Agriculture and Climatic Issues in South Asia

This book presents focussed information related to dynamic cropland transformation, agriculture development, climate change, and environment with the application of advance geospatial technology. It describes research using geospatial tools and techniques to develop the models, design, and planning for agricultural land use optimization especially in south Asian countries. It covers agriculture production, water scarcity, industrial development, natural resources, environmental degradation, and sustainable development.

Features:

- Provides the adaptation strategy from a multidisciplinary resilience perspective
- Addresses contemporary agricultural resilience to various climate change issues
- Develops novel approaches for sustainability with environmentally sound practices
- Discusses methodological and innovative approaches at local to global perspective
- Reports research using geospatial tools and techniques to develop the models, design, and planning for agricultural land use optimization

The book is aimed at researchers, professionals, and graduate students in GIS, environmental engineering, geography, agriculture, and climate studies.

Agriculture and Climatic Issues in South Asia

Geospatial Applications

Edited by Rukhsana Sarkar, Asraful Alam
and Azizur Rahman Siddiqui

CRC Press
Taylor & Francis Group
Boca Raton London New York

CRC Press is an imprint of the
Taylor & Francis Group, an **informa** business

First edition published 2024
by CRC Press
2385 NW Executive Center Drive, Suite 320, Boca Raton FL 33431

and by CRC Press
4 Park Square, Milton Park, Abingdon, Oxon, OX14 4RN

CRC Press is an imprint of Taylor & Francis Group, LLC

Library of Congress Cataloging-in-Publication Data
Names: Sarkar, Rukhsana, editor. | Alam, Asraful, 1990– editor. | Siddiqui, Azizur Rahman, editor.
Title: Agriculture and climatic issues in South Asia : geospatial applications / edited by Rukhsana Sarkar, Asraful Alam, and Azizur Rahman Siddiqui.
Description: First edition. | Boca Raton, FL : CRC Press, 2024. | Includes bibliographical references and index.
Identifiers: LCCN 2023032705 (print) | LCCN 2023032706 (ebook) | ISBN 9781032428338 (hardback) | ISBN 9781032456096 (paperback) | ISBN 9781003377825 (ebook)
Subjects: LCSH: Geospatial data—South Asia. | Environmental management—Geographic information systems. | Geoinformatics.
Classification: LCC G70.217.G46 A47 2024 (print) | LCC G70.217.G46 (ebook) | DDC 333.70285—dc23/eng/20231023
LC record available at https://lccn.loc.gov/2023032705
LC ebook record available at https://lccn.loc.gov/2023032706

ISBN: 978-1-032-42833-8 (hbk)
ISBN: 978-1-032-45609-6 (pbk)
ISBN: 978-1-003-37782-5 (ebk)

DOI: 10.1201/9781003377825

Typeset in Times
by Apex CoVantage, LLC

Contents

Chapter 6 Encroachment on Agricultural Land by Horizontal Development of Built-up Area: A Case Study of Sagar Island, East Coast of India..........76

Sabir Hossain Molla and Rukhsana Sarkar

Chapter 7 Consequences of Agricultural Land Transformation on the Natives of a Planned Town: A Case Study of New Town, Kolkata..........85

Mououda Khatun and Lakshmi Sivaramakrishnan

PART II Forest Vegetation, Disaster Resilience, and Sustainable Development

Chapter 13 Assessing Impact of Super Cyclone Amphan on Natural Vegetation in Sundarban Biosphere Reserve, India 174

Tania Nasrin, Md Nawaj Sarif, Mohd Ramiz, Mohammad Hashim, Sk Mohibul, Durgesh Dwivedi, Masood Ahsan Siddiqui, and Lubna Siddiqui

Chapter 14 Estimation of Surface Runoff by Soil Conservation Services–Curve Number Method in the Sanjai River Basin, Jharkhand 189

Arunashis Chandra and Swati Mondal

PART III   Application of Remote Sensing and GIS-Based Approach in Agricultural Practice and Climatic Issues

Chapter 17 Comparative Assessment of Landslide Risk Modelling by Bivariate Model in East Sikkim Himalaya, India..229

Sk Asraful Alam, Ramkrishna Maiti, and Sujit Mandal

Chapter 18 Using Geospatial Techniques for Detection of Land-Use/Land-Cover Changes Due to Major Hydropower Projects in Upper Beas River Basin, District Kullu, Himachal Pradesh, India..267

*Nishant Vaidya, Kesar Chand, Jagdish Chandra Kuniyal,
Suraj Kumar Singh, and Shruti Kanga*

*Sabirul Sk, Surajit Debarma, Asraful Alam, Lakshminarayan Satpati,
and R. Jagannathan*

Preface

Globally, agricultural planning, management, and sustainable development under the threat of climate change are essential to ensuring food security and economic growth, especially for low-income countries. These practices are responsible for producing affordable, wholesome food for a growing population. According to the United Nations, more than 836 million people worldwide continue to live in abject poverty, and most recent estimates show that 795 million people were undernourished between 2014 and 2016. The sustainability of agricultural practices is endangered by the greenhouse effect, one of the biggest contributors to global warming. Small landholdings, insufficient funding, and a lack of agrotechnological knowledge are obstacles to agriculture in developing nations. Under climatic change scenarios, agricultural planning and the use of agricultural technologies require precise spatiotemporal meteorological and crop data for accurate data analyses, forecasts, and their effective application in agricultural planning and management decisions, irrigation scheduling, crop stress management, and preparedness for disasters as well as for the sustainability of natural resources and ecosystems in various regions.

Remote sensing and geographic information systems (GIS) are highly applicable for the acquisition and management of huge spatiotemporal data by using satellite information, digital maps, and simulation models under changing climatic conditions, disaster management, and crop health status, including crop stress detection and damage assessment. A solution to conservation planning is a global concern in developing countries because of the rising population, rising food consumption, degrading state of the environment, and diminishing availability of freshwater.

This book provides precise and up-to-date information on environmental issues, agricultural development, climate change, and dynamic agricultural land alteration through the use of cutting-edge geospatial technology. The research for the book uses geospatial tools and techniques to create models, design, and plan for agricultural land use optimization, especially in South Asian nations, in order to inform decision-makers and planners on the best practices for agricultural management and sustainable livelihoods. A wide range of topics were covered in this book, including the growth of crops, changes in agricultural land use, climate change, disaster resilience, the loss of forest vegetation, urbanization, the depletion of natural resources, sustainable development, and the use of various statistics and GIS tools and techniques. This book is organized into three sections, each of which includes (1) mapping of cropland and agriculture transformation; (2) forest vegetation, disaster-resilient and sustainable development; and (3) application of remote sensing and GIS-based approaches in agricultural practice and climatic issues, which include various chapters collected from different countries worldwide, focused on cropland classification, monitoring cropland dynamics, cropping pattern planning for a flood-prone area, livestock-carrying capacity, land assessment for horticulture and agricultural drought, rural livelihood resilience and flood disaster and agriculture vulnerability, the impact of extreme cyclone events on agriculture, modeling of irrigation, land suitability analysis for different crops, and digital soil mapping using various geospatial technology. The gathered chapters aim to give planners and readers a thorough overview of the issues while also offering a multidisciplinary platform to investigate new areas in the fields of agriculture, climate change, and sustainable development.

Out of the three parts, which comprise 24 chapters provided by 67 authors, Part I consists of seven chapters focusing on various approaches and applications in the field of climatic issues using geospatial technology, a deep learning framework for extraction of crop and forest cover, crop pattern mapping using GIS and remote sensing, shrinking of cropland in Asansol Urban Agglomeration, fuzzy model–based spatiotemporal characterization of Dalbergiasissoo, encroachment on agricultural land by horizontal development of the built-up area, and consequences of agricultural land transformation on the natives of a planned town.

Part II has eight chapters discussing forest vegetation, disaster resilience, and sustainable development, including demographic vulnerability and women's participation to promote a disaster-resilient society based on disaster preparedness, evaluation of agricultural drought, sectorial vulnerability of climate change, social capital for reducing the effects of disaster on social vulnerabilities, impact of super cyclone Amphan on natural vegetation, and impact of land-use change on rural development.

Part III has nine chapters that are devoted to the use of remote sensing and a GIS-based approach in agricultural practice and climatic challenges, which examines and discusses the fuzzy machine learning model for mapping diseased sugarcane ratoon fields, landslide risk modeling by bivariate model, using geospatial techniques for detection of land-use/cover changes due to major hydro-power projects, forecasting the effects of urban expansion on the agricultural and forest landscape using the Multilayer Perceptron Neural Network–Markov chain model, soil erosion and digital soil mapping, geographical analysis of animal health facilities, flood susceptibility mapping based on morphometric parameters and one-dimensional simulation modeling, climatic impact on agricultural livelihood, and possible mitigation strategy of climate change and urban heat island using GIS and remote sensing in different regions. This book will be particularly helpful for students, academics, planners, and practitioners interested in geospatial technology–related issues in the sectors of agriculture, the environment, and development.

Rukhsana Sarkar, Asraful Alam, and Azizur Rahman Siddiqui

Acknowledgements

The completion of work is often a result of the direct or indirect contributions of many individuals. The process of completing this book was a great learning process and experience.

We express our appreciation to our colleagues and administration of *Aliah University, University of Calcutta,* and *Serampore Girls' College,* Kolkata, India, for providing a computer lab facility and necessary infrastructure to prepare the book.

We are sincerely indebted to the ICSSR (Ministry of Human Resource Development), New Delhi, India, for their sponsorship of Major Research Project 2021–2022 which enables us to learn more in the agriculture field and contribute to one chapter also.

We are pleased to express our gratitude to all our intellectual contributors for their valuable contributions that made it possible to complete this book.

We are immensely grateful to the publishing house, CRC Routledge, and all editors and team members, especially Dr. Gagandeep Singh and ISRS Indian Society of Remote Sensing, Dehradun, for enabling us to publish this edited book. The completion of this work would have been impossible without the publisher's support and guidance.

Rukhsana Sarkar, Asraful Alam, and Azizur Rahman Siddiqui
Kolkata, India

About the Editors

Rukhsana Sarkar is an assistant professor and former head of the Department of Geography at Aliah University, Kolkata. She has 12 years of teaching as well as research experience. She obtained her MA and PhD degrees in geography from Aligarh Muslims University. She was awarded nine academic awards and fellowships including the International Young Geographer Award 2009, the Aligaeh Muslim University Junior Research Fellowship award 2006–2007, and the University Grants Commission Research Fellowship 2007–2009 award. She has published more than 40 papers at the national as well as international levels in reputed journals and 23 chapters in edited books. Rukhsana has published six books and presented a number of research papers at the national and international levels. She has attended XXV International Federation of Surveyors Congress 2014, Malaysia, and International Conference on Geography & Geoinformatics for Sustainable Development, Bangkok, Thailand. Rukhsana has successfully completed one major research project sponsored by Indian Council of Social Science Research, Ministry of Education, New Delhi, in the agricultural field. Four doctorates have been successfully awarded under her supervision. She has served as a reviewer for many reputed international journals. Her specialization in research is agricultural development and planning, urban expansion and planning, environment, rural development, remote sensing, and geographic information systems. She has completed various training, workshops from different organizations, and a number of training courses held by *IIRS Indian Institute of Remote Sensing*, ISRO, Department of Space, Government of India. She is engaged in various professional activities and served the university in various posts. Presently she is working as project director on a major research project sponsored by ICSSR New Delhi, India.

Asraful Alam is an assistant professor and head of the Department of Geography, Serampore Girls' College, University of Calcutta, West Bengal, India. He received his MA and PhD degrees in geography from Aligarh Muslim University, Aligarh and Aliah University, Kolkata, India, respectively, and also completed a PG diploma in remote sensing and geographic information systems (GIS). Alam completed his post doctorate (PDF) from the Department of Geography, University of Calcutta, Kolkata, India. Previously, he was an assistant coordinator in the PG Department of Geography, Calcutta Women's College, University of Calcutta, Kolkata, India. Dr. Alam is one of the Project Director of collaborative major project on "Pradhan Mantri Ujjwala Yojana: An Impact Assessment in Relation to the Life of Women in Assam and West Bengal" sponsored by Indian Council of Social Science Research, Ministry of Education, New Delhi. His research interests include population geography, agricultural geography, climatology, health geography, remote sensing and GIS, and developmental studies. He has contributed various research papers published in various reputed national and international journals and edited book volumes. He has authored jointly edited books entitled *Habitat, Ecology and Ekistics: Case Studies of Human-Environment Interactions in India, Agriculture, Food and Nutrition Security: Case Study of Availability and Sustainability in India, Agriculture, Environment and Sustainable Development:—Experiences and Case Studies, Life and Living through Newer Spectrum of Geography, Self-Reliance (Atmanirbhar) and Sustainable Resource Management in India, Climate Change, Agriculture and Society— Approaches toward Sustainability and Population, Sanitation and Health: A Geographical Study Towards Sustainability*. He was a convener in the National Seminar on Self-reliance (Atmanirbhar), Sustainable Development and Environment 25–26 March 2022 sponsored by the Indian Council of Social Science Research (ICSSR) organized by the Department of Geography, Serampore Girls' College. Alam has served as an editorial board member in peer-reviewed international journals PLOS ONE, *Earth Science, Scientific Journal of Health Science Research*, and *Frontiers in Geochemistry*.

Azizur Rahman Siddiqui obtained MSc, MPhil, and DPhil degrees from Aligarh Muslim University. Siddiqui qualified for University Grants Commission National Eligibility Test June 1997 and qualified for Agricultural Research service National Eligibility Test October 1997 conducted by Indian Council of Agricultural Research, New Delhi. He also completed a postgraduate degree in geo-information science and earth observation and geo-hazard from the Indian Institute of Remote Sensing, April 2009. He has completed an National Natural Resource Management System training course and three international training courses sponsored by the United Nations. Siddiqui enjoyed a university fellowship for two years and was appointed as a permanent lecturer in P. G. College, Khurja Bulandshahar, C. C. S. University, Meerut, and later joined the Department of Geography, the University of Allahabad. He has supervised 17 PhD students and published 50 research papers in peer-reviewed journals and attended more than 77 national and international conferences. He is also the author of four books. He is the recipient of the Excellence Award for the year 2018 conferred by the University of Allahabad. He is the recipient of Dr. Sarvapalli Radha Krishnan award conferred by the Didiji Foundation (2022) and the Bhugol Bhushan award conferred by the Deccan Geographical Society (2022). He also delivered about 130 lectures in different University Grants Commission's Human Resource Development Centre and universities/colleges in India. Presently Siddiqui is International Advisor of Foreign Students, Chairman Data Cell, and Nodal officer National Institutional Ranking Framework. He holds the posts of Senior Vice President of Allahabad University Teachers Association and Secretary General of the Indian Institute of Geomorphologists.

Contributors

Siham Acharki
Department of Earth Sciences
Faculty of Sciences and Technologies of
    Tangier (FSTT)
Abdelmalek Essaadi University
Tetouan, Morocco

Shahana Akter
Environmental Science Discipline
Khulna University
Khulna, Bangladesh

Asraful Alam
Department of Geography
Serampore Girls' College
University of Calcutta
West Bengal, India

Najib Ansari
Department of Geography
Aliah University
Kolkata, India

Sk Asraful Alam
Department of Geography
Vidyasagar University
Midnapore, West Bengal, India

Brijmohan Bairwa
School of Earth Sciences
Banasthali Vidyapith
Tonk, Rajasthan, India

Divyadyuti Banerjee
Department of Geography
Jadavpur University
West Bengal, India

Md. Bariul Musabbir
Center for Environmental and Geographic
    Information Services
Dhaka, Bangladesh

Govindarajan Bhaskaran
Centre for Water Resources Management
University of Madras
Chennai, Tamil Nadu, India

Sayani Chakraborty
Department of Geography
University of Burdwan
Burdwan (W.B), India

Kesar Chand
G.B. Pant National Institute of Himalayan
    Environment
Himachal Regional Center
Mohal-Kullu, Himachal Pradesh, India

Arunashis Chandra
Department of Geography
Panskura Banamali College
West Bengal, India

K. K. Chattoraj
Department of Geography
Kazi Nazrul University
Asansol, West Bengal, India

Surajit Debarma
Department of Geography
University of Madras
Guindy, Chennai, India

Sneha Deka
Department of Geography
North-Eastern Hill University
Shillong, India

Sanu Dolui
Department of Geography
University of Burdwan
Burdwan (W.B), India

Durgesh Dwivedi
Department of Geography
Faculty of Natural Sciences
Jamia Millia Islamia
New Delhi, India

Asutosh Goswami
Department of Earth Sciences and Remote
    Sensing
JIS University
Agarpara, Kolkata, West Bengal, India

Somenath Goswami
Department of Geo-Informatics
Pandit Raghunath Murmu Smiriti
    Mahavidyalaya
Bankura University
Bankura, West Bengal, India

Mohammad Hashim
Department of Geography
Faculty of Natural Sciences
Jamia Millia Islamia
New Delhi, India

Sajjad Hussain
Department of Environmental Sciences
COMSATS University Islamabad,
    Vehari Campus
Pakistan

R. Jagannathan
Department of Geography
University of Madras
Guindy, Chennai, India

Shruti Kanga
School of Environment and Earth Sciences
Central University of Punjab,
    VPO-Ghudda
Bathinda, Punjab, India

Moududa Khatun
Department of Geography
Aliah University
Kolkata, West Bengal, India

Kishore S.
Calaver Technologies Pvt. Ltd.
Bengaluru, Karnataka, India

Anil Kumar
SG and Photogrammetry and Remote Sensing
    Department
Indian Institute of Remote Sensing
Dehradun, Uttarakhand, India

Arnab Kundu
Department of Geo-Informatics
Pandit Raghunath Murmu Smiriti Mahavidyalaya
Bankura University
Bankura, West Bengal, India

Jagdish Chandra Kuniyal
G.B. Pant National Institute of Himalayan
    Environment
Kosi-Katarmal-Almora, Uttrakhand, India

Santoshi Mahato
Department of Geo-Informatics
Pandit Raghunath Murmu Smiriti
    Mahavidyalaya
Bankura University
Bankura, West Bengal, India

Sunil Mahato
Department of Geo-Informatics
Pandit Raghunath Murmu Smiriti Mahavidyalaya
Bankura University
Bankura, West Bengal, India

Ramkrishna Maiti
Department of Geography
Vidyasagar University
Midnapore, West Bengal, India

Priyanka Majumder
Department of Earth Sciences and Remote
    Sensing
JIS University
Agarpara, Kolkata, West Bengal, India

Sujit Mandal
Department of Geography
Diamond Harbour Women's University
West Bengal, India

Sonakshi Mehrotra
Photogrammetry and Remote Sensing
    Department (PRSD)
Indian Institute of Remote Sensing (IIRS)
Dehradun, Uttarakhand, India

Sk Mohibul
Department of Geography
Faculty of Natural Sciences
Jamia Millia Islamia
New Delhi, India

Sabir Hossain Molla
Department of Geography
Aliah University
Kolkata, India

Swati Mondal
Department of Geography
Panskura Banamali College
Kanakpur, Panskura, West Bengal, India

Md. Mujibor Rahman
Environmental Science Discipline
Khulna University
Khulna, Bangladesh

Tania Nasrin
Department of Geography
Faculty of Natural Sciences
Jamia Millia Islamia
New Delhi, India

Md. Nawaj Sarif
Department of Civil Engineering
Indian Institute of Technology Delhi
Hauz Khas, New Delhi, India

Shruti Pancholi
Photogrammetry and Remote Sensing
    Department
Indian Institute of Remote Sensing
Dehradun, India

Mohd Ramiz
Department of Geography
Faculty of Natural Sciences
Jamia Millia Islamia
New Delhi, India

Swaagat Ray
Department of Geography
University of Calcutta
Kolkata, West Bengal, India

Ali Raza
School of Agricultural Engineering
Jiangsu University
Zhenjiang, P.R. China

Arijit Roy
SG and Disaster Management Sciences
    Department
Indian Institute of Remote Sensing
Dehradun, Uttarakhand, India

Sabirul Sk
Interdisciplinary Program in Climate Studies
Indian Institute of Technology Bombay
Mumbai, India

Pritom Saikia
G.B. Pant National Institute of Himalayan
    Environment
North East Regional Center
Itanagar, Arunachal Pradesh, India

Amiya Kumar Sarkar
Geography Department
Calcutta University
Kolkata, West Bengal, India

Rukhsana Sarkar
Department of Geography
Aliah University, Park Circus Campus
Kolkata, India

Lakshminarayan Satpati
UGC-Human Resource Development Centre
University of Calcutta
Kolkata, West Bengal, India

Megha Sharma
Department of Computer Science and Engineering
Amrita School of Computing
Amrita Vishwa Vidyapeetham
Bengaluru, Karnataka, India

Azizur Rahman Siddiqui
Department of Geography
University of Allahabad
Prayagraj, Uttar Pradesh, India

Lubna Siddiqui
Department of Geography
Faculty of Natural Sciences
Jamia Millia Islamia
New Delhi, India

Masood Ahsan Siddiqui
Department of Geography
Faculty of Natural Sciences
Jamia Millia Islamia
New Delhi, India

Govind Singh
Department of Geography
Jai Narain Vyas University
Jodhpur (Rajasthan), India

Sudhir Kumar Singh
K. Banerjee Centre of Atmospheric
   and Ocean Studies
University of Allahabad
Prayagraj, India

Suraj Kumar Singh
Suresh Gyan Vihar University
Jaipur, India

Lakshmi Sivaramakrishnan
Department of Geography
Jadavpur University
Kolkata, West Bengal, India

Supriya M.
Department of Computer Science
   and Engineering

Amrita School of Computing
Amrita Vishwa Vidyapeetham
Bengaluru, Karnataka, India

Neyha Rubab Syed
School of Energy and Environment
Power Engineering and Engineering
   Thermophysics
Southeast University
Nanjing, China

Nishant Vaidya
Department of Geography
Kumaun University, DSB, Campus
Nainital, Uttarakhand, India

Muhammad Zubair
School of Transportation
Southeast University
Nanjing, China

Part I

Mapping of Cropland and Agriculture Transformation

1 Agricultural Transformation and Climatic Issues Using Geospatial Technology
An Overview

*Rukhsana Sarkar, Najib Ansari, Asraful Alam,
and Azizur Rahman Siddiqui*

1.1 INTRODUCTION

Agricultural transformation and climatic issues are two crucial areas that require attention and innovative solutions in today's world. Geospatial technology plays a significant role in addressing these challenges and driving sustainable practices in agriculture. Geospatial technology also facilitates monitoring and evaluation of sustainable agricultural practices and policy implementation. By analyzing satellite imagery and other geospatial data, it is possible to assess the effectiveness of conservation practices, land-use policies, and the impact of agricultural activities on ecosystems. This knowledge is invaluable in formulating evidence-based policies and promoting sustainable agricultural practices at local, regional, and national levels.

The agricultural industry is varied and contradictory. According to estimates, 1.3 billion (or 19%) of the 7.1 billion people on the planet were directly employed in farming in 2012, although agriculture (which also includes the relatively tiny hunting, fishing, and forestry industries) only made up 2.8% of global revenue (World Bank, 2012). The majority of farmers in the world today may be found in middle- and low-income nations, where agriculture accounts for a sizable component of the economies of such nations and employs a sizable number of people. Despite being a remarkable accomplishment, the considerable growth in global food production over the past four decades has also led to severe environmental issues. These include faster conversion of crops to nonagricultural uses, pesticide risks and chemical fertilizers, desertification, and the cumulative impacts of salinization on land productivity and soil erosion. Over 4,000 plant and animal species are at risk due to agricultural intensification, which also causes genetic deterioration, species extinction, and the destruction of wildlife habitat (FAO, 2010).

Several studies have shown that significant progress has been made in reducing poverty and hunger as well as in enhancing the security of food and nutrition. Productivity growth and advancements in technology have contributed to better resource management and increased food security. But grave issues still exist. More than 2 billion people suffer from vitamin deficiencies, and 795 million people still experience starvation. Additionally, the sustainability of large-scale food systems may be jeopardized by rising demands on natural resources and climate change, both of which pose a threat to global food security (FAO, 2010; Rukhsana and Alam, 2021). It is astonishing how persistent and widespread hunger remains around the globe despite all of development's great achievements, including significant improvements in food production. More than 820 million people in the world experience daily hunger, and this number has been steadily rising over the past three years, according to the most recent State of Food Safety and Nutrition in the World Report (FAO, 2019). Furthermore, around 2 billion people experience some kind of food insecurity, including a lack of access to sufficient, healthy food. Particularly susceptible to famine are women, children, and

DOI: 10.1201/9781003377825-2

indigenous populations. Also on the topic of nutrition, there is an elevated risk of overweight and obesity, which is spreading quickly over the globe and reaching epidemic levels (Rukhsana and Alam, 2021).

The production of food will likely be severely impacted by climate change and global warming, especially in tropical and subtropical areas. Increasing temperatures have an impact on decreased agricultural yield because they limit crop duration and increase respiratory losses. India's agricultural output demonstrates that kharif crop production is more adversely affected. From 1901 through 2019, there was an average yearly temperature rise of 0.61°C/100 years, 1.0°C/100 years, and 0.22°C/100 years, respectively. Since 2000, yearly precipitation has diverged from the usual by around 20%. Increased winter temperatures also shorten the window of opportunity for rabi crops to fill their grains. As a result, both crops somewhat lose their nutritious value (Rukhsana et. al., 2021).

Producing enough food to feed a rising population is the biggest problem facing the planet in an era of frequent climate change. The globe now has a glut of food thanks to the use of high-yielding cultivars, advances in technology, irrigation water, and fertilizers. In order to have a robust global food system, accomplish sustainable agriculture, and govern and preserve the environment, developing nations must prioritize achieving food and nutrition security, which necessitates the restructuring of pertinent policies as one of the primary approaches to do so (Rukhsana and Alam, 2022; Molla and Rukhsana, 2023).

One way to increase food production and fulfill the rising demand of the global population is through proper land use (Rukhsana and Molla, 2022). The effectiveness of this method can be increased by using a variety of remote sensing techniques. The approach of remote sensing that has been utilized the most frequently is observation with the human eye. Any observation made using this method is frequently geo-referenced with the aid of modern technology and added to a geographic information systems (GIS) database. An important element of precision agriculture, a method to measure the field's greenness may be used to determine the production of different management zones (Brisco et al., 2013).

The capacity of nations to plan and monitor better exploitation and management of their land resources must be improved in order to maximize agricultural production without jeopardizing land and environmental security. As a result, land-use/land-cover changes have a significant impact on soil quality and cause land degradation, which in turn affects soil production (Biro et al., 2013). In addition to soil nutritional characteristics and certain pertinent climatic factors, land use/land cover has to be studied using remote sensing and GIS to determine if the land is suitable for cultivation (Bhagat et al., 2009; Samanta et al., 2011).

This chapter offers some fundamental knowledge on the mapping of cropland and agricultural transformation, forest vegetation, disaster-resilient development, and application of remote sensing and GIS-based approaches in agricultural practice and climatic issues using a systematic literature review. The book's optional material is summarized, and the concepts of agricultural transition and climate issues are explained.

1.1.1 Mapping of Cropland and Agriculture Transformation

The consequences of human activity and global environmental variation were significantly impacted by changes in land use and cover, which are also associated with interactions between humans and nature (Liu et al., 2014; Ning et al., 2018; Rukhsana and Molla, 2023). Cropland's dimensional patterns and temporal fluctuations must be well understood in order to understand the root causes and practical effects of agricultural landscapes. Dynamic crop mapping is essential to grasp the region's agrospatial variation (Rahman and Saha, 2009; Rukhsana and Molla, 2023; Molla and Rukhsana, 2022).

Social and economic advancement is fueled by agricultural productivity, land resource management, and food security (Huang et al., 2016; Lebourgeois et al., 2017). Crop variety maps may reveal crop trends, the location and proportion of different crops planted, and provide the foundation for yield prediction (Bolton and Friedl, 2013), disaster assessment, and water resource management (Vogels et al., 2019).

The availability of agriculture land is the major input in agricultural production and is indispensable and irreplaceable which is the best resource for economically boosting farmer wealth (Muyanga et al., 2013) and also a key driver of economic growth (Li, 2014). Agricultural land is arable land, contains perennial crops, and is utilized for perennial pastures. Arable land, often known as cropland, is defined by the Food and Agriculture Organization of the United Nations (FAO) as land that is momentarily fallow, temporarily under market or kitchen gardens, momentarily under temporary crops (double-cropped areas are counted once), or momentarily under meadows for mowing or pasture. It excludes any land left fallow owing to shifting agricultural practices. However, there is fierce competition over using limited and nonrenewable land, usually between the agricultural and nonagricultural sectors. This results in agricultural land transformation (ALT), drastically reducing the amount of available farmland and threatening the food supply. Ironically, countries that are developing (Azadi, 2011), which have large populations and high food consumption (Deloitte, 2013), have the greatest percentage of ALT. Therefore, handling ALT correctly is crucial to stabilize the food supply. A significant issue affecting agricultural land (AL) and ALT is the degradation of land quality brought on by inappropriate farming practices (Karunakaran, 2014).

Farmers frequently overexploit land by growing a high-value crop that is essentially unsuited for the characteristics of the land to maximize economic gains. (Singh and Nair, 2012; Kutywawo et al., 2012). This provides the farmer with a significant economic return. The condition of the land will deteriorate, compromising future food production in favor of instant financial profits. Urbanization was the ALT factor that produced the fastest and most permanent change (Schneider and Woodcook, 2012). It encouraged the development of a peri-urban area adjacent to the urban core in an agricultural area (Simon, 2008; Webster, 2001). By conventional location theory, the conversion of AL occurs due to the relatively higher rent produced by urban land use compared to rural land use (North, 1955; Irwin and Bockstael, 2002; Irwin and Bockstael, 2007). ALT is fast and usually unplanned in poor nations. Uncontrolled ALT immediately reduces food manufacture and the availability of ecosystem goods and services generated from various land types because most urban areas are bordered by productive AL (Imhoff et al., 1997). Controlling ALT is crucial for this reason.

Cropland availability has been threatened by social, physical, and climatic variables as a result of global urbanization (Hertel, 2011; Vinge, 2018). An urgent global issue is preserving agricultural property for present and upcoming generations (Hertel, 2011; Caldwell et al., 2017; FAO, 2021). Around the world, significant study is being done on the effects of urbanization, specifically on cropland resources, as seen in Europe (Perrin, 2013; Tan et al., 2009; Skog and Steinnes, 2016), the United States (Moroney and Castellano, 2018; Narducci et al., 2019), Canada (Qiu et al., 2015), and China (Duan et al., 2021; Zhang et al., 2016; Chien, 2015; Hu et al., 2018; Miao et al., 2021).

According to Paster (2004) and Tulloch et al. (2003), protecting cropland through land-use planning is a standard practice in most developed nations. To conserve farmland, in Japan, land-use master plans were created using four unique zones: AL zones, agricultural promotion zones, urbanization control zones, and urbanization promotion zones (Saizen et al., 2006). To preserve AL in the United States, both agricultural grid zoning and non-grid zoning strategies have been used (Paster, 2004).

Human modification of the earth's surface is reflected in land use/land cover activities (Gong et al., 2019a, 2019b). As a result of changes in land use/land cover, the planet's energy balance and biogeochemical cycles have been significantly altered Foley et al., 2005; Alkama and Cescatti, 2016; Turner et al., 2007). This has an impact on the properties of the extent of the land and the availability of ecosystem services. According to a recent study (Song et al., 2017, 2018), 60% of worldwide land changes are directly related to human activity. Kennedy et al.'s (2019) most recent estimate states that only 5% (6.96 million km^2) of the planet's terrestrial area is still undisturbed by human activity. Urbanization is characterized by migrated population to the urban extent from the rural world, and the expansion of urban land use (Wu et al., 2015) is one common manifestation of all these human-related modifications on global lands (Theobald et al., 2020).

The security of human food systems and sustainability depend heavily on cropland and other agricultural areas. A farming land base must be preserved because local, national, and international urbanization significantly impacts it. The sustainability and security of human food systems, habitats, agricultural businesses, and livelihoods depend heavily on farmland. Sustainable farmland management offers several other priceless ecosystem services, such as pollinator and wildlife habitat, carbon sequestration, nutrient cycling, water regulation, and amenity value, in addition to the provisioning value and benefits of farmland, such as the production of food and fibre (Power, 2010).

1.2 FOREST VEGETATION, DISASTER RESILIENCE, AND SUSTAINABLE DEVELOPMENT

Forest vegetation is a valuable natural resource that is used for wildlife conservation, economic output, and the preservation of water and soil (Gebrehiwot et al., 2014; Dixon et al., 1994; Keenan et al., 2013; Hanewinkel et al., 2013; McCarthy and Burgman, 1995). Additionally, forests contain significant carbon storage that is reliant on a balance between primary production and ecosystem respiration, procedures that adapt to climate variation (Peng et al., 2009; Reyer et al., 2014; Coomes et al., 2014; Hanewinkel et al., 2013; Michaletz et al., 2014; Margono et al., 2014). Transformation in forest vegetation could harm ecosystem variety and their ability to adapt at all scales, from local communities to the world (Lindner et al., 2010; Landres et al., 1999; Hanewinkel et al., 2013; Chazdon, 2008; Tian et al., 2011). A significant risk to society comes from global and regional policies that speed up the loss of biodiversity and the extinction of forests (Buizer et al., 2014; Ripple et al., 2014; Hanewinkel et al., 2013).

Disaster resilience is the ability of individuals, communities, organizations, and nations to deal with and recover from risks, shocks, or pressures without endangering long-term development possibilities. The Hyogo Framework for Action (2005) states that disaster resilience is determined at the international, regional, national, and local levels by people's capacity to organize themselves to learn from past disasters and reduce their risks from subsequent ones. Disaster resilience is a subset of the larger concept of resilience, which is defined as "the capacity of individuals, communities, states, and their institutions to absorb and recover from shocks while positively adapting and transforming their structures and means of living in the face of long-term changes and uncertainty" (OECD, 2013b, p. 1).

To better understand the concept of disaster resilience, we have to know the Sustainable Development Goal (SDG) of fulfilling the needs of disaster resilience. The core of sustainable development is the belief that three pillars promote sustainability: the economy, society, and the environment. The Brundtland Report's original definition of sustainable development states that it "meets the needs of the present without compromising the ability of future generations to meet their own needs." In developing countries, immediate pressures on the survival of the poorest citizens may cause attention to be focused on meeting their primary local requirements. However, as the Brundtland Report definition indicates, sustainable development's difficulty is meeting everyone's present-day requirements while preserving the environment for future generations. Actions in one area of the world impact how well people can meet their needs elsewhere. To promote more fair and reduced consumption of environmental resources, it is necessary to use and develop technology and social organization, such as through the creation of a circular economy (Stefen et al., 2015), given that the world's population already exceeds earth's bearing capacity and is expected to rise significantly (Stahel, 2016). Thus, the concept of sustainable development encompasses both spatial and temporal factors. The SDGs, therefore, aim to address the universal requirement for development that satisfies everyone's needs, whereas the Millennium Development Goals concentrate on action in underdeveloped nations.

We have finally included the 2030 Agenda for Sustainable Development to form more livable, resilient, and sustainable housing cities. In modern developing nations like India, the government has modernized their cities by implementing numerous city planning projects like the Swachh Bharat Mission, Smart Cities Mission, Pradhan Mantri Awas Yojana-Housing for All for the Urban Poor Population, Atal Mission for Rejuvenation and Urban Transformation, and many others. To complete urban resilience and sustainability, particular frames must be used to assess, implement, and adapt various planning and policy initiatives. Landscape and urban planning cannot accomplish resilience and sustainability in the future without sound strategic development planning (Zanotti et al., 2020; Ribeiro and Goncalves, 2019). Various planning and policy regarding urban land use have struggled in recent decades to achieve complete urban resilience and sustainable development in poor and developed countries (Keith et al., 2020; Jabareen, 2013; Zanotti et al., 2020). They applied the resilience framework to sustainability without considering the urbanization and flexibility of the local urban population and society.

International organizations and cooperation treaties have acknowledged the significance of research on sustainable forest vegetation and disaster resistance. To get cities involved in resilience building, the United Nations Office for Disaster Risk Reduction (UNDRR) created the "Cities are Resilient" campaign in 2010 (UNDRR, 2022). Furthermore, the Sendai Framework for Disaster Risk Reduction 2015–2030 emphasized the ability of disasters to be restored by using inclusive and integrated economic, educational, social, and political measures, both institutional and technological (UNDRR, 2022). The SDGs also emphasized increasing community resilience to natural disasters, focusing on local planning and administration (UN, 2022). Many conceptual frameworks have been proposed for the evaluation of catastrophe resilience. According to Cutter et al.'s Baseline Resilience Indicator for Communities concept, disaster resilience has numerous aspects, including physical, social, institutional, economic, and ecological elements. The Community Disaster Resilience Index is a conceptual framework created by Mayunga that considers social, economic, demographic, material, and natural capital (Mayunga, 2007). This methodology is capital based. Joerin et al. (2012) used the Climate-Related Disaster Community Resilience Framework to assess the physical, social, and economic resilience of individuals; Yoon et al. proposed the Community Disaster Resilience Index, which takes into account institutional, human, social, economic, and environmental elements, to assess the total catastrophe resilience of South Korea (Yoon et al., 2016).

1.3 APPLICATION OF REMOTE SENSING AND GIS-BASED APPROACH IN AGRICULTURAL PRACTICE AND CLIMATIC ISSUES

Agriculture and climate change are tied to one another. Agriculture is impacted by climate change in various ways, including variations in average temperatures, rainfall, and other factors. Lack of rain and changes in atmospheric components like carbon dioxide and ground-level ozone contribute to pests and diseases in AL. Globally, there is an increasing need for a spatially detailed geographic analysis of climate change and its effects. The fast development of computer-based GIS, open-source data/tools, and satellite imaging has been mirrored by the increasing trend of forecasting the impact of climate change (CC) on the agricultural and forestry industries. First, because GIS is frequently used to store, analyze, and display both spatial and nonspatial dimensions, spatial dimensions are given to address issues related to agriculture in mountain ecosystems. Second, digital elevation models (DEMs) made from satellite data are now conveniently accessible at a better-resolution grid through public geo-data portals, such as Bhuvan in India (https://bhuvan.nrsc.gov.in). This spatially explicit and regionally demonstrative data source aids in incorporating the spatial variability of landscape units and enables the grid-based operation of agro-ecosystem models. The Fourth Assessment Report of the Intergovernmental Panel on Climate Change (IPCC) extensively emphasized the use of crop models inside a GIS system for regional climate impact assessments (IPCC, 2013) (Easterling et al., 2007). GIS and agricultural models would enable the simulation of spatiotemporal plant processes, yield, and crop growth.

Numerous remote sensing techniques can be useful to make this process more effective. Observation using the human eye has been the most widely used remote sensing method. Any observation utilizing this technique is typically geo-referenced using contemporary technology and entered into a GIS database. The production of various management zones can be calculated using a method to estimate the greenness of the field, which is a key component of precise agriculture (Brisco et al., 2013). This method is based on the correlation between the reflection of red light and near-infrared light. The output from global circulation models on different climatic scenarios can be used to estimate crop performance over space using process-based crop models in current and future climate scenarios (Neelin et al., 2006; Lobell et al., 2008). Additionally, a rise in computer processing capacity has enabled regional or worldwide modeling of agricultural systems. The key components of integrated technologies, such as global circulation models, remote sensing, and GIS-based climate change scenarios, have substantially enhanced science's ability to handle the problem of crop model forecasts and improve crop planning in a changing environment.

From the microlevel to the worldwide survey of globally relevant crops, remote sensing data could be employed extensively in agricultural sector applications at various scales (Steven and Clark, 2013). Using remote sensing techniques, such as light detection and ranging (lidar) technology and thermal infrared (IR) sensors, it is now possible to measure the temperatures and heights of vegetation canopies to assess their biomass, chlorophyll, and nitrogen (N) contents at a certain moment (Berni et al., 2009; Deng et al., 2018). Additionally, remote sensing has been widely used to identify and map weeds in the agricultural sector (Lamb and Brown, 2001; Moran et al., 1997; Thorp and Tian, 2004; Zwiggelaar, 1998). It has been demonstrated that remote sensing is a useful tool for keeping track of agricultural practices. The spectral and temporal characteristics of the land surface resulting from human activities can be recorded and monitored at various spatial and temporal scales thanks to a wide variety of onboard sensors aboard an expanding number of civilian satellites. Remote sensing is crucial for dividing a large farm into management zones, and the use of GIS and GPS is required to satisfy the specific requirements of each zone. Thus, the first step in farming is to divide the land into management zones. The primary factors categorizing this area into zones include soil types, pH levels, pest infestations, nutrients available, soil moisture content, fertility requirements, weather forecasts, crop features, and hybrid responses.

Examining the available records will provide access to this data. Most farms regularly preserve documentation describing regional cropping practices, historical crop characteristics, and maps of soil surveys. This process can also use satellite and aerial images. Aerial and satellite images of the farm that were recently captured at different periods of the year or seasons can also be generated. The farmer can use this information to calculate the production of various management zones. Further, it is possible to identify each agricultural zone's growth and yield trends.

1.4   CONCLUSION

In conclusion, geospatial technology plays a crucial role in agriculture transformation and addressing climatic issues. By leveraging the power of remote sensing, GIS, and data analytics, farmers and policymakers can make informed decisions that optimize agricultural productivity, mitigate climate-related risks, and promote sustainable land management practices. Geospatial technology offers a powerful toolset to ensure a resilient and environmentally conscious agricultural sector in the face of evolving climatic challenges.

REFERENCES

Alkama R., and Cescatti, A. (2016). Biophysical climate impacts of recent changes in global forest cover. *Science* 351(6273), 600–604.

Azadi, H., Ho, P., and Hasfiati, L. (2011). Agricultural land conversion drivers: A comparison between less developed, developing and developed countries. *Land Degrad. Dev.* 22, 596–604.

Berni, J. A., Zarco-Tejada, P. J., Suarez, L., and Fereres, E. (2009). Thermal and narrowband multispectral remote sensing for vegetation monitoring from an unmanned aerial vehicle. *IEEE Trans. Geosci. Remote Sens.* 47, 722–738.

Bhagat, R. M., Singh, S., Sood, C., Rana, R. S., Kalia, V., Pradhan, S., Immerzeel, W., and Shrestha, B. (2009). Land suitability analysis for cereal production in Himachal Pradesh (India) using geographical information system. *J. Indian Soc. Remote.* 37(2), 233–240.

Biro, K., Pradhan, B., Buchroithner, M., and Makeschin, F. (2013). Land use/land cover change analysis and its impact on soil properties in the northern part of the Gadarif region, Sudan. *Land Degrad. Dev.* 24(1), 90–102.

Bolton D. K., and Friedl M. A. (2013). Forecasting crop yield using remotely sensed vegetation indices and crop phenology metrics. *Agric. For. Meteorol.* 173, 74–84.

Brisco, B., Schmitt, A., Murnaghan, K., Kaya, S., and Roth, A. (2013). Sarpolarimetric change detection for flooded vegetation. *Int. J. Digit. Earth.* 6, 103–114.

Buizer, M., Humphreys, D., and de Jong, W. (2014). Climate change and deforestation: The evolution of an intersecting policy domain. *Environ. Sci. Pol.* 35, 1–11.

Caldwell, W. J., Hilts, S., and Wilton, B. (eds.). (2017). *Farmland Preservation: Land for Future Generations, 2nd Edn.* University of Manitoba Press, Winnipeg Manitoba.

Chazdon, R. L. (2008). Beyond deforestation: Restoring forests and ecosystem services on degraded lands. *Science* 320, 1458–1460.

Chien, S. S. (2015). Local farmland loss and preservation in China: A perspective of quota territorialization. *Land Use Policy* 49, 65–74. doi: 10.1016/j.landusepol.2015.07.010

Coomes, D. A., Flores, O., Holdaway, R., Jucker, T., Lines, E. R., and Vanderwel, M. C. (2014). Wood production response to climate change will depend critically on forest composition and structure. *Glob. Chang. Biol.* 20, 3632–3645.

Deloitte. (2013). *The Food Value Chain a Challenge for the Next Century.* Deloitte, London, UK.

Deng, L., Mao, Z., Li, X., Hu, Z., Duan, F., and Yan, Y. (2018). UAV-based multi-spectral remote sensing for precision agriculture: A comparison between different cameras. *ISPRS J. Photogramm. Remote Sens.* 146, 124–136.

Dixon, R. K., Brown, S., Houghton, R. E. A., Solomon, A. M., Trexler, M. C., and Wisniewski, J. (1994). Carbon pools and flux of global forest ecosystems. *Science* 263, 185–189.

Duan, X., Meng, Q., Fei, X., Lin, M., and Xiao, R. (2021). The impacts of farmland loss on regional food self-sufficiency in Yangtze river delta urban agglomeration over last two decades. *Remote Sens.* 13, 3514. doi: 10.3390/rs13173514

Easterling, W. E., Aggarwal, P. K., Batima, P., Brander, K. M., Erda, L., Howden, S. M., Kirilenko, A., Morton, J., Soussana, J. F., Schmidhuber, J., and Tubiello, F. N. (2007). Food, fibre and forest products. In: Parry, M. L., Canziani, O. F., Palutikof, J. P., van der Linden, P. J., and Hanson, C. E. (eds.). *Climate Change 2007: Impacts, Adaptation and Vulnerability: Contribution of Working Group II to the Fourth Assessment Report of the Intergovernmental Panel on Climate Change.* Cambridge University Press, Cambridge, UK, pp. 273–313.

FAO. (2010). Characterisation of small farmers in Asia and the Pacific. Asia and Pacific Commission on Agricultural Statistics, twenty-third session, Siem Reap, 26–30 Apr.

FAO. (2019). State of Food Security and Nutrition in the World. www.fao.org/3/ca5162en/ca5162en.pdf (accessed on 20 November 2019).

Foley, J. A., DeFries, R., Asner, G. P., Barford, C., Bonan, G., Carpenter, S. R., Chapin, F. S., Coe, M. T., Daily, G. C., Gibbs, H. K., Helkowski, J. H., Holloway, T., Howard, E. A., Kucharik, C. J., Monfreda, C., Patz, J. A., Prentice, I. C., Ramankutty, N., and Snyder, P. K. (2005). Global consequences of land use. *Science* 309(5734), 570–574.

Food and Agriculture Organization of the United Nations (FAO). (2021). *The State of Food and Agriculture.* Rome: FAO.

Gebrehiwot, S. G., Bewket, W., and Bishop, K. (2014). Community perceptions of forest: Water relationships in the Blue Nile Basin of Ethiopia. *Geo. J.* 79, 605–618.

Gong, P., Chen, B., Li, X., Liu, H., Wang, J., Bai, Y., Chen, J., Chen, X., Fang, L., Feng, S., Feng, Y., Gong, Y., Gu, H., Huang, H., Huang, X., Jiao, H., Kang, Y., Lei, G., Li, A., Li, X., Li, X., Li, Y., Li, Z., Li, Z., Liu, C., Liu, C., Liu, M., Liu, S., Mao, W., Miao, C., Ni, H., Pan, Q., Qi, S., Ren, Z., Shan, Z., Shen, S., Shi, M., Song, Y., Su, M., Ping Suen, H., Sun, B., Sun, F., Sun, J., Sun, L., Sun, W., Tian, T., Tong, X., Tseng, Y., Tu, Y., Wang, H., Wang, L.,

Wang, X., Wang, Z., Wu, T., Xie, Y., Yang, J., Yang, J., Yuan, M., Yue, W., Zeng, H., Zhang, K., Zhang, N., Zhang, T., Zhang, Y., Zhao, F., Zheng, Y., Zhou, Q., Clinton, N., Zhu, Z., and Xu, B. (2019a). Mapping essential urban land use categories in China (EULUC-China): Preliminary results for 2018. *Sci. Bull.* 65(3), 182–187.

Gong, P., Liu, H., Zhang, M., Li, C., Wang, J., Huang, H., Clinton, N., Ji, L., Li, W., Bai, Y., Chen, B., Xu, B., Zhu, Z., Yuan, C., Suen, H. P., Guo, J., Xua, N., Lia, W., Zhao, Y., Yang, J., Yu, C., Wang, X., Fu, H., Yu, L., Dronova, I., Hui, F., Cheng, X., Shi, X., Xiao, F., Liu, Q., and Song, L. (2019b). Stable classification with limited sample: Transferring a 30-m resolution sample set collected in 2015 to mapping 10-m resolution global land cover in 2017. *Sci. Bull.* 64(6), 370–373.

Hanewinkel, M., Cullmann, D. A., Schelhaas, M. J., Nabuurs, G. J., Zimmermann, N. E. (2013). Climate change may cause severe loss in the economic value of European forest land. *Nat. Clim. Chang.* 3, 203–207.

Hertel, T. W. (2011). The global supply and demand for agricultural land in 2050: A perfect storm in the making? *Am. J. Agric. Econ.* 93, 259–275. doi: 10.1093/ajae/aaq189

Hu, Y., Kong, X., Zheng, J., Sun, J., Wang, L., and Min, M. (2018). Urban expansion and farmland loss in Beijing during 1980–2015. *Sustainability* 10, 3927. doi: 10.3390/su10113927

Huang, J., Sedano, F., Huang, Y., Ma, H., Li, X., Liang, S., Tian, L., Zhang, X., Fan, J., and Wu, W. (2016). Assimilating a synthetic Kalman filter leaf area index series into the WOFOST model to improve regional winter wheat yield estimation. *Agric. Forest. Meteorol.* 216, 188–202.

Imhoff, M. L., Lawrence, W. T., Stutzer, D. C., and Elvidge, C. D. (1997). A technique for using composite DMSP/OLS "city lights" satellite data to map urban area. *Remote Sens. Environ.* 61, 361–370.

IPCC. (2013). *Summary for policymakers*. In Climate Change 2013: The Physical Science Basis. Contribution of Working Group. Allen, J. Doschung, A. Nauels, Y. Xia, V. Bex, and P.M. Midgley, Eds., Cambridge University Press, pp. 3–29, doi:10.1017/CBO97 81107415324.004

Irwin, E. G., and Bockstael, N. E. (2002). Interacting agents, spatial externalities and the evolution of residential land use patterns. *J. Econ. Geogr.* 2, 31–54.

Irwin, E. G., and Bockstael, N. E. (2007). The evolution of urban sprawl: Evidence of spatial heterogeneity and increasing land fragmentation. *Proc. Natl. Acad. Sci. USA* 104, 33–46.

Jabareen, Y. (2013). Planning the resilient city: Concept and strategies for coping with climate change and environmental risk. *Cities* 31, 220–229. https://doi.org/10.1016/j.cities.2012.05.0040

Joerin, J., Shaw, R., Takeuchi, Y., and Krishnamurthy, R. (2012). Assessing community resilience to climate-related disasters in Chennai, India. *Int. J. Disaster Risk Reduct.* 1, 44–54.

Karunakaran, N. (2014). Cropping pattern and land degradation in Kasaragod, Kerala. *Rajagiri J. Soc. Dev.* 6, 5–20.

Keenan, T. F., Hollinger, D. Y., Bohrer, G., Dragoni, D., Munger, J. W., Schmid, H. P., and Richardson, A. D. (2013). Increase in forest water-use efficiency as atmospheric carbon dioxide concentrations rise. *Nature* 499, 324.

Keith, M., O'Clery, N., Parnell, S., and Revi, A. (2020). The future of the future city? The new urban sciences and a peak urban interdisciplinary disposition. *Cities* 105, Article 102820. https://doi.org/10.1016/j.cities.2020.102820

Kennedy, C. M., Oakleaf, J. R., Theobald, D. M., Baruch-Mordo, S., and Kiesecker, J. (2019). Managing the middle: A shift in conservation priorities based on the global human modification gradient. *Glob. Change. Biol.* 25(3), 811–826.

Kutywawo, D., Chemura, A., and Chagwesha, T. (2012). Soil Quality and Cropping Patterns as Affected by Irrigation Water Quality in Mutema Irrigation. In *13th WaterNet/WARFSA/GWP-SA International Symposium on Integrated Water Resource Management (IWRM)*, Global water Partnership-South Africa: Johannesburg, South Africa.

Lamb, D. W., and Brown, R. B. (2001). Remote-sensing and mapping of weeds in crops. *J. Agric. Eng. Res.* 78, 117–125.

Landres, P. B., Morgan, P., and Swanson, F. J. (1999). Overview of the use of natural variability concepts in managing ecological systems. *Ecol. Appl.* 9, 1179–1188.

Lebourgeois, V., Dupuy, S., Vintrou, É., Ameline, M., Butler, S., and Bégué, A. (2017). A combined random forest and OBIA classification scheme for mapping smallholder agriculture at different nomenclature levels using multisource data (simulated Sentinel-2 time series, VHRS and DEM). *Remote Sens.* 9(3), 259.

Li, J. (2014). Land sale venue and economic growth path: Evidence from China's urban land market. *Habitat Int.* 41, 307–313.

Lindner, M., Maroschek, M., Netherer, S., Kremer, A., Barbati, A., Garcia-Gonzalo, J., Seidl, R., Delzon, S., Corona, P., Kolström, M., Lexer, M., and Lexer, M. J. (2010). Climate change impacts, adaptive capacity, and vulnerability of European forest ecosystems. *For. Ecol. Manag.* 259, 698–709.

Liu, J., Kuang, W., Zhang, Z. et al. (2014). Spatiotemporal characteristics, patterns, and causes of land-use changes in China since the late 1980s. *J. Geogr. Sci.* 24, 195–210. https://doi.org/10.1007/s11442-014-1082-6

Lobell, D. B., Burke, M. B., Tebaldi, C., Mastrandrea, M. D., Falcon, W. P., Naylor, R. L. (2008). Prioritizing climate change adaptation needs for food security in 2030. *Science* 319, 607–610.

Margono, B. A., Potapov, P. V., Turubanova, S., Stolle, F., and Hansen, M. C. (2014). Primary forest cover loss in Indonesia over 2000–2012. *Nat. Clim. Chang.* 4, 730–735.

Mayunga, J. S. (2007). Understanding and applying the concept of community disaster resilience: A capital-based approach. *Summer Acad. Soc. Vulnerability Resil. Build.* 1, 1–16.

McCarthy, M. A., and Burgman, M. A. (1995). Coping with uncertainty in forest wildlife planning. *For. Ecol. Manag.* 74, 23–36.

Miao, Y., Liu, J., and Wang, R. Y. (2021). Occupation of cultivated land for urban: Rural expansion in China: Evidence from national land survey 1996–2006. *Land* 10, 1378. doi: 10.3390/land10121378

Michaletz, S. T., Cheng, D., Kerkhoff, A. J., and Enquist, B. J. (2014). Convergence of terrestrial plant production across global climate gradients. *Nature* 512, 39.

Molla, S. H., and Rukhsana. (2022). Spatio-temporal analysis of built-up area expansion on agricultural land in Mousuni Island of Indian Sundarban region. In: Rukhsana, and Alam, A. (eds.). *Agriculture, Environment and Sustainable Development.* Springer, Cham. https://doi.org/10.1007/978-3-031-10406-0_6

Molla, S. H., and Rukhsana. (2023). Mapping spatial dynamicity of cropping pattern and long-term surveillance of land-use/land-cover alterations in the Indian Sundarban region. *Arab. J. Geosci.* 16, 379. https://doi.org/10.1007/s12517-023-11444-8

Moran, M. S., Inoue, Y., and Barnes, E. M. (1997). Opportunities and limitations for image-based remote sensing in precision crop management. *Remote Sens. Environ.* 61, 319–346.

Moroney, J. L., and Castellano, R. S. (2018). Farmland loss and concern in the Treasure Valley. *Agric. Human Values* 35, 529–536. doi: 10.1007/s10460-018-9847-7

Muyanga, M., Jayne, T. S., and Burke, W. J. (2013). Pathways into and out of poverty: A study of rural household wealth dynamics in Kenya. *J. Dev. Stud.* 49, 37–41.

Narducci, J., Quintas-Soriano, C., Castro, A., Som-Castellano, R., and Brandt, J. S. (2019). Implications of urban growth and farmland loss for ecosystem services in the western United States. *Land Use Policy* 86, 1–11. doi: 10.1016/j.landusepol.2019.04.029

Neelin, J. D., Munnich, S. U., Meyerson, M. H., and Holloway, J. E. (2006). Tropical drying trends in global warming models and observations. *Proceedings of the National Academy of Sciences* 103(16), 6110–6115.

Ning, J., Liu, J., Kuang, W. et al. (2018). Spatiotemporal patterns and characteristics of land-use change in China during 2010–2015. *J. Geogr. Sci.* 28, 547–562. https://doi.org/10.1007/s11442-018-1490-0

North, D. C. (1955). Location theory and regional economic growth. *J. Polit. Econ.* 63, 243–258.

OECD. (2013b). What Does 'Resilience' Mean for Donors? An OECD Factsheet. OECD. www.oecd.org/dac/governancedevelopment/May%2010%202013%20FINAL%20resilience%20PDF.pdf

Paster, E. (2004). Preservation of agricultural lands through land use planning tools and techniques. *Nat. Res. J.* 44, 283–318.

Peng, C., Zhou, X., Zhao, S., Wang, X., Zhu, B., Piao, S., and Fang, J. (2009). Quantifying the response of forest carbon balance to future climate change in Northeastern China: Model validation and prediction. *Glob. Planet. Chang.* 66, 179–194.

Perrin, C. (2013). Regulation of farmland conversion on the urban fringe: From land-use planning to food strategies: Insight into two case studies in Provence and Tuscany. *Int. Plann. Stud.* 18, 21–36. doi: 10.1080/13563475.2013.750943

Power, A. G. (2010). Ecosystem services and agriculture: Tradeoffs and synergies. *Philos. Trans. R. Soc. B Biol. Sci.* 365, 2959–2971. doi: 10.1098/rstb.2010.0143

Qiu, F., Laliberté, L., Swallow, B., and Jeffrey, S. (2015). Impacts of fragmentation and neighbor influences on farmland conversion: A case study of the Edmonton-Calgary Corridor, Canada. *Land Use Policy* 48, 482–494. doi: 10.1016/j.landusepol.2015.06.024

Rahman, M. R., and Saha, S. K. (2009). Spatial dynamics of cropland and cropping pattern change analysis using Landsat TM and IRS P6 LISS III satellite images with GIS. *Geo Spat. Inf. Sci.* 12, 123–134.

Reyer, C., Lasch-Born, P., Suckow, F., Gutsch, M., Murawski, A., and Pilz, T. (2014). Projections of regional changes in forest net primary productivity for different tree species in Europe driven by climate change and carbon dioxide. *Ann. For. Res.* 71, 211–225.

Ribeiro, P. J. G., and Goncalves, L. A. P. J. (2019). Urban resilience: A conceptual framework. *Sustainable Cities and Society* 50, Article 101625. https://doi.org/10.1016/j.scs.2019.101625.

Ripple, W. J., Smith, P., Haberl, H., Montzka, S. A., McAlpine, C., and Boucher, D. H. (2014). Ruminants, climate change and climate policy. *Nat. Clim. Chang.* 4, 2–5.

Rukhsana, and Alam, A. (2021). Agriculture, food, and nutritional security: An overview. In: Rukhsana, and Alam, A. (eds.). *Agriculture, Food and Nutrition Security*. Springer, Cham. https://doi.org/10.1007/978-3-030-69333-6_1

Rukhsana, and Alam, A. (2022). Agriculture, environment and sustainable development: An overview. In: Rukhsana, and Alam, A. (eds.). *Agriculture, Environment and Sustainable Development*. Springer, Cham. https://doi.org/10.1007/978-3-031-10406-0_1

Rukhsana, Alam, A., and Mandal, I. (2021). Impact of microclimate on agriculture in India: Transformation and adaptation. In: Rukhsana, and Alam, A. (eds.). *Agriculture, Food and Nutrition Security*. Springer, Cham. https://doi.org/10.1007/978-3-030-69333-6_3

Rukhsana, and Molla, S. H. (2022). Investigating the suitability for rice cultivation using multi-criteria land evaluation in the Sundarban region of south 24 parganas district, West Bengal, India. *J. Indian Soc. Remote Sens.* 50, 359–372.

Rukhsana, and Molla, S. H. (2023). Soil site suitability for sustainable intensive agriculture in Sagar Island, India: A geospatial approach. *J. Coast. Conserv.* 27, 14. https://doi.org/10.1007/s11852-023-00943-1

Saizen, I., Mizuno, K., and Kobayashi, S. (2006). Effects of land-use master plans in the metropolitan fringe of Japan. *Landsc. Urban Plan.* 78, 411–421.

Samanta, S., Pal, D. P., and Pal, B. (2011). Land suitability analysis for rice cultivation based on multi-criteria decision approach through GIS. *Int. J. Sci. Emerg. Technol.* 2, 12–20.

Schneider, A., and Woodcook, C. E. (2012). Compact, dispersed, fragmented, extensive? A comparison of urban growth in twenty-five global cities using remotely sensed data, pattern metrics and census information. *Urban Stud.* 45, 659–692.

Simon, D. (2008). Urban environments: Issues on the peri-urban fringe. *Annu. Rev. Environ. Resour.* 33, 167–185.

Singh, P., and Nair, A. (2012). Environmental Sustainability of Cropping Patterns in Gujarat. IRMA Working Paper Series, Institute of Rural Management Anand: Gujarat, India.

Skog, K. L., and Steinnes, M. (2016). How do centrality, population growth and urban sprawl impact farmland conversion in Norway? *Land Use Policy* 59, 185–196. doi: 10.1016/j.landusepol.2016.08.035

Song, J., Huang, B., and Li, R. (2017). Measuring recovery to build up metrics of flood resilience based on pollutant discharge data: A case study in East China. *Water* 9, 619.

Song, X. P., Hansen, M. C., Stehman, S. V., Potapov, P. V., Tyukavina, A., Vermote, E. F., and Townshend, J. R. (2018). Global land change from 1982 to 2016. *Nature* 560(7720), 639.

Stahel, W. R. (2016). The circular economy. *Nature* 531, 435–438.

Stefen, W. et al. (2015). Planetary boundaries: Guiding human development on a changing planet. *Science* 347, 1259855.

Steven, M. D., and Clark, J. A. (2013). *Application of Remote Sensing in Agriculture*. Elsevier, Amsterdam, The Netherlands.

Tan, R., Beckmann, V., van den Berg, L., and Qu, F. (2009). Governing farmland conversion: Comparing China with the Netherlands and Germany. *Land Use Policy* 26, 961–974. doi: 10.1016/j.landusepol.2008.11.009

Theobald, D. M., Kennedy, C., Chen, B., Oakleaf, J., Baruch-Mordo, S., and Kiesecker, J. (2020). Earth transformed: Detailed mapping of global human modification from 1990 to 2017. *Earth Sys. Sci. Data* 12(3), 1953–1972.

Thorp, K. R., and Tian, L. F. (2004). A review on remote sensing of weeds in agriculture. *Precis. Agric.* 5, 477–508.

Tian, H., Melillo, J., Lu, C., Kicklighter, D., Liu, M., Ren, W., Xu, X., Chen, G., Zhang, C., Pan, S., Liu, J., and Liu, J. (2011). China's terrestrial carbon balance: Contributions from multiple global change factors. *Glob. Biogeochem. Cycles* 25, GB1007–GB1022.

Tulloch, D. L., Myers, J. R., Hasse, J. E., Parks, P. J., and Lathrop, R. G. (2003). Integrating GIS into farmland preservation policy and decision making. *Landsc. Urban Plan.* 63, 33–48.

Turner, B. L., Lambin, E. F., and Reenberg, A. (2007). The emergence of land change science for global environmental change and sustainability. *Proc. Natl. Acad. Sci.* 104(52), 20666–20671.

UN. (2022). Transforming Our World: The 2030 Agenda for Sustainable Development. https://sustainabledevelopment.un.org/post2015/transformingourworld/publication (accessed on 20 August 2022).

UNDRR. (2022). Making Cities Resilient 2030. www.undrr.org/event/launchmcr2030 (accessed on 20 August 2022).

Vinge, H. (2018). Farmland conversion to fight climate change? Resource hierarchies, discursive power and ulterior motives in land use politics. *J. Rural Stud.* 64, 20–27. doi: 10.1016/j.jrurstud.2018.10.002

Vogels, M. F., Jong, S. M., Sterk, G., and Addink, E. A. (2019). Mapping irrigated agriculture in complex landscapes using SPOT6 imagery and object-based image analysis: A case study in the Central Rift Valley, Ethiopia. *Int. J. Appl. Earth. Obs. Geoinf.* 75, 118–129.

Webster, D. (2001). *On the Edge: Shaping the Future of Peri-Urban East Asia.* Stanford University/Asia Pasific Research Center, Stanford, CA, USA.

World Bank. (2012). *World Development Report 2012: Gender Equality and Development.* World Bank, Washington, DC. doi: 10.1596/978-0-8213-8666-8

Wu, W., Zhao, S., Zhu, C., and Jiang, J. (2015). A comparative study of urban expansion in Beijing, Tianjin and Shijiazhuang over the past three decades. *Landsc. Urban Plan.* 134, 93–106.

Yoon, D. K., Kang, J. E., and Brody, S. D. (2016). A measurement of community disaster resilience in Korea. *J. Environ. Plan. Manag.* 59, 436–460.

Zanotti, L., Ma, Z., Johnson, J. L., Johnson, D. R., Yu, D. J., Burnham, M., and Carothers, C. (2020). Sustainability, resilience, adaptation, and transformation: Tensions and plural approaches. *Ecology and Society* 25(3), 4. https://doi.org/10.5751/ES-11642-250304

Zhang, Y., Chen, Z., Cheng, Q., Zhou, C., Jiang, P., Li, M., et al. (2016). Quota restrictions on land use for decelerating urban sprawl of mega city: A case study of Shanghai, China. *Sustainability* 8, 968. doi: 10.3390/su8100968

Zwiggelaar, R. (1998). A review of spectral properties of plants and their potential use for crop/weed discrimination in row-crops. *Crop Prot.* 17, 189–206.

2 A Deep Learning Framework for Extraction of Crop and Forest Cover from Multispectral Remote Sensing Images

Megha Sharma, Anil Kumar, Supriya M., and Kishore S.

2.1 INTRODUCTION

In the study of urban planning, ecological environment change, disaster management, or agriculture, land-cover classification of remote sensing images plays an essential function. Remote sensing technology advancements have made it possible to access vast remote sensing databases that were too big for manual processing. These include the United States Geological Survey (USGS) Earth Explorer, European Space Agency Sentinel Mission, or China High-Resolution Earth Observation System. The rapid growth in population directly impacts the pattern of global land use for agriculture, the amount of forest area, the proximity of various types of water bodies, etc. The literature presents the use of satellite remote sensing and geographic information system (GIS) technologies in mapping various land covers. The images collected from sensors could be analyzed and classified for a better understanding of the relationship between landscape dynamics and the ecosystem. For many different applications, including environment, forestry, hydrology, agriculture, and geology, satellite imagery and GIS maps for land-cover/land-use changes are essential. Programs for managing, planning, and monitoring natural resources depend on precise data regarding the local land cover.

Scientific research activities at landscape and regional dimensions have been made easier with the availability of satellite images in moderate to high-resolution formats. As remote sensing technology has advanced, remote sensing images data have become an integral part of the dynamic monitoring of changes in the world's resources and the use of land cover. Real-life applications like urban monitoring, fire detection, crop mapping, and estimating temperature or ocean salinity from remotely sensed multispectral images have great societal value. Remote sensing data processing deals with these applications successfully creating an impact on society. Among the various applications, land cover mapping has become the most important. Land use describes the use of the land surface by humans. The various uses of land include agricultural, transport, residential, commercial, and recreational (Campbell et al., 2011). The land cover indicates the visible features on the earth's surface including the vegetative cover, the water, soil, forest, and grassland. Land-cover classification of remote sensing data (Kussul et al., 2017) is a long-standing challenge because of its various features and heterogeneous study areas. It is important for many scientific, resource management, and policy-driven human activities (Cihlar, 2000) and helps in understanding the earth's biophysical systems.

When the electromagnetic energy from the sun falls on the earth's surface, the different features on the surface reflect, absorb, transmit, and emit this energy. Spectral reflectance measures

DOI: 10.1201/9781003377825-3

the percentage of electromagnetic energy reflected by a surface at a specific wavelength. Remote sensing technology uses this property to generate satellite images. The most important method in remote sensing for computerized analysis and pattern identification of satellite data is satellite image classification, which is based on several image structures. The classification of satellite images has been a conventional problem in machine learning applications, especially with the advent of neural networks (Unnikrishnan et al., 2018). Land-cover classification classifies pixels whose spectral characteristics are similar (Panda et al., 2005). The sensors can measure these interactions in the form of spectral reflectance. These pixels are further mapped to specific classes such as forest, water, grassland, and built-up area.

Consequently, a large number of traditional classification algorithms such as Maximum Likelihood Classifier (Murthy et al., 2003), Random Forest (Tatsumi et al., 2015), and Support Vector Machine (Zheng et al., 2015) have been proposed by researchers. Among them, significant results are shown by machine learning methods (Fausett, 2006; Kussul et al., 2006; Singh et al., 2022). Fuzzy methods have also seen extensive use in remote sensing applications like crop yield (Deve et al., 2017), mapping paddy fields (Rawat et al., 2021; Singh et al., 2020a; Rawat et al., 2022), urban mapping (Hamde et al., 2021; Akshay et al., 2020), burnt paddy field identification (Singh et al., 2020b), etc. Researchers have also explored soft computing, which combines the fundamentals of neural networks and fuzzy and genetic algorithms in their work (Senthil Kumar et al., 2017; Singh, Kumar et al., 2020, 2021). Hence, the development of reliable and efficient methods for automatic land-cover classification of these images is of prime importance.

The main aim of this research study has been based on exploring the capability of artificial neural networks (ANNs) and deep learning algorithms. The work presents the use of feature extractors convolution neural network (CNN), one-dimensional CNN (1D-CNN)–based models for the land-use/land-cover classification in the Haridwar region on multispectral Formosat-2. The objectives addressed in this work are as follows:

1. We develop a deep learning framework that utilizes the 1D-CNN for the wheat crop, eucalyptus, and dense forest classes identification.
2. We compare the classification results of ANN and 1D-CNN.

This chapter has been organized into six sections. Section 2.2 explains the literature survey done for this paper, Section 2.3 presents the experimental dataset, Section 2.4 describes the methodology, Section 2.5 provides experimental results and related discussion, and Section 2.6 concludes the outcomes.

2.2 LITERATURE SURVEY

The characteristic of supervised learning is the use of labeled datasets. In supervised learning, the algorithm learns from the training dataset by iteratively making predictions on the data and adjusting for the correct decision. This is why supervised learning models are more accurate than unsupervised learning models. In the present work, satellite imagery has been used as the input data. It uses labeled information about class membership of single pixels to build a model that can generalize the whole image (Camps-Valls, 2009). Supervised classification requires training of pixel data to teach the classifier to make correct decisions. In unsupervised learning, the goal is to get details from large volumes of new data. Here, the machine learning will determine what is different or interesting from the input set. The neural network approach for the analysis of data does not require any statistical data distribution (Sisodia et al., 2014). One of the categories of the feed-forward ANN is a multilayer perceptron. The traditional kind of neural network is a multilayer perceptron. The neural network consists of several processors, also known as nodes. These nodes are grouped in layers and run simultaneously. The raw input is sent to the first layer. Then, before passing its output to the tier below it, each succeeding layer receives input from the layer before it. The final output is processed in the last tier or layer. The nodes in the tiers before and following it are intricately related to one another. The ANN is also referred to as a feed-forward neural network because it only processes inputs moving forward. The key to the efficacy of neural networks is their high adaptability and quick learning. Those inputs which contribute more towards the correct output are given the highest weight. There can be one or more hidden layers providing levels of abstraction, and the predictions are made on the output layer, also called the visible layer (Bezdek et al., 1984).

A perceptron computes a linear combination of real-valued inputs using a vector of inputs and outputs 1 if the result exceeds a threshold and −1 otherwise (Mitchell, 2013). This is explained in Equation 2.1:

$$O\left(x_1,\ldots\ldots,x_n\right) = \begin{cases} 1, if \sum_{i=0}^{n} w_i x_i > 0 \\ -1, otherwise \end{cases} \tag{2.1}$$

where each w_i is a real-valued constant, or weight that determines the contribution of input x_i to the perceptron output. The conventional machine learning approaches first extract the relevant features and then apply shallow classification techniques. Deep learning is a state-of-the-art technology (Fakhfakh et al., 2017; Vinayakumar et al., 2019) in the remote sensing and geoscience field. This is primarily due to its capability for automated extraction of meaningful features (Tsagkatakis et al., 2019). Deep learning (LeCun et al., 2015) has had a significant impact on remote sensing image classification. Traditional approaches to remote sensing image classification involved manual feature extraction, which was time consuming and limited in terms of accuracy. With the advent of deep learning, these limitations have been overcome, and remote sensing image classification has become more accurate and efficient. One of the most significant advantages of deep learning in remote sensing image classification is its ability to handle large and complex datasets. With the increasing availability of remote sensing data, traditional approaches to image classification have become inadequate. Deep learning algorithms (Senthil Kumar et al., 2017) can handle large and complex datasets and learn from them, making them ideal for remote sensing applications. CNNs are explicit deep neural network architectures that have replaced the traditional matrix multiplication by using convolutions. CNN is a kind of deep learning method that uses filters with a specific window size that sweep the image or matrix to create feature maps at convolution layers (Singh, Kumar et al., 2021). Optimization is a mathematical discipline that determines the "best" solution in a quantitatively well-defined sense. The Adam optimization algorithm is an extension of the classical stochastic gradient descent algorithm to update network weights in an interactive manner based on training data. The name Adam is derived from adaptive moment estimation (Kingma et al., 2014). The power of the neural networks as described in Panda et al. (2005) and the various deep learning algorithms are finding great scope for satellite image analysis (Zhang et al., 2016).

2.3 STUDY AREA AND DATASET USED

The area under study is the Haridwar region which falls in the North Indian state of Uttarakhand. The proposed work uses Formosat-2 satellite image data. Formosat-2 is a Taiwan Earth Imaging Satellite that collects high-resolution panchromatic band with a spatial resolution of 2 m and a spatial resolution of multispectral bands of 8 m. There are various classes that have testing samples for testing the results. This region of Haridwar has all the important features of land cover, namely water, vegetation, and soil. Figure 2.1 shows the study area with various classes. In this chapter, the focus is on dense forests, wheat crops, and eucalyptus class. The coordinates of the location are 29°48′35.7″ N 78°14′58.9″ E. Significant research has been done in the past to classify this study area using supervised possibilistic c means (Singh et al., 2020a, 2019).

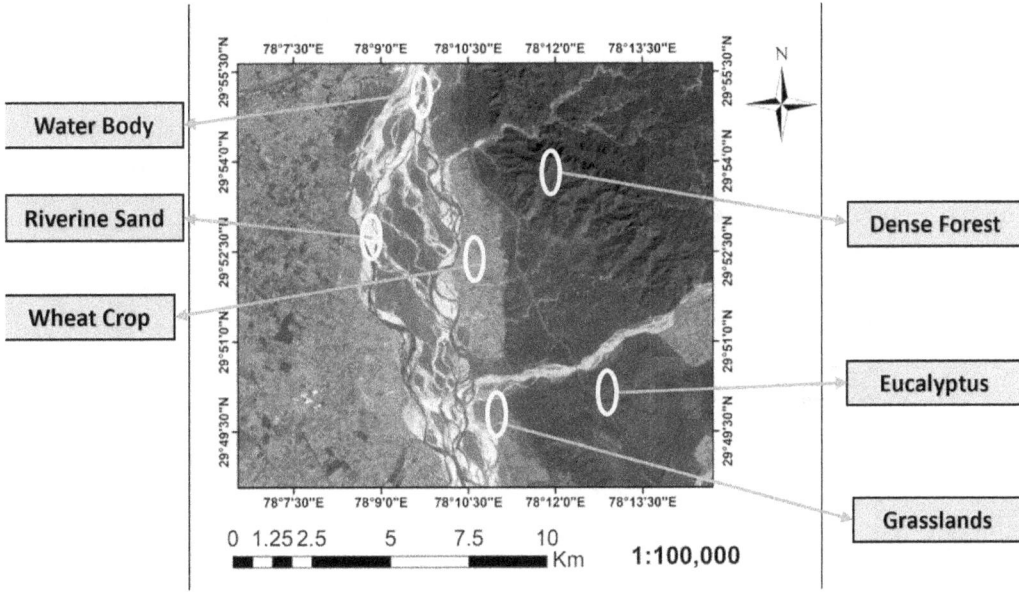

FIGURE 2.1   Study area—Haridwar region on the India map.

2.4   METHODOLOGY

Figure 2.2 shows the overall methodology of the proposed framework. Let X be the input image of size $(L \times M)$ and n dimensions, and let W be the filter (also called kernel) of size $(1 \times m)$ and n dimensions. We denote the convolution operation as $*$, and the element-wise multiplication as $\odot$. Then, the output Y of a 1D-CNN with a single filter can be computed as mentioned in Equation 2.2:

$$Y_{i,j} = \sigma \left(\sum_{p=1}^{n} \sum_{u=1}^{t} \sum_{v=1}^{m} W_{u,v,p} , X_{i+u-1,j+v-1,p} + b \right) \tag{2.2}$$

where σ is the activation function; b is the bias term; and i, j are the indices of the output feature map. Equation 2.2 represents the sliding window operation of the filter over the input image, where the filter is multiplied elementwise with the input image at each position, and the resulting values are summed up. The bias term is added to this sum, and the activation function is applied to the result to produce the output value. The filter W is a tensor of size $(1 \times m \times n)$ and represents the weights of the 1D-CNN. Each value of the filter tensor W is a learnable parameter, which is optimized during training to improve the accuracy of the model. The input image X is a tensor of size $(L \times M \times n)$, where each pixel in the image has n values corresponding to its color or intensity value in each channel. The sliding window operation is performed along the two spatial dimensions (L and M) and the n channels. In practice, a CNN can have multiple filters, each producing a different output channel. The outputs of these channels are stacked to form the final output tensor.

The convolution layer, pooling layer, dropout layer, and fully linked or dense layers make up the basic 1D-CNN structure used for classification. The feature extraction process makes use of the convolution and pooling layers. The dropout layer is a regularization method that is employed. It randomly drops some of the neurons so that the training does not get biased by relying only on a few neurons (Srivastava et al., 2014). Dropout of 0.5 is used to prevent overfitting. The number of

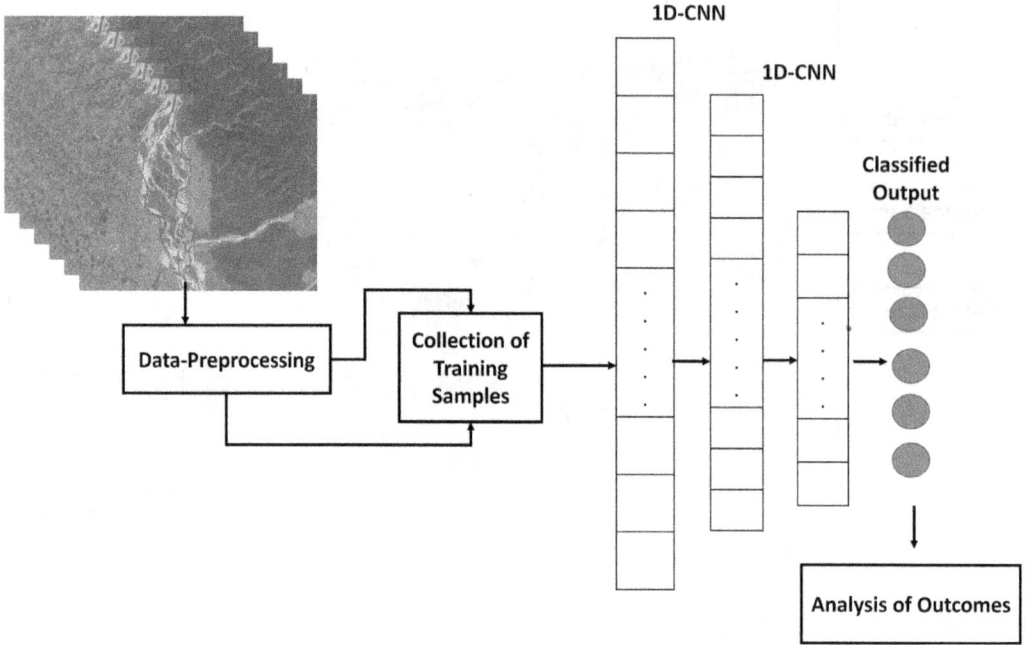

FIGURE 2.2 Framework of the proposed approach for the extraction of multiclass.

epochs = 30 and maxpooling 1D = 2 were considered. The maxpooling operation downsampled the input representation by taking the maximum value over the window refined by the pool size.

Stochastic gradient descent (SGD) is the fundamental algorithm for training deep neural networks. The present study uses state-of-the-art variations of SGD, namely, root mean square propagation, Adam, and adaptive gradient. The learning rate used was 0.001, and the window size was 2. Adaptive gradient adaptively sets the learning rate (Staib et al., 2019) according to a parameter. Parameters with high gradients or frequent updates should have a slower learning rate so that we do not overshoot the minimum value. Parameters with low gradients or infrequent updates have a faster learning rate to train them quickly. Stochastic gradient descent approximates the gradient descent search by updating weights incrementally, thus calculating the error for each individual record. It uses only a single sample to perform each iteration. The use of modular neural networks is suggested in the comparative analysis of the remote sensing classification techniques (Kussul et al., 2017), where a detailed study of the statistical as well as the neural network approaches is given. A combination of classification methods has been experimented with test results. The comparison of the performance of the combination of 1D-CNN with SGD and Adam was done with traditional ANN. The impact of the Adam optimizer was observed to see the variation in the output classified images.

2.5 EXPERIMENTAL RESULTS

Machine learning algorithms that learn, i.e. gradually optimise their internal parameters over time, are frequently represented by learning curves. This is what happens in deep learning neural networks (Sisodia et al., 2014) where algorithms learn from a training dataset incrementally. The learning happens in a manner that creates a small positive or negative change in a variable quantity or function. A learning curve is a plot that shows the performance of the model learning over the experience or time. In classification accuracy, better scores or larger numbers indicate more learning. Thus, the metric used to evaluate learning is maximizing.

The performance is the accuracy of the classification against the epoch. The epoch indicates the number of passes of the entire training dataset of the models. With each epoch, the dataset's internal model parameters are updated. The learning response shows a good fit with the testing or the cross-validation accuracy close to the training accuracy. We observe an initial increase in training and validation accuracy. After accuracy touches subsequent improvement, the performance remains flat till the end. The model runs for 30 epochs. The classification results are good when compared with the ground truth.

The models are run for various epochs with the best results obtained around 30 epochs. 1D-CNN classification gives better results as compared with ANN. Iterations play a very good part in increasing accuracy. The relationship between performance and iteration count is not always directly proportional. We see an initial increasing training and validation accuracy. After accuracy touches 0.91, the curve remains flat till the end. The model runs for 30 epochs. The classification results are good when compared with the ground truth. Table 2.1 displays the recorded accuracy and F1 score on running the different models. Observation shows that the Adam optimizer brought about the best results. Figure 2.3 depicts two classified images. The forest class is highlighted. When the ANN model is used on the input image, there were some areas that did not show the accurate forest area. Figure 2.3 shows its comparison with the 1D-CNN model which has classified the forest more accurately. The arrows highlight the specific areas where ANN shows incorrectly estimated areas while 1D-CNN accurately highlights the forest cover. Figure 2.4 depicts extracted wheat class. When the ANN model was used on the input image, there were some areas that did not show the accurate wheat area. In ANN, the wheat area was incorrectly classified as compared to the ground truth. Figure 2.4 shows its comparison with the 1D-CNN which has classified the wheat more accurately. Figure 2.5 depicts the eucalyptus class identification. The arrows highlight the specific areas where ANN shows an underestimated eucalyptus class, while 1D-CNN accurately highlights the eucalyptus cover next to the river basin.

The performance of 1D-CNN shows several advantages compared to general ANN. 1D-CNNs can learn hierarchical representations of the input data. The first layer captures low-level features, while subsequent layers capture more abstract features that are built on top of the lower-level features. 1D-CNNs are robust to noise in the input data. This is because the convolution operation smooths the input, effectively reducing the impact of noise. Table 2.1 shows the comparison of class-wise accuracy metrics of ANN and 1D-CNN. Table 2.1 signifies the impact of incorporating 1D-CNN in extracting dense forest, wheat, and eucalyptus classes in terms of recall, precision,

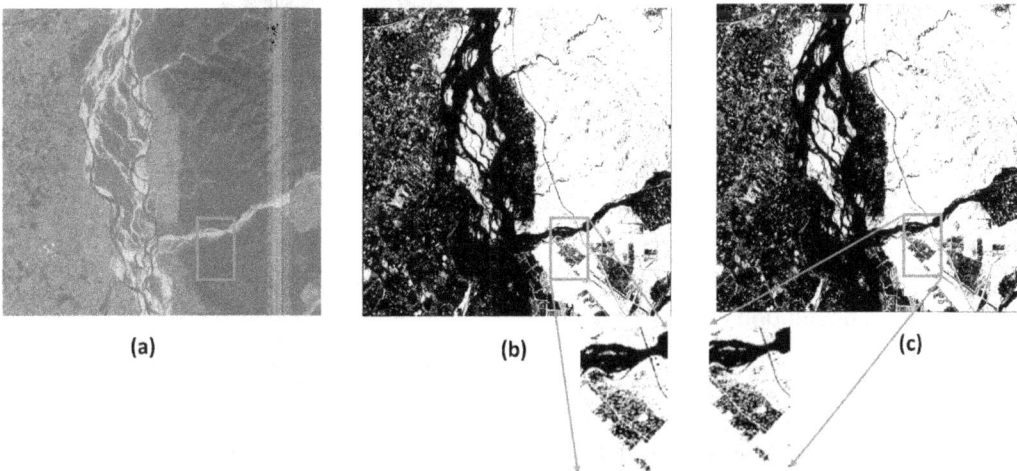

FIGURE 2.3 Comparison between the classified images for the forest class as obtained by ANN and 1D-CNN models: (a) original image, (b) 1D-CNN, (c) ANN.

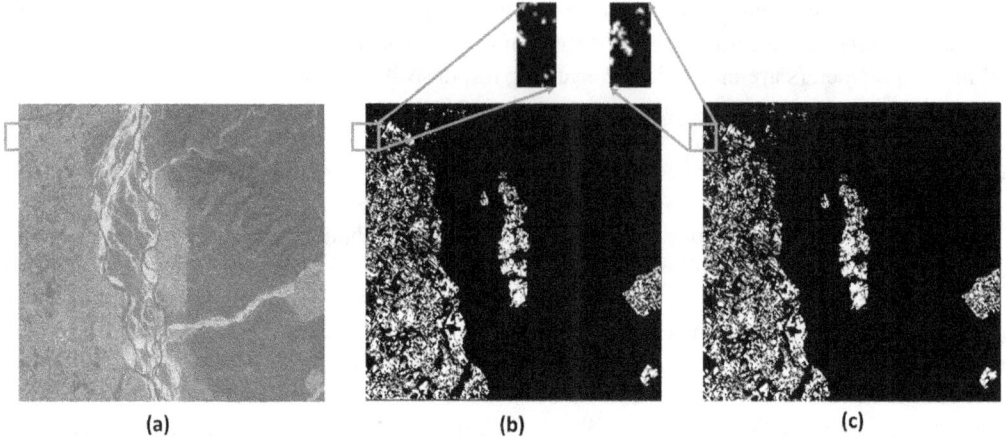

FIGURE 2.4 Comparison between the classified images for the wheat class as obtained by ANN and 1D-CNN models: (a) original image, (b) 1D-CNN, (c) ANN.

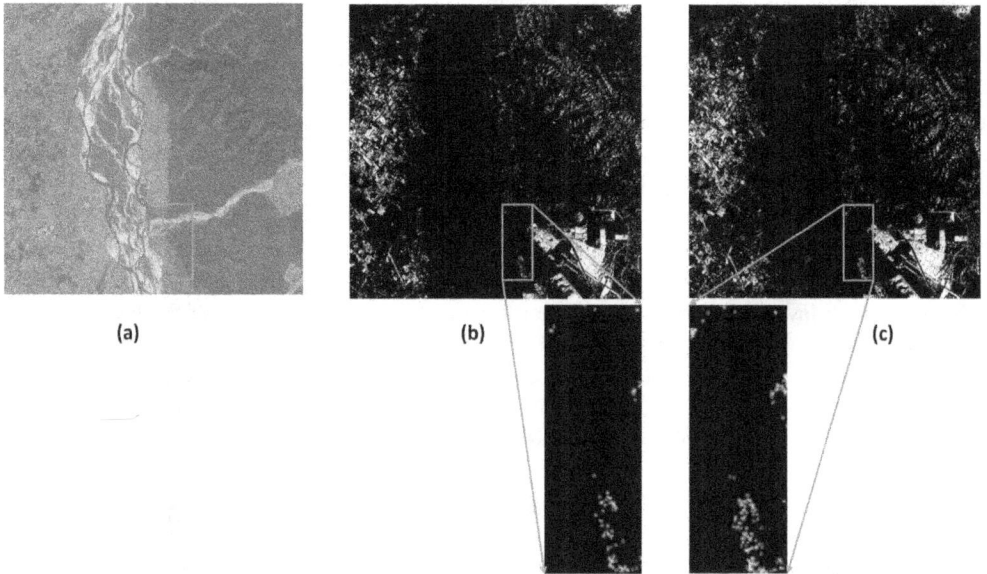

FIGURE 2.5 Comparison between the classified images for the eucalyptus class as obtained by ANN and 1D-CNN models: (a) original image, (b) 1D-CNN, (c) ANN.

TABLE 2.1

Comparison of Class-Wise Accuracy Metrics of Artificial Neural Network and One-Dimensional Convolution Neural Network

	Class-Wise Accuracy (%)					
	ANN			1D-CNN		
Classes/Metrics	Recall	Precision	F1-Score	Recall	Precision	F1-Score
Denseforest	**99.85**	96.65	98.22	99.22	**97.79**	**98.50**
Wheat	**99.27**	94.99	97.08	95.56	**99.97**	97.71
Eucalyptus	95.28	87.67	91.32	**98.86**	91.48	**95.03**

Note: ANN, artificial neural network; 1D-CNN, one-dimensional convolution neural network.

and F1 score. The 1D-CNN shows significant improvement compared to ANN. In 1D-CNN, the classification was crisp in comparison to ANN as the pixels could be sharply differentiated in the output images.

2.6 CONCLUSION

In this chapter, we have analyzed a framework for measuring and monitoring wheat crop, eucalyptus, and dense forest classes of the study area. In this work, ANN and deep neural networks including 1D-CNN were used to classify and extract water, wheat, and eucalyptus classes. The tests have been carried out using various optimizers like SGD and Adam. The classification results have produced an F1 score of 0.95 for ANN and 0.98 for CNN. The 1D-CNN with Adam optimizer produced crisper boundaries and was successful in extracting the water, wheat, and eucalyptus classes. The output images show that the 1D-CNN model with the Adam optimizer could sharply differentiate the classes, resulting in better classification. Overall, the proposed approach has been successful in extracting water, wheat, and eucalyptus classes. We will investigate the proposed approach for extracting features from high-dimensional images at a large scale.

REFERENCES

Akshay, S., Mytravarun, T., Manohar, N., & Pranav, M. (2020). Satellite image classification for detecting unused landscape using CNN. In 2020 International Conference on Electronics and Sustainable Communication Systems (ICESC) (pp. 215–222).

Bezdek, J. C., Ehrlich, R., & Full, W. (1984). FCM: The fuzzy c-means clustering algorithm. Computers & Geosciences, 10 (2–3), 191–203.

Campbell, J. B., & Wynne, R. H. (2011). Introduction to remote sensing. Guilford Press.

Camps-Valls, G. (2009). Machine learning in remote sensing data processing. In 2009 IEEE International Workshop on Machine Learning for Signal Processing (pp. 1–6).

Cihlar, J. (2000). Land cover mapping of large areas from satellites: Status and research priorities. International Journal of Remote Sensing, 21 (6–7), 1093–1114.

Deve, S. N., & Geetha, P. (2017). Land use modeling for sustainable crop development using fuzzy approach. In 2017 International Conference on Communication and Signal Processing (ICCSP) (pp. 1473–1476).

Fakhfakh, R., Ammar, A. B., & Amar, C. B. (2017). Deep learning-based recommendation: Current issues and challenges. International Journal of Advanced Computer Science and Applications, 8 (12), 59–68.

Fausett, L. V. (2006). Fundamentals of neural networks: Architectures, algorithms and applications. Pearson Education India.

Hamde, N. S., Kumar, A., & Maithani, S. (2021). Fuzzy machine learning approach for transitioned building footprints extraction using dual-sensor temporal data. SN Applied Sciences, 3 (4), 1–11.

Kingma, D. P., & Ba, J. (2014). Adam: A method for stochastic optimization. arXiv preprint arXiv:1412.6980.

Kussul, N., Lavreniuk, M., Skakun, S., & Shelestov, A. (2017). Deep learning classification of land cover and crop types using remote sensing data. IEEE Geoscience and Remote Sensing Letters, 14 (5), 778–782.

Kussul, N., Skakun, S., & Kussul, O. (2006). Comparative analysis of neural networks and statistical approaches to remote sensing image classification. International Scientific Journal of Computing, 5 (2), 93–99.

LeCun, Y., Bengio, Y., & Hinton, G. (2015). Deep Learning. Nature, 521 (7553), 436–444.

Mitchell, T. M. (2013). Machine learning, Indian edition. McGraw Hill Education.

Murthy, C., Raju, P., & Badrinath, K. (2003). Classification of wheat crop with multi-temporal images: Performance of maximum likelihood and artificial neural networks. International Journal of Remote Sensing, 24 (23), 4871–4890.

Panda, B., & Panda, B. (2005). Remote sensing principles and applications. Viva Books Pvt. Limited.

Rawat, A., Kumar, A., Upadhyay, P., & Kumar, S. (2021). Deep learning-based models for temporal satellite data processing: Classification of paddy transplanted fields. Ecological Informatics, 61, 101214.

Rawat, A., Kumar, A., Upadhyay, P., & Kumar, S. (2022). A comparative study of 1D-convolutional neural networks with modified possibilistic c-mean algorithm for mapping transplanted paddy fields using temporal data. Journal of the Indian Society of Remote Sensing, 50 (2), 227–238.

Senthil Kumar, A., Kumar, A., Krishnan, R., Chakravarthi, B., & Deekshatalu, B. (2017). Soft computing in remote sensing applications. Proceedings of the National Academy of Sciences, India Section A: Physical Sciences, 87 (4), 503–517.

Singh, A., & Bruzzone, L. (2022). Mono-and dual-regulated contractive-expansive-contractive deep convolutional networks for classification of multispectral remote sensing images. IEEE Geoscience and Remote Sensing Letters, 19, 1–5.

Singh, A., & Kumar, A. (2019). Introduction of local spatial constraints and local similarity estimation in possibilistic c-means algorithm for remotely sensed imagery. Journal of Modeling and Optimization, 11 (1), 51–56.

Singh, A., & Kumar, A. (2020a). Fuzzy-based approach to incorporate spatial constraints in possibilistic c-means algorithm for remotely sensed imagery. International Journal of Intelligent Information and Database Systems, 13 (2–4), 307–318.

Singh, A., & Kumar, A. (2020b). Identification of paddy stubble burnt activities using temporal class-based sensor-independent indices database: Modified possibilistic fuzzy classification approach. Journal of the Indian Society of Remote Sensing, 48, 423–430.

Singh, A., Kumar, A., & Upadhyay, P. (2020). Modified possibilistic c-means with constraints (MPCM-S) approach for incorporating the local information in a remote sensing image classification. Remote Sensing Applications: Society and Environment, 18, 100319.

Singh, A., Kumar, A., & Upadhyay, P. (2021). A novel approach to incorporate local information in possibilistic c-means algorithm for an optical remote sensing imagery. Egyptian Journal of Remote Sensing and Space Science, 24 (1), 151–161.

Sisodia, P. S., Tiwari, V., & Kumar, A. (2014). A comparative analysis of remote sensing image classification techniques. In 2014 International Conference on Advances in Computing, Communications and Informatics (ICACCI) (pp. 1418–1421).

Srivastava, N., Hinton, G., Krizhevsky, A., Sutskever, I., & Salakhutdinov, R. (2014). Dropout: A simple way to prevent neural networks from overfitting. Journal of Machine Learning Research, 15 (1), 1929–1958.

Staib, M., Reddi, S., Kale, S., Kumar, S., & Sra, S. (2019). Escaping saddle points with adaptive gradient methods. In International Conference on Machine Learning (pp. 5956–5965).

Tatsumi, K., Yamashiki, Y., Torres, M. A. C., & Taipe, C. L. R. (2015). Crop classification of upland fields using random forest of time-series Landsat 7 ETM+ data. Computers and Electronics in Agriculture, 115, 171–179.

Tsagkatakis, G., Aidini, A., Fotiadou, K., Giannopoulos, M., Pentari, A., & Tsakalides, P. (2019). Survey of deep-learning approaches for remote sensing observation enhancement. Sensors, 19 (18), 3929.

Unnikrishnan, A., Sowmya, V., & Soman, K. (2018). A two-band convolutional neural network for satellite image classification. In International Conference on Communications and Cyber Physical Engineering 2018 (pp. 161–170).

Vinayakumar, R., Soman, K., & Poornachandran, P. (2019). A comparative analysis of deep learning approaches for network intrusion detection systems (N-IDSS): Deep learning for N-IDSS. International Journal of Digital Crime and Forensics (IJDCF), 11 (3), 65–89.

Zhang, L., Zhang, L., & Du, B. (2016). Deep learning for remote sensing data: A technical tutorial on the state of the art. IEEE Geoscience and Remote Sensing Magazine, 4 (2), 22–40.

Zheng, B., Myint, S. W., Thenkabail, P. S., & Aggarwal, R. M. (2015). A support vector machine to identify irrigated crop types using time-series Landsat NDVI data. International Journal of Applied Earth Observation and Geoinformation, 34, 103–112.

3 Crop Pattern Mapping Using GIS and Remote Sensing
A Case Study from Multan District of Pakistan

Neyha Rubab Syed, Ali Raza, Muhammad Zubair,
Siham Acharki, Sajjad Hussain, and Sudhir Kumar Singh

3.1 INTRODUCTION

The sustainability of the agriculture system is the most significant factor in both developing and developed countries. Utilizing the latest technology and techniques to map and monitor cropping patterns can greatly contribute to achieving this goal. Therefore, it is the responsibility of everyone to practice good agricultural methods which will ensure the production of safe and high-quality agricultural products. For attaining sustainable agriculture of an area, mapping and monitoring of cropping patterns including their land use/land cover (LULC) are necessary (Hussain et al. 2023). The process of data modelling involves transforming information requirements into a database design that can be fully implemented. This is achieved through three levels of abstraction—conceptual, logical, and physical—which represent the outcomes of the database design process. The focus of this process is on the logical data model, which builds on the data context established in the conceptual model and introduces logical data structuring. To translate the designed data model into a physical database, a specific database management system (DBMS) can be used to organize the database implementation, such as the moderate resolution imaging spectroradiometer (MODIS) approach (Jin et al. 2011)

Long-term Normalized Difference Vegetation Index (NDVI) data can provide insights into changes in bioclimatic zones' geographical distribution, indicating variations in land-use patterns and large-scale circulation (He et al. 2017). Although remote sensing time-series data's effectiveness in monitoring vegetation seasons is acknowledged (Hussain et al. 2022a), only a few strategies have been devised for deriving seasonality factors from this data (Jönsson and Eklundh, 2004). Crop yield assessment and NDVI data projections are broadly applied for vegetation monitoring (Hu et al. 2023; Benedetti and Rossinni, 1993; Kohli et al. 2020).

The agricultural and hydrological fields are influenced by the broad range of applications offered by remote sensing (RS) and GIS (Raza et al. 2023a, 2023b). Many researchers have directed their attention toward LULC studies due to their significant impact on the local ecology and vegetation (Bisht et al. 2013). Changes in LULC are critical for a variety of applications, including temperature, erosion, and land planning (Sajjad et al. 2019; Useya and Shengbo, 2019). NDVI values reached their highest when peak season came near October. In this investigation, we concentrate on NDVI because of its potential to detect subtle modifications in vegetation. The results of this study may help alleviate food insecurity in the country. In addition, NDVI is a strong indicator of the conditions for plant development and the extent of plant cover in an investigation. According to Faisal et al. (2020), if a region is covered in vegetation, its NDVI value is a positive number that rises as the vegetation cover improves. The application of remote sensing is particularly useful in agriculture and horticulture to estimate crop yield and vegetation strength, assess water demand, create crop mapping of vegetation, monitor changes in land cover, and implement irrigation and precision farming techniques (Mallupattu and Sreenivasula, 2013; Hussain et al. 2022b; Mamatkulov et al. 2021; Hussain et al. 2020).

The objectives of current research are mapping and monitoring of cropping patterns in the Multan district located in South-Punjab, Pakistan, by using a logical modelling approach and NDVI technique. The study aimed to map and monitor the cropping patterns of a selected study area using different tools (GIS and remote sensing) and analyzing them both qualitatively and quantitatively. The survey-based questionnaire was also used to perform analysis and evaluate study site features for characterization and assessment. This chapter illustrates that to achieve these objectives on a larger scale, the government should encourage development of techniques for mapping and monitoring cropping patterns in order to attain a well-developed agriculture sector.

3.2 SITE DESCRIPTION AND METHODOLOGY

3.2.1 SELECTED SITE

The Multan district lies in the southern part of Punjab, Pakistan, as shown in Figure 3.1. It spans over 227 km² and is located at 30° 11′52″ N, 71° 28′11″ E. Geographically, the location of Multan is in the center of the country. Multan has a hot desert climate with very hot summers and relatively cool winters; average annual rainfall is 175 mm, and the normal annual rainfall is 186 mm (Raza et al. 2020). The lowest temperature is around −1°C, while the warmest is around 52°C. Multan

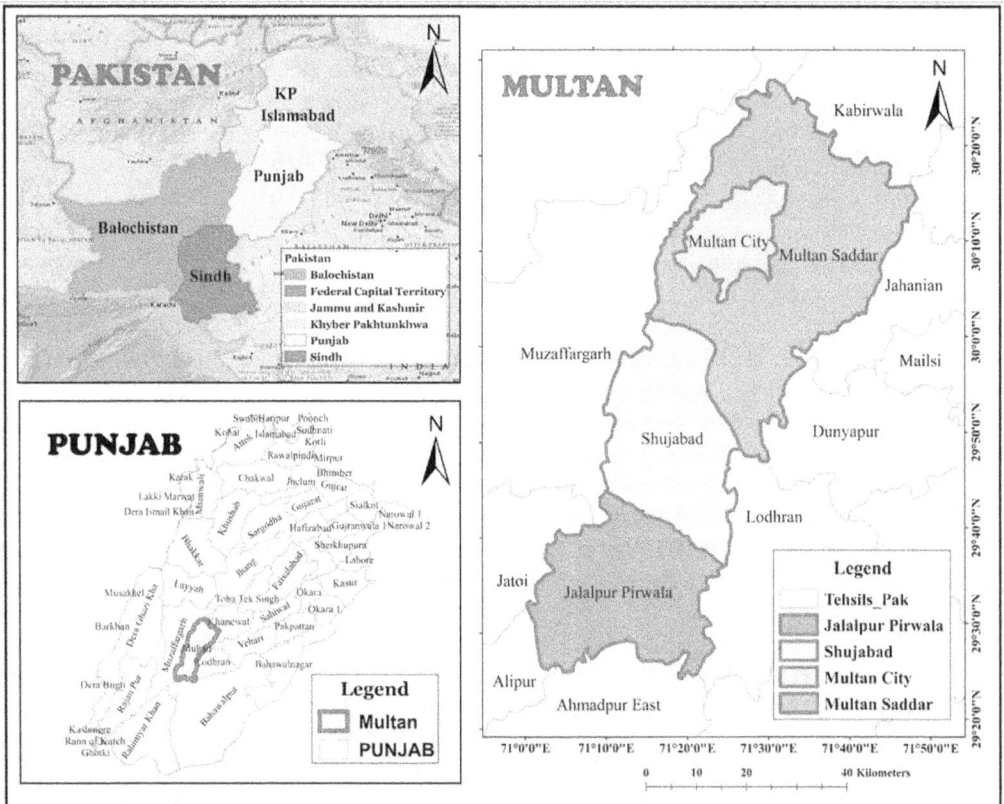

FIGURE 3.1   Study area of Multan district.

has an annual mean temperature of 25.6°C. June is the hottest month with an average temperature of 35.5°C. With an average temperature of 13.2°C, January is the coldest month of the year (Wang et al. 2022). There are several farms in the Multan area. More than 80% of the people rely on agriculture for their livelihood, and the area's topography is nearly plain, flat, and perfect for agriculture. Mangoes, citrus, guava, date palm, pomegranate, falsa, jujube, bananas, strawberries, and grapes are among the fruit grown. Major crops include wheat, cotton, sugarcane, rice, maize, sunflower, fodder crops, etc. Minor crops include rapeseed and mustard, pulses (i.e., chickpea, lentil, mung bean, and mush bean), etc. Mangoes from Multan are well known across Pakistan and beyond. The farms in the Multan district rely heavily on water from the River Chenab.

3.2.2 DATA ACQUISITION

The MOD13Q1 (MOD13Q1: h24v05, h24v06) and Landsat (4–5, 7, and 8) datasets were obtained from US NASA and United States Geological Survey websites, respectively. The current study used 24 datasets of MODIS (MOD13Q1) at 250-m resolution for a 20-year period (15 April 2000–30 December 2020) in order to map and monitor the cropping patterns over the study area. Furthermore, 16 datasets of MODIS (MOD13Q1) were used for a 1-year period (15 April 2020–30 December 2020) to derive phenological parameters. Table 3.1 summarizes the dataset used in the current study.

According to satellite data, the Multan district is located in path 150 row 39 and path 150 row 40 (Hu et al. 2023). LULC maps were prepared from satellite image data. The Landsat 4–5 images for rabi season (2000), Landsat 7 for kharif season (2000), and Landsat 8 for both kharif and rabi seasons (2016, 2020) were used in this study. Moreover, field survey was conducted to collect information from local people regarding various aspects of crops. In addition, a questionnaire-based study was carried out in the field to ascertain farmers' perspectives on cropping patterns and their preferences in crop selection. A total of 60 farmers were interviewed using a questionnaire consisting of 12 questions (see Appendix 3.1).

3.2.3 METHODOLOGY

The first step was to acquire data from US NASA (MODIS data) and USGS (Landsat data) websites. The MODIS data obtained were in HDF (Hierarchical Data Format), and Landsat data were in TIFF (Tag Image File Format). The obtained MODIS data were re-projected from original sinusoidal projection to geographic projection (WGS84 datum) before further processing. Later, stacking, mosaicking, and subset to area of interest (AOI) were done using ArcMap 10.8 for further analysis. The NDVI technique was applied to map cropping pattern, crop identification, and NDVI of rabi and kharif seasons of the years 2000, 2004, 2008, 2012, 2016, and 2020 using MODIS (MOD13Q1) data. Cropping patterns are mapped using NDVI matrices and calculated through raster calculator using the ARC map tool spatial analysis in ArcMap 10.8. LULC maps are extracted by the supervised classification of Landsat (4–5, 7, and 8) images. The quantitative analysis of remote sensing image data is commonly performed using supervised classification. This technique involves creating thematic maps from satellite imagery.

TABLE 3.1
Remote Sensing Dataset Used in Current Study

Satellite Image	Resolution (m)	Year	Source	File Format
Terra MODIS (MOD13Q1)	250	2000 to 2020	https://lpdaac.usgs.gov/tools/data-pool/	HDF
Landsat (4–5, 7, 8)	30	2000, 2016, 2020	https://earthexplorer.usgs.gov/	TIF

A thematic map is an informative representation of an image that displays the spatial distribution of a particular theme. The image classification process involves several steps, including generating a feature space image, defining training sites, and performing the classification process. The land of the Multan district was classified into five categories: water, settlements, barren lands, crops, and vegetation. Crop phenological parameters are the start of the season (SOS), end of the season (EOS), total length of the growing season (LOS), and Amplitude (AMP) (Upadhyay et al. 2008). SOS indicates the crop type that is sowed in a specific season; EOS also indicates the specific crop type; LOS shows the growth of the specific crop during the season; and AMP shows the crop yield, or production of the crop, in the growing season. Derivation of phenological parameters was done by using the logical modelling approach with the help of ArcMap (V10.8). This program was used for estimating growing seasons of crops from satellite time-series as well as for calculating phenological metrics from the data. A schematic illustration of the methodology applied in the current study is shown in Figure 3.2.

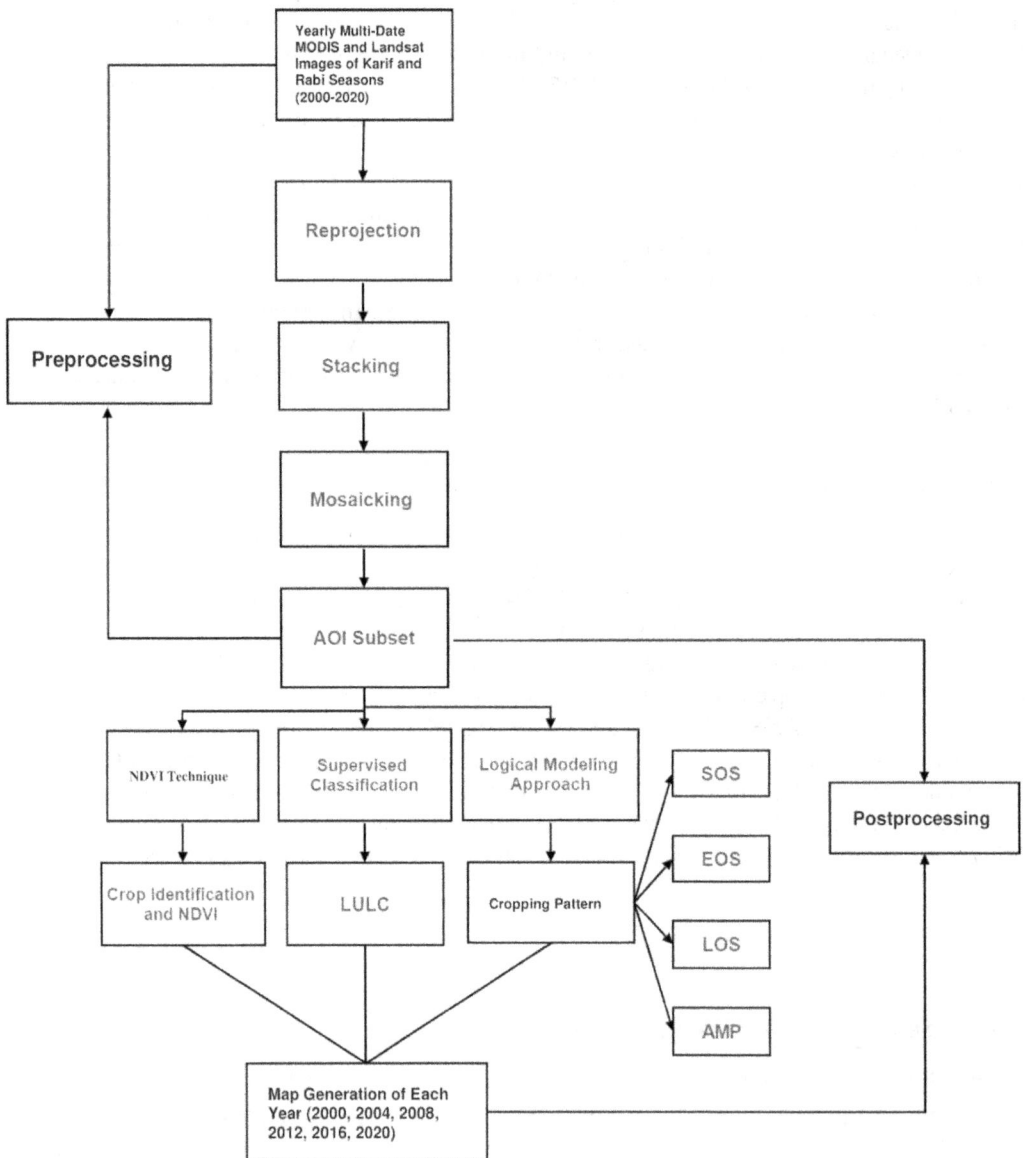

FIGURE 3.2   Methodology used.

3.3 RESULTS

It is well known that Pakistan has long been considered for agriculture production. The country faces shortages in crop production and could not meet the demands of the people because of population pressure and the urbanization phenomenon. The availability of water will determine if production will be able to meet demand in coming years. Major crops that are grown in the Multan district are cotton and wheat, which are rotated round the year. This study has demonstrated that rotation of crops has a significant effect on production and quality.

3.3.1 CROP PRODUCTION

It was observed that most farmers base their preferences on inflation rate and financial conditions. Order of preference of cropping for most of the farmers was, viz., cotton preferable to wheat, then rice, then sugarcane and corn are preferable to sunflower, as shown in Figure 3.3. According to Pakistan Statistical Bureau, the statistics area allocated to different crops for cultivation is depicted in Figure 3.4. It can be seen that cotton and wheat occupy most of the agricultural area. It depicts that maximum area, i.e. 27%, is covered by wheat, and minimum area is covered by sunflower and sugarcane, i.e. 4%. It was observed that wheat and cotton were preferred due to their alternative harvesting periods and good profit ratios. Farmers prefer rice because of its good yield and high price in market. The preference towards sugarcane is due to its one time sowing and three times harvesting quality.

The availability of water to the fields is through canal system and tube-well. Due to this switch process, the farmers face a lot of problems. Water is not the only issue; power-cut problems have affected farmers equally. The process of seed sowing is done by different methods. Seeds are sowed manually by farmers with small budgets, and the drill method has been adopted by high-budget farmers. The fertilizers frequently used in the Multan district were DAP and urea (Figure 3.3). It can be observed that many fertile lands are becoming barren due to the hazardous waste discharge by the industrial factory in Multan district. Moreover, many housing societies are occupying the agricultural lands and building housing projects over them, impacting the farmers' investment as they switch towards alternatives that are much more expensive.

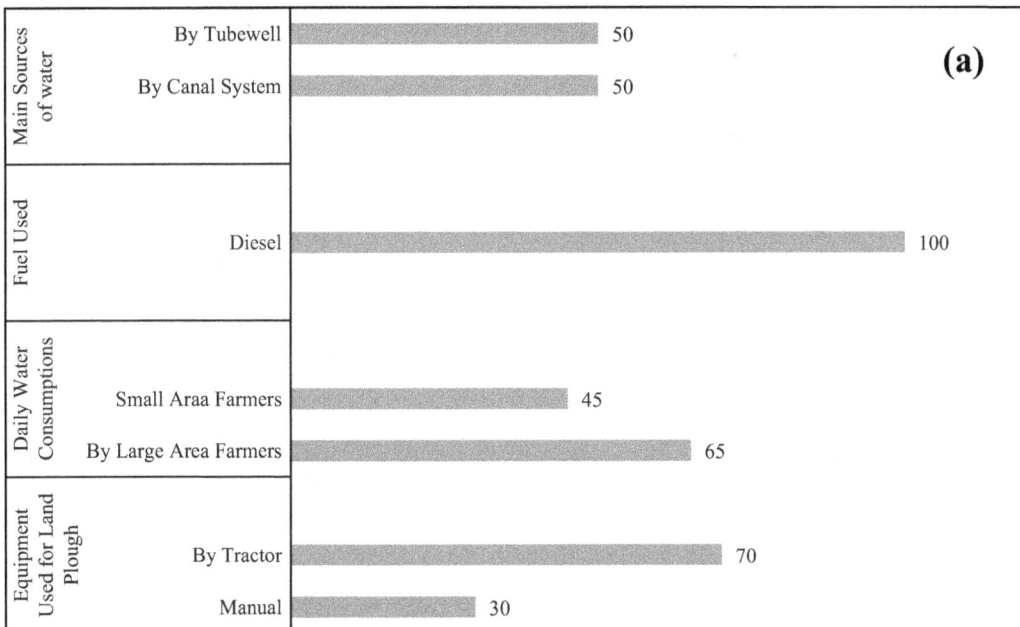

FIGURE 3.3 (a) Water availability and usage. (b) Various seasonal crops and their supply to market. (c) Sustainability of crops and fertilizer preferences for the farmers in the Multan district.

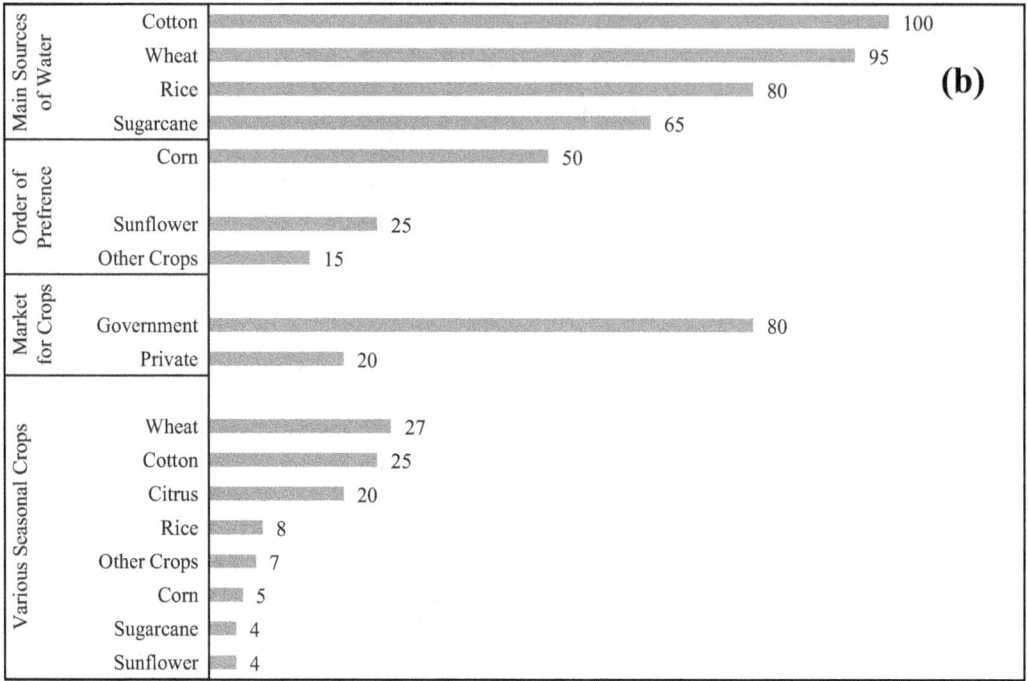

(b)

Main Sources of Water
- Cotton 100
- Wheat 95
- Rice 80
- Sugarcane 65

Order of Prefrence
- Corn 50
- Sunflower 25
- Other Crops 15

Market for Crops
- Government 80
- Private 20

Various Seasonal Crops
- Wheat 27
- Cotton 25
- Citrus 20
- Rice 8
- Other Crops 7
- Corn 5
- Sugarcane 4
- Sunflower 4

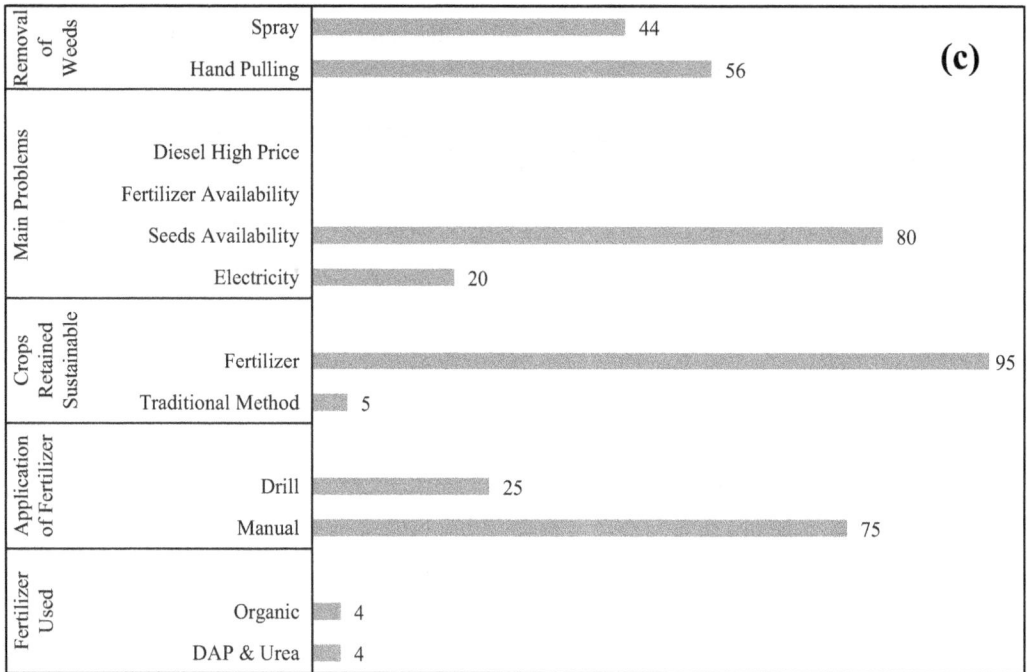

(c)

Removal of Weeds
- Spray 44
- Hand Pulling 56

Main Problems
- Diesel High Price
- Fertilizer Availability
- Seeds Availability 80
- Electricity 20

Crops Retained Sustainable
- Fertilizer 95
- Traditional Method 5

Application of Fertilizer
- Drill 25
- Manual 75

Fertilizer Used
- Organic 4
- DAP & Urea 4

FIGURE 3.3 (Continued)

3.3.2 CROP IDENTIFICATION

The precise location of agricultural crops in the growing environment may be determined with the use of crop pattern maps using the kharif and rabi seasons. Crops were identified based on phenological metrics and field survey. The results shown in Figure 3.5 clearly mention that cotton, rice,

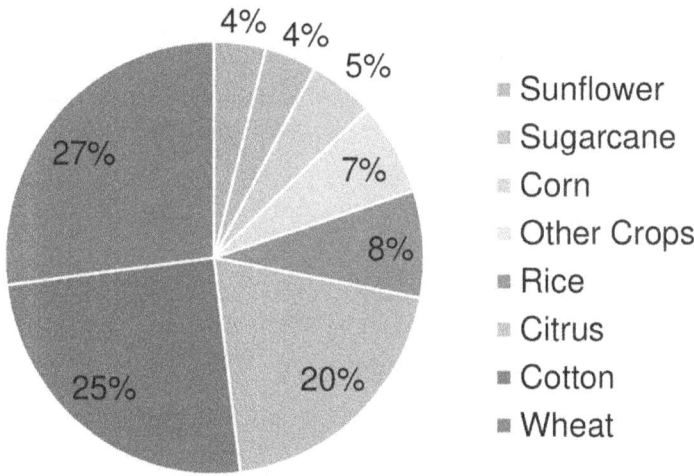

FIGURE 3.4 Pie chart depiction of cropping area in the Multan district.

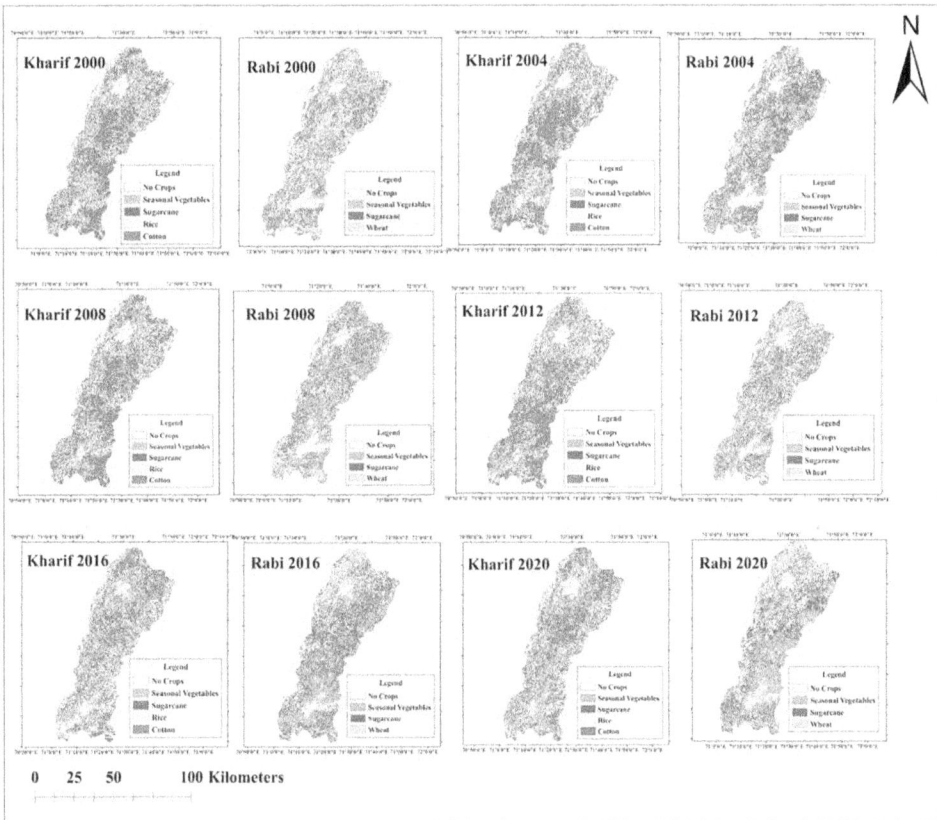

FIGURE 3.5 The crops of kharif and rabi seasons of the Multan district from 2000 to 2020.

and different seasonal vegetables are mostly cultivated during the kharif season. During the rabi season, wheat, sugarcane, and different seasonal vegetables are cultivated. All these results were computed by satellite data combined with field data which can be seen in Table 3.2. The mapping and monitoring of cropping patterns in the Multan area using remote sensing and GIS were a significant

TABLE 3.2
Different Crop Parameters of the Multan District

Serial Number	Crop	Year	Productivity ($\times 10^3$ tons)	Area ($\times 10^3$ ha)	Yield (kg/ha)
1	Wheat	2000	502	195.86	2,561.13
		2004	535	195.46	2,736.17
		2008	590	199.91	2,948.06
		2012	503	173.205	2,902
		2016	616	189.79	3,242.87
2	Cotton	2000	618.24	171.58	1,836.09
		2004	858.55	193.84	2,257.11
		2008	670.77	167.13	1,924.533
		2012	774.57	169.15	2,195.38
		2016	495.28	141.23	1,681.31
		2020	296.4	105.22	1,260.29
3	Rice	2000	14.13	9.307	1,516.41
		2004	14.62	10.117	1,443.63
		2008	27.31	16.99	1,604.85
		2012	16.19	9.712	1,664.73
		2016	47.6	23.067	2,060.88
		2020	59.9	28.327	1,970.59
4	Sugarcane	2000	153.3	3.88	39,430.35
		2004	100.6	2.43	41,365.02
		2008	121.49	2.83	42,839.05
		2012	186.1	3.642	51,038.35
		2016	328.64	6.07	54,078.54
		2020	162.96	2.43	57,026.61

TABLE 3.3
Crop Area Distribution

Years	Wheat (ha)	Cotton (ha)	Rice (ha)	Sugarcane (ha)
2000	195.86	171.58	9.307	3.88
2004	195.46	193.84	10.117	2.43
2008	199.91	167.13	16.99	2.83
2012	173.205	169.15	9.712	3.642
2016	189.79	141.23	23.067	6.07
2020	N/A	105.22	28.327	2.43

challenge. The findings of this research have important implications for the mapping and tracking of cropping patterns. Additionally, crop distribution over the selected area is given in Table 3.3.

3.3.3 Normalized Difference Vegetation Index during Rabi and Kharif Seasons

Figure 3.6 and Figure 3.7 show the NDVI maps of kharif and rabi seasons of the years 2000–2020, respectively, which are drawn using ArcMap 10.8. Figure 3.6 depicts that the NDVI value was 0.68

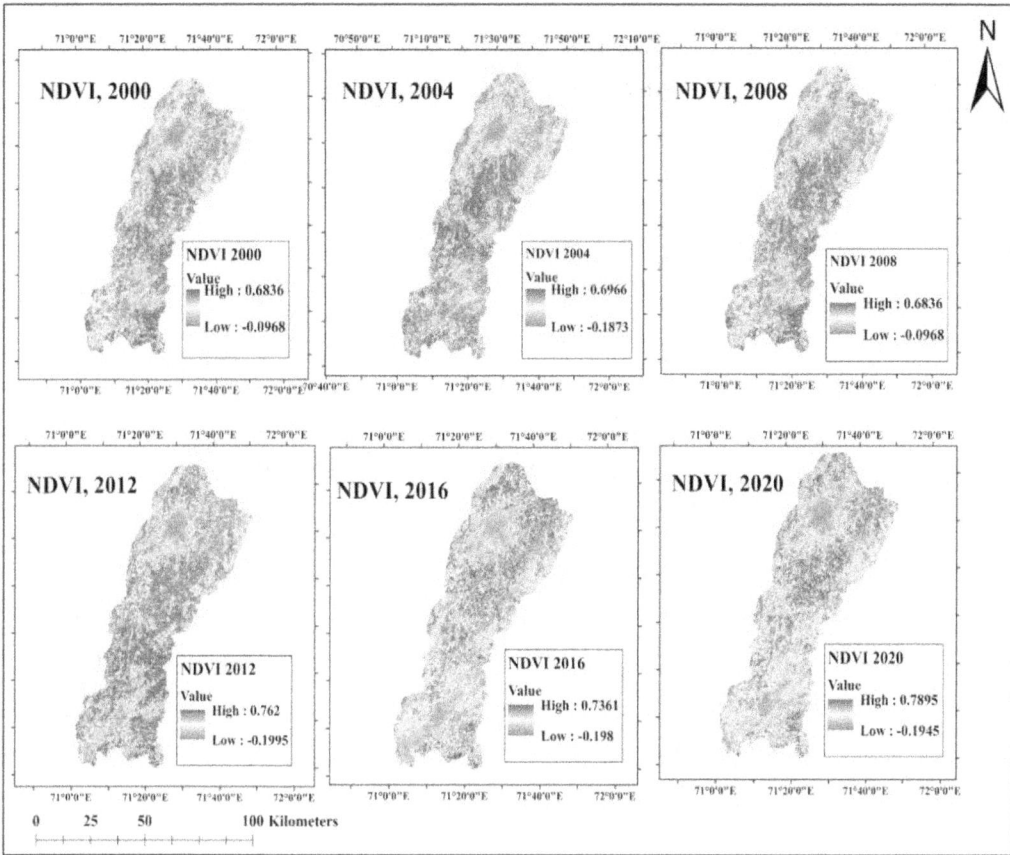

FIGURE 3.6 Normalized Difference Vegetation Index values of the kharif season from 2000 to 2020 in the Multan district.

in 2000 while it is high, i.e. 0.8 during the year 2020 in kharif season. Figure 3.7 depicts that the NDVI value was 0.72 in 2000 while it increases up to 0.8 in 2020 in rabi season. Similarly, Figure 3.8 and Figure 3.9 present graphical representations of NDVI values from 2000 to 2020 for both kharif and rabi seasons.

The health of plants can be determined through the measurement of NDVI, which is based on the reflection of light at specific frequencies. When certain waves of light are absorbed and others are reflected, NDVI can be used as an indicator of plant health. Visible light is strongly absorbed by chlorophyll, an important health indicator, while near-infrared light is strongly reflected by the cellular structure of leaves. A higher NDVI value indicates the presence of more vegetation. Alternatively, there is a possibility of little or nearly no vegetation in case of the lowest NDVI (Hussain and Karuppannan, 2023; Chen et al. 2023).

The most common tool for quantifying plant life is NDVI. Extremely low NDVI values are characteristic with arid environments like rocks, sand, or snow. Shrublands and grasslands have NDVI levels in the moderate range, whereas temperate and tropical rainforests have NDVI values in the high range (Khan et al. 2022; Baqa et al. 2022). Figure 3.10 depicts the cropping area covered by different crops in both cropping seasons. Areas of crops support the result produced as crop identification and length of the seasons. Figure 3.11 depicts that maximum area was occupied by wheat and minimum area was occupied by sugarcane in the Multan district from 2000 to 2020. Similar results were determined by Hussain et al. (2020) which validate our study's results.

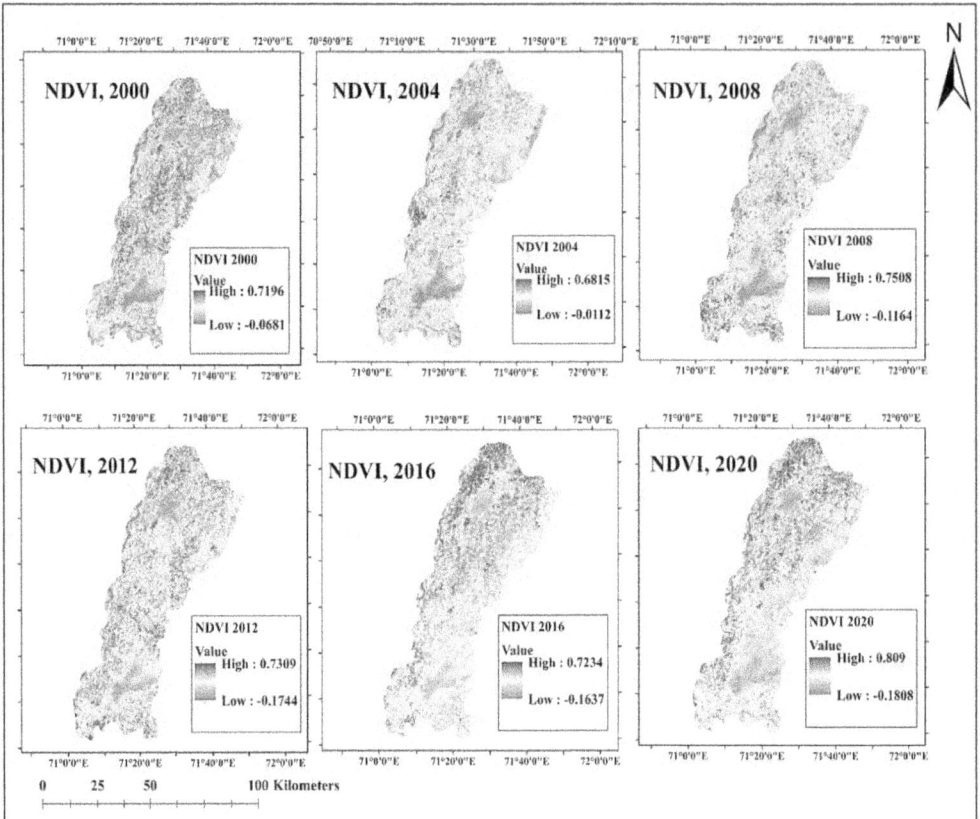

FIGURE 3.7 Normalized Difference Vegetation Index values of the rabi season from 2000 to 2020 in the Multan district.

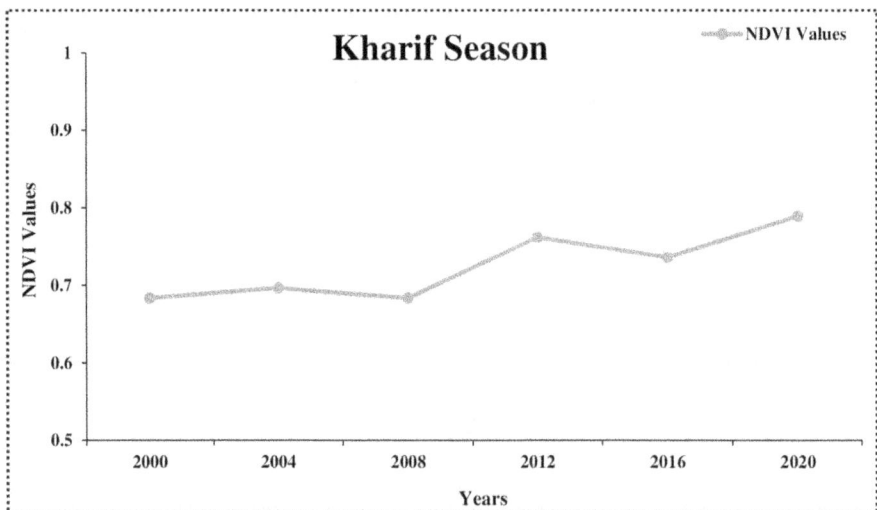

FIGURE 3.8 Normalized Difference Vegetation Index values during the kharif season in the Multan district.

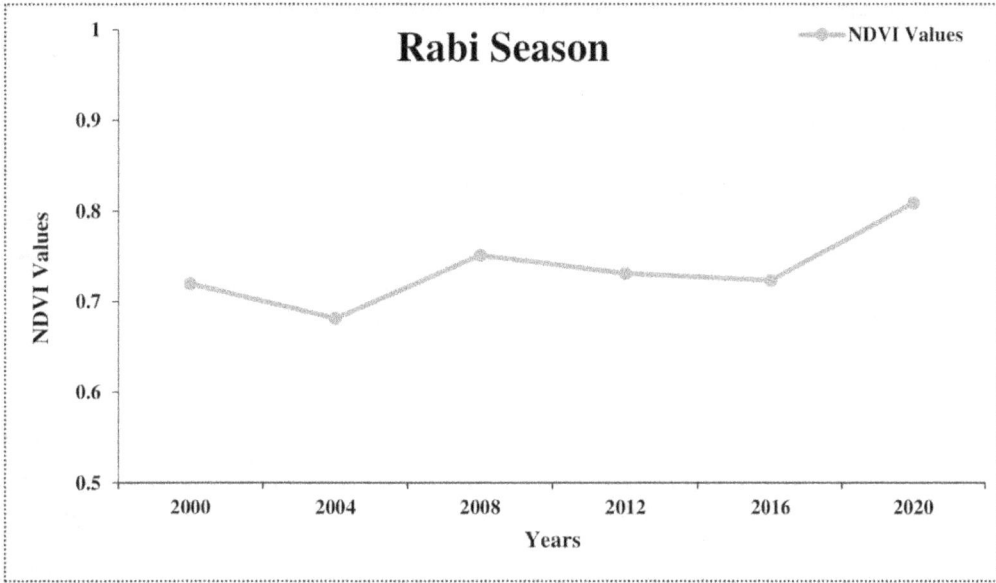

FIGURE 3.9 Normalized Difference Vegetation Index values during the rabi season in the Multan district.

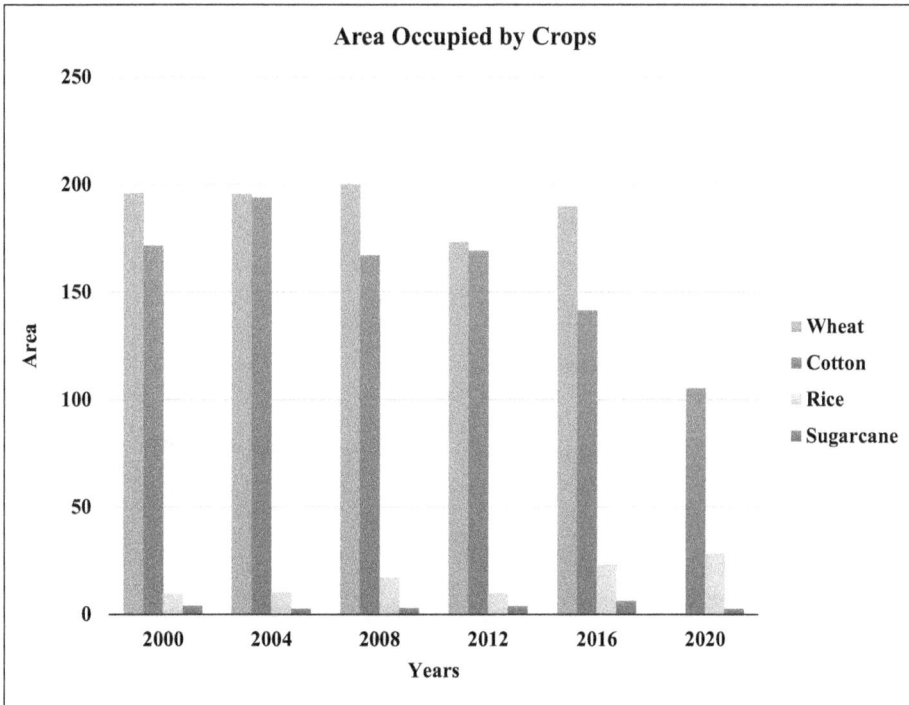

FIGURE 3.10 The area occupied by different crops.

3.3.4   LAND-USE/LAND-COVER MAPPING

The LULC changes were estimated for the years 2000, 2016, and 2020 for both rabi and kharif seasons and are shown in Figure 3.11. It is evident from the LULC maps of the year 2000 that a large area was covered by crops and vegetation as compared to settlements. But with the passage of time, it can be seen clearly in the LULC maps of 2016 that agriculture is now replaced with settlements such as building, roads, etc. due to an increase in population and the increasing demand for shelter. In addition, urbanization is a major reason for increasing settlements in the Multan district. The water spread tanks and reservoirs also decrease because of increasing residential area due to a significant increase in population. The LULC map of the year of 2020 shows an increase in crops and vegetation due to the implementation of afforestation projects and spreading awareness among people in the area.

3.3.5   CROPPING PATTERN

Generally, it was observed that the kharif season started in May and lasted till November. The kharif season's major crops are cotton and rice, respectively. The rabi season started in December and ends in June. The major crop of rabi season is wheat. The area having cotton in the kharif season is mostly occupied by wheat in the rabi season. AMP is the crop yield per season. Therefore,

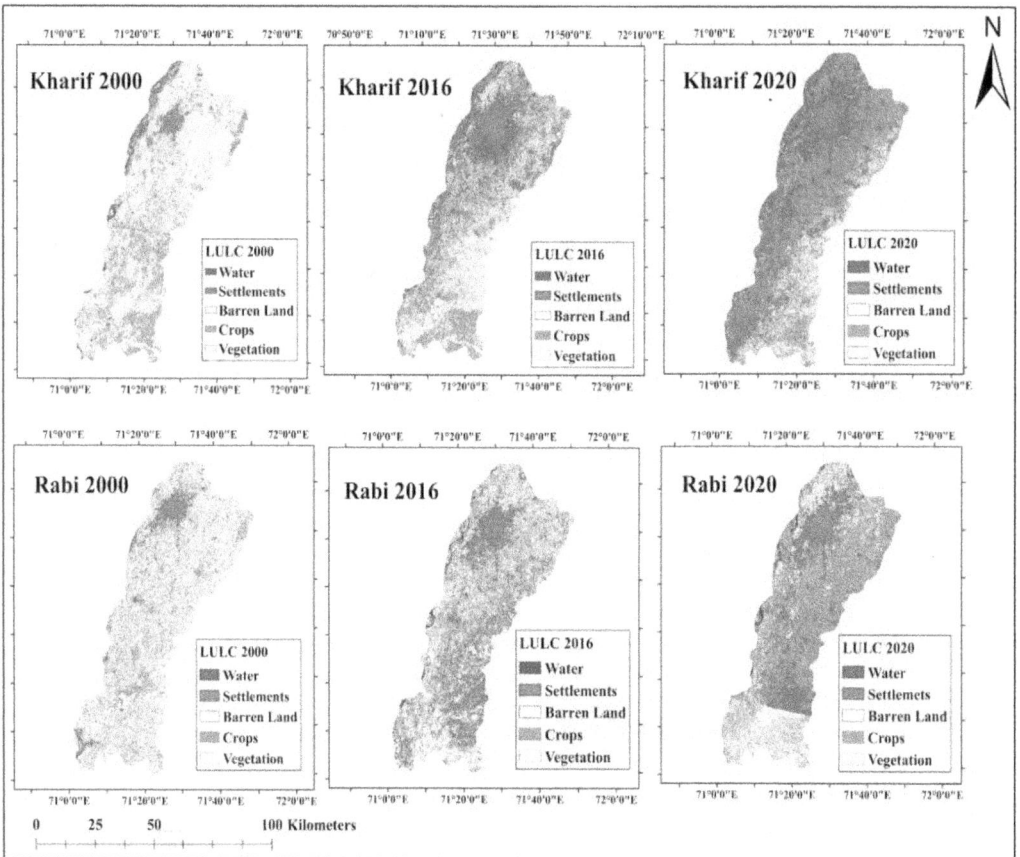

FIGURE 3.11   Land-use/land-cover maps of kharif and rabi seasons of the Multan district from the years 2000, 2016, and 2020.

AMP describes the number of crops. All these parameters are described in the phenology metrics in Figure 3.12. It is observed that the maximum NDVI value during kharif season was 0.8281 (EOS) and minimum was −0.1966 (LOS). Maximum NDVI value during the rabi season was 0.8697 (SOS) and minimum was −0.1943 (EOS). Moreover, Figure 3.13 indicates trends in different crop phenology metrics for kharif and rabi seasons in the Multan district for 2020.

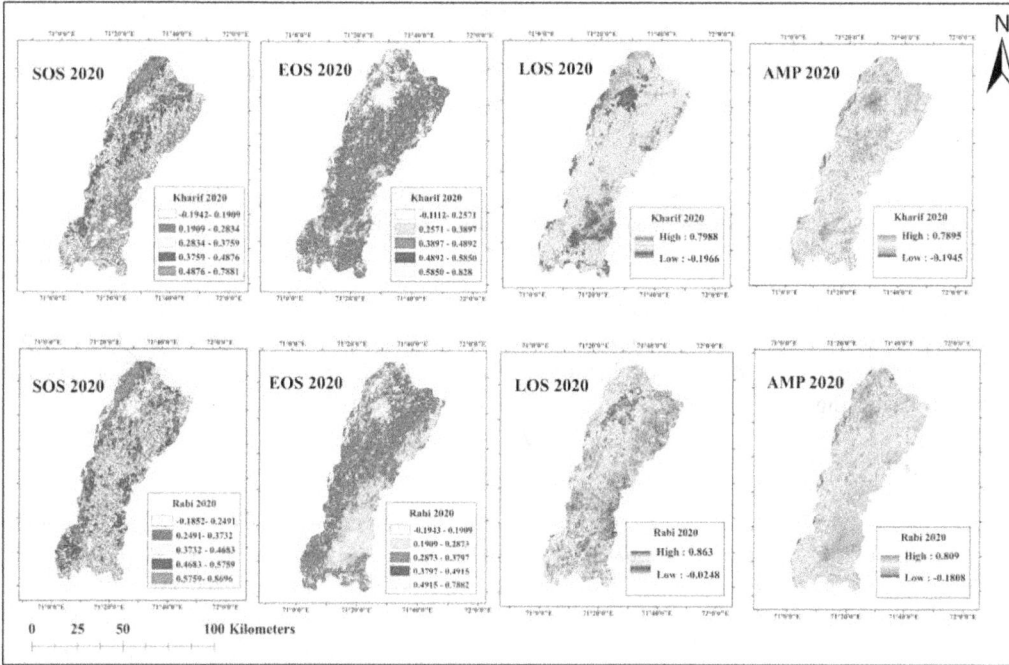

FIGURE 3.12   Variation of phenological metrics over studied region for kharif and rabi seasons.

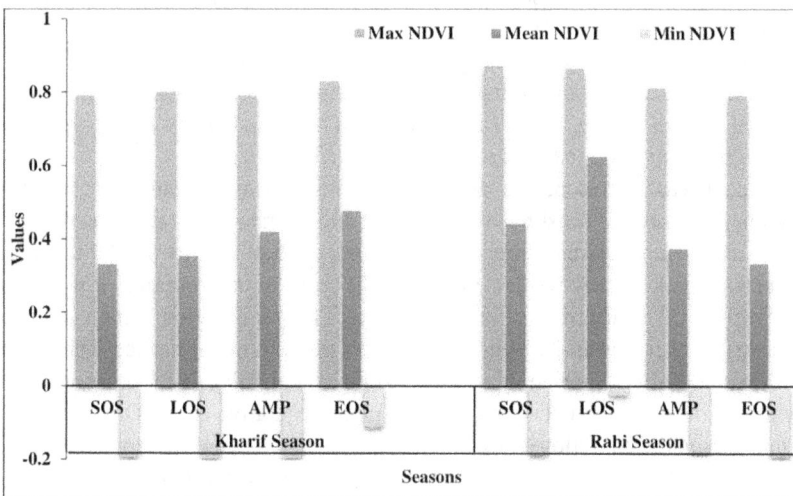

FIGURE 3.13   Minimum, maximum, and mean values of phenological metrics for kharif and rabi seasons.

TABLE 3.4

Results of Various Normalized Difference Vegetation Index Variables during Kharif and Rabi Seasons

Name	Kharif Season				Rabi Season			
	SOS	EOS	LOS	AMP	SOS	EOS	LOS	AMP
X_{Max}	0.7882	0.8281	0.7988	0.7895	0.8697	0.7882	0.863	0.809
X_{DMax}	22–4–20	16–12–20	16–11–20	15–10–20	2–12–20	9–5–20	5–3–20	26–4–20
X_{Min}	−0.1943	−0.1112	−0.1966	−0.1945	−0.1853	−0.1943	−0.0248	−0.1808
X_{Mean}	0.3299	0.4746	0.353	0.4188	0.4388	0.3319	0.6223	0.3733
X_{SD}	0.1171	0.1307	0.1313	0.131	0.1258	0.1175	0.1574	0.1171

Note: AMP, amplitude; EOS, end of season; LOS, length of season; SOS, start of season; XMax, maximum; XDmax, maximum date; XMin, minimum; Xmean, mean value; XSD, standard deviation.

Table 3.4 shows the maximum and minimum NDVI during rabi and kharif seasons of the year 2020. It shows that NDVI value was maximum in the EOS, i.e. 0.8281, whereas minimum NDVI value was observed to be −0.1966 during LOS in the kharif season. During the rabi season, maximum NDVI was observed during SOS, i.e. 0.8697, and minimum NDVI was observed during EOS, i.e. −0.1943.

3.4 DISCUSSION

The study's primary objective was to comprehensively map and monitor the cropping pattern of the Multan district, Pakistan, from 2000 to 2020. During a 20-year period, the study found substantial changes in settlements, vegetation, agriculture, and water supplies. In comparison to settlements, the majority of the land was initially covered with vegetation and crops. Rapid urbanization and population growth later resulted in an increase in settlements, such as housing schemes, buildings, and roads, etc. The water area has shrunk as a result of the steady transformation of aquatic areas into urban areas in order to accommodate the enormous population (Mallupattu and Sreenivasula Reddy 2013; Khan et al. 2020; Hasan et al. 2021). Mallupattu and Sreenivasula Reddy (2013) developed LULC maps using a satellite dataset from 1976 to 2003. The results indicated that cropland and aquatic area converted into residential area with the passage of time due to population transformation. Similarly, Khan et al. (2022) used a dataset (1993–2018) to determine LULC variation and found that barren land increased nearly 12% due to reduction in cropland and water bodies. Moreover, Hasan et al. (2021) reported that water body and forest area both declined, by 9.20% and 3.86%, respectively. Furthermore, 5.18% of Bangladesh's total land area was converted into residential area. This is due to overpopulation and unsustainable urban growth.

During the kharif season, cotton, rice, and a variety of seasonal vegetables are widely cultivated. Wheat, sugarcane, and several seasonal vegetables, on the other hand, are grown during the rabi season. Satellite and field data were used to calculate all of these results. NDVI technique, phenological metrics, and a field survey were used to identify crops in this research. It was found that the kharif season, which is dominated by cotton (25%), has a longer duration, which is due to the early start of the season, which causes the vegetative stage's peak to occur earlier. The rabi season, on the other hand, had a shorter duration. It was caused by a clear delay in the start of the rabi season, which resulted in a delay in the occurrence of the peak vegetative stage. During the rabi season, wheat is the major crop (27%), followed by other crops. Wheat and cotton crops have the maximum area of 189.79 ha and 105.22 ha, respectively. Other crops, i.e. rice and sugarcane show the minimum rest of the area, 30.757 ha. Similarly, Nema et al. (2018) studied wheat, gram, sugarcane, and other crops. Wheat crop had the largest area (84.90%), gram crop had the second largest area (10.23%), and other crops had the minimum rest of the area. Wheat was found to be the most common crop in the Hoshangabad district of Madhya Pradesh, India, covering 2,264,902 ha, followed by gram

(31,895 ha), sugarcane (6,025.6 ha), and other crops (8,857.6 ha). Jamil and Sajjad (2016) studied and mapped 17 cropping patterns in Bijnor, Uttar Pradesh state. Net cultivated area accounted for about 79% of the overall geographical area. Sugarcane as a single crop, rice-wheat, and sugarcane-wheat were the most common cropping patterns, accounting for 33%, 11%, and 8% of the district's total geographical area, respectively. It can be deduced from obtained results that NDVI increased from 0.62 to 0.8 in both rabi and kharif seasons. The identification of spatial distribution of crops makes it easier to understand the cropping patterns of the area with respect to crop productivity. For extracting the seasonal data from the NDVI time series, the method used is based on value thresholds, presuming that a specific phenology phenomenon has started when NDVI values exceed a given threshold (Ali et al. 2022; Bradley et al. 1994). In previous studies, different modelling approaches were applied to map and monitor cropping patterns, such as threshold model, Fourier analysis method, and wavelet transform. Sakamoto et al. (2005) developed a new method for remotely identifying paddy rice phenological phases. The study applied wavelet and Fourier transforms. Wavelet transform uses three types of mother wavelets (Daubechies, Symlet, and Coiflet). Validation results indicated that the wavelet transform outperformed the Fourier transform. Zhang and Zhang (2007) employed the Fourier analysis method to study crop-planting patterns in the North China plain. The findings revealed that at the county level, the estimated area and statistics correlated. Furthermore, the crop area predicted using MODIS data and the statistics at the provincial level were nearly identical. Liu et al. (2018), based on time series of vegetation index data, developed a phenology-based approach to map cropping patterns. The proposed method for retrieving phenological metrics was based on the well-known 'threshold model'. Cropping patterns were mapped using MODIS Enhanced Vegetation Index (EVI) data collected over an eight-day period. The suggested technique extracts cropping patterns with a reasonable overall accuracy (84%) according to the results. The present study utilized 16 datasets of MODIS to extract four phenological parameters (SOS, EOS, LOS, and AMP) using the logical modelling approach. The results showed that NDVI value was maximum in the EOS (0.8281), whereas the minimum NDVI value was observed (−0.1966) during LOS in the kharif season. During the rabi season, maximum NDVI was observed during SOS, i.e. 0.8697, and minimum NDVI was observed during EOS, i.e. −0.1943. The results of the current study agreed with the previous studies using GIS and remote sensing tools (Hu et al. 2023; Hussain and Karuppannan, 2023; Hussain et al. 2022b; Hussain et al. 2023, 2020).

3.5   CONCLUSION

This research study mapped and monitored cropping patterns of the Multan district, located in Pakistan, comprehensively. This study employed NDVI, logical modelling approach, and supervised classification techniques. According to the findings of the study, cotton, rice, and seasonal vegetables are mostly cultivated in the kharif season, while wheat, sugarcane, and a variety of seasonal vegetables are cultivated in the rabi season. In addition, NDVI increased from 0.62 to 0.8 in both kharif and rabi seasons. As the world's population grows, so does the need for food, putting further strain on agriculture. The comparison of LULC in 2000, 2016, and 2020 shows that rapid urbanization and population growth have resulted in a significant increase in settlements and crop area over the research period. Crop phenology indicates that cotton dominated the kharif season, while wheat dominated the rabi season throughout the year. Moreover, NDVI values during the kharif and rabi seasons ranged from maximum 0.8281 (EOS) to minimum −0.1966 (LOS) and maximum 0.8699 (SOS) to minimum −0.1943 (EOS), respectively. This research illustrates that mapping and monitoring of cropping patterns on a regular basis are essential to maintain a healthy agricultural system. Trends in cropping patterns are the key indicators for the nature of regional and local climate change. For assessing cropping patterns, a hybrid strategy that combines field data with remotely sensed map products is a viable choice. Therefore, the crop fields and satellite data are found to be quite beneficial in mapping the cropping patterns in the Multan district. The agricultural research institute of the government might use the study's findings to improve crop strategy and plan agricultural research and extension activities for crop diversification.

NOTE

Work Acknowledgement: We are grateful to the editors and the anonymous reviewers for their insightful comments and suggestions for enhancing this manuscript.
Funding: The current study received no external funding.
Conflict of interest: There is no conflict of interest present in current manuscript.
Author contributions: A.R. and M.Z. conceptualized the current study; A.R. and N.R.S. finalized the research draft; the remaining coauthors edited and reviewed the draft.

REFERENCES

Ali, M., Yar, P., Khan, S., Muhammad, S., Hussain, W., Hussain, K., Hussain, G., Aneva, I. Y., Yue Phin Tng, D. and Bussmann, R. W. (2022). Land use and land cover modification and its impact on biodiversity and the ecosystem services in district Kurram, Pakistan. *Boletín Latinoamericano y del Caribe de Plantas Medicinales y Aromáticas, 21*(3).

Baqa, M. F., Lu, L., Chen, F., Nawaz-ul-Huda, S., Pan, L., Tariq, A., Qureshi, S., Li, B. and Li, Q. (2022). Characterizing spatiotemporal variations in the urban thermal environment related to land cover changes in Karachi, Pakistan, from 2000 to 2020. *Remote Sensing, 14*(9), 2164.

Benedetti, R. and Rossinni, P. (1993). On the use of NDVI profiles as a tool for agricultural statistics: The case study of wheat yield estimate and forecast in Emilia Romagna. *Remote Sensing of Environment., 45*, 311–332.

Bisht, P., Kumar, P., Yadav, M., Ravat, J. S., Sharma, M. P. and Hooda, R. S. (2013). Spatial dynamics for relative contribution of cropping pattern analysis on environment by integrating remote sensing and GIS. *International Journal of Plant Production, 4*(3), 299–304.

Bradley C. R., Jesslyn F. B., Darrel V., Thomas R. L., James W. M. and Donald O. O. (1994). Measuring phenological variability from satellite imagery. *Vegetation Science, 5*, 703–704.

Chen, X., Wang, Y., Chen, Y., Fu, S. and Zhou, N. (2023). NDVI-based assessment of land degradation trends in Balochistan, Pakistan, and analysis of the drivers. *Remote Sensing, 15*(9), 2388.

Faisal, B. R., Rahman, H., Sharifee, N. H., Sultana, N., Islam, M. I., Habib, S. A. and Ahammad, T. (2020). Integrated application of remote sensing and GIS in crop information system: A case study on aman rice production forecasting using MODIS-NDVI in Bangladesh. *AgriEngineering, 2*(2), 264–279.

Hasan, M., Islam, R., Saifur Rahman, M., Ibrahim, M., Shamsuzzoha, M., Khanam, M. and Zaman, M. (2021). Analysis of land use and land cover changing patterns of Bangladesh using remote sensing technology. *American Journal of Environmental Sciences, 17*(3), 71–81.

He, R., Jin, Y., Kandelous, M. M., Zaccaria, D., Sanden, B. L., Snyder, R. L., Jiang, J. and Hopmans, J. W. (2017). Evapotranspiration estimate over an almond orchard using Landsat satellite observations. *Remote Sensing, 9*(5), 436.

Hu, Y., Raza, A., Syed, N. R., Acharki, S., Ray, R. L., Hussain, S., Dehghanisanij, H., Zubair, M. and Elbeltagi, A. (2023). Land use/land cover change detection and NDVI estimation in Pakistan's Southern Punjab Province. *Sustainability, 15*(4), 3572.

Hussain, S., Qin, S., Nasim, W., Bukhari, M. A., Mubeen, M., Fahad, S., Raza, A., Abdo, H. G., Tariq, A., Mousa, B. G. and Mumtaz, F. (2022a). Monitoring the dynamic changes in vegetation cover using spatiotemporal remote sensing data from 1984 to 2020. *Atmosphere, 13*(10), 1609.

Hussain, S. and Karuppannan, S. (2023). Land use/land cover changes and their impact on land surface temperature using remote sensing technique in district Khanewal, Punjab Pakistan. *Geology, Ecology, and Landscapes, 7*(1), 46–58.

Hussain, S., Mubeen, M., Akram, W., Ahmad, A., Habib-ur-Rahman, M., Ghaffar, A., Amin, A., Awais, M., Farid, H. U., Farooq, A. and Nasim, W. (2020). Study of land cover/land use changes using RS and GIS: A case study of Multan district, Pakistan. *Environmental Monitoring and Assessment, 192*, 1–15.

Hussain, S., Mubeen, M. and Karuppannan, S. (2022b). Land Use and Land Cover (LULC) change analysis using TM, ETM+ and OLI Landsat images in district of Okara, Punjab, Pakistan. *Physics and Chemistry of the Earth, Parts a/b/c, 126*, 103117.

Hussain, S., Mubeen, M., Nasim, W., Fahad, S., Ali, M., Ehsan, M. A. and Raza, A. (2023). Investigation of irrigation water requirement and evapotranspiration for water resource management in Southern Punjab, Pakistan. *Sustainability, 15*(3), 1768.

Jamil, M. and Sajjad, H. (2016). Deriving cropping system efficiency pattern using remote sensing and GIS: A case study of Bijnor district, India. *International Journal of Advancement in Remote Sensing, GIS and Geography*, 4, 27–40.

Jin, Y., Randerson, J. T. and Goulden, M. L. (2011). Continental-scale net radiation and evapotranspiration estimated using MODIS satellite observations. *Remote Sensing of Environment*, 115(9), 2302–2319.

Jönsson P. and Eklundh L. (2004). Timesat: A program for analyzing time-series of satellite sensor data. *Computers and Geosciences*, 30, 833–845.

Khan, R., Li, H., Basir, M., Chen, Y. L., Sajjad, M. M., Haq, I. U., Ullah, B., Arif, M. and Hassan, W. (2022). Monitoring land use land cover changes and its impacts on land surface temperature over Mardan and Charsadda Districts, Khyber Pakhtunkhwa (KP), Pakistan. *Environmental Monitoring and Assessment*, 194(6), 409.

Khan, S. M., Ullah, S., Sun, T., Rehman, A. U. and Chen, L. (2020). Land-use/land-cover changes and its contribution to urban heat island: A case study of Islamabad, Pakistan. *Sustainability*, 12(9).

Kohli, G., Lee, C. M., Fisher, J. B., Halverson, G., Variano, E., Jin, Y., Carney, D., Wilder, B. A. and Kinoshita, A. M. (2020). Ecostress and CIMIS: A comparison of potential and reference evapotranspiration in riverside county, California. *Remote Sensing*, 12(24), 4126.

Liu, J., Zhu, W., Atzberger, C., Zhao, A., Pan, Y. and Xin, H. (2018). A phenology-based method to map cropping patterns under a wheat-maize rotation using remotely sensed time-series data. *Remote Sensing*, 10(80).

Mallupattu, P. K. and Sreenivasula Reddy, J. R. (2013). Analysis of land use/land cover changes using remote sensing data and GIS at an urban area, Tirupati, India. *The Scientific World Journal*, 2013, 1–6.

Mamatkulov, Z., Safarov, E., Oymatov, R., Abdurahmanov, I. and Rajapbaev, M. (2021). Application of GIS and RS in real time crop monitoring and yield forecasting: A case study of cotton fields in low and high productive farmlands. *EDP Sciences*.

Nema, S., Awasthi, M. K. and Nema, R. K. (2018). Spatial crop mapping and accuracy assessment using remote sensing and GIS in tawa command. *International Journal of Current Microbiology and Applied Sciences*, 7(5), 3011–3018.

Raza, A., Hu, Y., Acharki, S., Buttar, N. A., Ray, R. L., Khaliq, A., Zubair, N., Zubair, M., Syed, N. R. and Elbeltagi, A. (2023a). Evapotranspiration Importance in Water Resources Management Through Cutting-Edge Approaches of Remote Sensing and Machine Learning Algorithms. In *Surface and Groundwater Resources Development and Management in Semi-Arid Region: Strategies and Solutions for Sustainable Water Management* (pp. 1–20). Cham: Springer International Publishing.

Raza, A., Khaliq, A., Hu, Y., Zubair, N., Acharki, S., Zubair, M., Syed, N. R., Ahmad, F., Iqbal, S. and Elbeltagi, A. (2023b). Water Resources and Irrigation Management Using GIS and Remote Sensing Techniques: Case of Multan District (Pakistan). In *Surface and Groundwater Resources Development and Management in Semi-arid Region: Strategies and Solutions for Sustainable Water Management* (pp. 137–156). Cham: Springer International Publishing.

Raza, A., Shoaib, M., Faiz, M. A., Baig, F., Khan, M. M., Ullah, M. K. and Zubair, M. (2020). Comparative assessment of reference evapotranspiration estimation using conventional method and machine learning algorithms in four climatic regions. *Pure and Applied Geophysics*, 177, 4479–4508.

Sajjad, S., Mannan, A., Liu, J., Zhongke, F., Khan, T., Saeed, S., Mukete, B., Yong, S., Yongxiang, F., Ahmad, A., Amire, M., Ahmed, S. and Shr, S. (2019). Application of land-use/land cover changes in monitoring and projecting forest biomass carbon loss in Pakistan. *Global Ecology and Conservation*, 17.

Sakamoto, T., Yokozawa, M., Toritani, H., Shibayama, M., Ishitsuka, N. and Ohno, H. (2005). A crop phenology detection method using time-series MODIS data. *Remote Sensing of Environment*, 96, 366–374.

Upadhyay, G., Ray, S. S and Panigrahy, S. (2008). Derivation of crop phenological parameters using multi-date spot-vgt-NDVI data: A case study for Punjab. *Journal of Indian Society of Remote Sensing*, 36, 37–50.

Useya, J. and Shengbo, C. (2019). Exploring the potential of mapping cropping patterns on smallholder scale cropland using sentinel-1 SAR data, *Chinese Geographical Science*, 29(4), 626–639.

Wang, J., Raza, A., Hu, Y., Buttar, N. A., Shoaib, M., Saber, K., Li, P., Elbeltagi, A. and Ray, R. L. (2022). Development of monthly reference evapotranspiration machine learning models and mapping of Pakistan: A comparative study. *Water*, 14(10), 1666.

Zhang, J. and Zhang, Y. (2007). Remote sensing research issues of the national land use change program of China. *ISPRS Journal of Photogrammetry and Remote Sensing*, 62(6), 461–472.

APPENDIX 3.1

Questionnaire for Mapping and Monitoring Cropping Patterns

Name _____

Age _____

Gender _____

Date _____

Q1: Which equipment is used for land plough? a. Manual b. Tractors
_____

Q2: How is the water consumed on daily basis?
_____
_____

Q3: Which fuel is used?
_____

Q4: What are the main sources of water?
_____
_____

Q5: What are the various seasonal crops sowed in your field?
_____
_____

Q6: Which are the main markets for crops?
_____
_____

Q7: What is the order of your preferences towards crops?
_____
_____

Q8: Which fertilizers do you use normally?
_____
_____

Q9: How do you apply the fertilizers?
_____
_____

Q10: How are the crops retained sustainable?
_____
_____

Q11: How are the weeds removed?
_____
_____

Q12: What kinds of problems do you face normally?
_____
_____

4 Shrinking of Cropland Using Geospatial Approach
A Case Study of Asansol Urban Agglomeration

Najib Ansari and Rukhsana Sarkar

4.1 INTRODUCTION

Agriculture is essential in developing nations for eradicating rural poverty, improving food security, lowering income inequities, and providing environmental services (Byerlee et al., 2009; Rukhsana and Alam, 2022). Climate change and food security are complex issues that span political boundaries (Islam and Kieu, 2020). The ability of the planet's resources to satisfy the rising food demand is already in jeopardy. Worldwide agriculture is under pressure to feed an increasing number of people while also being limited by factors such as degradation, food insecurity, land availability, biodiversity loss, and climate change (Dawson et al., 2014).

The rate of urbanisation has dramatically grown during the past 200 years. Only 2% of the population resided in the city in 1800; this number increased to 12% in 1900 and is predicted to reach 75% by 2030 (Triantakonstantis and Mountrakis, 2012). In recent decades, India has also witnessed this scenario of urban transition (Sarkar and Chouhan, 2019; Hasnine and Rukhsana, 2023). From 11.4% in 1901 to 31.16% in 2011, the population increased dramatically (Bose and Chowdhury, 2020). According to Rukhsana and Hasnine (2020), this has a negative impact on the socioeconomic system, natural ecosystem, and environment. To create a sustainable urban region, environmental and ecological sustainability is necessary (Hasnine and Rukhsana, 2020; Rukhsana and Hasnine, 2018).

Lack of agricultural land is the main obstacle to the development of countries like India that rely heavily on agriculture. This issue is getting worse every day as a result of the population's rapid growth, which is displacing existing resources and land. Therefore, increasing agricultural production is the greatest approach to fulfil the demands of the mounting population (Halder, 2019). About 40% of India's gross national product comes from agriculture, which provides a living for over 70% of its citizens. Since India frequently suffers severe weather, such as floods and droughts, the environment has a significant influence on the productivity of the agricultural sector. According to Gandhi and Savalia in 2014 and Dissanayake et al. (2019), soil is the most important natural resource for food production and socioeconomic growth of humans. If we want to increase food production and ensure food sufficiency, we need to produce crops in environments that are suitable for them. The most important thing to do to accomplish this is to carry out a land suitability assessment (Molla et al., 2020; Rukhsana and Molla, 2023; Moisa et al., 2022).

The Normalized Difference Vegetation Index (NDVI), which is obtained from Landsat images, is the technique that is most frequently utilised. To discriminate between vegetated and non-vegetated environments, researchers have used NDVI (Singh, 2018; Nero, 2017; Sathyakumar et al., 2019). According to Yang et al. (2014), the NDVI is directly correlated with climate change, particularly with variations in rainfall. Plant species with more than normal rainfall have higher NDVI values (Wang et al., 2003). The various vegetation indices, especially NDVI, typically serve as

DOI: 10.1201/9781003377825-5

indicators of ecosystem change and vegetation growth (Nemani et al., 2003). The NDVI is used to measure how green a tree is. It is the most used index for analysing the phenomenon of dense vegetation (Goward et al., 1985).

Weng et al. (2008) found a strong connection between land surface temperature (LST) and urban biophysical traits as NDVI, Normalized Difference Water Index (NDWI), Normalized Difference Built-up Index (NDBI), and modified Normalized Difference Water Index. The urban heat island result is the term used to describe how LST in metropolitan regions is frequently greater than in non-urban locations (Chen et al., 2006). Additionally, various studies by various scholars have been done to illustrate the connection between LST, (Sandholt et al., 2002; Raynolds et al., 2008; Estoque and Murayama, 2017) land-use/land-cover changes, greenness (NDVI) and impervious land (NDBI). A distinctive technique for getting LST data is thermal infrared (TIR) remote sensing on both a regional and global scale since sensors in this spectral area are able to gather up energy that is directly released from the land surface (Yu et al., 2014). The LST satellite-derived images are used for land-use classification, hydrological examination in urban growth, thermal environment, and urban heat island research, or even on a broader scale (Shi et al., 2015). The use of LST to analyse the connection between urban thermal trends, spatial organisation, and surface features is a crucial method for studying urban climate, which is helped by the TIR bands of data from space-based sensors' remote sensing (Chen et al., 2006).

Rainfall has an effect on the LST as well (Schultz and Halpert, 1995), and LST can be lowered when there has been a lot of rain and there is vegetation cover (Wan et al., 2004). Due to the growth of urban areas, the natural landscape has been significantly altered into a built-up environment, which changes the thermal characteristics of the land surface (Mohajerani et al., 2017) and ultimately leads to an increase in the LST (Li et al., 2020). Remote sensing has become a crucial and effective instrument for tracking and analysing changes to the earth's natural elements. The examination of land-cover data over a wide spatial area using satellite data is cost effective. It has been extensively used in long-term vegetation change detection as well as environmental and climatic analysis (Nega et al., 2019; Zhang et al., 2010). According to Lo et al. (1997), the thermal properties of the various forms of land cover are related to the distinctive LST. It has been well established in the literature that the LST and NDVI are related (Chen et al., 2006, 2014).

According to certain studies, the temperature vegetation index was developed to look into how land changes affected LST (Jiang and Tian, 2010). The spatial distribution of LST in a small city in Korea or China has been the subject of several studies (Babalola and Akinsanola, 2016). Furthermore, earlier studies (Karimi et al., 2017; Pal and Ziaul, 2017) only created NDBI or NDVI to look into the impact of land changes over LST. Most studies only examined small towns or one or two factors when evaluating the influence of LST on land use. This study looked at four land-use/land-cover (LULC) indices to determine how land changes affected LST in a large, heavily populated urban area.

Another study focuses on examining how Sagar Island's agricultural land in the Indian Sundarban region has been impacted by the growth of built-up regions over time and space. The study makes use of a geographic information systems (GIS) environment and remotely sensed data to do this. Researchers can efficiently and practically evaluate the changes in LULC through time and space by using these tools. They may analyse and map the LULC dynamics using this method, and the historical data are helpful for environmental monitoring (Lambin et al., 2001; Molla and Rukhsana, 2022). The computer-assisted examination of geometric properties of specific land uses, such as compactness and contiguousness, is another area where GIS is quite effective. Combining the two disciplines can produce a beneficial symbiotic relationship for both fields (Abdel Rahman et al., 2016; Molla and Rukhsana, 2022). The focusses of this study are to (1) mark out the shrinkage of the farmland area of the Asansol Urban Agglomeration (AUA), (2) evaluate the spatiotemporal dynamic of the AUA LULC from 1991 to 2021, and (3) investigate the link between LST and LULC indexes such as the NDVI and NDWI.

4.2   STUDY AREA

Asansol, the second-largest urban agglomeration (UA) in West Bengal after Kolkata, is located in Paschim Burdwan's far western region. In terms of population size, Asansol ranks 39th in India. The study region extends longitudinally and latitudinally from 87°09′35″ E to 86°47′40″ E and 23°35′12″ N to 23°46′37″ N (Figure 4.1), respectively. The AUA is made up of the four blocks Barabani, Jamuria, Raniganj, and Salanpur, where there is only one municipal corporation, Asansol. There are two municipalities: Kulti and Raniganj. There are 12 census towns, 1 outgrowth, and 1 township area. The total population was 1,243,414 according to the 2011 Indian census. This UA is geographically located at the meeting point of the Ganga plain and the Chottanagpur plateau in the west and east, correspondingly. The southern and western urban areas are bounded by the Ajay and Damodar rivers, respectively. This area has a well-developed transportation infrastructure. This region is connected to both north and south India via the railway lines and GT road. Along the NH2 and the railway, linear habitation is growing. The best non-coking coal reserve in the country is located near Asansol, which also happens to be the location of the greatest coal reserve in the country. The Raniganj Coalfield, which helped the subdivision become more industrialised, was the origin of the Indian coal industry.

Although the coal industries were discovered in the beginning of the eighteenth century, the area did not become established until the middle of the nineteenth century. Forests in the district's western portion were cleared as a result of industrial development, but in the district's eastern

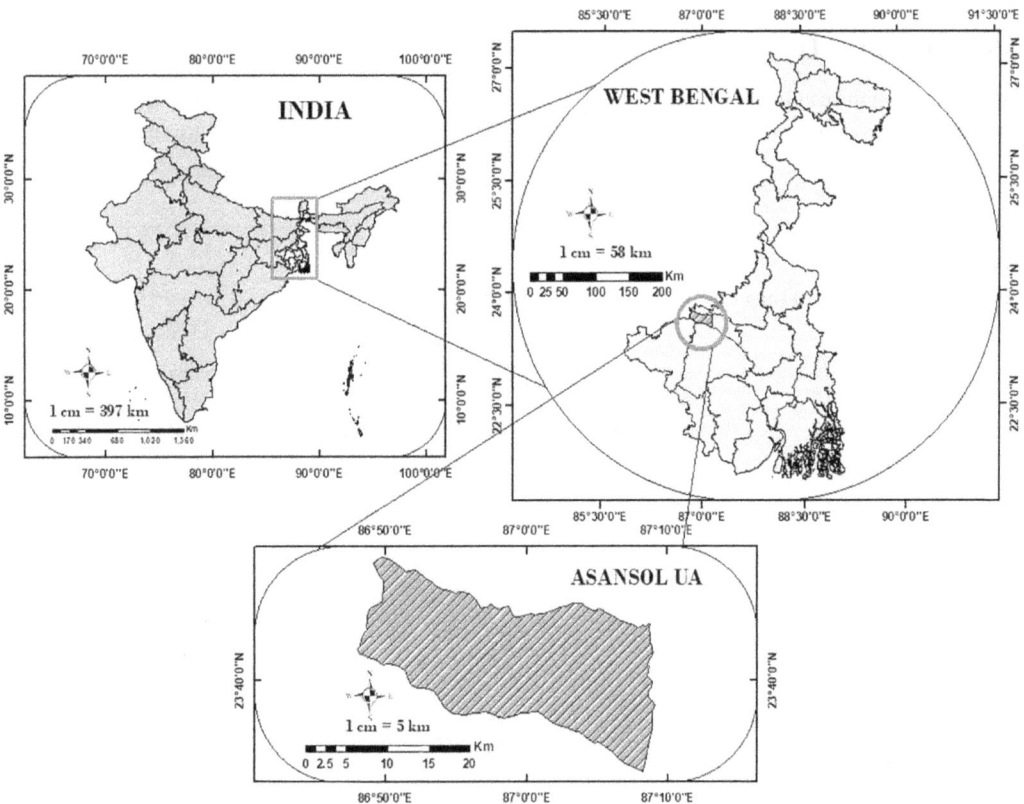

FIGURE 4.1   The location of the study area, Asansol Urban Agglomeration.

portion, forests still exist in Kanksa and the nearby Faridpur and Ukhra areas. Asansol features sandy soil which is typical, consisting of granitic rocks and sandstones. On the other hand, Asansol has a mixture of quartz veins, conglomeratic, and pegmatites sandstones that have weathered into coarse, abrasive soil intermingled with rock pieces. In this area, one can find the red and yellow Ultisols, which are poor in phosphate, calcium, nitrogen, and other plant nutrients. The Asansol Sub-division is renowned for its mineral resources, which include manganese coal, silica bricks, calcium carbonate, abrasives, glass sands, bauxite, iron ore, moulding sands, building supplies, laterite, etc.

4.3 MATERIAL AND METHODS

4.3.1 SATELLITE IMAGES

This study used multi-temporal satellite imagery from 1991, 2001, 2011, and 2021 to analyze land transformation and LULC classification and LST, vegetation index, built-up index, and water index. Three images from the Landsat Thematic Mapper (TM) were obtained for 1991, 2001, and 2011, and one Landsat Operational Land Imager (OLI) image was obtained for 2021 from the United States Geological Survey (https://earthexplore.usgs.gov). The data were collected throughout the same season of the year, i.e. February to April. For NDVI, LST, and NDWI, January to February data are taken for the entire period. Samples obtained from Google Earth Pro helped with the ground validation of the LULC maps for three decades. Table 4.1 lists all the specific details on the satellite images.

4.3.2 REFERENCE DATA

The classification result was validated using certain reference data. For a more precise classification outcome in this study area for the reference year, the data were taken from a topographical map

---

TABLE 4.1
List of Multi-Sensor Data Collected from Landsat

			Date of Acquisition		
			AUA		
Number	Satellite	Spatial Resolution (m)	LULC	LST, NDVI, NDWI	Map Projection
1	Landsat 4 and 5 TM	30	10 April 1991	24 January 1991	UTM-WGS84, Polar Stereographic for the continent of Antarctica
2	Landsat 4 and 5 TM	30	25 April 2001	19 January 2001	UTM-WGS84, Polar Stereographic for the continent of Antarctica
3	Landsat 4 and 5 TM	30	20 March 2011	31 January 2011	UTM-WGS84, Polar Stereographic for the continent of Antarctica
4	Landsat 8 OLI/TIRS	15–30	27 February 2021	27 February 2021	UTM-WGS84, Polar Stereographic for the continent of Antarctica

Note: AUA, Asansol Urban Agglomeration; LST, land surface temperature; LULC, land use/land cover; NDVI, Normalized Difference Vegetation Index; NDWI, Normalized Difference Water Index; OLI, Operational Land Imager; TIRS, Thermal Infrared Sensor; TM, Thematic Mapper; UTM, Universal Transverse Mercator; WGS, World Geodetic System.

that was managed by the Survey of India and Google Earth. LST, NDVI, and NDWI were also calculated in ArcGIS (v.10.1). We also used QGis 3.16.3 version for more accuracy of NDVI, NDWI, and LST.

4.3.3 Pre-processing of Satellite Images

ArcGIS 10.1 was used to import the multi-temporal satellite images and create Standard False Color Composite using the image interpreter toolkit (False Color Composites). The area of interest was sub-set using geo-referenced multi-temporal satellite imagery using the Universal Transverse Mercator projection technique and a Survey of India topographical map at a scale of 1:50,000 with the aid of the World Geodetic Survey (WGS-84), Datum 45 N. These four satellite images were properly registered with the help of a Ground Control Point (GCP). Much of the GCP was gathered from this area's cropland, built-up areas, water bodies, railway stations, and the intersection of the roads in order to facilitate well identification (Amuti and Luo, 2014). The reference year for registering GCP was determined using the Landsat-8 (2021) image, while the remaining three images were registered using a second-degree polynomial model (Islam et al., 2018). The images were further pre-processed using radiometric correction, geometric correction, and picture enhancement techniques in Erdas software (2014). After that, a GeoTIFF file was created for further usage, exporting the images.

4.3.4 Land-Use/Land-Cover Classification

The trendiest technique for extracting data from satellite images is LULC classification (Srivastava et al., 2012). Supervised and unsupervised classifications are the two LULC classification methods that are most frequently used (Lang et al., 2008). Four fixed land classes, e.g. vegetation cover, cropland, built-up area, and other class, were noted based on the Digital Number (DN) value of each pixel for various land classifications using 258 signature points for these four land-use classes, out of which 100 accuracy points were collected through the grid system and verified with the help of Google Earth Pro software. All satellite imageries' spectral signatures were collected using pixel-contained polygons, and training samples were chosen for each terrain class using polygons. Afterwards, Supervised LULC classification was perform using the Gaussian Maximum Likelihood Classifier Algorithm in Erdas 2014 version and ArcGIS (v.10.1) software, and the classification was validated using accuracy evaluation. The accuracy findings of the land transformation have a range that is acceptable over 0.8 (Pal and Ziaul, 2017). After the images had been validated, a theme layer was created by superimposing the four different land classes on top of one another and subtracting one class from another for each of the two different years. The 'From-To' method was then applied to calculate the amount of land that was transferred from one class to a different class, and four change maps (1991–2001, 2001–2011, 2011–2021, and 1991–2021) were created to provide information on how the land use and land cover have changed between the classed photographs.

4.3.5 Analysis of the Directional Zone and Intensity of Agricultural Shrinkage

Cropland shrinkage of AUA was calculated for the years 1991, 2001, 2011, and 2021 by superimposing these images (1991–2001, 2001–2011, 2011–2021, and 1991–2021), and the cropland areal shrinkage was extracted in ArcGIS (v.10.8.1) and Erdas (v2014). We have now added a radar diagram to indicate the distance and direction of agricultural loss from the AUA city centre. Radar graph is a useful tool for displaying the spatial distribution of specific urban land uses using sectorial and concentric circle diagrams (Yin et al., 2011). As a result, we chose a 10-km buffer made up of eight consecutive concentric circles at an angle of 45°. Thus, we have adopted 67 km of buffer from the core city centre to detect the eight zones of urban built-up expansion, namely, north (N), east (E), south (S), west (W), north-east (NE), south-east (SE), south-west (SW), and North-west (NW).

4.3.6 ACCURACY ASSESSMENT

An accurate classification assessment is required when maps of LULC are produced by remote sensing (Jensen, 1996), even though the resolution is very poor in satellite imagery (Thapa, 2009). For assessing the accuracy of the seven LULC classes, random selections of LULC maps were made.

An essential component of the LULC categorisation and mapping process was utilised. The post-classification accuracy assessment is a method used to evaluate the quality of the classified maps (Manandhar et al., 2009). The quality of the produced maps is measured by classification accuracy, making it easier to determine whether a map is suitable for a given usage. A categorised map should have an accuracy level of at least 80% for accurate interpretability and identification (Anderson et al., 1976). Numerous studies have already assessed the accuracy of LULC maps using methodologies like the kappa coefficient, error matrices, and index-based techniques (Manandhar et al., 2009; Shahfahad et al., 2020; Talukdar et al., 2020). This kappa coefficient is a popular and trustworthy approach for indicating accuracy (Foody, 1992; Ma and Redmond, 1995).

The accuracy of LULC maps is assessed in this study using the kappa coefficient method for AUA using about 165 arbitrarily chosen points. The points were chosen in a manner that equally represents each LULC class and the entire study region. Because of the inaccessibility of the recent LULC data and field visit here, we used Google Earth images software as subsidiary information for accuracy assessment.

4.3.7 NORMALIZED DIFFERENCE VEGETATION INDEX

At the local, regional, and international levels, the use of remotely sensed vegetation data has become crucial for processes of vegetation evaluation, monitoring, and management. One of the most popular indices created from the red and near-infrared bands is the NDVI. The output of the normalisation process ranges from −1 to 1. (Rouse et al., 1974). Positive values indicate the presence of vegetation, while values near 1 indicate that the vegetation is dense. NDVI can be expressed as

$$\text{NDVI} = \frac{\text{NIR Band} - \text{R Band}}{\text{NIR Band} + \text{R Band}} \tag{4.1}$$

The NDVI method was used in this study to calculate the vegetation index. The vegetation index was obtained from the NDVI after manual thresholding:

$$NDWI = \frac{\text{Green Band} - \text{NIR Band}}{\text{Green Band} + \text{NIR Band}} \tag{4.2}$$

The NDWI method is frequently used to identify open water features. It uses the green and near-infrared bands and is expressed using the formula in Equation 4.2. NDWI is used to show the water index of an area. The range of the NDWI's value is −1 to 1. Values between 0 and 1 denote water bodies, whereas values near 1 denote water bodies with a high density (Ghosh and Singh, 2022).

4.3.8 LAND SURFACE TEMPERATURE

According to Qin et al., the study used the thermal infrared bands of various satellite images to derive LST (Nishara et al., 2021). Spectral reflectance measuring between 10.4 and 12.5 µm was used to find the ground surface radiation. There were four main steps in the LST retrieval (Mondal, 2021; Coluzzi et al., 2022).

Conversion of Digital Number Values to Spectral Radiance

The scaling method was used to convert the digital numbers, or the pixel values of the satellite images into spectral radiance as presented in Equation 4.3 for Landsat TM and ETM+ and Equation 4.4 for Landsat OLI.

$$D\lambda = \left(\frac{DMax.\lambda - DMin.\lambda}{QCal.Max - QCal.Min} \right) \times \left(\left(QCal - QCal.Min \right) \right) + DMin.\lambda \qquad (4.3)$$

$$D\lambda = ML \times Q_{Cal} + \Delta D \qquad (4.4)$$

where $D\lambda$ equals spectral radiance (W sr^{-1} m^{-3}); $DMax.\lambda$ equals spectral radiance scaled to QCal.Max (i.e., DN value 255); $DMin.\lambda$ equals spectral radiance scaled to QCal.Min (i.e., DN value 1); QCal equals pixel values of satellite images (Digital Number); QCal.Max equals quantitised and calibrated maximum pixel value that corresponds to $DMax.\lambda$; QCal.Min equals quantitised and calibrated minimum pixel value that corresponds to $DMin.\lambda$; ML equals multiplicative scaling factor for the radiance of the specific spectral band (x) obtained from the metadata of the dataset (i.e., RADIANCE_MULT_BAND_x); and ΔD equals additive scaling factor for the radiance of the spectral band (x) retrieved from the image's dataset (i.e., RADIANCE_ADD_BAND_x).

- Conversion of Spectral Radiance to Temperature (in Kelvin)

In this phase, the Planck Radiance Function is used to convert spectral radiation into brightness temperature (TB), or the top of the atmosphere (TOA). The standard formula shown in Equation 4.5 is used to determine the brightness temperature (TB), This is also related to the apparent surface temperature that is detected by the satellite sensor (Coluzzi et al., 2021):

$$TB = \frac{K2}{Log\left(\frac{K1}{L\lambda} + 1 \right)} \qquad (4.5)$$

where $K1$ and $K2$ are the calibration constants of thermal bands obtained from the image's metadata (Table 4.2), $L\lambda$ is the spectral radiance (W sr^{-1} m^{-3}), and TB is the brightness temperature (in Kelvin).

- Conversion of Temperature (in Kelvin) to Degrees Celsius

TABLE 4.2
Landsat Thermal Band Calibration Constant

S/No.	Satellite	Thermal Band	K1 (W m^{-2} sr^{-1} μm^{-1})	K₂ (Kelvin)
1	Landsat 4 TM	Band 6	671.62	1284.3
2	Landsat 7 ETM+	Band 6	666.09	1282.71
3	Landsat 8 OLI	Band 10	774.89	1321.08
4	Landsat 8 OLI	Band 11	480.89	1201.14

At this phase, surface temperature will be converted from degrees Kelvin (K) to degrees Celsius (C) using Equation 4.6:

$$T_{B\ (°C)} = T_B(\text{in Kelvin}) - 273.15 \tag{4.6}$$

• Assessment of LST

Equation 4.7 was used to calculate LST for the study from at-sensor brightness temperature:

$$\text{LST}\left(°\text{C}\right) = \frac{\text{TB}}{1 + \left(\lambda \times \dfrac{\text{TB}}{\rho}\right) \times \text{Ln}\left(\varepsilon\right)} \tag{4.7}$$

where *TB* is the sensor brightness temperature; λ is the emitted radiance wavelength (11.5 µm); ρ equals $h \times (c/\sigma) = 1.438 \times 10^2$ mk (h = Planck's constant [6.626×10^{34} JS], c = velocity of light [2.998×10^8 m/s] and k = Boltzmann constant [1.38×10^{23} j/k]) and ε is the surface emissivity calculated using Equation (4.8):

$$\text{Surface emissivity } (\varepsilon) = 0.004\left(\text{Pv}\right) + 0.986 \tag{4.8}$$

where *Pv* is the proportion of vegetation retrieved using Equation 4.9:

$$\text{Pv} = \left| \frac{\left(\text{NDVI} - \text{NDVI}min\right)}{\left(\text{NDVI}max - \text{NDVI}min\right)} \right|^2 \tag{4.9}$$

where NDVI is the Normalized Difference Vegetation Index, NDVI*min* is the minimum value of NDVI, and NDVI*max* is the maximum value of NDVI.

4.3.9 Correlation and Regression Analyses

The LST component of urban climate was then compared to spectral indices, which include NDVI, LST, and NDWI, using a linear correlation (Pearson) analysis (Brinkmann et al., 2012). The LST values for the four-time epochs under discussion were extracted from each year's image pixel at the exact locations of the various spectral indices. After that, the study's model was built using these points as input data, and the dynamics of land cover change were investigated in connection to LST.

4.4 RESULTS AND DISCUSSION

4.4.1 Land-Use and Land-Cover Status and Their Changing Scenario

The LULC classification of AUA shows the four LULC classes: cropland, natural vegetation, built-up area, and other class (Table 4.3). A vivid illustration of the spatial distribution of LULC change has been given by the four historical eras' classified images (Figure 4.2) and their rate of change for each land class in AUA (Figure 4.3). The focus of this paper is to better understand the cropland pattern throughout the study period. If we ignore the other classes, then the findings showed that cropland was the most common type of land covering greater than one-quarter of the area (28.00%) followed by natural vegetation (10.06%) and built-up area (8.10%) during 1991 (Table 4.3). Nonetheless, there was a significant change in the land-use classifications in 2001, within ten years, when cropland area

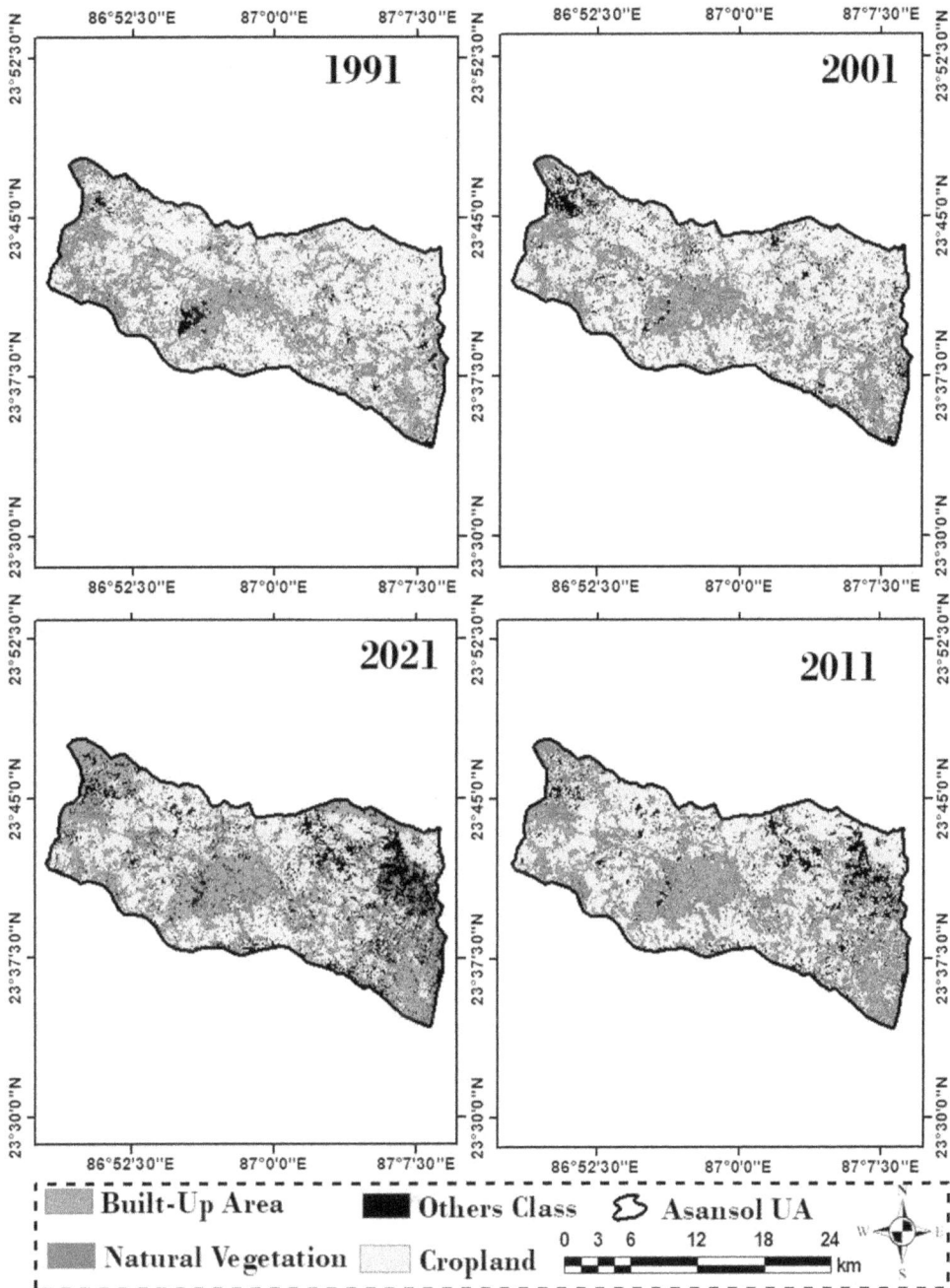

FIGURE 4.2 Land-use and land-cover maps of the Asansol Urban Agglomeration illustrating the dynamics of land use through four distinct time periods: 1991, 2001, 2011, and 2021.

reduced to 24.91%. This demonstrates that the majority of the agricultural area surrounding the main town was developed owing to overpopulation. This is pushed by in-migration from the rural area and increasing demands of housing for habitation in the western and south-eastern portions along the Damodor River. While in 2011 the cropland changed a lot and natural vegetation decreased, by about 22.35% and 7.387%, respectively, on the other side, the built-up area continuously grew from 8.52% to 12.05%. In 2021, the built-up area stands more or less equal in proportion to cropland in

TABLE 4.3

Percentage of Area under Different Land Use/Land Cover in Different Periods

LULC Category	1991 Area	1991 Percentage of Area (%)	2001 Area	2001 Percentage of Area (%)	2011 Area	2011 Percentage of Area (%)	2021 Area	2021 Percentage of Area (%)
Natural vegetation	90.67	10.06	78.27	8.69	70.93	7.87	68.51	7.60
Cropland	252.23	28.00	224.38	24.91	201.34	22.35	174.06	19.32
Built-up area	73.01	8.10	76.80	8.52	108.53	12.05	161.56	17.93
Other class	485.02	53.84	521.47	57.88	520.14	57.73	496.74	55.14
Total	900.93	100.00	900.92	100.00	900.94	100.00	900.87	100.00

Note: LULC, land use/land cover.

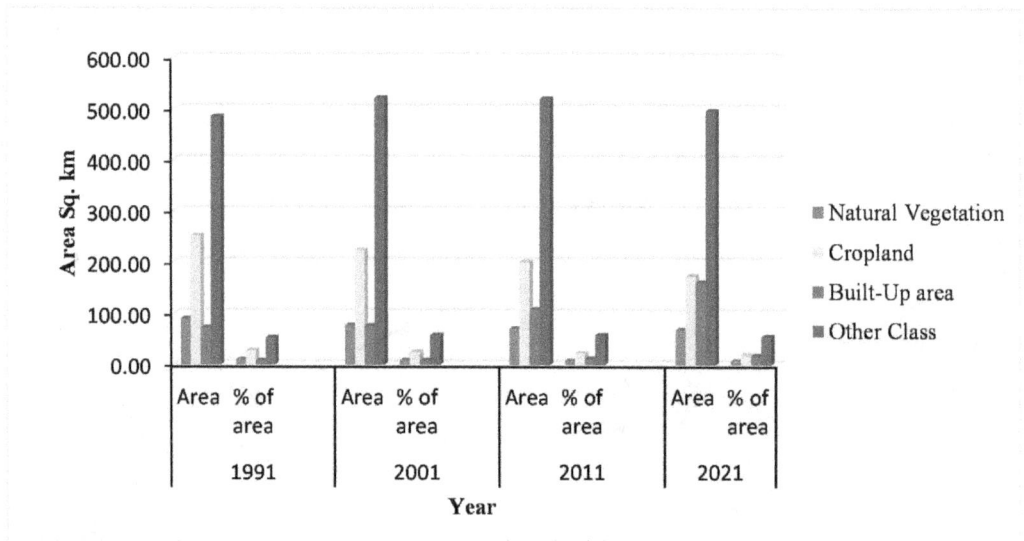

FIGURE 4.3 The percentage of area covered by each land use/land cover in 1991, 2001, 2011, and 2021.

terms of percentage, although it was more than three times less as compared to cropland in 1991. This illustrates that urban growth is rising quickly caused by population pressure. The overall accuracy of supervised classification was high enough to support the AUA transformation scenarios, coupled with kappa values of 0.89, 0.84, 0.89, and 0.83 for four reference years (1991, 2001, 2011, and 2021).

4.4.2   ANALYSIS OF LAND TRANSFORMATION BETWEEN 1991 AND 2021

The land transformation map and matrix both show various classes between natural vegetation, cropland, and built-up areas. This clear representation indicates that the majority of these components were transformed by the built-up region.

4.4.3   SPATIAL SHRINKAGE OF THE CROPLAND AREA

The cropland area of AUA from 1991 to 2021 decreased 27.85 km² during the years 2001–2011; this decreased 23.04 km² which is more than in the previous decade (Table 4.4). The annual rate of cropland area reduction during the whole period of 2001–2011 followed the decreasing trend (Figure 4.4) due to roadway construction, building construction, and deforestation due to natural population growth. An important factor in the growth of urban settlements in AUA was the conversion of agriculture due to industrial development and entrepreneur investment in a variety of sectors. As a result of the low cost of land, a number of urban sprawls have grown towards the south-eastern portion of the study area.

Another remarkable change is that AUA was construct as a halt station alongside G.T. Road in the past. Because of the abundance of minerals in the areas surrounding AUA, the development of IISCO in Burnpur has made Asansol a significant urban centre in the West Barddhaman district. This growth has been largely attributed to the establishment of rail lines and the availability of raw materials. As a result, in the south-western portion of the research region, another dominant urban sprawl has emerged. On the other hand, in the south-east direction, Raniganj has better transport and connectivity which has played a key role in urban in-migration which directly transformed the peripheral area of the core city into a built-up area. Also, because of the strong competition for land and the lower cost of land compared to the city's core and centre, people are more eager to choose areas on the periphery, which has led to the growth of the urban area. Particularly, middle-class societies would find it appropriate to live in this area. The important highlight from Table 4.4 is that the cropland was extremely reduced in the period 1991–2021 by 78.17 km. The annual rate of reduction of the cropland throughout this three decades averaged 1.50 km/year. This creates an environmental challenge.

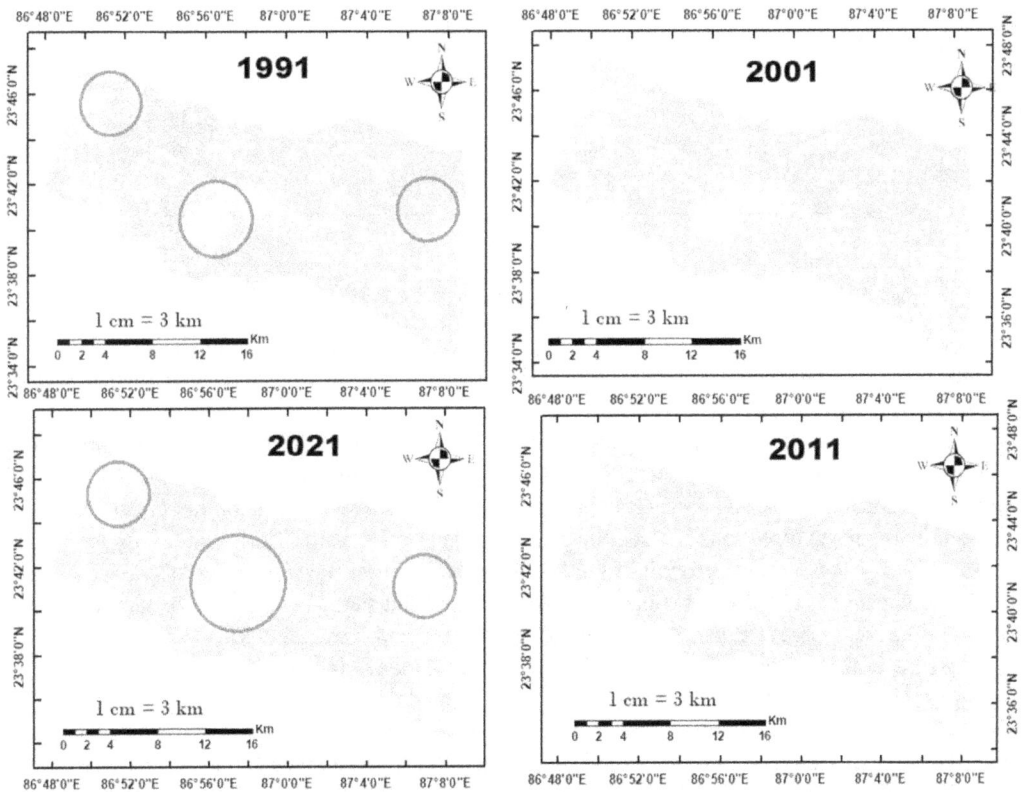

FIGURE 4.4 Spatio-temporal shrinkage of cropland from 1991 to 2021.

TABLE 4.4
Shrinkage of Cropland Area in Last 30 Years

Year	Cropland (km²)	Change (km)			
		1991–2001	2001–2011	2011–2021	1991–2021
1991	252.23	−27.85			
2001	224.38		−23.04		
2011	201.34			−27.28	
2021	174.06				−78.17
Percentage of change (km²/year)		−1.1	−1.03	−1.35	−1.5

TABLE 4.5
Directional Shrinkage of Cropland (km²) of Asansol Urban Agglomeration from 1991 to 2021

Direction	Year				Change			
	1991	2001	2011	2021	1991–2001	2001–2011	2011–2021	1991–2021
North	13.21	11.77	12.39	10.89	−1.44	0.63	−1.50	−2.32
North-east	10.87	9.99	10.53	7.61	−0.88	0.54	−2.91	−3.26
East	49.99	47.25	39.70	34.86	−2.74	−7.55	−4.84	−15.13
South-east	61.20	53.00	37.65	33.72	−8.20	−15.35	−3.93	−27.49
South	16.07	14.21	13.71	12.94	−1.86	−0.50	−0.77	−3.13
South-west	14.03	11.24	9.60	8.33	−2.79	−1.64	−1.27	−5.70
West	20.34	21.41	18.77	16.58	1.06	−2.63	−2.19	−3.76
North-west	66.36	55.38	58.87	48.99	−10.98	3.49	−9.87	−17.36
Total	252.07	224.24	201.22	173.93				

4.4.4 ZONE-WISE DIRECTIONAL SHRINKAGE OF THE CROPLAND AREA

The cropland of AUA decreased by 69% during the period 1991–2021. In order to have a thorough understanding of urban expansion patterns, we plotted line diagrams from Asansol's earliest locales (Ward no. 6) to the key eight directions for this study. The results suggest that the farmland shrank most in the south-east due to the placement of the industrial zone, built-up area, and brickwork building, and then in the north-west and east along the railway line and G.T. Road (Table 4.5 and Figure 4.5). During the period 1991–2021, the south-east, north-west, east, south-west, west, north-east, south, and north directions showed the greatest change. The results of this study suggested that AUA could experience urban growth in the future in the south-eastern, north-western, and eastern directions because it has a good transport infrastructure and better infrastructure facilities compared to other regions. As a result, Asansol has grown into an industrial city.

The focus of this study is to determine LST, vegetation index, and correlation variables utilising multi-temporal Landsat 5 TM, Landsat 8 OLI, and TIRS data. This chapter also gives the NDWI which is defined as the water index of this particular region. The final outputs of the mean value of the NDVI and LST have been correlated. The LST and NDVI have been found for the years 1991, 2001, 2011, and 2021. Figures 4.3 and 4.4 have exposed the spatial distribution of estimated LST, respectively, for the same years. The maximum and minimum temperatures and statistics of LST

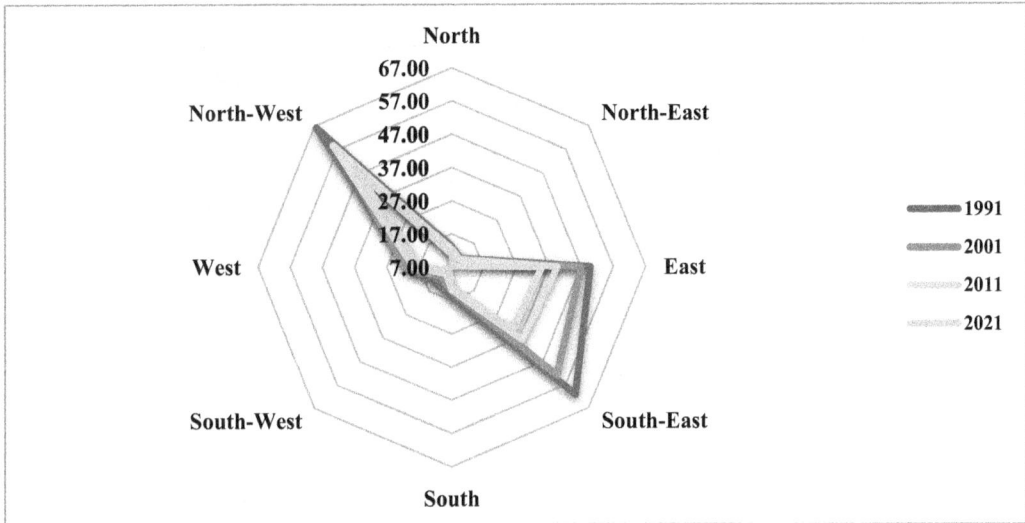

FIGURE 4.5 Directional shrinkage of cropland (km²) from 1991 to 2021.

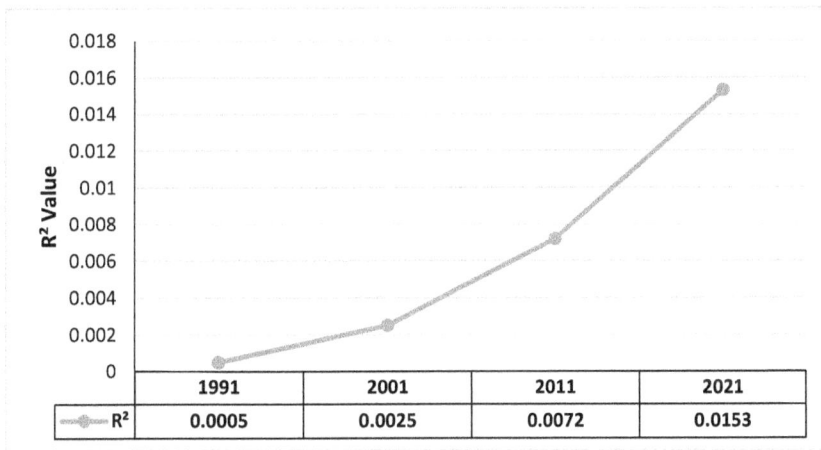

FIGURE 4.6 Change in R^2 value from 1991 to 2021.

and NDVI are 31.25°C and 15.65°C, respectively, for the year 1991; 35.27°C and 15.18°C for the year 2001; 28.78°C and 17.02°C for 2011; and 44.45°C and 23.87°C for 2021 (Figure 4.8). The mean value of LST was recorded as 10.909°C in 1991 and 12.584°C in 2017. Here maximum and minimum values of LST increased from 1991 to 2021. The NDVI of 1991 has maximum and minimum values −0.3 and 0.52, respectively, for 2001 the maximum and minimum NDVI values are −032 and 0.60. On the other hand, in 2011, the NDVI value is −0.14 (minimum) and 0.4 (maximum) with a mean value of 0.863 (Figure 4.9). For 2021, the minimum NDVI value is −0.52 and maximum is 0.74. The maximum value of NDVI slightly increased, but maximum variation was found in the minimum value of NDVI. LST is negatively correlated with NDVI (vegetation cover).

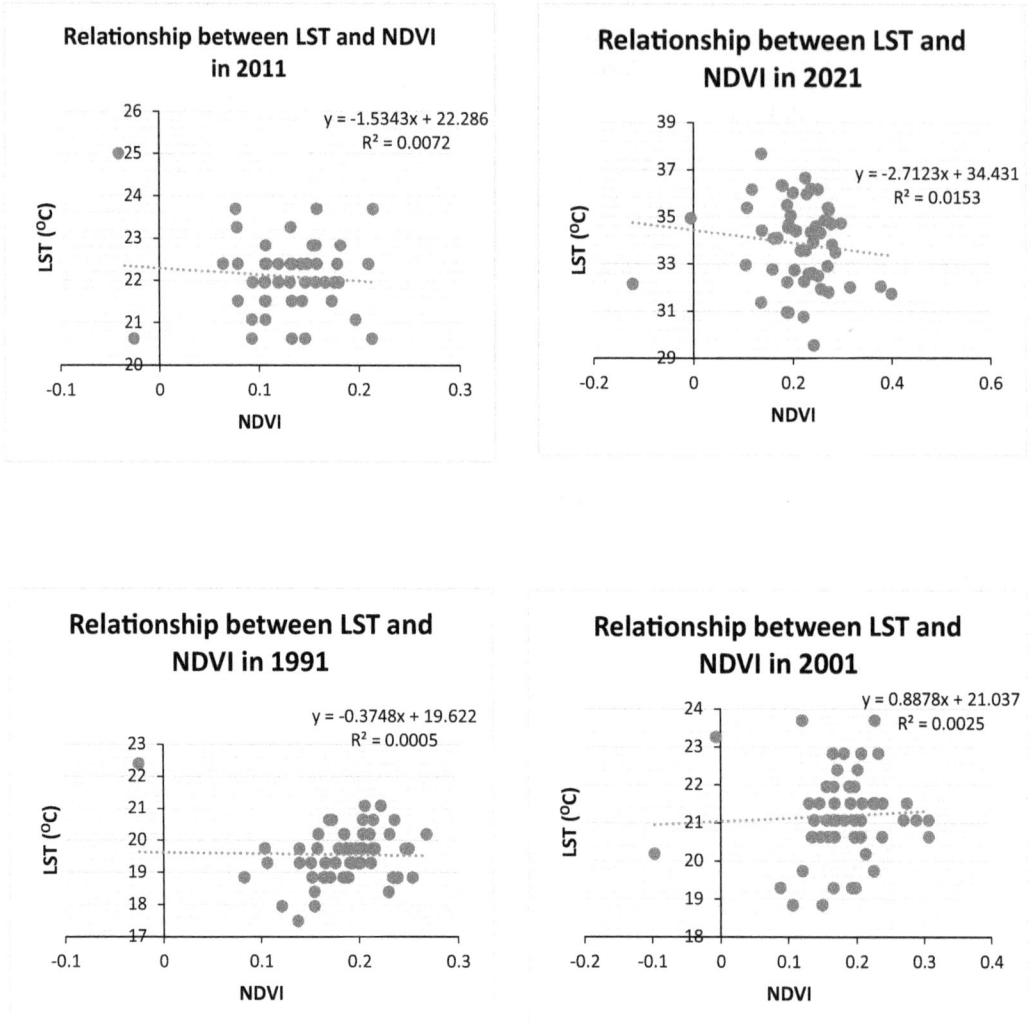

FIGURE 4.7 Correlation between land surface temperature and Normalized Difference Vegetation Index for the years 1991, 2001, 2011, and 2021.

Figures 4.8, 4.9, and 4.10 show spatial distributions of the LST, NDVI, and NDWI indices derived from the Landsat 5 TM and Landsat 8 OLI data, 1991, 2001, 2011, and 2021. The best methods for determining the region's temperature are LST and NDVI combinations, and NDVI serves as research validation. The connection between LST and NDVI highlights the status of vegetation in the study area. LST mean value increased from 10.909°C in 1991 to 12.584°C in 2017. LST and NDVI mean values have been correlated using the Pearson correlation coefficient. Figures 4.5 and 4.6 give the statistical relationship between LST and NDVI in the years of 1991, 2001, 2011, and 2021, respectively.

The vegetation proportion and the coefficient of determination value $R^2 = 0.0005$ for 1991, $R^2 = 0.0025$ for 2001, $R^2 = 0.0072$ for 2011, and finally for 2021 $R^2 = 0.0153$ (Figures 4.6 and 4.7) are strongly correlated with the brightness temperature. Intensity of the LST gradually increases in AUA, West Bengal. High LST and low NDVI are observed in the south-eastern and north-eastern parts of the study area in the year 2021, especially part of Jamuria block and most of Raniganj block (Figure 4.4). So, the

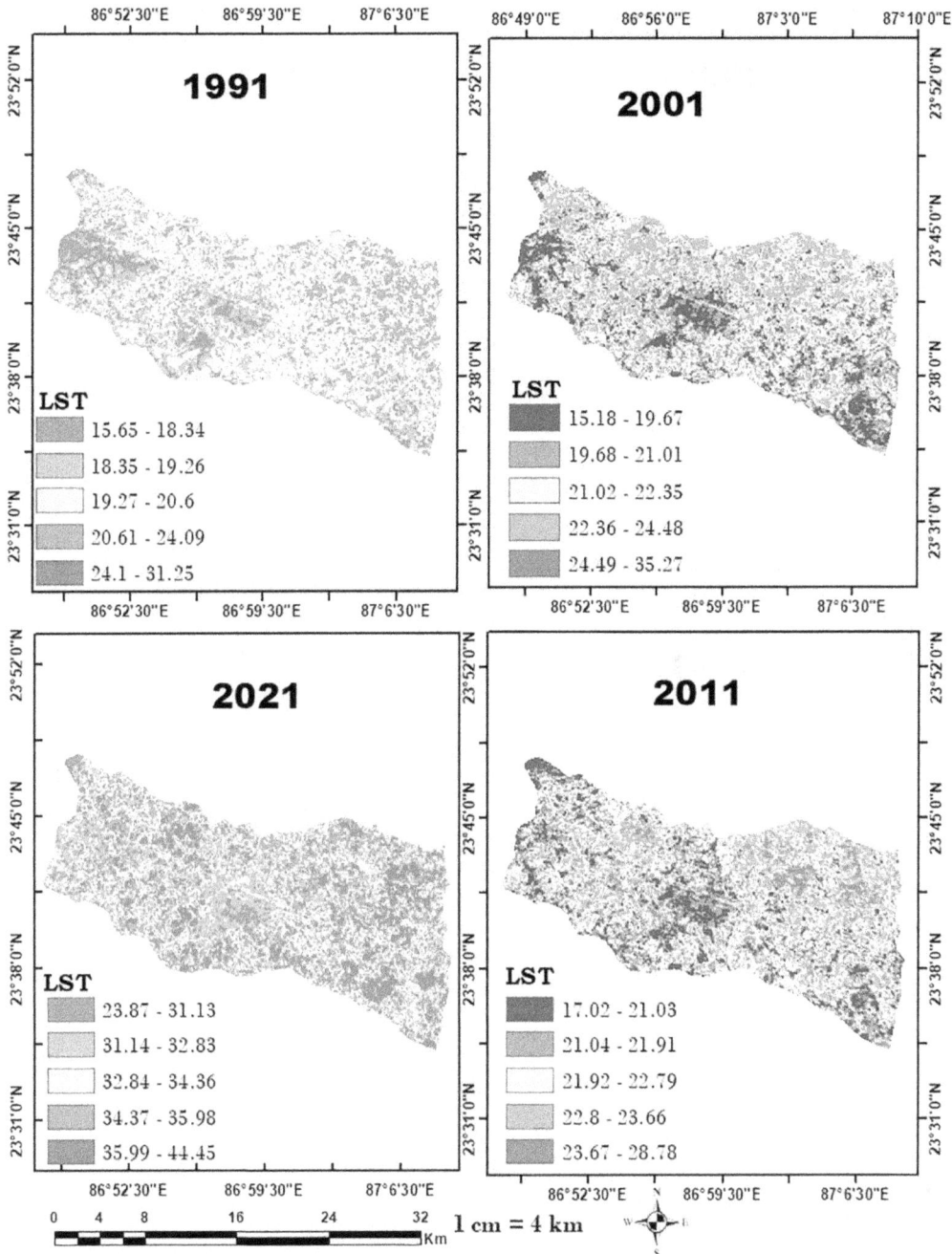

FIGURE 4.8   Status of land surface temperature (°C) in 1991, 2001, 2011, and 2021.

vegetation health of these regions is negatively affected by high LST. This chapter also calculated the NDWI, which defined the density of a water body in the region. The minimum and maximum values of NDWI are −0.303 and 0.524 for 1991. And for 2001 the maximum and minimum NDWI values are −0.543 and 0.387. The minimum and maximum values for 2011 are −0.322 and 0.233. For 2021, the NDWI minimum and maximum values are −0.633 and 0.57 (Figure 4.10). The NDWI value for 1991–2011 rapidly decreased because of vegetation and agricultural land shrinkage.

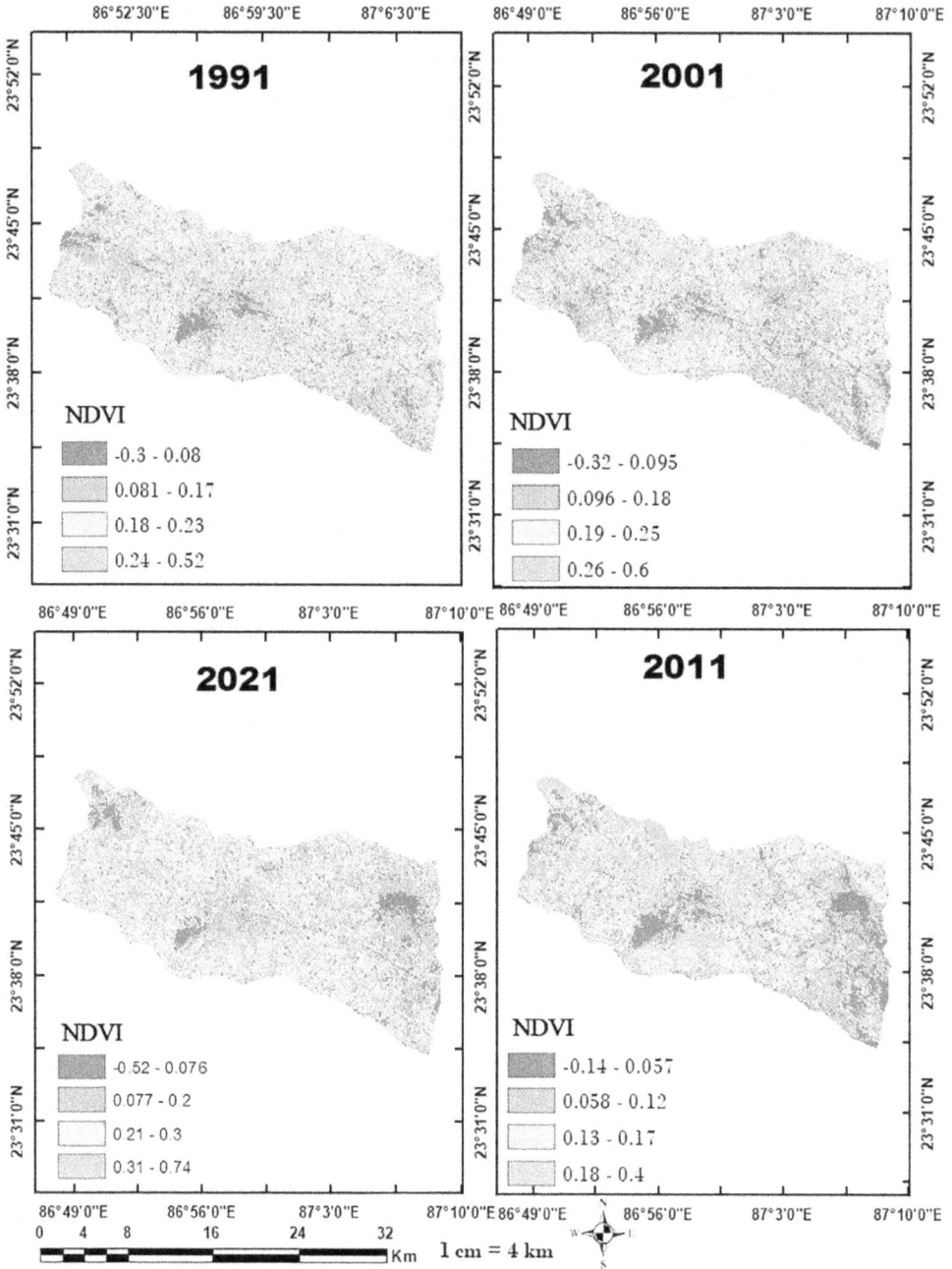

FIGURE 4.9 Status of Normalized Difference Vegetation Index in 1991, 2001, 2011, and 2021.

This study was carried out to assess the influence of LULC transformation on the LST in AUA of Paschim Barddhaman district of West Bengal during February to April of the years 1991, 2001, 2011, and 2021. It was useful to consult earlier research on LULC changes and LST in order to carry out the study and evaluate its results. The analysis indicated an increase in built-up areas and a declining trend in natural vegetation and agriculture, which is consistent with earlier findings (Choudhury et al., 2019; Chatterjee et al., 2022; Dutta & Guchhait, 2022; Maity et al., 2022). Choudhury et al. (2019) reported that dense and scattered vegetation had rapid shrinkage from 1993

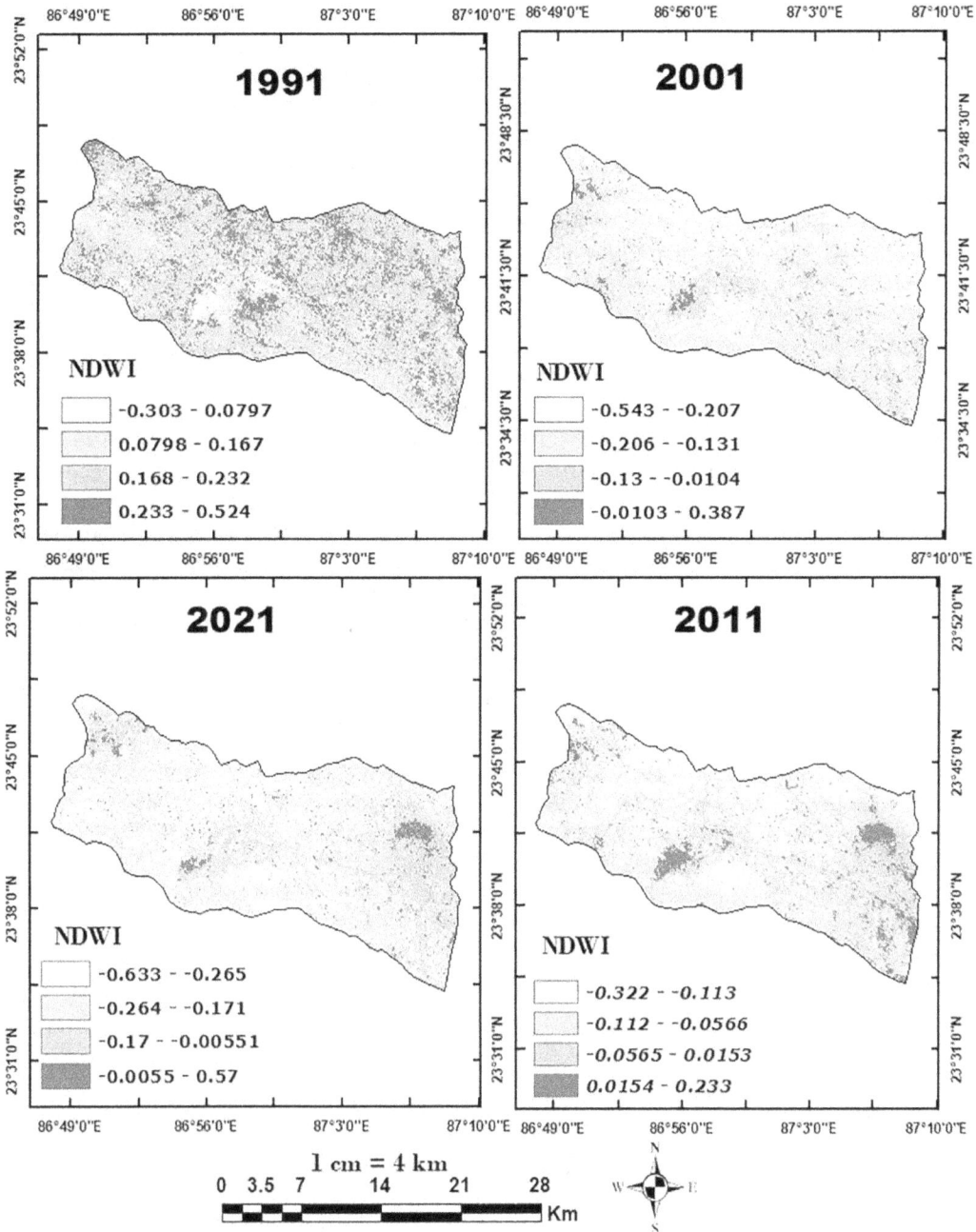

FIGURE 4.10 Status of Normalized Difference Water Index (°C) in 1991, 2001, 2011, and 2021.

to 2015 in the Asansol-Durgapur Development Region: 6.90% to 4.97% for dense vegetation and 43.23% to 28.37% for scattered vegetation. Settlement rapidly increased from 9.01% to 19.93%, about a 121.19% increase throughout these 22 years. According to Chatterjee et al. (2022), LULC changes were made for 21 years, or between 1996 and 2017, to examine the dynamics in each LULC category in the Paschim Bardhaman District of West Bengal. A total of 5,870 ha of agricultural land and 1,333.4 ha of water bodies have been lost, which, correspondingly, accounts for 8.9% and 37.7% of net loss, in the following research. On the other hand, 3,866 acres, or 10.8% of the total study area, have seen a considerable rise in built-up areas. Maity et al. (2022) found that in comparison to the preceding ten years, the built-up area of Asansol municipal corporation (AMC) rose by 14.28 km^2 from 1991 to 2008 and by 16.34 km^2 from 2008 to 2018. It is also evident that the urban expansion between 2008 and 2018 played a big role due to the inevitable increase of the population and a variety of employment alternatives. The growth of urban areas and the investments made by entrepreneurs in numerous sectors were key factors in the transformation of the natural landscape in AMC. Dutta and Guchhait (2022) noticed in their studies of Paschim Bardhhaman that the built-up area increased from 11.39% from 1.41% in the period 1990–2010. Beside that, the vegetation decreased from 29.5% to 27.1%, and agricultural land was reduced to 58.32% in 2010 from 63.58% in 1990. The main causes of the new urban momentum are the growth of the sponge iron industry in the Durgapur region, the revitalisation of trade and transportation caused by the construction of the Kona Expressway, and the indiscriminate movement of rural residents into urban areas, urban peripheries, and rural service centres at convenient transportation nodes.

Additionally, a significant amount of the water bodies, low-lying land, and vegetation cover were developed, affecting the environment's biodiversity and natural habitat (Alphan, 2003). The AUA's population grew as a result of rising birth rates, rural-to-urban migration, social and economic development, and urban expansion. People moving from the countryside to the city was the main factor in the greater population growth rate in the AUA. In 1991, the urban migrated population was 2,49,574, and in the 2011 census it reached 4,39,394, about a 76.05% increase in migration toward the AUA. As a result of ongoing over-exploitation of the scarce natural resources, significant urbanisation has already taken place in order to meet the needs of a growing population. The AUA urban area did not grow in all directions equally; instead, the edge of the city had the greatest urban growth.

4.5 CONCLUSION

The AUA in West Bengal has been studied using an integrated geospatial technique to analyse the spatio-temporal analysis of cropland dynamics. The overall findings suggested that this city's landscape has changed significantly, primarily towards the south-east direction. But the major concern is that the expansion is with the encroachment of cropland to a built-up area around AUA. Consequently, the south-east sector of the study region has revealed a built-up area, followed by the north-west and eastern sections. Thus, it is critically necessary to design numerous aspects of construction of a built-up area in order to safeguard the agricultural and vegetation area in the area surrounding it as well as inside the residential zone. The region needs to plant more trees both inside and in the periphery of the city to increase the green infrastructure needed to combat an urban heat island in the future. Due to the reduction of agricultural land and vegetation land, many changes in the temperature of the city can be observed. In the case of AUA, it can be seen that the LST minimum value in 1991 was 15.65°C which reached 23.87°C in 2021. The maximum LST value was 31.25°C in 1991 which became 44.45°C in 2021. This rise in temperature has turned the city into an urban heat island. One of the reasons behind this is the decrease in vegetation and agricultural land. Cropland and vegetation land in 1991 were 252.23 km^2 and 90.67 km^2, respectively, which decreased to 174.06 km^2 and 68.51 km^2 in 2021. On the other hand, the built-up area increased from 73.01 to 161.56 km^2. Cropland has decreased by about 1.5 km^2 per year in the last 30 years. And built-up area increased by 121.28% encroaching on cropland and vegetation. This is proved by using the NDVI and NDWI methods. It can be seen that the NDVI minimum value in 1991 was −0.3 which decreased to −0.14 in 2011, while the maximum NDVI value in 1991 was 0.52 which decreased to 0.4 in 2011. Not only that, the NDWI value has seen a sharp decline from 0.524 to 0.233, from 1991 to 2011.

However, the shift of the landscape towards populated areas is a significant worry for every metropolis in developing countries. The most prominent element that directly affects LULC in AUA is rapid population expansion in all directions. Therefore, the AUA's local government and planning committee have a significant responsibility to plan appropriately in accordance with AUA's land-use concerns. Unless this tendency of urban built-up growth is reversed, there will soon be another vulnerable issue in this area. To ensure that future development will be habituated to sustainable practises, we should prioritise sound water management, green infrastructure development, solid waste management, and regular atmosphere monitoring.

REFERENCES

Abdel Rahman, M.A.E., Natarajan, A., and Hegde, R. (2016). Assessment of land suitability and capability by integrating remote sensing and GIS for agriculture in Chamarajanagar district, Karnataka, India. Egypt J Remote Sens Space Sci 19:125–141.

Alphan, H. (2003). Land-use change and urbanization of Adana, Turkey. Land Degrad Dev 14:575–586. https://doi.org/10.1002/ldr.581

Amuti, T., and Luo, G. (2014). Analysis of land cover change and its driving forces in a desert oasis land of Xinjiang, northwest China. Solid Earth 5(2):1071–1085. https://doi.org/10.5194/se-5-1071-2014

Anderson, J.R., Hardy, E.E., Roach, J.T., and Witmer, R.E. (1976). A landuse/landcover classification system for use with remote sensor data. Geological Survey Professional Paper 964.

Babalola, O.S., and Akinsanola, A.A. (2016). Change detection in land surface temperature and land use land cover over Lagos Metropolis, Nigeria. J Remote Sens GIS. https://doi.org/10.4172/2469-4134.1000171

Bose, A., and Chowdhury, I.R. (2020). Monitoring and modeling of spatio-temporal urban expansion and land-use/land-cover change using Markov chain model: A case study in Siliguri Metropolitan area, West Bengal, India. Model. Earth Syst Environ 6:2235–2249. https://doi.org/10.1007/s40808-020-00842-6

Brinkmann, K., Schumacher, J, Dittrich, A, Kadaore, I, and Buerkert, A. (2012). Analysis of landscape trans-formation processes in and around four West African cities over the last 50 years. Landsc Urban Plan 105(1–2):94–105. https://doi.org/10.1016/j.landurbplan.2011.12.003

Byerlee, D., De Janvry, A., and Sadoulet, E. (2009). Agriculture for development: Toward a new paradigm. Annual Review of Resource Economics 1:15–31.

Chatterjee, S., Dutta, S., Dutta, I., and Das, A. (2022). Ecosystem services change in response to land use land cover dynamics in Paschim Bardhaman district of West Bengal, India. Remote Sens Appl Soc Environ 27. https://doi.org/10.1016/j.rsase.2022.100793

Chen, J., Chang, K., Karacsonyi, D., and Zhang, X. (2014). Comparing urban land expansion and its driv-ing factors in Shenzhen and Dongguan, China. Habitat Int 43:61–71. https://doi.org/10.1016/j.habit atint.2014.01.004

Chen, X.L., Zhao, H.M., Li, P.X., Yin, Z.Y. (2006). Remote sensing image-based analysis of the relationship between urban heat island and land use/cover changes. Remote Sens Env 104(2):133–146. https://doi.org/10.1016/j.rse.2005.11.016

Choudhury, D., Das, K., and Das, A. (2019). Assessment of land use land cover changes and its impact on variations of land surface temperature in Asansol-Durgapur Development Region. Egyptian Journal of Remote Sensing and Space Science 22(2):203–218. https://doi.org/10.1016/j.ejrs.2018.05.004

Coluzzi, R., Bianchini, L., Egidi, G., Cudlin, P., Imbrenda, V., Salvati, L., and Lanfredi, M. (2022). Density matters?: Settlement expansion and land degradation in peri-urban and rural districts of Italy. Environ Impact Assess Rev 92:106703. https://doi.org/10.1016/j.eiar.2021.106703

Dawson, P.T., Perryman, N.A., and Osborne, T. (2014). Modelling impacts of climate change on global food security. Climatic Change 134(3):1–12. http://link.springer.com/article/10.1007%2Fs10584-014-1277-y (DOI 10.1007/s10584-014-1277-y)

Dissanayake, D., Morimoto, T., and Ranagalage, M. (2019). Accessing the soil erosion rate based on RUSLE model for sustainable land use management: A case study of the Kotmale watershed, Sri Lanka. Model Earth Syst Environ 5:291–306.

Dutta, S., and Guchhait, S.K. (2022). Assessment of land use land cover dynamics and urban growth of Kanksa block in Paschim Barddhaman district, West Bengal. GeoJournal 87(2):971–990. https://doi.org/10.1007/s10708-020-10292-3

Estoque, R.C., and Murayama, Y. (2017). Monitoring surface urban heat island formation in a tropical moun-tain city using Landsat data (1987–2015). ISPRS J Photogramm Remote Sens 133:18–29.

Foody, G.M. (1992). On the compensation for chance agreement in image classification accuracy assessment. Photogramm Eng Remote Sens 58(10):1459–1460.

Gandhi, G., and Savalia, S.G. (2014). Soil-site suitability evaluation for mustard in calcareous soils of Girnar toposequence in Southern Saurashtra region of Gujarat. J Oilseed Brassica 5(2):128–133.

Ghosh, P., and Singh, K.K. (2022). Spatiotemporal dynamics of urban green and blue spaces using geospatial techniques in Chandannagar city, India. GeoJournal 87(6):4671–4688. https://doi.org/10.1007/s10708-021-10524-0

Goward, S.N., Tucker, C.J., and Dye. D.G. (1985). North American vegetation patterns observed with the NOAA-7 advanced very high resolution radiometer. Vegetatio 64(1):3–14. https://doi.org/10.1007/BF00033449

Halder, J.C. (2019). Modeling the effect of agricultural inputs on the spatial variation of agricultural efficiency in West Bengal, India. Model Earth Syst Environ 5:1103–1121.

Hasnine, M., and Rukhsana. (2020). An analysis of urban sprawl and prediction of future urban town in urban area of developing nation: Case study in India. J Indian Soc Remote Sens 48(6):909–920. https://doi.org/10.1007/s12524-020-01123-6

Hasnine, M., and Rukhsana. (2023). Spatial and temporal analysis of land use and land cover change in and around Kolkata city, India, using geospatial techniques. J Indian Soc Remote Sens. https://doi.org/10.1007/s12524-023-01669-1

Islam, K., Jashimuddin, M., Nath, B., and Nath, T. K. (2018). Land use classification and change detection by using multi-temporal remotely sensed imagery: The case of Chunati wildlife sanctuary, Bangladesh. Egypt J Remote Sens Space Sci 21(1):37–47. https://doi.org/10.1016/j.ejrs.2016.12.005

Islam, M.S., and Kieu, E. (2020). Tackling regional climate change impacts and food security issues: A critical analysis across ASEAN, PIF, and SAARC. Sustainability 12:883. https://doi.org/10.3390/su12030883

Jensen, J.R. (1996). Introductory Digital Image Processing: A Remote Sensing Perspective, 2nd edn. Prentice-Hall Inc., Upper Saddle River.

Jiang, J., and Tian, G. (2010). Analysis of the impact of land use/land cover change on land surface temperature with remote sensing. Proc Environ Sci 2(5):571–575. https://doi.org/10.1016/j.proenv.2010.10.062

Karimi, A., Pahlavani, P., and Bigdeli, B. (2017). Land Use Analysis on Land Surface Temperature in Urban Areas Using a Geographically Weighted Regression and Landsat 8 Imagery, a Case Study: Tehran, Iran. ISPRS: International Archives of the Photogrammetry, Remote Sensing and Spatial Information Sciences, XLII-4/W4, 117–122. https://doi.org/10.5194/isprs-archives-XLII-4-W4-117-2017

Lambin, E.F., Turner, B.L., Geist, H.J., Agbola, S.B., Angelsen, A., Bruce, J.W., Coomes, O.T., Dirzo, R., Fischer, G.W., Folke, C., George, P.S., Homewood, K., Imbernon, J., Leemans, R., Li, X., Moran, E.F., Mortimore, M.J., Ramakrishnan, P.S., Richards, J.F., Skånes, H., Steffen, W., Stone, G.D., Svedin, U., Veldkamp, T.A., Vogel, C., and Xu, J. (2001). The causes of land-use and land-cover change: Moving beyond the myths. Global Environmental Change-human and Policy Dimensions 11:261–269.

Lang, R., Shao, G., Pijanowski, B.C., and Farnsworth, R.L. (2008). Optimizing unsupervised classifications of remotely sensed imagery with a data-assisted labeling approach. Computers & Geosciences 34:1877–1885.

Li, Y., Schubert, S., Kropp, J.P., and Rybski, D. (2020). On the influence of density and morphology on the urban heat island intensity. Nat Commun 11(1):1–9. https://doi.org/10.1038/s41467-020-16461-9

Lo, C.P., Quattrochi, D.A., and Luvall, J.C. (1997). Application of high resolution thermal infrared remote sensing and GIS to assess the urban heat island effect. International Journal of Remote Sensing 18:287–303.

Ma, Z., and Redmond, R.L. (1995). Tau coefficients for accuracy assessment of classification of remote sensing data. Photogramm Eng Remote Sens 61(4):435–439.

Maity, B., Mallick, S.K., and Rudra, S. (2022). Spatiotemporal dynamics of urban landscape in Asansol municipal corporation, West Bengal, India: A geospatial analysis. GeoJournal 87(3):1619–1637. https://doi.org/10.1007/s10708-020-10315-z

Manandhar, R., Odeh, I.O.A., and Acnev, T. (2009). Improving the accuracy of land use and land cover classification of Landsat data using post-classification enhancement. Remote Sensing 1:330–344.

Mohajerani, A., Bakaric, J., Jefrey-Bailey, T. (2017). The urban heat island effect, its causes, and mitigation, with reference to the thermal properties of asphalt concrete. J Environ Manage 197:522–538. https://doi.org/10.1016/j.jenvman.2017.03.095

Moisa, M.B., Tiye, F.S., Dejene, I.N., and Gemeda, D.O. (2022). Land suitability analysis for maize production using geospatial technologies in the Didessa watershed, Ethiopia. Artif Intell Agric 6:34–46. https://doi.org/10.1016/j.aiia.2022.02.001

Molla, S.H., and Rukhsana. (2022). Spatio-Temporal Analysis of Built-Up Area Expansion on Agricultural Land in Mousuni Island of Indian Sundarban Region. In: Agriculture, Environment and Sustainable Development: Experiences and Case Studies (pp. 91–104). Cham: Springer International Publishing.

Molla, S.H., Rukhsana, and Alam, A. (2020). Land suitability appraisal for the growth of potato cultivation: A study of Sagar island, India. Sustainable Development Practices Using Geoinformatics:111–126. © 2021 Scrivener Publishing LLC.

Mondal, D. (2021). Basic service provisioning in peri-urban India: A regional perspective from Kolkata Metropolis. Indian J Human Dev 15(1):97–116. https://doi.org/10.1177/09737030211000930

Nega, W., Hailu, B.T., and Fetene, A. (2019). An assessment of the vegetation cover change impact on rainfall and land surface temperature using remote sensing in a subtropical climate. Ethiopia Remote Sens Appl Soc Environ 16:100266. https://doi.org/10.1016/j.rsase.2019.100266

Nemani, R.R., Keeling, C.D., Hashimoto, H., Jolly, W.M., Piper, S.C., Tucker, C.J., Myneni, R.B., and Running, S.W. (2003). Climate-driven increases in global terrestrial net primary production from 1982 to 1999. Science 300(5625):1560–1563. https://doi.org/10.1126/science.1082750

Nero, B.F. (2017). Urban green space dynamics and socioenvironmental inequity: Multi-resolution and spatio-temporal data analysis of Kumasi, Ghana. International Journal of Remote Sensing 38(23):6993–7020. https://doi.org/10.1080/01431161.2017.1370152

Nishara, V.P., Sruthi Krishnan, V., and Firoz, C.M. (2021). Geo-Intelligence-based Approach For Sustainable Development of Peri-Urban Areas: A Case Study of Kozhikode City, Kerala (India). In: Singh, T.P., Singh, D., and Singh, R.B. (eds.) Geo-Intelligence For Sustainable Development: Advances in Geographical and Environmental Sciences. Springer, Singapore. https://doi.org/10.1007/978-981-16-4768-0_3

Pal, S., and Ziaul, S. (2017). Detection of land use and land cover change and land surface temperature in English Bazar Urban Centre. Egypt J Remote Sens Space Sci 20(1):125–145. https://doi.org/10.1016/j.ejrs.2016.11.003

Raynolds, M.K., Comiso, J.C., Walker, D.A., and Verbyla, D. (2008). Relationship between satellite-derived land surface temperatures, arctic vegetation types, and NDVI. Remote Sens Environ 112(4):1884–1894. https://doi.org/10.1016/j.rse.2007.09.008

Rouse, J., Haas, R.H., Schell, J.A., and Deering, D.W. (1974). Monitoring vegetation systems in the great plains with ERTS. NASA Special Publication 351(1974):309.

Rukhsana, and Alam, A. (2022). Levels of Agriculture Development and Crop Diversification: A District-Wise Panel Data Analysis in West Bengal. In: Rukhsana, and Alam, A. (eds.) Agriculture, Environment and Sustainable Development. Springer, Cham. https://doi.org/10.1007/978-3-031-10406-0_7

Rukhsana, and Hasnine, M. (2018). Modelling of potential sites for residential development at South East Peri-Urban of Kolkata.

Rukhsana, and Hasnine, M. (2020). Population Pressure and Urban Sprawl in Kolkata Metropolitan Area. In Rukhsana et al. (eds.) Habitat, Ecology and Ekistics, Advances in Asian Human-Environmental Research (pp. 163–178). Springer Nature Switzerland AG. (1). https://doi.org/10.1007/978-3-030-49115-4_9

Rukhsana, and Molla, S.H. (2023). Soil site suitability for sustainable intensive agriculture in Sagar Island, India: A geospatial approach. J Coast Conserv 27:14. https://doi.org/10.1007/s11852-023-00943-1

Sandholt, I., Rasmussen, K., and Andersen, J. (2002). A simple interpretation of the surface temperature/vegetation index space for assessment of surface moisture status. Remote Sens Env 79(2–3):213–224. https://doi.org/10.1016/S0034-257(01)00274-7

Sarkar, A., and Chouhan, P. (2019). Dynamic simulation of urban expansion based on cellular automata and Markov chain model: A case study in Siliguri metropolitan area, West Bengal. Modelling Earth System and Environment 5:1723–1732. https://doi.org/10.1007/s40808-019-00626-7

Sathyakumar, V., Ramsankaran, R., and Bardhan, R. (2019). Linking remotely sensed urban green space (UGS) distribution patterns and socio-economic status (SES): A multi-scale probabilistic analysis based in Mumbai, India. GIScience Remote Sens 56(5):645–669. https://doi.org/10.1080/15481603.2018.1549819

Schultz, P.A., and Halpert, M.S. (1995). Global analysis of the relationships among a vegetation index, precipitation and land surface temperature. Remote Sens 16(15):2755–2777. https://doi.org/10.1080/01431169508954590

Shahfahad, Maurya, M., Kumari, B., tayyab, M., Paarcha, A., Asif, et al. (2020). Indices based assessment of built-up density and urban expansion of fast-growing Surat city using multi-temporal Landsat data sets. GeoJournal. https://doi.org/10.1007/s10708-020-10148-w2020

Shi, T., Huang, Y., Wang, H., Shi, C.E., and Yang, Y.J. (2015). Influence of urbanization on the thermal environment of meteorological station: Satellite-observed evidence. Adv Clim Chang Res 6:7–15.

Singh, K.K. (2018). Urban green space availability in Bathinda city, India. Environmental Monitoring and Assessment 190(11):671. https://doi.org/10.1007/s10661-018-7053-0

Srivastava, P.K., Han, D., Rico-Ramirez, M.A., Bray, M., and Islam, T. (2012). Selection of classification techniques for land use/land cover change investigation. Adv Space Res 50:1250–1265.

Talukdar, S., Singha, P., Shahfahad, Mahato, S., Praveen, B., and Rahman, A. (2020). Dynamics of ecosystem services (E.S.s) in response to land use land cover (LU/LC) changes in the lower Gangetic plain of India. Ecol Indic 112:106121.

Thapa, R.B. (2009). Spatial Process of Urbanization in Kathmandu Valley Nepal (p. 153). University of Tsukuba Graduate School of Life and Environmental Sciences, University of Tsukuba, Japan.

Triantakonstantis, D., and Mountrakis, G. (2012). Urban growth prediction: A review of computational models and human perceptions. J Geograph Inf Syst 4(6):555–587. https://doi.org/10.4236/jgis.2012.46060

Wan, Z., Wang, P., and Li, X. (2004). Using MODIS land surface temperature and normalized difference vegetation index products for monitoring drought in the southern great plains, USA. Int J Remote Sens 25(1):61–72. https://doi.org/10.1080/0143116031000115328

Wang, J., Rich, P.M., and Price, K.P. (2003). Temporal responses of NDVI to precipitation and temperature in the central great plains, USA. Int J Remote Sens 24(11):2345–2364. https://doi.org/10.1080/01431160210154812

Weng, Q., Liu, H., Liang, B., and Lu, D. (2008). The spatial variations of urban land surface temperatures: Pertinent factors, zoning effect, and seasonal variability. IEEE J Sel Top Appl Earth Obs Remote Sens 1:2. https://doi.org/10.1109/JSTARS.2008.917869

Yang, X., Yang, T., Ji, Q., He, Y., and Ghebrezgabher, M.G. (2014). Regional-scale grassland classification using moderate-resolution imaging spectrometer datasets based on multistep unsupervised classification and indices suitability analysis. J Appl Remote Sens 8(1):083548. https://doi.org/10.1117/1.JRS.8.083548

Yin, J., Yin, Z., Zhong, H., Xu, S., Hu, X., Wang, J., and Wu, J. (2011). Monitoring urban expansion and land use/land cover changes of Shanghai metropolitan area during the transitional economy (1979–2009) in China. Environmental Monitoring and Assessment 177:609–621. https://doi.org/10.1007/s10661-010-1660-8

Yu, X., Guo, X., and Wu, Z. (2014). Land surface temperature retrieval from Landsat 8 TIRS-comparison between radiative transfer equation-based method, split window algorithm and single channel method. Remote Sens 6:9829–9852.

Zhang, X.X., Wu, P.F., and Chen, B. (2010). Relationship between vegetation greenness and urban heat island effect in Beijing city of China. Procedia Environ Sci 2:1438–1450. https://doi.org/10.1016/j.proenv.2010.10.157

5 Fuzzy-Model-Based Spatio-Temporal Characterisation of *Dalbergia sissoo* in Doon Valley
Post-Classification Approach

Sonakshi Mehrotra, Anil Kumar, and Arijit Roy

5.1 INTRODUCTION

The physical components of the surface of the land, such as trees, water bodies, and open spaces, are referred to as land cover (Treitz & Rogan, 2004). Spatially explicit land-cover information and summary statistics are essential to making decisions about natural resource management at local, national, and international levels (Lu et al., 2004). In ecological studies, biophysical and habitat data are collected across large geographic and temporal scales, a task for which remote sensing (RS) is particularly useful (Kerr & Ostrovsky, 2003). RS and the availability of field resources can be used to identify different habitats in land-cover classification (Franklin & Wulder, 2002). Detailed knowledge of land-cover changes can also assist in predicting individual species distributions (Saveraid et al., 2001). Medium-resolution Landsat data has been useful to quantify changes at the species level (Meng et al., 2020; Molla et al., 2023) in homogeneous areas. However, classification at the species level is challenging in heterogeneous areas using medium-resolution datasets due to lower spatial resolution and mixing of signatures. The use of high-resolution data for species-level classification has been common for the past few years (Herrick et al., 2014; Lin et al., 2015). Technological developments and an increase in the availability of high-resolution satellite data can be advantageous in the classification and detection of changes at the species level. PlanetScope (PS) provides daily 3-m spatial resolution data which can be beneficial for different aspects of ecological studies. Using PS data for species-level classification has given promising results in both forest (Kluczek et al., 2023; Rösch et al., 2022) and agricultural species classification (Sabir & Kumar, 2022; Sivaraj et al., 2022).

The selection of an appropriate classification approach is important to obtain true land-cover information. Different approaches, including semi-supervised and supervised, have been used to generate land-cover maps (Alshari & Gawali, 2021). Though widely used, these hard classification techniques presume the purity of a pixel which does not exist in the real world. These algorithms ignore the mixed pixel problem by combining the spectra of all pixels of the training set of a feature (Lu & Weng, 2007). A mixed pixel refers to a pixel composed of more than one element. Algorithms based on fuzzy logic are useful in this situation by handling mixed pixel problems (Singhal et al., 2021). Fuzzy-based approaches are based on membership values ranging from 0 to 1. Dunn (1973) developed the fuzzy *c* means (FCM) algorithm which divides a pixel into different membership values corresponding to the classes of the image. This algorithm was improved by James C. Bezdek (1981) by showing that the sum of the membership values of each pixel should be one. To handle the limitations of FCM, Krishnapuram and Keller (1993) introduced the concept of possibilistic *c* means

DOI: 10.1201/9781003377825-6

(PCM) which could deal with the concept of belongingness to the class. Li et al. (2003) developed a modified possibilistic *c* means (MPCM) approach to handling the shortfalls of both FCM and PCM. A single class could be classified using this method without optimising parameters. The use of the conventional mean approach as the concept of training parameters is another challenge (Jose & Kumar, 2021). They use the mean of the samples to generate the training sample. This lessens the individual influence of each sample and renders it unable to handle heterogeneity within the class. A modified training approach, individual sample as mean (ISM), has been experimented with in different forests (Mehrotra et al., 2022) and agricultural-based (Sivaraj et al., 2022) species extraction studies to understand its impact on the classification output while handling heterogeneity within the class.

The concept of change detection (CD) refers to observing an object or phenomenon at different times to identify differences in its state (Singh, 1989). There are multiple techniques of CD and research on these techniques is an active topic (Jianya et al., 2008). To select appropriate CD techniques, it is essential to determine the direction of the change. It is challenging to select an appropriate method for a given project as the digital CD is subject to spectral, thematic, temporal, and spatial constraints (Lu et al., 2004). Some cases like image differencing only detect change or no change, while other methods like the post-classification method generate a complete matrix of the directionality of change. Classified output is highly dependent upon the quality and quantity of data used in the post-classification process (Goswami et al., 2022) which is one of the most widely used quantitative CD methods. This approach uses a separate classification of multi-temporal images into thematic maps followed by pixel-by-pixel comparisons (Lu et al., 2004). It minimises the impact of atmospheric effects between multi-temporal images. The technique has been successfully used to determine changes in different areas (Bhatt et al., 2015; Mishra et al., 2020; Suribabu et al., 2012).

The goals of this study were threefold. The first was to study the phenological cycle of *D. sissoo* over 2018 and 2021. This was done using one of the most widely used vegetation indices (VI) called the Normalized Difference Vegetation Index (NDVI). The second was to classify *D. sissoo* over the study period using multi-temporal satellite data. This has been achieved using 28 and 32 images for the years 2018 and 2021, respectively. The final goal was to find the change in the area occupied by the species over the study period and identify its probable causes. This study combines a supervised approach of fuzzy-based classification with a change matrix to obtain the change in the target class present in the study area over two time periods.

5.2 MATHEMATICAL CONCEPTS

5.2.1 Vegetation Index

The spectral dimensionality of a dataset can be reduced while maintaining its temporal dimensions by using VIs. This study uses the NDVI which works with red and near-infrared (NIR) bands (Equation 5.1) (Rouse et al., 1974). It is one of the most commonly used VIs for vegetation detection and ranges from −1 to 1.

$$NDVI = \frac{NIR - Red}{NIR + Red} \tag{5.1}$$

5.2.2 Classification Algorithm Used

This study uses the MPCM approach. This algorithm can handle noise and has a fast clustering ability (Li et al., 2003). It can map a single class of interest from a temporal database. The objective function is given in Equation 5.2:

$$J_{MPCM}(u,v) = \sum_{i=1}^{c}\sum_{k=1}^{N}\mu_{ki}^{m}D_{ki^2} + \eta_i \sum_{k=1}^{N}\left(\mu_{ki}log\mu_{ki} - \mu_{ki}\right) \tag{5.2}$$

where $v = \{v_1, v_2, \ldots, v_n\}$ represents clusters of vector centres;
u = matrix with membership values with dimension N ×C;
N = number of data points;
c = number of clusters;
μ_{ki} = typicality value or possibilistic value of class x_k in class i, $0 \le \mu_{ki} \le 1$;
m = weight component that controls the degree of fuzziness; it is independent when there is only one class;
D_{ki} = squared Euclidean distance between cluster centre and individual value given in (Equation 5.3):

$$D_{ki}^2 = \left\| x_k - v_i \right\| \forall i, k > 1 \tag{5.3}$$

η_i = distribution parameter and controls the shape and size of the cluster. It is given in (Equations 5.4 and 5.5):

$$\eta_i = \frac{\sum_{k=1}^{N} \mu_{ki,FCM}^m D_{ki}^2}{\sum_{k=1}^{N} \mu_{ki,FCM}^m} \tag{5.4}$$

$$\mu_{ki} = \left[\sum_{j=1}^{c} \left(\frac{D_{ki}^2}{D_{kj}^2} \right)^{\frac{1}{m-1}} \right]^{-1} \forall k; i \tag{5.5}$$

where $\mu_{ki,FCM}$ is the terminal membership value FCM.
For MPCM, the possibilistic values (μ_{ki}) (Equation 5.6):

$$\mu_{ki} = exp\left(-\frac{D_{ki^2}}{\eta_i} \right) \forall i; k \tag{5.6}$$

5.2.3 TRAINING PARAMETER CONCEPT

This study uses a newly developed training parameter approach called the ISM approach. Traditional classifiers use the mean of samples for classification which reduces the impact of each sample on the output. This approach uses all samples individually as the mean in the algorithm rather than taking the mean of the samples (Singhal et al., 2021). Mathematically, the traditional approach uses $\bar{x}$ while the ISM approach uses $x_1, x_2 x_3, \ldots x_n$ individually as means to train the algorithm (Figure 5.1). This approach reduces the heterogeneity within the target class, thereby giving a more homogeneous output. It has been tested in different crops (Jose & Kumar, 2021; Sabir & Kumar, 2022; Sivaraj et al., 2022) and land-use/land-cover classification (Suman et al., 2022) and is found to reduce heterogeneity in the output.

5.2.4 ACCURACY ASSESSMENT

Accuracy assessment of soft output of fuzzy classification can be challenging. Root Mean Square Error (RMSE) and Fuzzy ERror Matrix (FERM) have been used to indirectly assess the accuracy of the fuzzy classified output (Upadhyay et al., 2014). This study uses the approach of mean membership difference (MMD) for accuracy assessment. In this approach, a difference in the means of the membership values within a class is used to measure accuracy. MMD values within the class tend to be 0, while values between different classes tend to be closer to 1 to show high accuracy of classified output. A high value of MMD between different classes shows better separation (Singh et al., 2021).

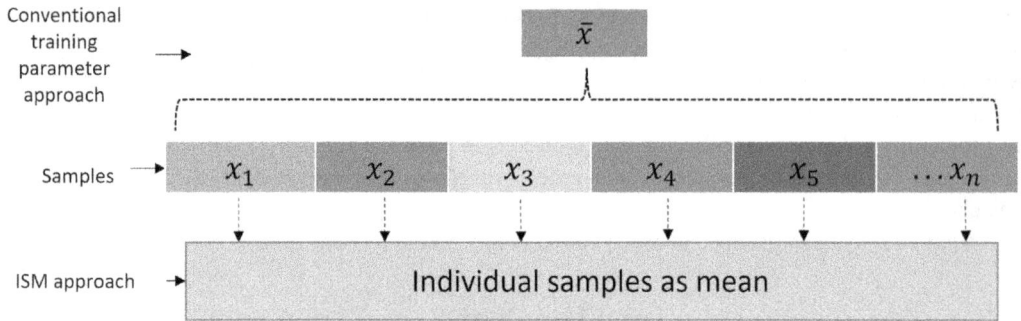

FIGURE 5.1 Conventional versus individual samples as mean training parameter approach.

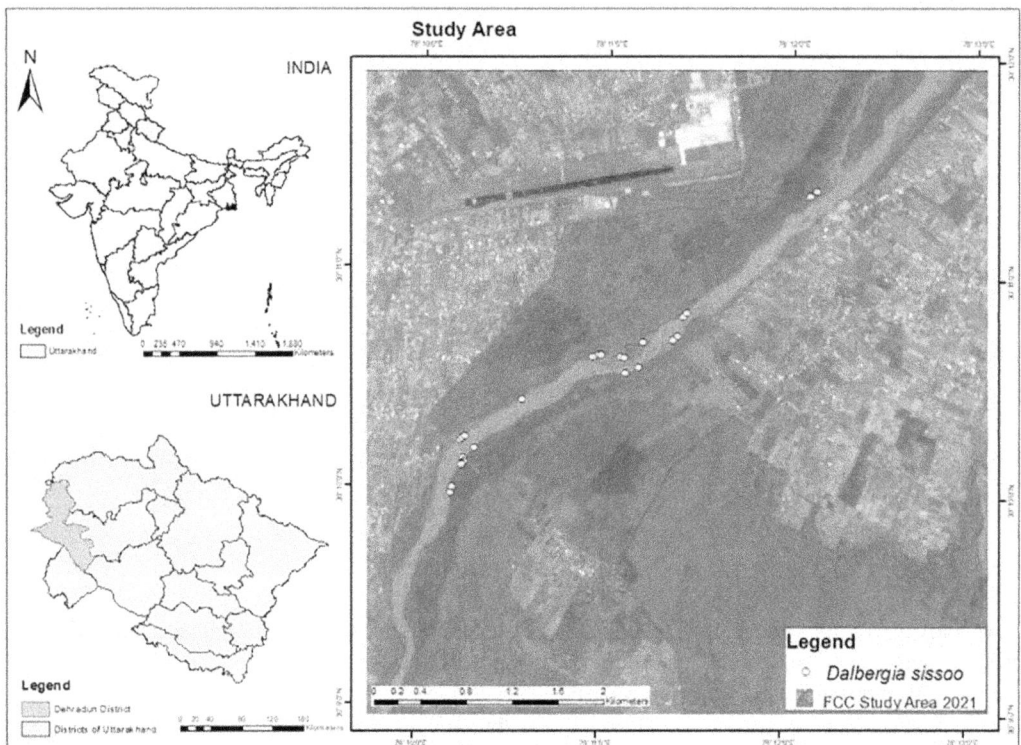

FIGURE 5.2 Study area map with data points of *Dalbergia sissoo* spread around the Jakhan Rao.

5.3 STUDY AREA

The study site lies around a 7-km-long stretch of Jakhan Rao, Dehradun, Uttarakhand. It is located between 30°11′49.2551″ N, 78°9′45.8038″ E and 30°8′46.9330″ N, 78°12′44.3179″ E (Figure 5.2). The Jakhan Rao is an ephemeral river and a tributary of the song river (Ahmad et al., 2018). Mining for sand, gravel, and boulders is done on the riverbed (Chandra & Kainthola, 2019). The area is a Sal mixed moist deciduous forest with patches of *Acacia catechu* associated with *Dalbergia sissoo*. In this area, *D. sissoo* is found along the Jakhan Rao close to the riverbank in pure and mixed patches. A deciduous tree, *D. sissoo* or Shisham belongs to the Papilionaceae family. It is an example of a pioneer species in the riverine succession of the Gangetic plains of India (Champion &

Seth, 1968). The species grows on newly laid down traces of rivers, fresh embankments, and freshly exposed soil with good drainage. The phenological stages of the species based on literature and surveys are given in Table 5.1 (Chhetri et al., 2020; Lodhiyal et al., 2002).

5.4 MATERIALS AND METHODS

5.4.1 FIELD DATA

Fieldwork around Jakhan Rao, in the Doon Valley, was conducted in January, May, and October 2022. During January, *D. sissoo* was observed in its leaf fall stage with two distinct classes, namely, DS1 and DS2 (Figure 5.3). DS1 referred to smaller trees with no leaves, and DS2 referred to taller trees which retained some leaves during the leaf fall stage. In this study, both DS1 and DS2 were taken together as one class. Different stages of a mature *D. sissoo* in an urban area are shown in Figure 5.4.

5.4.2 SATELLITE DATA

Four-band PS data, with a spatial resolution of 3 m, was used for this study. The four available bands are blue (455–515 nm), green (500–590 nm), red (590–670 nm), and NIR (780–860 nm). The four-band product was selected for its availability in both years 2018 and 2021. Surface reflectance products for the years 2018 and 2021, with a cloud cover of less than 5%, were downloaded based on their availability. The details of the dataset used are given in Table 5.2.

TABLE 5.1
Phenological Stages of *Dalbergia sissoo*

Phenological Stage	Months
Leaf flush	February–April
Flowering	March–April
Fruiting	April–May
Leaf fall	October–February

FIGURE 5.3 *Dalbergia sissoo* observed in January 2022. (a) DS1 shows a leafless patch of young Shisham trees. (b) DS2 shows a leafless patch of older Shisham trees.

5.4.3 Method

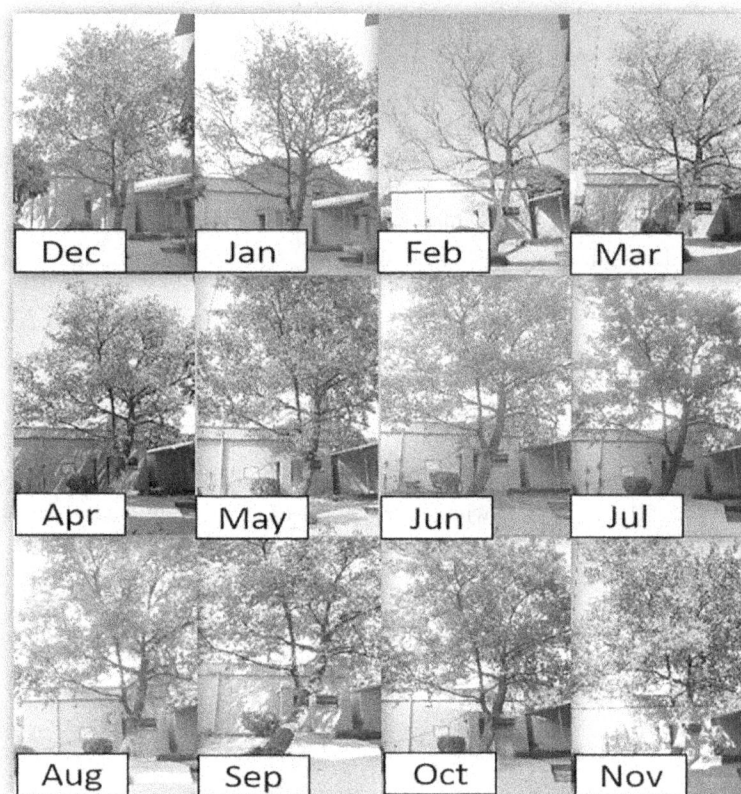

FIGURE 5.4   Different stages of *Dalbergia sissoo* between December 2021 and November 2022.

TABLE 5.2
Dates Used for the Years 2018 and 2021

	2018		2021	
January	03, 09, 18, 29	(4)	09, 31	(2)
February	09, 14, 20	(3)	08, 17, 24	(3)
March	06, 16, 23	(3)	04, 11, 18, 26, 31	(5)
April	01, 15, 30	(3)	08, 13, 25	(3)
May	10, 20	(2)	09, 14, 26	(3)
June	12	(1)	02, 08	(2)
July	—	—	1	(1)
August	—	—	16	(1)
September	17, 30	(2)	27	(1)
October	06, 12, 18, 27	(4)	09, 16, 30	(3)
November	08, 18	(2)	02, 08, 17, 29	(4)
December	02, 11, 20, 29	(4)	04, 11, 21, 31	(4)
Total		28		32

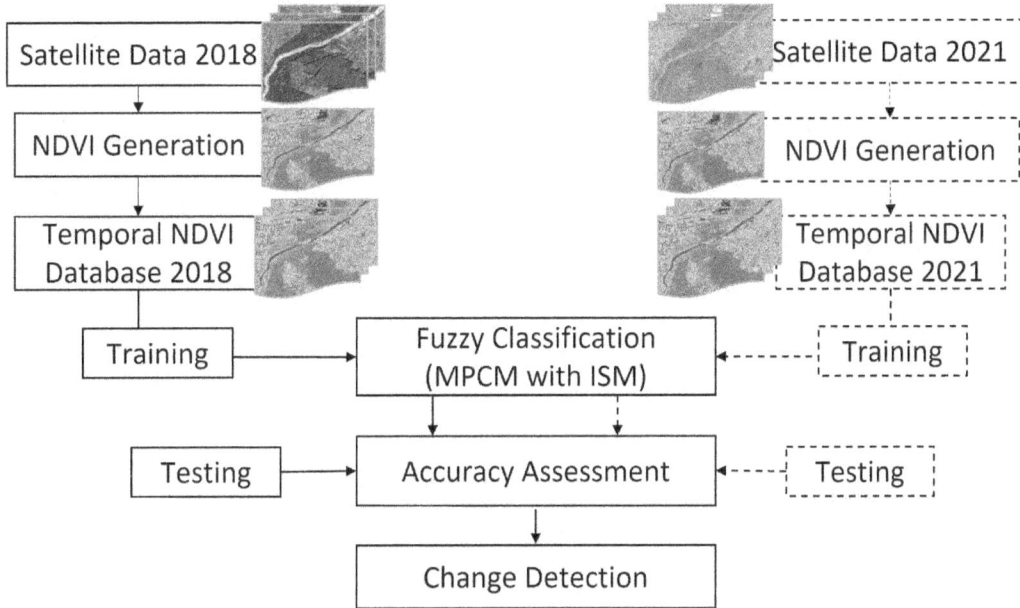

FIGURE 5.5 The methodology adopted for this research work.

The method followed included pre-processing of the images, classification, accuracy assessment, and CD. Figure 5.5 shows the method followed. Software programs used were ERDAS Imagine 2018, QGIS 3.28.1, and an in-house Sub-Pixel Multispectral Image Classifier (SMIC) tool.

5.4.4 PRE-PROCESSING

Pre-processing was done in ERDAS Imagine 2018. Subsets of all images were created, and NDVI was computed for each image. NDVI was rescaled to 0 to 255. The temporal NDVI databases for the years 2018 and 2021 were created using 28 and 32 images, respectively. This temporal database was exported in the generic binary format.

5.4.5 CLASSIFICATION

Supervised classification was performed using an in-house SMIC tool with the ISM approach of MPCM fuzzy classifiers (Kumar et al., 2006). 25 and 40 patches were considered to train and test the algorithm in 2018 and 2021, respectively. The training and testing ratio was taken as 7:3, where 70% of the data was used to train the algorithm and 30% was used to test it. MMD between training and testing data of the target class was used for accuracy assessment. The soft output was converted to hard output by selecting the appropriate α cut value on output membership values for each database based on the testing samples for representing the target class of *D. sissoo*.

5.4.6 CHANGE DETECTION

The CD was performed using QGIS 3.28.1. The single-layer classified outputs of 2018 and 2021 were considered as the before and after images, respectively. CD matrix was generated using the post-classification process in the semi-automatic classification plugin (Congedo, 2021). It was considered that during the study period, some of the patches of *D. sissoo* of 2018 could have either grown or vanished in 2021 due to the dynamic nature of the river. During the study, it was also observed that some patches of *D. sissoo* vanished within the year 2021 (i.e., they were present till August of 2021 and vanished afterwards). To identify the patches which vanished within 2021, an additional experiment was done where training samples were selected from the patches which disappeared in September.

5.5 RESULTS AND DISCUSSION

5.5.1 PHENOLOGICAL CYCLE OF *D. SISSOO* IN 2018 AND 2021 BASED ON NDVI

The NDVI profile of *D. sissoo* in 2018 and 2021 is shown in Figure 5.6. It can be observed that the NDVI increased between February and April 2018 corresponding to the flushing stage and gradually reduced between September and December 2018 corresponding with the leaf fall stage. Less available images in 2018 resulted in a gap in understanding the phenological stages between July and September. A gradual rise in the NDVI values was observed between July 2021 and September 2021.

5.5.2 THE CLASSIFIED OUTPUT OF 2018 AND 2021

The classified output is shown in Figure 5.7. It can be seen that more area is occupied by *D. sissoo* in 2021 in comparison to that occupied in 2018. *D. sissoo* covers an approximate area of 7.53 ha in the 2021 dataset (Table 5.3). The accuracy assessment of the classified output was done using MMD (Table 5.4). MMD values for both study years were close to 0 when tested within the same class.

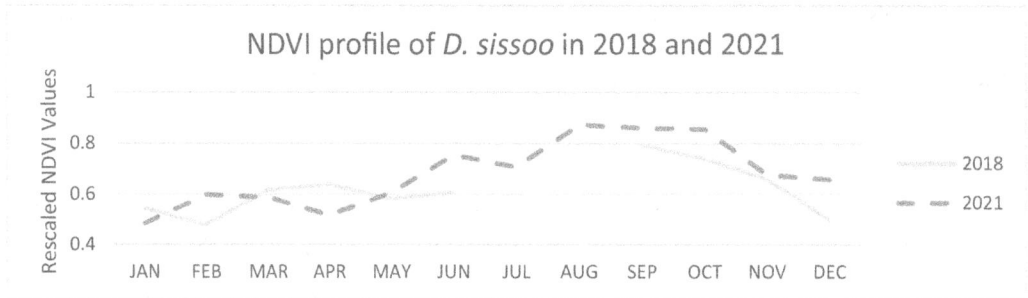

FIGURE 5.6 Normalized Difference Vegetation Index profile of *Dalbergia sissoo* in 2018 and 2021.

TABLE 5.3
Area Covered by *Dalbergia sissoo* in 2018 and 2021

Year	Area Covered by D. sissoo in Hectares
2018	3.43
2021	7.53

TABLE 5.4
Mean Membership Difference Values within the Class in 2018 and 2021

Year	Mean Membership Difference Within Class
2018	0.011765
2021	0.013725

5.5.3 CHANGE DETECTION

An increase in the area of *D. sissoo* was observed in 2021 in comparison to 2018 (Figure 5.7). It was observed that in 2021, about 2.46 ha of the area of target species was lost, a 6.56-ha area was gained, while 2,814.96 ha of the study site remained unchanged (Table 5.5). It was observed that *D. sissoo* nearly doubled in the three years.

Training sets from a few patches were used to identify the different patches that disappeared in 2021. Figure 5.9 shows the change in the presence of the target species within one year. It was observed that small patches of *D. sissoo* present till August 2021 vanished in September 2021. The sudden dip at the 21st date in NDVI observed in the figure corresponds to September 2021 indicating the disappearance of identified patches (Figure 5.8). The results were verified by comparing the FCC of August and September 2021.

TABLE 5.5

Change matrix of *Dalbergia sissoo* in 2021

Serial Number	2021	2018	Analysis	Pixel Sum	Area (m²)	Area (ha)
1	NT	NT	Unchanged NT	3127732	28149588.00	2814.96
2	NT	DS	Decrease	2742	24678	2.47
3	DS	NT	Increase	7291	65619	6.56
4	DS	DS	Unchanged DS	1072	9648	0.96

Note: DS, Dalbergia sissoo; NT, non-target class.

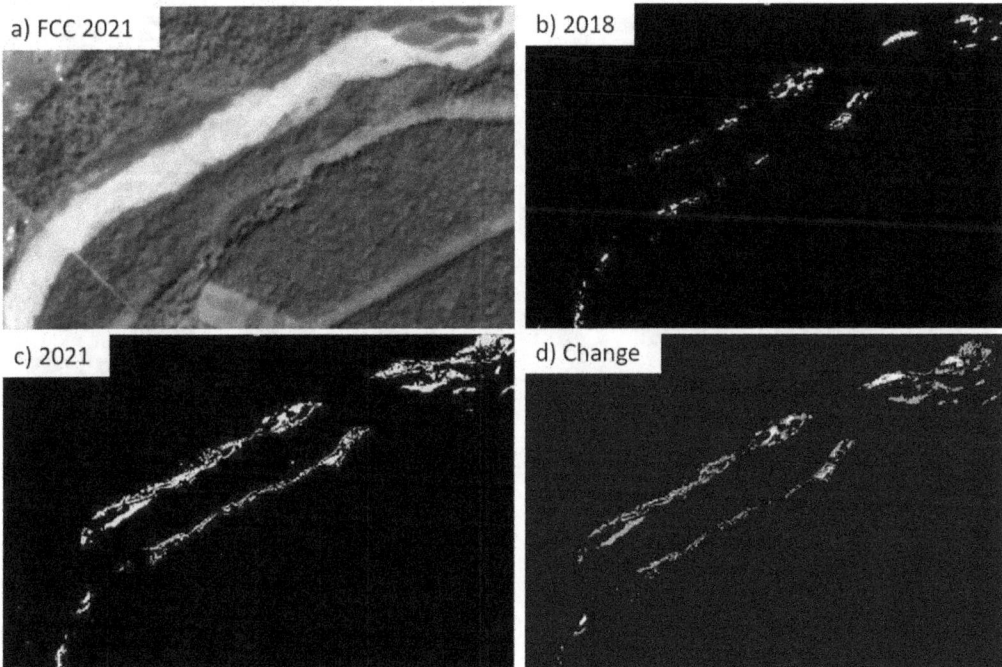

FIGURE 5.7 Classified output with change detection. (a) FCC 2021 shows a part of the study area. (b) Output-2018. (c) Output-2021. (d) CD output of *Dalbergia sissoo* where black shows non-target class, red shows decrease, green shows increase, and white shows no change in the target class.

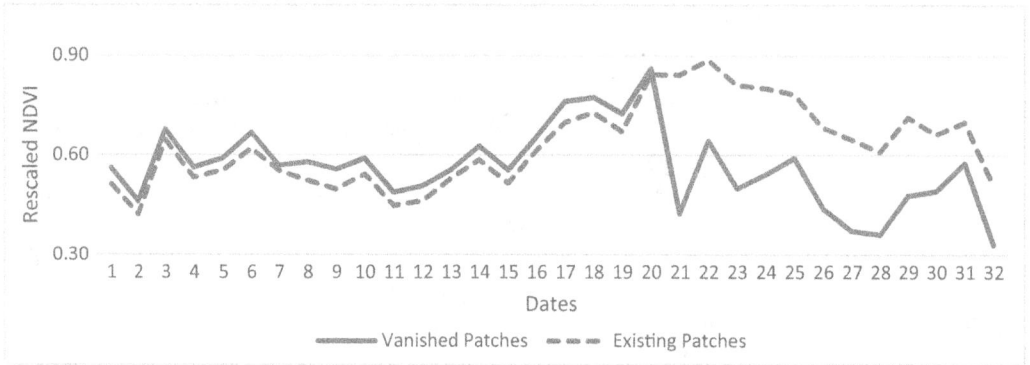

FIGURE 5.8   Normalized Difference Vegetation Index profile of *Dalbergia sissoo* patches in 2021.

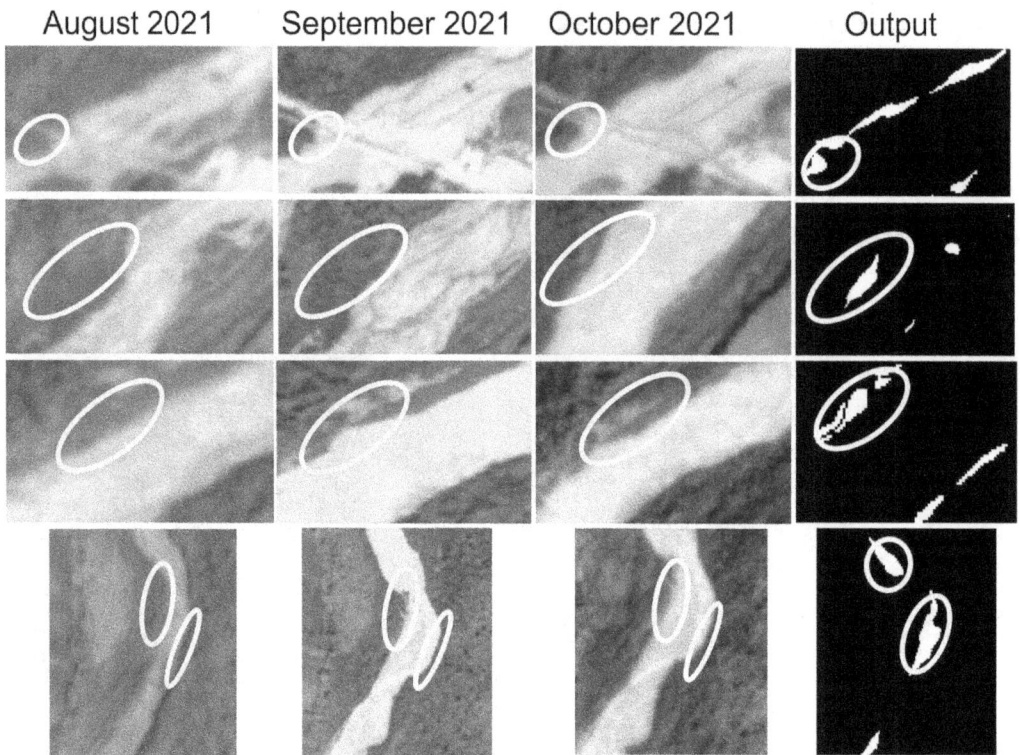

FIGURE 5.9   Output showing vanished patches in 2021 where patches present in August 2021 disappeared in September 2021.

5.5.4   Discussion

In this study, classification using fuzzy-based MPCM with the ISM approach showed promising results with good MMD values. The selection of an appropriate training sample was essential when using the ISM training parameter approach. This is because contrary to the conventional mean approach, ISM considers each sample individually and thus the classification output gets the impact of each training sample. Fuzzy-based soft classifications using MPCM have proven to be advantageous in land-use/land-cover mapping (Singh et al., 2020), teak forest and eucalyptus classification (Chhapariya et al., 2022), and psyllium husk medicinal crop mapping (Sabir & Kumar, 2022). Specific crop mapping for psyllium husk (Sabir & Kumar, 2023) and pigeon pea (Sivaraj et al., 2022) using the ISM approach have shown better results than the conventional method. In both cases, the MMD was lower when using the ISM approach, indicating that the ISM approach outperformed the conventional mean approach of training parameters and resulted in better classification outputs.

In this study, more temporal images in 2021 provide a more comprehensive picture of the phenological stages of the target species than those available in 2018. The increased NDVI between February and April 2018 corresponds to the flushing stage, and its gradual reduction between September and December 2018 corresponds with the leaf fall stage. A gradual rise in the NDVI values observed between July 2021 and September 2021 could probably be due to the monsoon season in the study area. It should be noted that the target species occupy a small area of the whole study site. An approximately 2,500 ha area is composed of different species of vegetation. These include the dominant species of *Shorea robusta* and *Tectona grandis*. *D. sissoo* is a non-dominant species in the region found around the river covering a small portion of the study area. It was also observed that the area around Jakhan Rao is dynamic which can cause changes in the coverage of *D. sissoo* within one year as well. Daily Climate Hazards Group InfraRed Precipitation with Station data were used to note rainfall over the study area (Funk et al., 2015). Heavy precipitation observed during August and September 2021 could have led to bank erosion and hence caused the disappearance of patches of *D. sissoo*. PS data provide promising results in detecting small patches which vanished in September 2021. Regular PS data availability over the research sites can be helpful to spot subtle changes happening in the area and support early management-based action.

5.6 CONCLUSION

The ability to detect and quantify changes over areas and long periods provides valuable information for decision-making and policy development. From a methodological viewpoint, this study shows that it is effective to classify a non-dominant species using ground truth data, ISM training parameter concept, and fuzzy-based MPCM algorithm. The methodology followed can be experimented with using different species in forest and agricultural landscapes to identify species-level changes occurring in the regions. This study uses an innovative approach of the CD where multi-temporal images of two years were used to detect changes in the species coverage. Through this study of the change of area coverage of *D. sissoo* around the Jakhan Rao over three years, it can be seen that the area of *D. sissoo* expanded continuously. Overall, this study presents an important contribution to the field of remote sensing and highlights the need for continued research in monitoring and quantifying changes in species in forested regions. More research in this domain is required to identify and quantify the reasons for changes in species cover in the region. Further research can include detailed studies on understanding the dynamics of Jakhan Rao and its impact on species around it.

REFERENCES

Ahmad, R., Chauhan, P., & Srivastava, S. K. (2018). Groundwater study in eastern Dehradun valley and adjoining areas using remote sensing and GIS observations. *Himalayan Geology, 39*(2), 212–222.

Alshari, E. A., & Gawali, B. W. (2021). Development of classification system for LULC using remote sensing and GIS. *Global Transitions Proceedings, 2*(1), 8–17. https://doi.org/10.1016/j.gltp.2021.01.002

Bezdek, J. C. (1981). *Pattern recognition with fuzzy objective function algorithms*. Plenum Press.

Bhatt, A., Ghosh, S. K., & Kumar, A. (2015). Automated change detection in satellite images using machine learning algorithms for Delhi, India. *International Geoscience and Remote Sensing Symposium (IGARSS), 2015-Novem*, 1678–1681. https://doi.org/10.1109/IGARSS.2015.7326109

Champion, H. G., & Seth, S. K. (1968). *A revised survey of the forest types of India*. Natraj Publishers.

Chandra, K., & Kainthola, H. (2019). *Scheme of mining for Sand, Bajri and Boulder in Jakhan River Dehradun Forest Division Area: 96.50 ha.* KainGeotech Geological, Geotechnical, Mining & Environmental Consulting.

Chhapariya, K., Kumar, A., & Upadhyay, P. (2022). Handling non-linearity between classes using spectral and spatial information with kernel based modified possibilistic c-means classifier. *Geocarto International*, *37*(6), 1704–1721. https://doi.org/10.1080/10106049.2020.1797186

Chhetri, R., Kumar, P., & Megha. (2020). Phenological trend of tree species at Forest Research Institute, Dehradun, India. *Research Journal of Agriculture and Forestry Sciences*, *8*(2), 1–8.

Congedo, L. (2021). Semi-automatic classification plugin: A python tool for the download and processing of remote sensing images in QGIS. *Journal of Open Source Software*, *6*(64), 3172. https://doi.org/10.21105/joss.03172

Dunn, J. C. (1973). A fuzzy relative of the ISODATA process and its use in detecting compact well-separated clusters. *Journal of Cybernetics*, *3*(3), 32–57. https://doi.org/10.1080/01969727308546046

Franklin, S. E., & Wulder, M. A. (2002). Remote sensing methods in medium spatial resolution satellite data land cover classification of large areas. *Progress in Physical Geography*, *26*(2), 173–205. https://doi.org/10.1191/0309133302pp332ra

Funk, C., Peterson, P., Landsfeld, M., Pedreros, D., Verdin, J., Shukla, S., Husak, G., Rowland, J., Harrison, L., Hoell, A., & Michaelsen, J. (2015). The climate hazards infrared precipitation with stations—A new environmental record for monitoring extremes. *Scientific Data*, *2*, 1–21. https://doi.org/10.1038/sdata.2015.66

Goswami, A., Sharma, D., Mathuku, H., Gangadharan, S. M. P., Yadav, C. S., Sahu, S. K., Pradhan, M. K., Singh, J., & Imran, H. (2022). Change detection in remote sensing image data comparing algebraic and machine learning methods. *Electronics*, *11*(3), 1–26. https://doi.org/10.3390/electronics11030431

Herrick, C., Palace, M. W., Finnell, D. R., Garnello, A., Sullivan, F., Anderson, S. M., & Varner, R. K. (2014). Use of high resolution UAS imagery to classify sub-arctic vegetation types. *AGU Fall Meeting Abstracts*, *2014*, B31F-0089.

Jianya, G., Haigang, S., Guorui, M., & Qiming, Z. (2008). A review of multi-temporal remote sensing data change detection algorithms. *The International Archives of the Photogrammetry, Remote Sensing and Spatial Information Sciences*, *37*(B7), 757–762.

Jose, N., & Kumar, A. (2021). Handling heterogeneity through "individual sample as mean" approach—A case study of Isabgol (psyllium husk) medicinal crop. *Remote Sensing Applications: Society and Environment*, *25*(August 2021). https://doi.org/10.1016/j.rsase.2021.100671

Kerr, J. T., & Ostrovsky, M. (2003). From space to species: Ecological applications for remote sensing. *Trends in Ecology and Evolution*, *18*(6), 299–305. https://doi.org/10.1016/S0169-5347(03)00071-5

Kluczek, M., Zagajewski, B., & Zwijacz-Kozica, T. (2023). Mountain tree species mapping using Sentinel-2, PlanetScope, and Airborne HySpex Hyperspectral Imagery. *Remote Sensing*, *15*(3). https://doi.org/10.3390/rs15030844

Krishnapuram, R., & Keller, J. M. (1993). A possibilistic approach to clustering. *IEEE Transactions on Fuzzy Systems*, *1*(2), 98–110. https://doi.org/10.1109/91.227387

Kumar, A., Ghosh, S. K., & Dhadhwal, V. K. (2006). Sub-pixel land cover mapping: SMIC system. *ISPRS Int. Sym. "Geospatial Databases for Sustainable Development"*, Goa, India.

Li, K., Huang, H. K., & Li, K. L. (2003). A modified PCM clustering algorithm. *International Conference on Machine Learning and Cybernetics*, *2*(November), 1174–1179. https://doi.org/10.1109/icmlc.2003.1259663

Lin, C., Popescu, S. C., Thomson, G., Tsogt, K., & Chang, C. I. (2015). Classification of tree species in overstorey canopy of subtropical forest using QuickBird images. *PLoS One*, *10*(5), 1–23. https://doi.org/10.1371/journal.pone.0125554

Lodhiyal, N., Lodhiyal, L. S., & Pangtey, Y. P. S. (2002). Structure and function of Shisham forests in central Himalaya, India: Dry matter dynamics. *Annals of Botany*, *89*(1), 41–54. https://doi.org/10.1093/aob/mcf004

Lu, D., Mausel, P., Brondízio, E., & Moran, E. (2004). Change detection techniques. *International Journal of Remote Sensing*, *25*(12), 2365–2401. https://doi.org/10.1080/0143116031000139863

Lu, D., & Weng, Q. (2007). A survey of image classification methods and techniques for improving classification performance. *International Journal of Remote Sensing*, *28*(5), 823–870. https://doi.org/10.1080/01431160600746456

Mehrotra, S., Kumar, A., Roy, A., Kushwaha, S. P. S., & Singh, R. P. (2022). Studying dual-sensor time-series remote sensing data for *Dalbergia sissoo* mapping in a lesser Himalayan area. *Journal of Applied Remote Sensing*, *16*(3), 1–20. https://doi.org/10.1117/1.jrs.16.034521

Meng, Y., Cao, B., Mao, P., Dong, C., Cao, X., Qi, L., Wang, M., & Wu, Y. (2020). Tree species distribution change study in Mount Tai based on Landsat remote sensing image data. *Forests*, *11*(2), 1–14. https://doi.org/10.3390/f11020130

Mishra, P. K., Rai, A., & Rai, S. C. (2020). Land use and land cover change detection using geospatial techniques in the Sikkim Himalaya, India. *Egyptian Journal of Remote Sensing and Space Science*, *23*(2), 133–143. https://doi.org/10.1016/j.ejrs.2019.02.001

Molla, G., Addisie, M. B., & Ayele, G. T. (2023). Expansion of eucalyptus plantation on fertile cultivated lands in the North-Western highlands of Ethiopia. *Remote Sensing*, *15*(3), 1–16. https://doi.org/10.3390/rs15030661

Rösch, M., Sonnenschein, R., Buchelt, S., & Ullmann, T. (2022). Comparing PlanetScope and Sentinel-2 imagery for mapping mountain pines in the Sarntal Alps, Italy. *Remote Sensing*, *14*(13), 1–24. https://doi.org/10.3390/rs14133190

Rouse, J. W. J., Haas, R. H., Schell, J. A., & Deering, D. W. (1974). Monitoring vegetation systems in the great plains with erts. *Proceedings of the 3rd ERTS Symposium*, *1*, 309–317.

Sabir, A., & Kumar, A. (2022). Harvesting information extraction using Sentinel-2 and CubeSat temporal data for medicinal psyllium husk crop. *Journal of Geomatics*, *16*(1), 45–54.

Sabir, A., & Kumar, A. (2023). Study of integrated optical and synthetic aperture radar-based temporal indices database for specific crop mapping using fuzzy machine learning model. *Journal of Applied Remote Sensing*, *17*(1), 1–15. https://doi.org/10.1117/1.jrs.17.014502

Saveraid, E. H., Debinski, D. M., Kindscher, K., & Jakubauskas, M. E. (2001). A comparison of satellite data and landscape variables in predicting bird species occurrences in the Greater Yellowstone ecosystem, USA. *Landscape Ecology*, *16*(1), 71–83. https://doi.org/10.1023/A:1008119219788

Singh, A. (1989). Review article: Digital change detection techniques using remotely-sensed data. *International Journal of Remote Sensing*, *10*(6), 989–1003. https://doi.org/10.1080/01431168908903939

Singh, A., Kumar, A., & Upadhyay, P. (2020). Modified possibilistic c- means with constraints (MPCM-S) approach for incorporating the local information in a remote sensing image classification. *Remote Sensing Applications: Society and Environment*, *18*(April), 100319. https://doi.org/10.1016/j.rsase.2020.100319

Singh, A., Kumar, A., & Upadhyay, P. (2021). A novel approach to incorporate local information in possibilistic c-means algorithm for an optical remote sensing imagery. *Egyptian Journal of Remote Sensing and Space Science*, *24*(1), 151–161. https://doi.org/10.1016/j.ejrs.2020.06.001

Singhal, M., Payal, A., & Kumar, A. (2021). Importance of individual sample of training data in modified possibilistic c-means classifier for handling heterogeneity within a specific crop. *Journal of Applied Remote Sensing*, *15*(3), 1–18. https://doi.org/10.1117/1.jrs.15.034507

Sivaraj, P., Kumar, A., Koti, S. R., & Naik, P. (2022). Effects of training parameter concept and sample size in possibilistic c-means classifier for pigeon pea specific crop mapping. *Geomatics*, *2*(1), 107–124. https://doi.org/10.3390/geomatics2010007

Suman, S., Kumar, D., & Kumar, A. (2022). Fuzzy based convolutional noise clustering classifier to handle the noise and heterogeneity in image classification. *Mathematics*, *10*(21). https://doi.org/10.3390/math10214056

Suribabu, C. R., Bhaskar, J., & Neelakantan, T. R. (2012). Land use/cover change detection of Tiruchirapalli city, India, using integrated remote sensing and GIS tools. *Journal of the Indian Society of Remote Sensing*, *40*(4), 699–708. https://doi.org/10.1007/s12524-011-0196-x

Treitz, P., & Rogan, J. (2004). Remote sensing for mapping and monitoring land-cover and land-use change-an introduction. *Progress in Planning*, *61*(4), 269–279.

Upadhyay, P., Ghosh, S. K., & Kumar, A. (2014). A brief review of fuzzy soft classification and assessment of accuracy methods for identification of single land cover. *Studies in Surveying and Mapping Science*, *2*(Mlc), 1–13. www.as-se.org/ssms

6 Encroachment on Agricultural Land by Horizontal Development of Built-up Area
A Case Study of Sagar Island, East Coast of India

Sabir Hossain Molla and Rukhsana Sarkar

6.1 INTRODUCTION

The matter of overpopulation is a crucial and pressing global issue (UN DESA, 2015). This issue urges sufficient social and economic resources, including enough land for urban development and economic activities such as housing, industries, businesses, health-care facilities, and schools. Unfortunately, this requirement often leads to the depletion of valuable cultivable land in the surrounding areas of a town (Magsi and Torre, 2012; Liu et al., 2014). Agricultural lands play a crucial role as buffer zones within town and natural habitats, providing food, fiber, and essential natural resources while mitigating the impact of urbanization on the environment and wildlife (Doygun, 2009). However, the diminution in agricultural lands can have severe consequences on biodiversity and the disappearance of aboriginal plant species. In addition, the increasing prevalence of non-porous surfaces can be ascribed to the growing inclination towards urbanization, which comes at the cost of agricultural land (Turok and Mykhnenko, 2007; Prokop et al., 2011). Due to this fact, an increasing requirement is emerging for regional planning bodies to enhance their supervision of land transformation and safeguarding of farming areas (Gravert and Wiechmann, 2016; Li and Yeh, 2001).

This research focuses on studying how the expansion of built-up areas over time and space has impacted the agricultural land on Sagar Island in the Indian Sundarban region. To achieve this, the study utilizes remotely sensed data and a geographic information systems (GIS) environment. By employing these tools, researchers can effectively and practically assess the changes in land use/land cover (LULC) over time and space. This approach enables them to analyze and map the dynamics of LULC, providing historical data that are useful for environmental monitoring (Lambin et al., 2001; Molla & Rukhsana, 2022). Additionally, GIS is highly appropriate for the computer-aided analysis of geometric characteristics of definite land uses, such as compactness and contiguousness. Combining both disciplines can result in a symbiotic relationship that benefits each field (Abdelrahman et al., 2016; Molla & Rukhsana, 2022). Numerous scholars have addressed the challenge of precisely tracking alterations in the coverage and utilization of land in various scenarios (Singh, 1989; El Bastawesy et al., 2008; Muchoney and Haack, 1994; Almutairi and Warner, 2010).

Sagar Island, the largest among the Indian Sundarban islands, experiences the greatest population pressure, resulting in frequent expansion of settlement areas. With a population density of 893 individuals per square kilometer, the holy site of Sagar Island attracts both the Gangasagar Mela and the Kapil Muni Ashram as a place of pilgrimage, leading to the development of numerous business activities alongside agriculture as the primary source of income for its inhabitants. As the

                    DOI: 10.1201/9781003377825-7

population on Sagar Island continues to increase, the result is a noticeable expansion in the residential areas (Hajra et al., 2017), resulting in substantial changes to the original terrain and triggering massive environmental, ecological, and societal consequences (Haregeweyn et al., 2012). Thus, the goal of this study was to utilize an integrated GIS tool, remote sensing technique, and Random Forest method to examine the spatial and temporal built-up expansion on agricultural lands between 2000 and 2020.

6.2 STUDY AREA

Sagar Island, also known as Ganga Sagar, is the selected study area located at the conflux of the Ganga and the Bay of Bengal, situated 100 km south of Kolkata (Rukhsana & Molla, 2023; Mukherjee et al., 2019). The island is situated within a delta region that is influenced by tides and composed of alluvial deposits transported by the Ganga and Brahmaputra Rivers and their associated tributaries (Mukherjee et al., 2019) (Figure 6.1). The process of reclaiming Sagar Island from the Sundarban mangrove wetlands of the western Ganga-Brahmaputra delta was first started in 1811 (Bandyopadhyay, 1997). The elevation of the islands being examined ranges from 2.10 to 2.75 m above the mean sea level Purkait, 2009). The village, or mouza, serves as the main management unit for financial resource allocation and strategic decision-making in Sagar Island. The island is home to 42 villages that are divided amongst eight gram panchayats. On average, each village seizes an area of 673 ha, with a population of 4,847 individuals, 820 households, and 418 cultivators (Mandal et al., 2018). Sagar Island is confronted with several environmental challenges, such as severe coastal erosion (Bandyopadhyay, 1997; Gopinath, 2010), the incursion of saline water (Majumdar and Das, 2011), and recurrent cyclones and depressions over the past 120 years (Chand et al., 2012).

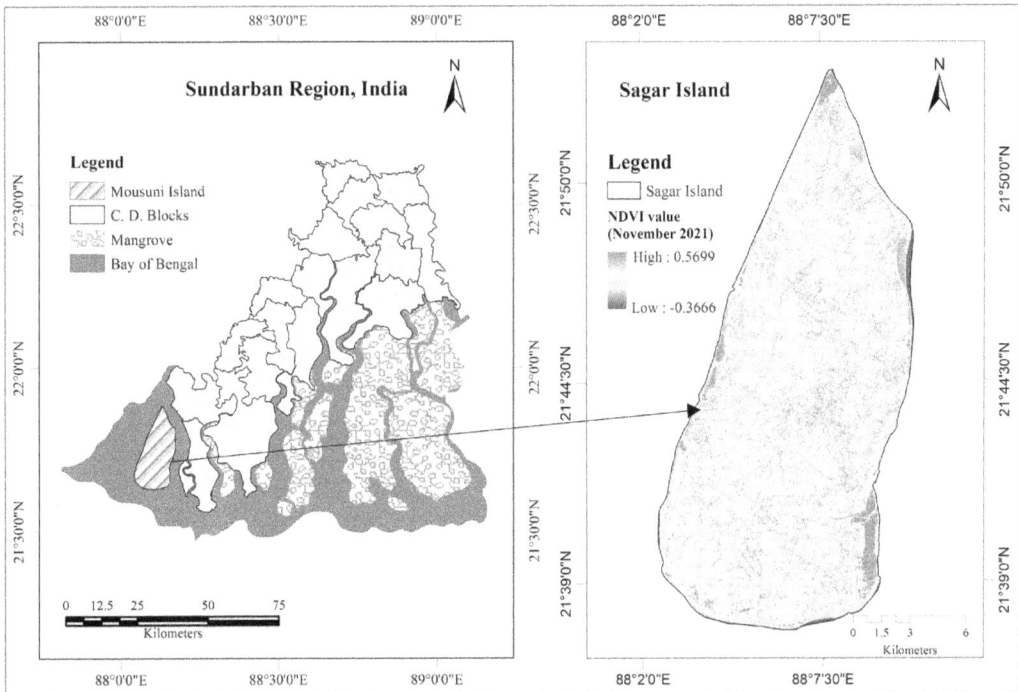

FIGURE 6.1   Location and Normalized Difference Vegetation Index value of the study area.

6.3   DATA TYPE AND METHODOLOGY

6.3.1   SATELLITE IMAGERIES AND ANCILLARY DATA

In this research, details on LULC for Sagar Island, Indian Sundarban region, were extracted using Landsat Thematic Mapper (TM) of November 2000 and Landsat 8 (Operational Land Imager [OLI]) of 2020 (Table 6.1). Satellite data were obtained by downloading from the USGS Earth Explorer website (https://earthexplorer.usgs.gov). The Survey of India's (SOI) topographic map (79 C/4) with a scale of 1:50,000 was utilized to geo-reference the data. Using GPS, ground truth data were collected and utilized for image classification as well as for evaluating overall accuracy.

6.3.2   IMAGE PRE-PROCESSING, CLASSIFICATION, AND ACCURACY ASSESSMENT

The pre-processing step involved merging a total of six bandsets (excluding thermal band) for Landsat 5 TM and eight bandsets (excluding thermal bands and band 1) for Landsat 8 OLI, using the open-source Quantum Geographic Information System (QGIS) version 3.2.0 and the Semi-Automatic Classification Plugin (SCP) (Congedo, 2016). The resulting imagery was then merged and clipped through the shapefile of Sagar Island on ArcGIS Environment.

The training set for image classification was developed successfully using Google Earth historical data and field observation after pre-processing. Random Forest, a powerful ensemble supervised algorithm developed by Breiman (2001), was used for image classification of all Landsat images in ArcGIS 10.5 software. At least 350 training samples were selected for each image classification by drawing polygons around representative classes. The Landsat images were primarily classified into six classes, including mangrove, vegetation, water body, mudflats, agricultural land, and built-up, but later, they were reclassified into agricultural land, built-up, and others categories using the Reclassification technique in ArcGIS 10.5 to fulfil the study's purpose.

To evaluate the accuracy output, the classified LULC categories were compared with referenced data (Lillesand et al., 2015). The evaluation involved computing overall accuracy (O_A) (Equation 6.2), kappa coefficient (k) (Equation 6.1), user accuracy (U_A), and producer accuracies (P_A) after consultation with Lillesand et al. (2015):

$$Kappa\ Statistics\left(k\right) = \frac{N\sum_{i=1}^{n}m_{ii} - \sum_{i=1}^{n}G_iC_i}{N^2 - \sum_{i=1}^{n}G_iC_i} \tag{6.1}$$

$$Overall\ Accuracy\ \left(OA\right) = \frac{\sum_{i=1}^{n}m_{ii}}{N} \tag{6.2}$$

where m_{ii} denotes counts samples in cluster I (I = 1, 2, ..., n); G_i represents overall classification in category I; C_i denotes the reference data for all categories; and N is the total number of samples across all categories.

TABLE 6.1

Concise Metadata Summary for Chosen Landsat Datasets in Land-Use Classification

Year	Source	Satellite (sensor)	Acquisition Date	Path/Row	Resolution (m)
2000	EarthExplorer	Landsat 5 TM	9 November 2000	138/045	30 × 30
2020	EarthExplorer	Landsat 8 OLI	16 November 2020	138/045	30 × 30

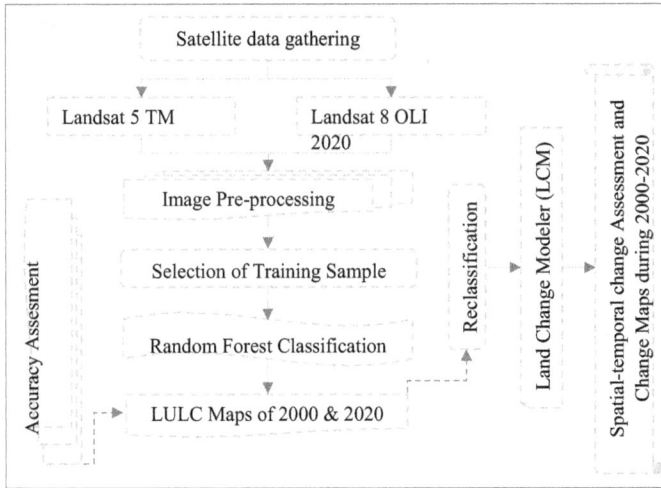

FIGURE 6.2 Workflow for processing, classifying, and measuring changes in Landsat image using QGIS, ArcGIS, and TerrSet software.

6.3.3 Quantifying the Expansion of Built-up Land

The TerrSet software's Land Change Modeler (LCM) was employed to examine the changes in the built-up areas' spatio-temporal dynamics. This was achieved by creating land-use maps from the Landsat TM and Landsat OLI images taken in 2000 and 2020, respectively. LCM is a collection of tools for LULC change analysis and modelling, initially developed by Eastman (2006). The methodology for this study is outlined in Figure 6.2.

6.4 RESULTS AND DISCUSSION

The study focused on the encroachment of agricultural land caused by built-up expansion on Sagar Island in the Indian Sundarban region. The increase in population on the island, driven by the development of agriculture, aquaculture, and fishing industries, resulted in demand for new housing, communication, education, health facilities, and employment. These demands have led to significant land-use changes on the island, with agricultural land being transformed to built-up areas to reconcile the growing population's needs (Mondal et al., 2019).

6.4.1 Classification Results and Change Detection

The study found that the overall accuracies of the Landsat 5 TM images (2000) and Landsat 8 OLI images (2020) were 83.73% and 85.21%, respectively, with kappa co-efficient values of 75.31% and 79.62%. Congalton and Green (2002) proposed that a kappa value exceeding 0.8 represents a robust agreement, whereas values ranging from 0.6 to 0.8 signify a considerable level of agreement between the reference data and the classified maps.

In Figure 6.3, the LULC maps of the island for the years 2000 and 2020 are presented, showing a decrease in agricultural area from 13,736.25 to 11,298.06 ha; an increase in built-up area from 1,427.40 to 5,477.58 ha; and a decrease in other categories including mangrove, vegetation, water body, and mudflats area from 8,757.81 to 7,145.82 ha during the 2000–2020 period, as outlined in Table 6.2. The study's findings indicate a considerable boost in built-up area, with a rise of 4,050.18 ha from 2000 to 2020, as demonstrated in Figure 6.3. This has resulted in a rise in the proportion of built-up area within the total land, from 5.97% to 22.90%, as shown in Table 6.2.

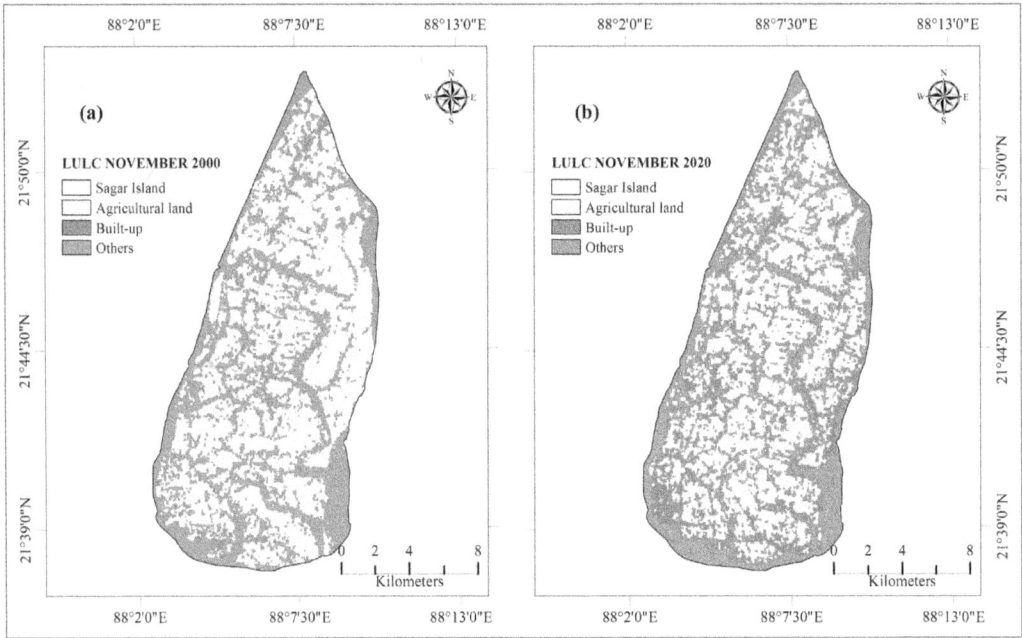

FIGURE 6.3   Structural changes in agriculture and built-up areas between (a) 2000 and (b) 2020.

Conversely, there has been a decrease of 2,438.19 hain agricultural land and 1,611.99 hain other land, as illustrated in Figure 6.3. From 2000 to 2020, the Sundarban region and Sagar Island have undergone numerous development-related activities, such as reclaiming forest land for settlements and agriculture, creating new residential structures while developing road networks to enhance connectivity and cater to the needs of the tourism industry (construction of a bridge connecting Sagar Island to the mainland in 2018), and embankments to prevent erosion and seawater ingress. These activities have influenced changes in the LULC dynamics of the area (Ghosh et al., 2015).

6.4.2   THE MAGNITUDE AND SHIFTING OF BUILT-UP GROWTH TOWARDS AGRICULTURAL LAND

The analysis revealed that over the study period from 2000 to 2020, there was a nearly fourfold increase in built-up areas, rising from 1,427.40 ha to 5,477.58 ha, as shown in Table 6.2. The built-up area of Sagar Island experienced a growth of 283.75% from its early built-up area in 2000, equivalent to 16.93% or 4,050.18 ha of the total study area, as denoted in Figure 6.4. On average, 202.51

TABLE 6.2
LAND-USE AREA COVERAGES

Land Use	2000		2000		2000–2020	
	Area (ha)	Percentage of Total Area (%)	Area (ha)	Percentage of Total Area (%)	Net Change (area)	Percentage Change (%)
Agriculture land	13,736.25	57.42	11,298.06	47.23	−2,438.19	−10.19
Built-up	1,427.40	5.97	5,477.58	22.90	4,050.18	16.93
Others	8,757.81	36.61	7,145.82	29.87	−1,611.99	−6.74
Total	23,921.46	100.00	23,921.46	100.00		

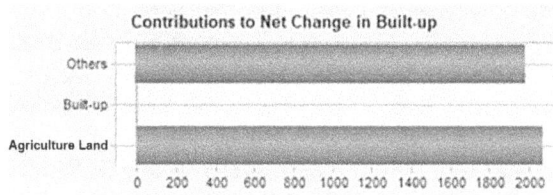

FIGURE 6.4 Built-up area net change contributions (in hectares) from 2000 to 2020.

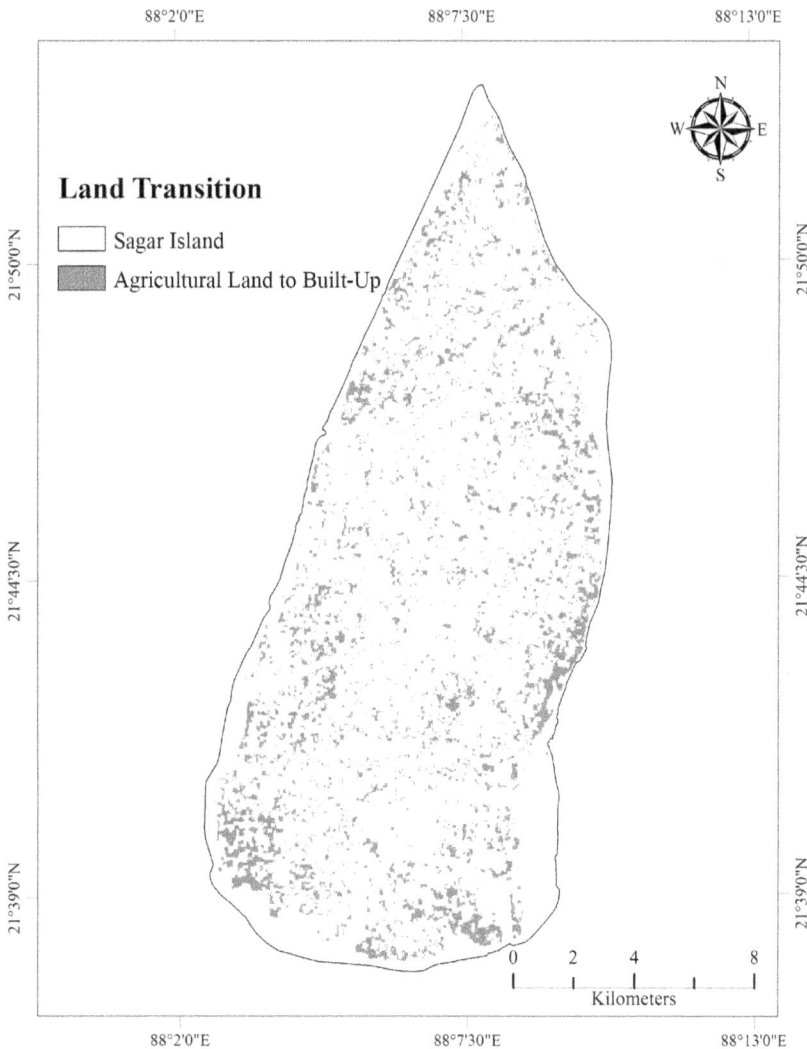

FIGURE 6.5 Expansion of built-up agricultural land from 2000 to 2020.

ha of the study area was converted to built-up land annually during the past 20 years. The development of the built-up area occurred mainly towards the south to south-west and central-eastern part of the island, where significant roads connected to surrounding areas, as depicted in Figure 6.5. The rise in built-up areas was driven by the swift population growth and their need for new housing units and improved transportation infrastructure. This growth took off without a thoughtful approach to planning and managing land use.

The study found that the newlyconstituted built-up area accounts for 16.93% (4,050.18 ha), with 8.64% (2,067.45 ha) originating from the conversion of agricultural land (Figure 6.4). The proportion of agricultural land has exhibited a steady decline from 57.42% (13,736.25 ha) to 47.23% (11,298.06 ha), as indicated in Table 6.2. Specifically, the outcome showed a loss of 2,438.19 ha of the original agricultural land (13,736.25 ha), which equates to 121.91 ha/year. The exhaustive loss of agricultural lands owing to built-up growth can be attributed to several factors. Firstly, the expansion of built-up areas typically follows major roads that often run through agricultural lands, or agricultural roads built to facilitate farmers' access to their farms and orchards (Al Rawashdeh and Saleh, 2006). Secondly, the inadequacy of legislation and land-use policies that govern building or the unjust application of such regulations and laws (Al Tarawneh, 2014) could also contribute to this phenomenon. Such conversion of agricultural land into other purposes, such as built-up areas, can potentially cause conflicts regarding land use and food availability (Magsi et al., 2017).

6.5   CONCLUSION AND SUGGESTIONS

The current investigation confirms that the utilization of GIS and remote sensing techniques can be efficient means for analyzing alterations in land usage. The development of built-up areas on Sagar Island is resulting in the loss of agricultural land, constituting a threat to the country's food security. Over a 20-year timeframe (2000–2020), urban land expansion eventuated in the conversion of 2,067.45 haof agricultural land. The built-up area of Sagar Island has increased by 283.75% from the original built-up area of 1,427.40 hain 2000. The presence of built-up areas on agricultural land suggests that urbanization does not necessarily entail an increase in population density on a limited and scarce piece of land. However, the higher demand for goods and services associated with an increased population requires the government to provide fundamental needs such as food, housing, transportation, education, healthcare, and public spaces. Conversely, urbanization may exacerbate issues of cleanliness, pollution, transportation, water and sewerage, and conflicts, among others. The loss of agricultural land may lead to people becoming landless and jobless, resulting in reduced agricultural production and increased dependence on imports, leading to higher prices for agricultural products. Therefore, competent land-use zoning is essential to prevent the expansion of urban areas over agricultural land, and building on such land should be stringently divergent. The government may recognize alternative sites on barren land for the expansion of new housing projects. Furthermore, educating the general public about the negative impacts of unplanned built-up expansion is crucial.

REFERENCES

Abdelrahman, M., Natarajan, A. V., & Hegde, R. (2016). Assessment of land suitability and capability by integrating remote sensing and GIS for agriculture in Chamarajanagar district, Karnataka, India. *The Egyptian Journal of Remote Sensing and Space Science*, *19*, 125–141.

Almutairi, A., & Warner, A. T. (2010). Change detection accuracy and image properties: A study using simulated data. *RemoteSensense*, *2*(1), 1508–1529. doi: 10.3390/rs2061508

Al Rawashdeh, S., & Saleh, B. (2006). Satellite monitoring of urban spatial growth in the Amman area, Jordan. *Journal of Urban Planning and Development*, *132*(4), 211–216.

Al Tarawneh, W. M. (2014). Urban sprawl on agricultural land (Literature survey of causes, effects, relationship with land use planning and environment): A case study from Jordan (Shihan Municipality Areas). *Journal of Environment and Earth Science*, *4*(20), 97–124.

Bandyopadhyay, S. (1997). Natural environmental hazards and their management: A case study of Sagar Island, India. *Singapore Journal of Tropical Geography*, *18*(1), 20–45.

Breiman, L., (2001). Random forests. *Machine Learn*, *45*(1), 5–32.

Chand, B. K., Trivedi, R. K., Dubey, S. K., & Beg, M. M. (2012). *Aquaculture in Changing Climate of Sundarban: Survey Report on Climate Change Vulnerabilities, Aquaculture Practices &Coping Measures in Sagar and Basanti Blocks of Indian Sundarban*. Kolkata, India: West Bengal University of Animal & Fishery Sciences, 198pp.

Congalton, R. G., & Green, K. (2002). *Assessing the Accuracy of Remotely Sensed Data: Principles and Practices.* QGIS Development Team. Open Source Geospatial Foundation Project: CRC Press.

Congedo, L. (2016). Semi-automatic classification plugin documentation. doi: http://dx.doi.org/10.13140/RG.2.2.29474.02242/1

Doygun, H. (2009). Effects of urban sprawl on agricultural land: A case study of Kahramanmaraş, Turkey. *Environmental Monitoring and Assessment, 158*, 471–478.

Eastman, J. R. (2006). *Idrisi Andes User's Manual [M].* Worcester, MA: Clark Labs, Clark University.

El Bastawesy, M., Khalaf, F., & Arafat, F. (2008). The use of remote sensing and GIS for the estimation of water loss from Tushka lakes, South Western Desert, Egypt. *Journal of African EarthSciences, 52*(3), 73–80.

Ghosh, A., Schmidt, S. K., Fickert, T., & Nüsser, M. (2015). The Indian Sundarban mangrove forests: History, utilization, conservation strategies and local perception. *Diversity, 7*, 149–169.

Gopinath, G. (2010). Critical coastal issues of Sagar Island, east coast of India. *Environmental Monitoring and Assessment, 160*, 555–561.

Gravert, A., & Wiechmann, T. (2016). Climate change adaptation governance in the Ho Chi Minh city region. In Katzschner, A., Waibel, M., Schwede, D., Katzschner, L., Schmidt, M., Storch, H. (Eds.), *Sustainable Ho Chi Minh City: Climate Policies for Emerging Mega Cities* (pp. 19–33). Cham: Springer International Publishing.

Hajra, R., Ghosh, A., & Ghosh, T. (2017). Comparative assessment of morphological and landuse/landcover change pattern of Sagar, Ghoramara, and Mousani island of Indian Sundarban delta through remote sensing. *Environment and Earth Observation: Case Studies in India* (pp. 153–172). Cham: Springer.

Haregeweyn, N., Fikadu, G., Tsunekawa, A., Tsubo, M., & Meshesha, D. T. (2012). The dynamics of urban expansion and its impacts on land use/land cover change and small-scale farmers living near the urban fringe: A case study of Bahir Dar, Ethiopia. *Landscape and Urban Planning, 106*(2), 149–157.

Lambin, E. F., Turner, B. L., Geist, H. J., Agbola, S. B., Angelsen, A., Bruce, J. W., Coomes, O. T., Dirzo, R., Fischer, G. W., Folke, C., George, P. S., Homewood, K., Imbernon, J., Leemans, R., Li, X., Moran, E. F., Mortimore, M. J., Ramakrishnan, P. S., Richards, J. F., Skånes, H., Steffen, W., Stone, G. D., Svedin, U., Veldkamp, T. A., Vogel, C., & Xu, J. (2001). The causes of land-use and land-cover change: Moving beyond the myths. *Global Environmental Change-Human and Policy Dimensions, 11*, 261–269.

Li, X.,& Yeh, A. G.-O. (2001). Zoning land for agricultural protection by the integration of remote sensing, GIS, and cellular automata. *Photogrammetric Engineering and Remote Sensing, 67*(4), 471–478.

Lillesand, T. M., Kiefer, R. W., & Chipman, E. J. (2015). *Remote Sensing and Image Interpretation* (7th ed.). Hoboken, NJ: John Wiley & Sons.

Liu, Y., Fang, F., & Li, Y. (2014). Key issues of land use in China and implications for policy making. *Land Use Policy, 40*, 6–12.

Magsi, H., & Torre, A. (2012). Social network legitimacy and property right loopholes: Evidences from an infrastructural water project in Pakistan. *Journal of Infrastructure Development, 4*(2), 59–76.

Magsi, H., Torre, A., Liu, Y., & Sheikh, M. J. (2017). Land use conflicts in the developing countries: Proximate driving forces and preventive measures. *The Pakistan Development Review*, 19–30.

Majumdar, R. K., & Das, D. (2011). Hydrological characterization and estimation of aquifer properties from electrical sounding data in Sagar Island Region, South 24 Parganas, West Bengal, India. *Asian Journal of Earth Sciences, 4*(2), 60.

Mandal, S., Satpati, L. N., Choudhury, B. U., & Sadhu, S. (2018). Climate change vulnerability to agrarian ecosystem of small island: Evidence from Sagar Island, India. *Theoretical and Applied Climatology, 132*, 451–464.

Molla, S. H., & Rukhsana. (2022). Spatio-temporal analysis of built-up area expansion on agricultural land in Mousuni Island of Indian Sundarban region. In *Agriculture, Environment and Sustainable Development: Experiences and Case Studies* (pp. 91–104). Cham: Springer International Publishing.

Mondal, I., Thakur, S., Ghosh, P., De, T. K., & Bandyopadhyay, J. (2019). Land use/land cover modeling of Sagar Island, India using remote sensing and GIS techniques. In *Emerging Technologies in Data Mining and Information Security: Proceedings of IEMIS 2018, Volume 1* (pp. 771–785). Springer Singapore.

Muchoney, D. M., & Haack, B. (1994). Change detection for monitoring forest defoliation. *Photogrammetric Engineering andRemote Sensing, 60*, 1243–1251.

Mukherjee, N., Siddique, G., Basak, A., Roy, A., & Mandal, M. H. (2019). Climate change and livelihood vulnerability of the local population on Sagar Island, India. *Chinese Geographical Science, 29*, 417–436.

Prokop, G., Jobstmann, H., & Schoenbauer, A. (2011). Report on best practices for limiting soil sealing and mitigating its effects. Study contracted by the European Commission, DG Environment, Technical Report-2011–50, Brussels, Belgium, 231 pp.

Purkait, B. (2009). Coastal erosion in response to wave dynamics operative in Sagar Island, Sundarban delta, India. *Frontiers of Earth Science in China*, *3*, 21–33.

Rukhsana, & Molla, S. H. (2023). Soil site suitability for sustainable intensive agriculture in Sagar Island, India: A geospatial approach. *Journal of Coastal Conservation*, *27*(2), 14.

Singh, A. (1989). Digital change detection techniques using remotely sensed data. *International Journal of RemoteSensing*, *10*, 989–1003.

Turok, I., & Mykhnenko, V. (2007). The trajectories of European cities, 1960–2005. *Cities*, *24*, 165–182.

UN DESA. (2015). *World Population Prospects: The 2015 Revision*. New York: United Nations, Department of Economic and Social Affairs, Population Division.

7 Consequences of Agricultural Land Transformation on the Natives of a Planned Town
A Case Study of New Town, Kolkata

Moududa Khatun and Lakshmi Sivaramakrishnan

7.1 INTRODUCTION

Presently, India is the largest populated country in the world. It shares about 11% of the world's urban population, and it is expected to increase to 13% by the year 2030 (MHUPA, 2016). There will be a simultaneous increase in the number of million-plus cities as well. The noble concept of Ebenezer Howard's 'Garden City' shows hope for exporting a huge proportion of people to new urban areas in the open countryside (Howard, 1902; Hobson, 1999). In India also, decentralisation through planned satellite cities is considered as a solution to multiple problems of large cities (Ranjan et al., 2014). Thus, many planned towns are emerging and will come up to cope with the increasing urban population. Consequently, the transformation of agricultural land for urban development is increasing (Abd EL-kawy et al., 2019; Pandey and Seto, 2015) and is a serious issue. Such transformation tends to increase social tensions among the natives due to unexpected changes in livelihood (Phuc et al., 2014).

Many studies are focusing on urbanisation and associated land-use transformation which happen at a cost to the natural environment (Easterling and Peterson, 1996; Weng, 2001; Bekele, 2005; Hegazy and Kaloop, 2015; Tolessa et al., 2017). Besides, among the different types of land-use changes, urbanisation of agricultural land is predominant (Muller and Middleton, 1994). Such land-use transformation leads to the loss of agricultural land (Shuaib, 2018; Fazal, 2000; Seto and Kaufmann, 2003), reduction of agricultural productivity (Yan et al., 2009), and other associated problems (Tong and Chen, 2002; Grau et al., 2003; Lambin et al., 2003). Urbanisation and consequent conversion of agricultural land have been widely studied in different parts of the world like China (Hou et al., 2021; Lu et al., 2011), Hyderabad district, Pakistan (Peerzado, 2019), Egypt (Abd EL-kawy et al., 2019), Nile Delta (Abd-Elmabod, 2019), and India (Tripathi and Rani, 2018; Gumma et al., 2017). Land-use change also led to the transformation of employment opportunities as well as the overall socio-economic living of the locals due to the loss of agricultural land (Wang et al., 2010; Sardar, 2013). Therefore, there is a need to focus on the effects of agricultural land transformation on the locals in the newly developed planned towns.

This chapter discusses the long-term effect of agricultural land transformation on the natives in New Town, Kolkata (NTK). After 30 years of plan implementation, what natives have lost and what they have gained due to the transformation and change of the basic amenities are discussed. Along with these, the study assesses what problems they are facing. The mental state of the land losers at the time of land acquisition and the present state of mind has been deliberated to understand their adjustment to the changes they receive. Finally, a proposal is given to tackle the problem.

DOI: 10.1201/9781003377825-8

7.2 STUDY AREA

NTK is a planned town of Kolkata and is located very close to Kolkata Municipal Corporation (Figure 7.1). The area is bounded by Salt Lake City on its western side, at the southern end there is a wide-open space of East Kolkata Wetland, and at the northern end there is Netaji Subhas Chandra Bose International Airport. It was planned during the 1990s to support Kolkata by increasing the housing stock supply by creating new residential units for people of different income levels and by establishing new central business districts (CBDs) to reduce the mounting pressure on the existing CBD of Kolkata (WBHIDCO, 2012). The whole NTK extends over 93.3 km². Out of that, the plan has already been implemented on about 60.36 km², covering 23 *mouzas* and is considered a 'Projected Area' or planned area.

This planned area is mostly a low-lying agricultural area. Thus, many natives have lost their agricultural land and have had to change their means of livelihood. They have to adapt to the changing land use and living. This superimposed and sudden change has an obvious impact on their living style. This truly agrarian society was forced to adopt a semi-urban character within a short period.

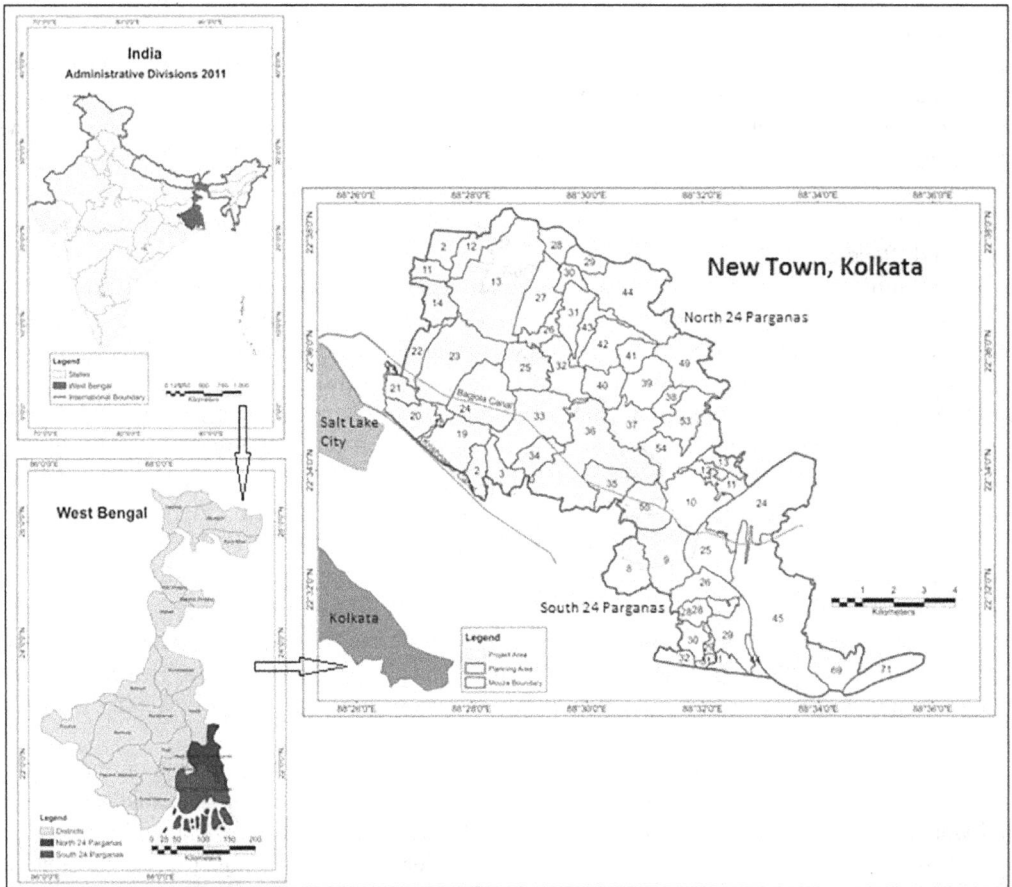

FIGURE 7.1 Location map of the study area.

Source: Census of India, 2011 and WBHIDCO, 2013.

7.3 DATABASE AND METHODOLOGY

The study is predominantly based on primary data, which were collected in the year 2019 using a well-designed schedule. For the survey, five sample *mouzas* were taken into consideration where both the existing settlement and adjoined planned urban units co-existed. Thus, based on the afore-mentioned criteria, Thakdari, Jatragachhi, and Patharghata of North 24 Parganas and Kochpukur and Jotbhim of South 24 Parganas have been selected for the detailed household survey of the natives. The household survey has been conducted in 5% of the total households of each of the aforementioned mouza. Thus, a total of 262 households and 1,366 populations have been randomly surveyed comprising existing settlements. Moreover, their perception regarding the basic amenities before and after plan implementation has been ranked with Ridit Analysis (Flora, 1974; Zimmerman, 1974; Selvin, 1977; Mantel, 1979). Seven basic amenities like availability of electricity, quality and availability of water, availability of education facility, availability of the medical facility, availability of shopping facility, transport facility, and security of living have been taken into consideration. Thereafter, respondents' perceptions regarding these basic amenities were recorded on a 5-point Likert scale (i.e., Excellent, Good, Average, Poor and Very Poor). Finally, a ranking of selected basic facilities has been done by Ridit Analysis both before plan implementation (BPI) and after plan implementation (API) based on the following methods (Wu, 2007).

Computation of Ridit for the reference dataset

- The population to serve has been selected as a reference dataset. Here, in the Likert scale survey, the reference dataset includes the total responses of the survey.
- The frequency (f_j) has been computed for each category of responses, where $j = 1 \ldots .n$.
- Mid-point accumulated frequency (F_j) has been computed for each category of responses:

$$F_j = \frac{1}{2} f_j + \sum_{k=1}^{j-1} f_k$$

where $2 \ldots n$.

- Ridit value R_j for each category of responses in the reference dataset:

$$R_j = \frac{F_j}{N}$$

where $j = 1 \ldots n$ and $N = 1862 (7 \times 266)$.

- N is the total number of responses from the Likert scale. The expected value of R for the reference dataset is always 0.5 by definition.

Computation of Ridit and mean Ridit for comparison datasets

The comparison dataset is composed of the frequencies of responses for each category (Services) of a Likert-scale item. Since there are 7 (m) Likert-scale items in this illustration, there are seven comparison datasets.

- Ridit value (r_{ij}) has been computed for each category of scale items:

$$r_{ij} = \frac{R_j \times r_{ij}}{\pi_i}$$

- π_{ij} is the frequency of category j for the ith scale item, and π_i is a short form for the summation of frequencies for scale item i across all categories, i.e.

$$\pi_i = \sum_{k=1}^{n} \pi_{ik}$$

- Mean Ridit (P_i) for each Likert-scale item has been computed:

$$P_i = \sum_{k=1}^{n} r_{ik}$$

7.4 DISCUSSION

The rational goal of transformation of agricultural land to NTK for decentralisation of the crowded city of Kolkata ultimately led to the development of some serious issues. Planned settlement has a social and spatial influence of new settlement on the central city and the natives (Wang et al., 2010). This chapter has been arranged in such a way so that the losses, gains, and problems of the natives can be assessed properly. Thus, firstly it focusses on what they have lost. The second portion focusses on what they have gained and then on the problems or sufferings of the natives. It also focusses on the changing psychological state of the land losers. Finally, a proposal is given to tackle the problem.

7.5 WHAT NATIVES HAVE LOST

Planning of NTK has been done in such a way that a maximum number of natives can save their dwelling units. Despite that, about 79.32% of the total surveyed households are land losers, who have lost their land during the development of NTK. Most of them squandered the compensation money without proper utilisation.

Notably, the lands they have lost are predominantly agricultural (Figure 7.2). Consequently, natives' means of livelihood, their occupation, has been changed drastically (Table 7.1 and Figure 7.3). Census data for 1991 and 2011 show a *mouza*-wise percentage share of agricultural and non-agricultural activity in the projected area of NTK (Figure 7.4). Data from 1991 represent the scenario of BPI, and data from 2011 show the scenario of API. Almost in all the *mouza* percentage share of agricultural activity has been reduced, and non-agricultural activity has increased drastically from 1991 to 2011. Primary data show that during BPI, cultivation was the prominent occupation, and

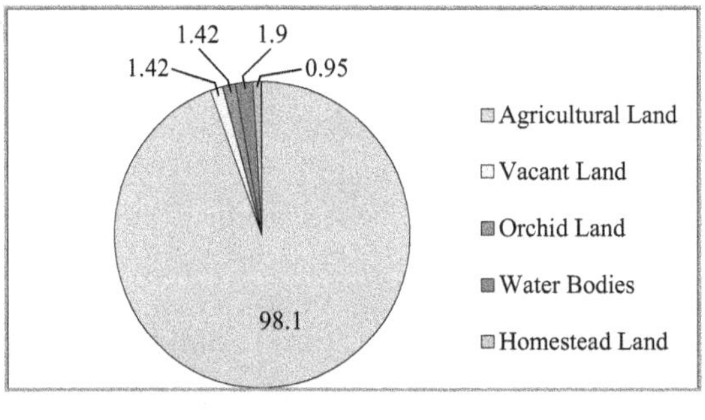

FIGURE 7.2 Type of land acquired from sampled household.

Source: Field Survey.

TABLE 7.1

Percentage Share of Dominant Occupation of the Total Population

Dominant Occupation	Population (in Percentage)	
	Before Plan Implementation	After Plan Implementation
Cultivation	60.14	0.23
Agriculture labour	6.76	0
Fishing	6.08	6.48
Non-agriculture labour	4.39	19.68
Household industry worker	2.03	2.31
Government job	4.05	4.63
Private enterprise	0.00	11.34
Self-employment	12.16	35.65
Security guard	0.68	10.42
Driver	2.70	9.03
Other	1.01	0.23
Total population	**100 (*n* = 1,366)**	**100 (*n* = 1,366)**

Source: Field Survey.

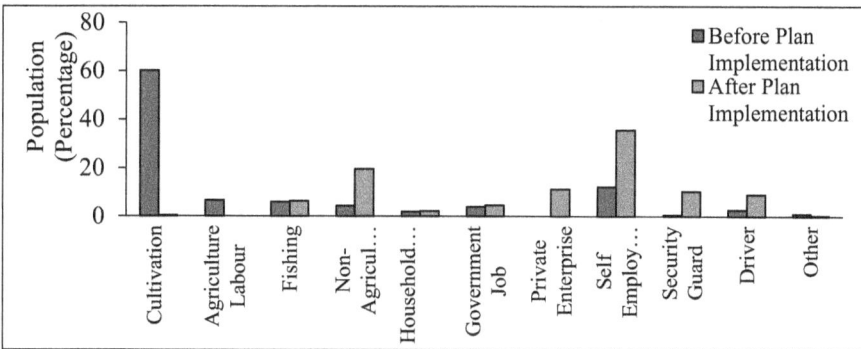

FIGURE 7.3   Share of dominant occupation of the total population in percentage.

Source: Field survey.

about 60% of the total workers were engaged in cultivation. After plan implementation, due to the acquisition of agricultural lands, domains are either converted into a built-up area or left vacant for future development. Thus, these are not available to local people for practising agricultural activities. They have to change their occupation for the sake of survival. Thus, presently, cultivation activity as well as agricultural labourer which were the dominant occupations BPI became near to absent in the study area. However, self-employment and non-agricultural labour became the dominant occupations after implementation of the plan.

7.6   WHAT NATIVES HAVE GAINED

Although, they have lost their land and their livelihood, in return they receive compensation. Maximum land losers (98.55%) got money as compensation based on the then-market price of their

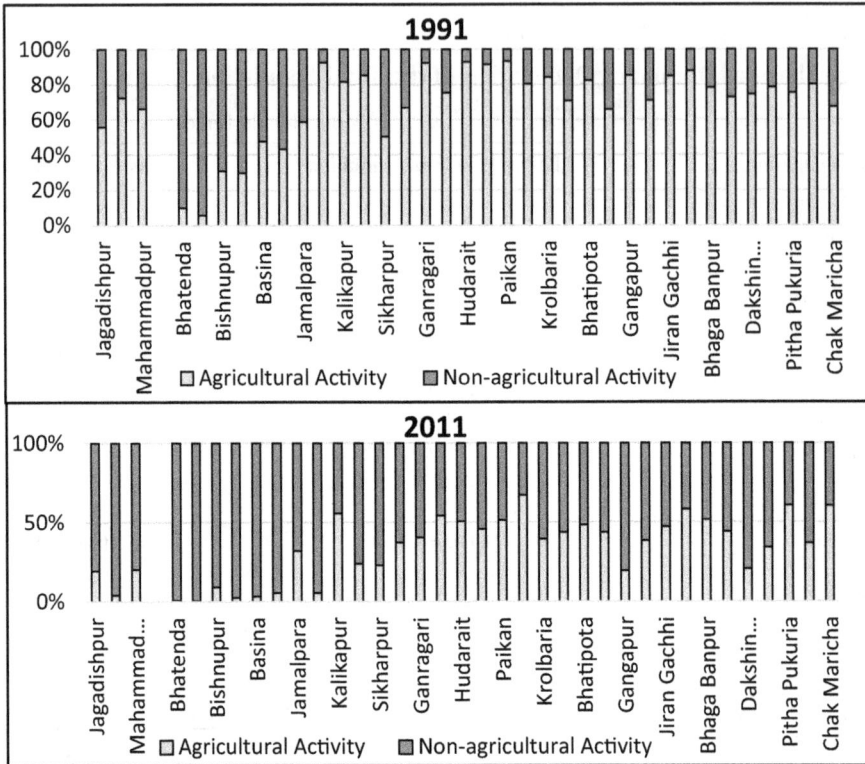

FIGURE 7.4 Changing of agricultural and non-agricultural activity in the *mouzas* of projected areas in New Town, Kolkata (1991 and 2011).

Source: Census of India, 1991, 2011.

land. Although the compensation amount varies with the quality of land, land losers receive a handsome sum of compensation money at the time of acquisition. Natives also get the diversification of employment opportunities in the newly developed planned area.

Besides, they have gained the development of infrastructure around their settlement. The plan has not been implemented in the residential areas of the existing settlement, even though the waves of change of the planned area touch the existing settled area. Thus, it has been noticed from the study area that along with the implementation of the plan, the availability of basic facilities changes. This change has been assessed with the changing perceptions of the respondents about the basic amenities of the respondents (Table 7.2). Transport facility, in BPI, ranked seventh and now, in API, ranked first. After the development of NTK, transportation facilities including mode, flow, network, and overall infrastructure improved considerably around the existing settled area. The availability of electricity also improved substantially. Almost all the surveyed households have electric connections and interestingly the problem of power-cut throughout the year reduced with the development of the area. Although the quality and quantity of education, shopping, and medical facilities improved in the planned area, they are not affordable for most of the locals. Thus, ranks remained more or less the same in BPI and API. However, the quality and availability of water and security status have deteriorated after the plan implementation. In BPI, the quality and availability of water ranked first. Respondents reported that they had adequate clean water for drinking and other purposes throughout the year. Now, it ranked seventh as they are not getting adequate water throughout

TABLE 7.2

Ranking of Preference of Basic Services Before and After Plan Implementation

Item No.	Services (m)	Rank	
		Before Plan Implementation	After Plan Implementation
1	Transport	7	1
2	Electric	6	2
3	Education	4	3
4	Shopping	3	4
5	Medical	5	5
6	Security	2	6
7	Water	1	7

Source: Computed by researcher based on Tables 7.10, 7.11, 7.12, and 7.13 in Appendix 7.1.

the year. For drinking purposes, either they have to purchase clean water or need to compromise on water quality. Likewise, they feel insecure after the huge in-migration of residents as well as workers in the planned area.

7.7 PROBLEMS OF NATIVES

Expectantly, the newly developed town NTK emerged with many employment opportunities in different sectors, including highly paid jobs. Roy (2005), in his paper for the Annual Conference of HUDCO Chair 2005, explains the development of NTK became the impetus for rural surroundings and villages by creating employment opportunities. Practically natives were not well prepared to grab high-paid jobs.

Most of the sample population are either illiterate or literate without formal education or have a low level of education (Table 7.3). Overall education level of the local people is relatively low. Thus, after the land acquisition, when they had to change their occupation, they could not grab well-paid job opportunities and had to do low-paid unskilled jobs. The types of work they are doing after the plan implementation present an interesting picture (Table 7.4). Home maid, mason, tailor, water supplier, vegetable vendor, the tea stall owner, hotel owner, priest, shop owner, idol maker, plumber, painter, private tutor, any kind of mechanic, carpenter, decorator, any kind of small business, house rent, gardening, horticulture, housekeeper, aaya at school, pantry boy, bus conductor, etc. are some major kinds of jobs they used to exercise after plan implementation. Notably, these are low-paid unskilled jobs and mainly serve the population living in the planned area. The mean monthly per capita incomes of the natives, i.e. both the land losers and non-land losers are low and more or less similar monthly per capita income (Table 7.5). Besides, the costly basic amenities in NTK like education, medical, market, etc. are attractive and accessible but not affordable for the natives. Utilisation of those facilities by the natives may lead to overburdened and impediments to economic conditions.

7.8 THE MENTAL STATE OF THE LAND LOSERS

Thereafter, land losers' psychological state at the time of land acquisition and present has been enquired about (Table 7.6). A number of people handed over land unhappily (49.76%), followed by

TABLE 7.3

Literacy and Educational Attainment by Gender of the Surveyed Population Excluding Minors

Category	Gender		Total
	Male	Female	
Illiterate	6.82	15.85	11.12
Literate without formal schooling	5.34	5.56	5.44
Pre-primary	11.87	9.31	10.65
Primary	11.72	12.91	12.29
Upper primary	25.37	20.10	22.86
M.P./secondary education	16.32	17.97	17.11
H.S./higher secondary education	9.79	10.13	9.95
Graduation	10.39	7.19	8.86
Post-graduation (P.G.)	1.04	0.65	0.86
Professional degree	1.04	0.33	0.7
Diploma	0.30	0.00	0.16
Total	100 ($n = 674$)	100 ($n = 612$)	100 ($n = 1,286$)

Source: Field Survey

TABLE 7.4

Type of Works under Dominant Occupation Observed among Natives

Item No.	Dominant Occupation	Type of Work Found During Survey
1	Does not work	Minors and those who are not engaged in any type of work
2	Home maker	Women who are engaged in domestic work or household work without a pecuniary return
3	Cultivation	Persons engaged in the cultivation of owned and leased land
4	Agriculture labourer	Labour in agricultural activities
5	Fishing	Works related to fishing activity
6	Non-agriculture labour	Labour engaged in other than agricultural work
7	Household industry worker	Working under any kind of household industry, specifically bidi roller in the study area
8	Government job	Any kind of job in any government sector
9	Private enterprise	Any kind of work under private enterprise, including gardening, horticulture, housekeeper, aaya at school, pantry boy, bus conductor
10	Self-employment	All self-employed persons, including hawker, home maid, mason, tailor, water supplier, vegetable vendors, tea stall owner, hotel owner, priest, shop owner, idol maker, plumber, painter, private tutor, any kind of mechanic, carpenter, decorators, any kind of small business, house rent
11	Security guard	Persons working as a watchman or security provider
12	Driver	Taxi driver, auto driver, toto driver, van driver, rickshaw driver
13	Other	All other occupations not included

Source: Selected by the researcher from field survey

TABLE 7.5
Monthly Per Capita Incomes in the Study Area

Household	Items	Per Capita Income (Rs.)
Total sampled household	Monthly average	2,688.86
	Minimum	750.00
	Maximum	13,333.00
Sampled households, who lose their land	Monthly average	2,689.00
	Minimum	750.00
	Maximum	8,400.00
Sampled households, who do not lose their land	Monthly average	2,675.00
	Minimum	833.00
	Maximum	13,333.00

Source: Field survey.

TABLE 7.6
Psychological or Mental Status of Land Losers

Items	Category	Percentage of Households (%)
Mental state at the time of land acquisition	Happily	0
	Willingly	2.37
	Neutrally	13.74
	Unhappily	49.76
	Forcefully	34.12
	Total	100 (n = 211)
Present mental state	Too satisfied	0.00
	Satisfied	0.47
	Neutral	18.48
	Quite depressed	52.13
	Too depressed	28.91
	Total	100 (n = 211)

Source: Field survey.

forcefully, neutrally, and willingly. No one handed over land happily among the sampled household. Those who handed over land willingly believed that they would get a better opportunity. Respondents disclose that at the time of land acquisition, they were promised by local leaders to get a job or other kind of facilities in exchange for their land, but no one gets such kind of compensation. However, their psychological state after land acquisition shows that about 52.16% of land losers were quite depressed, followed by too depressed (28.91%), neutral (18.48%), and satisfied (0.47%), and no one was too satisfied. One of the respondents told his distress story that after losing the land his father was depressed and died of a heart attack. The majority of respondents were annoyed that though the value of the money they received was quite good at that time, increasing the price of their land made them more upset. The land value is collected based on the respondent's experience, which has further averaged for each *mouza*. Land value increased abruptly in all the sampled *mouza* and the average of NTK from 1990 to 2017 (Table 7.7). Thus, direct land losers recalled the good situation (when they had owned land), imagined how much property they could have now (if

TABLE 7.7

Change of the Land Value in '000' Rupee per Bigha

Mouza	Land Value in '000' Rupee per Bigha			
	1990	2000	2010	2020
Jotbhim	7–10	30–35	14,000–16,000	24,000–30,000
Jatragachhi	30–40	150–200	30,000–50,000	60,000–90,000
Kochpukur	40–50	120–150	12,000–15,000	20,000–40,000
Patharghata	10–20	50–100	60,000–10,000	10,000–15,000
Thakdari	20–25	120–150	10,000–12,000	40,000–50,000
Average	**50**	**220**	**4,600**	**7,600**

Source: Field survey.

the land was not being acquired), and regretted the present situation (as they remain as a serving population of the NTK).

After the acquisition of agricultural land, along with land losers, non-land losers too had to change their occupations for the sake of survival. The adaptation of occupation happens based on their level of education, availability, and feasibility. However, those who were not the landowner and engaged in agricultural activity used to work as labourers in the agricultural field BPI. API, they got an opportunity to diversify occupations, and this change made them happy. On the other hand, BPI, the land losers were the owners and felt superior in the area. Now, they feel frustrated because working as a serving population offends their self-respect. Although, Sengupta (2022) suggested that natives can still practice agricultural activity in vacant spaces through government initiatives with proper planning which is practically difficult because of increasing land value. The price of land increased beyond their expectations which augmented their bitterness. During the field survey, they acknowledged that they used to imagine the property they could have if they have some land at present. Of course, most of the land losers received compensation in return for their land, but soon after receiving the money, the amount was squandered without being utilised properly. It is also an unavoidable fact that improved infrastructural facilities are primarily developed for the newly settled society of NTK. These facilities are physically accessible but financially unaffordable for the native people.

7.9 A PROPOSAL FOR TACKLING THE PROBLEM

The loss, gain, and different problems will be more or less common after the development of the planned city (Table 7.8). Thus, it is very necessary to find out some solution so that in the future these kinds of problems can be circumvented. Therefore, here is a suggestion that can sort out the problem of economic stability of the land losers (Table 7.9). Compensation money can be provided in different phases until the land losers survive. Thus, when they are not in a position to change their occupation and are unable to earn income, they can enjoy it as a pension. For instance, a person of Thakdari contributed about 1 Bigha for the development of the city and got Rs. 3,000/- per *Katha* which turns to a total of Rs. Rs. 60,000/- much higher than the contemporary land price of Rs. 20,000–25,000 per *Bigha*. The compensation money they received was about three times greater than the market value. The developer could provide compensation money at the time of acquisition and provide some money at regular intervals as pension only to the land losers until they survive. If someone receives Rs. 60,000/- at the time of acquisition and only Rs. 5,000/- per month, then overall that person will receive Rs. 1,860,000/- (i.e. [5,000 × 30 × 12] + 60,000). This amount is also much less than the present value of their land. This estimation is just an instance that can be considered

TABLE 7.8
Summary of Losses, Gains, and Problems of Natives

Aspects	Type	Land Losers	Non-Land Losers
Lost	The land has been acquired	Acquired lands are dominantly agricultural lands	—
	Compensation money	Received compensation money squandered quickly	—
	Occupation	They have to change their occupation as most of the natives were engaged in agricultural activities.	
	Water quality	There is a paucity of adequate clean water for drinking and other purpose throughout the year.	
	Overall security	Because of the huge in-migration of residents in the area, they feel insecure.	
Gained	Compensation money	They receive a handsome sum of compensation money at the time of acquisition.	—
	Occupation	Diversification of occupation takes place.	
	Infrastructural facility	Electricity availability, transport facility, educational institutes, medical facility, and many shopping malls have been improved in NTK.	
	Employment opportunities	The NTK has developed huge employment opportunities in different sectors.	
Problems	Low level of education	Native populations dominantly are either illiterate or literate without formal education or educated up to the upper primary level.	
	Low-paid unskilled job	Mostly engaged in unskilled jobs with low monthly per capita income.	
	Become helping the population of the NTK	Most of the work they are doing is mainly serving the newly settled population of NTK.	
	Infrastructural facilities are not affordable	Costly education, medical, and shopping facilities are not affordable for the natives.	

Source: Compiled by the researcher.

TABLE 7.9
Estimation of Compensation Money That Can Be Given

Calculation for 1 *Bigha* Land (1 *Bigha* = 20 Katha)			
Situation	**Condition**		**Total Amount**
Actual compensation received	At the rate of Rs. 3,000 per *Katha* (1 *Bigha* = 20 *Katha*)		Rs. 60,000
They could have (i.e. present value of the land)	The average rate of 1 *bigha* land (based on Table 7.7)		Rs. 45,000,000
If a pension is provided along with compensation	Pension at a rate of a minimum of Rs. 5000 per month for 30 years	Rs. 5,000 × 30 × 12 1,800,000	Rs. 1,860,000
	Compensation money	At a rate of Rs. 3,000 per *Katha* 60,000 (1 *Bigha* = 20 *Katha*)	

Source: Compiled by the researcher.

by the developers according to their feasibility. Such pension kind allowances can provide economic stability as well as can reduce the emotional scratch of the land losers up to some extent.

Besides, compulsory inclusion of land losers or their direct family members in the development process based on their level of education in every project of a newly developed area can generate more employment opportunities for the land losers' families. This will bring economic stability to the land losers' families.

However, it is also necessary to enhance their level of education, otherwise they will be segregated as a service village only and regional inequality will emerge. Therefore, along with employment opportunities, their level of education should be enhanced by providing adequate and affordable institutions. The problem related to infrastructure including water quality should be improved by implementing a water treatment plant for the existing settlement too. Local administration can take responsibility.

7.10 CONCLUSION

The development of planned towns excludes existing settlements, which consequently developed physical inequality (differences in land use and infrastructure). Now if native people remain as a serving population for the planned area, social inequality will also emerge and remain for a long time. The vicious cycle begins of low level of education, low-paid jobs, attraction to an expensive lifestyle, overburdened, and impediments to their economic stability. Such socio-economic and infrastructural polarisation must be considered by planners and policymakers to cater for these issues in future.

REFERENCES

Abd EL-kawy, O. R., Ismail, H. A., Yehia, H. M., and Allam, M. A. (2019, December). Temporal detection and prediction of agricultural land consumption by urbanization using remote sensing. *Egyptian Journal of Remote Sensing and Space Science*, 22(3), 237–246. doi: https://doi.org/10.1016/j.ejrs.2019.05.001

Abd-Elmabod, S. K. (2019). Rapid urbanisation threatens fertile agricultural land and soil carbon in the Nile delta. *Journal of Environmental Management*, 252. doi: https://doi.org/10.1016/j.jenvman.2019.109668

Bekele, H. (2005). *Urbanization and Urban Sprawl*. Retrieved from www.scribd.com/document/356976883/Bekele-2005-Urbanization-and-Urban-Sprawl

Census of India (1991). Retrieved from http://censusindia.gov.in/

Census of India (2011). Retrieved from http://censusindia.gov.in/

Easterling, D. R., and Peterson, T. C. (1996). Influence of land use/land cover on climatological values of the diurnal temperature range. *Journal of Climate*, 9, 2941–2944.

Fazal, S. (2000). Urban expansion and loss of agricultural land—a GIS based study of Saharanpur City, India. *Environment & Urbanization*, 12(2), 133–149. Retrieved from https://journals.sagepub.com/doi/pdf/10.1177/095624780001200211

Flora, J. D. (1974). *A note on Ridit Analysis* (Technical Report). Ann Arbor, MI: Hlighway Safety Research Institute. Retrieved from https://deepblue.lib.umich.edu/bitstream/handle/2027.42/1530/30927.0001.001.pdf?sequence=2

Grau, H. R., Aide, T. M., Zimmerman, J. K., Thomlinson, J. R., Hemler, E., and Zou, X. (2003). The ecological consequences of socioeconomic and land-use changes in postagriculture Puerto Rico. *BioScience*, 53(12), 1159–1168.

Gumma, M. K., Mohammad, I., Nedumaran, S., Whitbread, A., and Lagerkvist, C. J. (2017). Urban sprawl and adverse impacts on agricultural land: A case study on Hyderabad, India. *Remote Sensing*, 9(11), 1136. https://doi.org/10.3390/rs9111136

Hegazy, I. R., and Kaloop, M. R. (2015). Monitoring urban growth and land use change detection with GIS and remote sensing techniques in Daqahlia governorate Egypt. *International Journal of Sustainable Built Environment*, 4(1), 117–124. Retrieved from http://dx.doi.org/10.1016/j.ijsbe.2015.02.005.

Hobson, J. (1999). Development Planning Unit, University College London. Retrieved from www.ucl.ac.uk/bartlett/development/sites/bartlett/files/migrated-files/wp108_0.pdf.

Hou, D., Meng, F., and Prishchepov, A. V. (2021). How is urbanization shaping agricultural land-use? Unraveling the nexus between farmland abandonment and urbanization in China. *Landscape and Urban Planning*, 214. Retrieved from www.sciencedirect.com/science/article/abs/pii/S016920462100133X

Howard, E. (1902). *Garden Cities of Tomorrow*. London: S. Sonnenschein & co. ltd., pp. 1–20.

Lambin, E. F., Geist, H. J., and Lepers, E. (2003). Dynamics of land-use and land-cover change in tropical regions. *Annual Review of Environment and Resources*, 28, 205–241.

Lu, Q., Liang, F., Bi, X., Duffy, R., and Zhao, Z. (2011). Effects of urbanization and industrialization on agricultural land use in Shandong Peninsula of China. *Ecological Indicators*, 11(6), 1710–1714. https://doi.org/10.1016/j.ecolind.2011.04.026

Mantel, N. (1979). Ridit analysis and related ranking procedures—use at your own risk. *American Journal of Epidemiology*, *109*(1), 25–29.

MHUPA. (2016). *India Habitat III National Report 2016*. New Delhi: Ministry of Housing and Urban Poverty Alleviation. Retrieved from http://mhupa.gov.in/writereaddata/1560.pdf.

Muller, M. R., and Middleton, J. (1994). A Markov model of land-use change dynamics in the Niagara Region, Ontario, Canada. *Landscape Ecology*, *9*, 151–157.

Pandey, B., and Seto, K. C. (2015). Urbanization and agricultural land loss in India: Comparing satellite estimates with census data. *Journal of Environmental Management*, *148*, 53–66. https://doi.org/10.1016/j.jenvman.2014.05.014

Peerzado, M. B. (2019). Land use conflicts and urban sprawl: Conversion of agriculture lands into urbanization in Hyderabad, Pakistan. *Journal of the Saudi Society of Agricultural Sciences*, *18*(4), 423–428. https://doi.org/10.1016/j.jssas.2018.02.002

Phuc, N. Q., Westen, A. V., and Zoomers, A. (2014). Agricultural land for urban development: The process of land conversion in Central Vietnam. *Habitat International*, *41*, 1–7. https://doi.org/10.1016/j.habitatint.2013.06.004

Ranjan, N., Prakash, P., Nathani, N., and Khan, M. R. (2014). Sustainable development strategies for satellite towns in India (Case Study—Hajipur). In *Sustainable Constructivism: Traditional vis-à-vis Modern Architecture*, edited by J. C. Mishra, 91–97. New Delhi: Execellent Publishing House.

Roy, U. K. (2005). Development of New Townships: A catalyst in the growth of rural fringes of Kolkata Metropolitan Area (KMA). *Annual Conference of HUDCO Chair*. Kolkata.

Sardar, J. (2013). Land use change and perception mapping of New Town, Rajarhat, North 24 Parganas, West Bengal. *International Journal of Remote Sensing & Geoscience*, *2*(6), 28–32.

Selvin, S. (1977). A further note on the interpretation of Ridit analysis. *Amearican Journal of Epidemiology*, *105*(1), 16–20.

Sengupta, P. (2022). Alternative rehabilitation program against agricultural land acquisition for new development projects—A case study of New Town, Kolkata, West Bengal, India. In *Sustainable Urbanism in Developing Countries*, edited by U. Chatterjee, A. Biswas, J. Mukherjee, & D. Mahata, 1–15. Boca Raton: Taylor & Francis Group.

Seto, K. C., and Kaufmann, R. K. (2003). Modeling the drivers of urban land use change in the Pearl River Delta, China: Integrating remote sensing with socioeconomic data. *Land Economics*, *79*, 106–121. https://doi.org/10.2307/3147108

Shuaib, M. A. (2018). Impact of rapid urbanization on the floral diversity and agriculture land of district Dir, Pakistan. *Acta Ecologica Sinica*, *38*(6), 394–400. https://doi.org/10.1016/j.chnaes.2018.04.002

Tolessa, T., Senbeta, F., and Kidane, M. (2017). The impact of land use/land cover change on ecosystem services in the central highlands of Ethiopia. *Ecosystem Services*, *23*, 47–54.

Tong, S. T. Y., and Chen, W. (2002). Modeling the relationship between land use and surface water quality. *Journal of Environmental Management*, *66*, 377–393.

Tripathi, S., and Rani, C. (2018). The impact of agricultural activities on urbanization: Evidence and implications for India. *International Journal of Urban Sciences*, *22*(1), 123–144. DOI: 10.1080/12265934.2017.1361858

Wang, L., Kundu, R., and Chen, X. (2010). Building for what and whom? New town development as planned suburbanization in China and India. *Research in Urban Sociology*, *10*, 319–345.

WBHIDCO. (2012). *Landuse and Development Control Plan for New Town Planning Area*. Retrieved from www.wbhidcoltd.com: www.wbhidcoltd.com/uploads/reports/report11.pdf

WBHIDCO. (2013). *Present Land Use Map & Register*. New Town, Kolkata: West Bengal Housing Infrastructure Development Company Ltd. Retrieved from www.wbhidcoltd.com/uploads/announcement/Ann80Notice_LUMR.pdf

Weng, Q. (2001). A remote sensing—GIS evaluation of urban expansion and its impact on surface temperature in the Zhujiang Delta, China. *International Journal of Remote Sensing*, *22*(10), 1999–2014. Retrieved from www.tandfonline.com/doi/pdf/10.1080/713860788?needAccess=true

Wu, C. H. (2007). On the application of Grey Relational Analysis and RIDIT analysis to Likert Scale surveys. *International Mathematical Forum*, *2*(14), 675–687. Retrieved from chrome-extension://efaidnbmnnnibpcajpcglclefindmkaj/www.m-hikari.com/imf-password2007/13-16-2007/chienhowuIMF13-16-2007.pdf

Yan, H., Liu, J., Huang, H. Q., and Cao., M. (2009). Assessing the consequence of land use change on agricultural productivity in China. *Global and Planetary Change* (Elsevier), *67*, 13–19.

Zimmerman, S., and Johnston, D. A. (1974). Nonparametric tests and Ridits. *Journal of Periodontal Research*, 193–206.

APPENDIX 7.1

TABLE 7.10
Ridit for the Reference Data Set (Preference of Basic Services) Before Plan Implementation

Item No.	Services (m)	Response (n)					Total
		Excellent	Good	Average	Poor	Very Poor	
1	Electric	0	34	126	79	27	266
2	Water	5	94	149	15	3	266
3	Education	0	20	197	45	4	266
4	Medical	0	16	186	56	8	266
5	Shopping	1	37	203	22	3	266
6	Transport	1	5	59	137	64	266
7	Security	3	102	106	50	5	266
	fj	10	308	1,026	404	114	1,862
	Fj	5	164	831	1,546	1,805	
	R_j	0.0027	0.0881	0.4463	0.8303	0.9694	

Source: Computed by the researcher.

TABLE 7.11
Ridit for Comparative Dataset and Relative Ranking of Basic Services Before Plan Implementation

Item No.	Services (m)	r_{ij} of Response (n)					Total (P_i)	Rank
		Excellent	Good	Average	Poor	Very Poor		
1	Electric	0.0000	0.0113	0.2114	0.2466	0.0984	0.5677	6
2	Water	0.0001	0.0311	0.2500	0.0468	0.0109	0.3389	1
3	Education	0.0000	0.0066	0.3305	0.1405	0.0146	0.4922	4
4	Medical	0.0000	0.0053	0.3121	0.1748	0.0292	0.5213	5
5	Shopping	0.0000	0.0123	0.3406	0.0687	0.0109	0.4325	3
6	Transport	0.0000	0.0017	0.0990	0.4276	0.2332	0.7615	7
7	Security	0.0000	0.0338	0.1778	0.1561	0.0182	0.3860	2

Source: Computed by the researcher.

TABLE 7.12

Ridit for the Reference Dataset (Preference of Basic Services) After Plan Implementation

Item No.	Services (m)	Response (n)					Total
		Excellent	Good	Average	Poor	Very Poor	
1	Electric	35	177	46	5	3	266
2	Water	2	78	114	60	12	266
3	Education	10	182	59	15	0	266
4	Medical	6	148	80	29	3	266
5	Shopping	16	152	68	27	3	266
6	Transport	40	180	39	4	3	266
7	Security	2	119	105	38	2	266
	f_j	111	1,036	511	178	26	1,862
	F_j	55.5	629	1,402.5	1,747	1,849	
	R_j	0.0298	0.3378	0.7532	0.9382	0.9930	

Source: Computed by the researcher.

TABLE 7.13

Ridit for Comparative Dataset and Relative Ranking of Basic Services After Plan Implementation

Item No.	Services (m)	r_{ij} of Response (n)					Total (P_i)	Rank
		Excellent	Good	Average	Poor	Very Poor		
1	Electric	0.0039	0.2248	0.1303	0.0176	0.0112	0.3878	2
2	Water	0.0002	0.0991	0.3228	0.2116	0.0448	0.6785	7
3	Education	0.0011	0.2311	0.1671	0.0529	0.0000	0.4522	3
4	Medical	0.0007	0.1879	0.2265	0.1023	0.0112	0.5286	5
5	Shopping	0.0018	0.1930	0.1925	0.0952	0.0112	0.4938	4
6	Transport	0.0045	0.2286	0.1104	0.0141	0.0112	0.3688	1
7	Security	0.0002	0.1511	0.2973	0.1340	0.0075	0.5902	6

Source: Computed by the researcher.

Part II

Forest Vegetation, Disaster Resilience, and Sustainable Development

8 Analysis of Demographic Vulnerability and Women's Participation to Promote Disaster-Resilient Society Based on Disaster Preparedness Behavior of South-West Coastal Bangladesh

Shahana Akter and Md. Mujibor Rahman

8.1 INTRODUCTION

Bangladesh has been classified as one of the nations that is most susceptible to natural disasters because of its proximity to frequent and intense meteorological phenomena like cyclones and their related storm surges (Karim and Mimura, 2008). Increasing natural disasters are a consequence of rapid urbanization, commercialization, global warming, and economic expansion (Gencer and Gencer, 2013). In recent years, more and more people have begun to recognize that natural catastrophes constitute a hindrance to achieve sustainable development objectives (Akter and Rahman, 2023). Various natural disasters, such as cyclones, river erosion, seawater intrusion, and floods, etc. have plagued the south-western coast of Bangladesh for many years (Saha, 2015). Due to its position, the south-west is especially susceptible to natural disasters (Cutter and Finch, 2008). Many individuals in rural areas lose their jobs and homes when a natural catastrophe strikes. They also frequently go without necessities like food, water, and sanitation (Sanderson, 2000). Natural catastrophes, such as cyclones, saline intrusion, river erosion, storm surge, and floods, impose a serious warning to the coast of the south-western region of Bangladesh (Afjal et al. 2012). There is a long history of human effort made to prevent catastrophe and lessen the impact of catastrophic events (Akter and Rahman, 2023).

Disaster response behaviors might help lessen negative impacts of natural calamities. Socio-demographic considerations positively or negatively influence disaster response behavior. Recent studies have shown that these areas are the most vulnerable, and it is estimated that a large amount of economic loss occurs frequently. Disastrous weather events, riverbank degradation, salinization, and other catastrophic events such as floods and high tides cannot be avoided, but their losses can be minimized by disaster response behavior that shows people have the ability to protect their health and property from disaster.

Factors affecting disaster response behavior include age, gender, income, education level, marital status, occupation, religious status, birth rate and mortality rate, average family age, and previous

DOI: 10.1201/9781003377825-10

disaster experience. They also include perception of risk, perception of readiness, perception of importance, optimism and normalization bias, self-efficacy, collective effectiveness, mortality, position of control, anxiety, societal values, individual accountability, social accountability, personal style, and accessible resources (Najafi et al., 2015). The most important factors influencing disaster response behavior are demographics. How demographics have an influence in disaster response behavior remains unclear. It is essential to determine the influence of demographic factors such as sex, age, employment, educational level, prior natural calamities training, housing varieties, and neighborhood on disaster response behavior (Najafi et al., 2015).

Based on the results of certain investigations, males are more involved with disaster response behavior than females. Likewise, many studies have found that disaster preparedness behavior (DPB) grows with age, but other research suggests that seniors are even less inclined to engage in disaster readiness (Najafi et al., 2015). Therefore, this study aims to identify traditional adaptation mechanisms and represents the global idea of the participation of demographic determinants in disaster risk reduction. It is hoped that the study's findings may shed new light on disaster management. It will also be useful to the authorities in formulating and adapting mitigation measures and introducing policy interventions.

8.2 MATERIALS AND METHODS

8.2.1 GENERAL INFORMATION AND SELECTION OF RESEARCH AREA

Natural disasters are particularly dangerous in Bangladesh's south-western coastline region. In southern Bangladesh, the municipalities of Khulna, Bagerhat, as well as Satkhira are regarded the most disaster-prone. Due to the severity of natural catastrophes like as tropical cyclones, salinization, river erosion, and high tides, the hamlet of Borunpara of Gangarampur Union in Batiaghata Upazila in Khulna district was chosen as the research area among the three districts. A good level of communication is also crucial.

Delegation from Gangarampur Borunpara is a village in Khulna's Batiaghata Upazila. Borunpara village is located approximately 22 km from Khulna district. This union has a total size of around 34.4 km². The population is estimated to be around 24,000 people. Borunpara is located between 22°37'59" and 22°38'59" N longitude and 89°30'40" and 89°31'23" E latitude (Figure 8.1). There is a total of 462 households in the area. In this community, there are 160 households in total. The majority of the people in this village work in agriculture.

The maps of the study area are given in Figure 8.1.

8.2.2 ANALYSIS OF DEMOGRAPHIC VULNERABILITY AT HOUSEHOLD LEVEL

It is critical to identify the demographic vulnerability at the household level of the research area in order to meet the study's objectives. Several demographic characteristics are examined for the study of demographic vulnerability, such as age group, male-to-female ratio, education and employment position, income level, and so on. Natural hazard incidence, natural hazard frequency, and disaster preparedness are all factors to consider. The DPB scores for the study's participants, knowledge scores concerning to disaster preparedness, show the contribution of each variable to predict relative vulnerability status among research participants at the household level. Similar studies are found in Primer (n.d.)

8.2.3 ASSESSING WOMEN'S PARTICIPATION IN PROMOTING DISASTER-RESILIENT SOCIETY

Several studies were carried out to measure women's role in fostering disaster-resilient societies. Community capacity for hazard resilience, particularly among women, is critical in the development of human capital for DPB stages. Independent sample t-tests for differences between the groups and one-way ANOVA for differences between the groups show the participation of women in DPB.

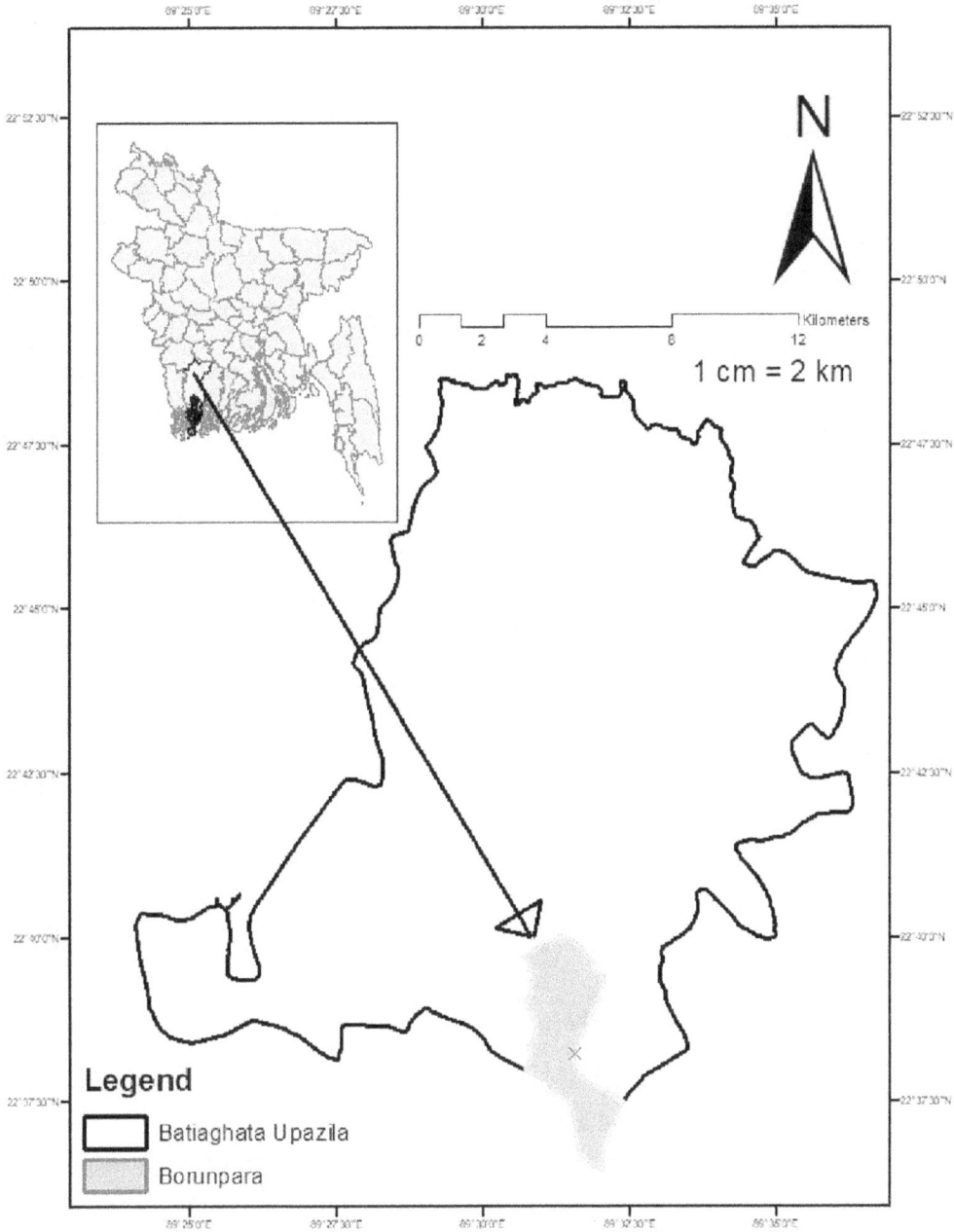

FIGURE 8.1 Study area map of Borunpara village of Batiaghata Upazilla.

8.2.4 Vulnerability Assessment in Respect to Disaster Preparedness Behavior to Promote Community Capacity for Disaster Risk Reduction

Community hazard resilience ability for disaster preparedness was assessed in this study. To determine degrees of resilience, the Hazard Resilience Capacity for Disaster Preparedness (GOAL Toolkit, 2015) was employed. The toolkit's score ranges from one to five. The greater the number, the more resilient you are. Information on risk zoning, hazard ranking and risk valuation, and disaster vulnerability score was gathered via a questionnaire survey and focus group discussion (disaster vulnerability score [DVS]). Hazard ranking and risk valuation scores of 800–1,000 indicate a very high-risk zone, and the vulnerable group is recognized by this evaluation.

8.2.5 Data Collection: Sampling, Analysis, Interpretation

The village of Borunpara, under the Gangarampur Union in Batiaghata Upazilla, was selected as a good representative of the disaster area. A household was used as a sampling unit. Households were randomly selected from the villagers. The sample unit is considered a household member over the age of 30 years with extensive knowledge and experience in disasters. A total of 160 households from the village of Borunpara were recorded. Correlation matrices of the major socio-demographic characteristics of the village of Borunpara and all other statistical analyses were performed by using SPSS, Minitab, and MS Office (2013) statistical software. To estimate the sample size, Kothari was consulted, and it was discovered that there were 104 people included in the samples for the Borunpara village:

$$n = \frac{z^2 pqN}{e^2 (N-1) + z^2 pq}$$

where N is the total number of people, n is the number of people in the sample, p is the percentage of people in the sample, q is one-half of p, z is the standard deviation at the confidence level in question, and e is the allowable deviation from the true value (95% confidence):

$$z = 1.96; p = 0.25; q = (1 - 0.25) = 0.75; e = 0.05$$

The population of Borunpara, $N = 160$. The estimated sample size for Borunpara village is 104 after plugging in the numbers into the equation and doing the math.

The home survey method for Borunpara village has a total number of participants of 104. The method of sampling that was used for this survey is one that is completely at random. This is done to eliminate any possibility of bias and to ensure that the samples are accurate representations of the population.

8.3 RESULTS AND DISCUSSION

8.3.1 Demographic Vulnerability at Household Level

8.3.1.1 Disaster Preparedness Behavior Scores for the Study Participants

The DPB scores of the individuals in the study are shown in Table 8.1. The data have been arranged on the basis of public readiness indexing. The demographic determinants like gender, age, occupation, education, and income level have been compared with community of practices. The DPB score of less than or equal to five was achieved by 90% of the overall participants, where in the previous research it was equal to eight. It also shows that 58.7% of participants have taken no measures to prepare for a potential tragedy during periods of 2021. None of the 58.7% of responders had taken any measures to ensure their safety. So, through time it is increasing.

8.3.2 Knowledge Scores Concerning to Disaster Preparedness

Nine individual knowledge-based questions about disaster preparedness (Figure 8.3) were asked to every respondent of the villages to understand their knowledge level in disaster occurrence and the level of vulnerability at household level. The knowledge scoring is based on public readiness

TABLE 8.1
Disaster Preparedness Behavior Respondent Ratings

Disaster Preparedness Behavior Score	Frequency	Percentage	Cumulative Percentage
0	61	58.7	58.7
1	11	10.57	69.27
2	7	6.73	76
3	10	9.62	86
4	5	4.81	90.43
5	2	1.9	98.33
6	2	1.9	94.23
7	6	5.61	100.0

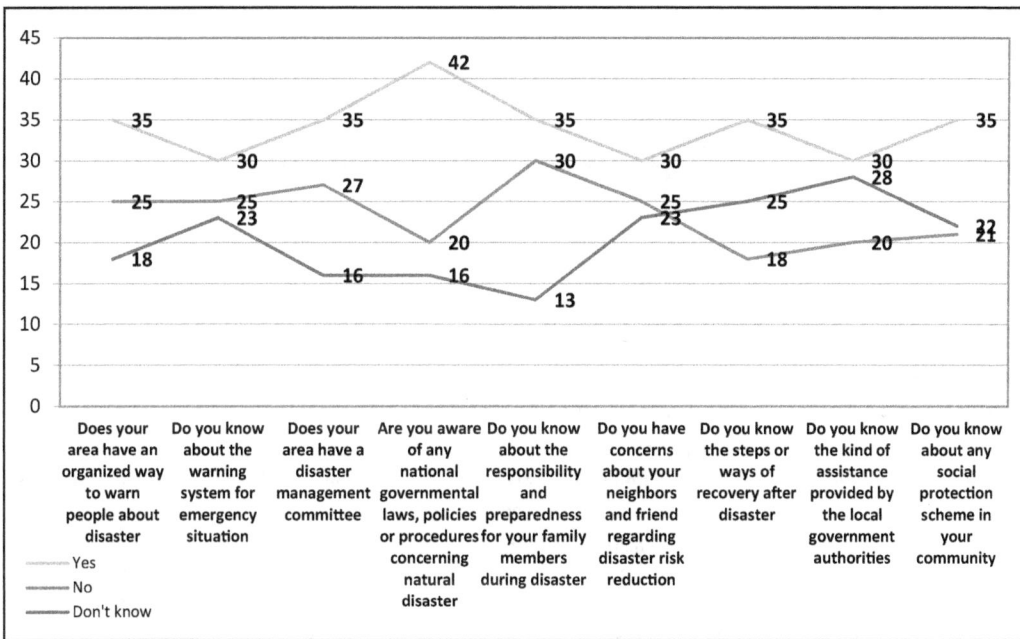

FIGURE 8.2 Frequency distribution of individuals regarding knowledge-based questions on disaster preparedness by multiple response.

Source: Field survey (2021).

indexing which helps to measure the knowledge factors during emergency situations. Frequency of the respondents of villages is being estimated for further scoring.

Frequency is converted to the percentage for knowledge scoring. According to the survey questionnaire, 39% of the respondents answered yes which represents that, among the 100%, the 39% respondents was much aware about disaster preparedness behaviour according to the mentioned eight questionnaire, whereas 46% answered no which represents that they had zero knowledge about disaster preparedness behaviour regarding knowledge-based questions, and the rest of the 15% of respondents could not answer the questions or they did not have any precautions regarding the preparedness behaviour of disaster considering knowledge-based questions (Figure 8.2).

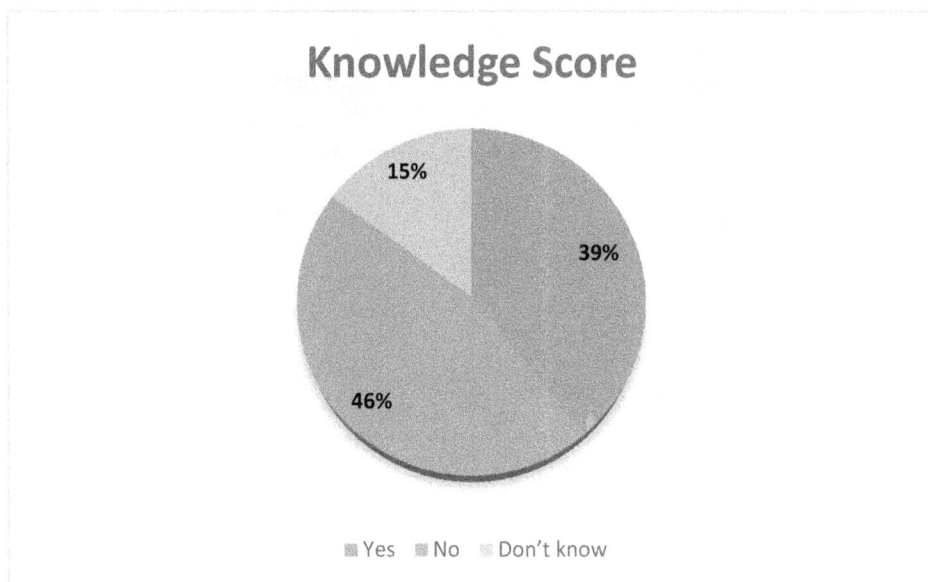

FIGURE 8.3   Proportion or Percentage (%) of individuals obtaining knowledge score.

Source: Field survey (2021).

TABLE 8.2

Comparisons between Groups Using an Independent *t*-Test

Variable	Groups	N (sample size)	Mean of DPB Score	SD	t (t-test)	t-test (total)	Sig. (2-tailed)	90% CI (confidence interval)
Age	21–45 years	95	1.51	8.09	8.62	5.06	0.010	0.9002 to 0.1259
	46–65 years	9	8.20	1.09	8.44		0.070	8.5602 to 0.1602
Sex	Male	87	1.39	1.91	8.11	4.23	0.037	0.7716 to 0.0245
	Female	17	8.41	8.7	8.12		0.050	8.8205 to 0.0030
Education	No education	12	1.0000			6.329		
	Primary	73	1.91	8.21	−0.341		0.734	0.3932 to −0.5560
	Secondary	19	0.575	1.22	6.67		0.000	−0.09897 to −0.1.8588
Occupation	Farmer	49	0.5556	1.21	−7.965	−6.86	0.000	−1.0790 to −1.8099
	Labor	29	1.73	1.84	−1.006		0.319	0.2649 to −0.7955
	Businessman	7	8.84	8.60	1.169		0.265	8.422 to −0.7305
	Service holder	3	1.50	8.12	−0.333		0.795	18.5593 to −19.5593
	Housewife	16	3.18	3.09	1.267		0.235	3.2594 to −0.8958
Marital status	Single	11	3.75	8.12	8.33	−0.87	0.052	3.5235 to −0.0235
	Married	86	1.39	1.98	−3.20		0.002	−0.2306 to −0.9766
	Others (widow, divorced)	7	0.000					
Income level	<5,000	21	1.0000			−8.83		
	5,000–10,000	45	1.0357	1.63	−5.417		0.000	−0.6103 to −1.3183
	10,000–15,000	28	3.1290	8.43	8.584		0.015	8.0212– 0.2369
	15,000–20,000	7						
	>20,000	3	0.0000	0.00000				

8.3.3 Assessing Women's Participation for Promoting Disaster-Resilient Society

8.3.3.1 Comparisons between Groups Using an Independent *t*-Test

Any statistically significant difference between two groups were determined using the *t*-test of independence (Table 8.2). Females were shown to be more active participants in DPB than males in this study. Moreover, it was shown that the subjects who have high education, high income, and high occupation value are more prepared compared to those who had no such values. At the time of the survey's administration, those with jobs had considerably higher DPB scores than those without. Assessment also shows that women are more concern about the effects, preparedness, and participation of disaster among all stakeholders like men. The score on the DPB scale did not correspond with marital status, etc.

8.4 COMPARISONS BETWEEN GROUPS USING ONE-WAY ANALYSIS OF VARIANCE

To determine if there were statistically significant differences between the groups, a one-way analysis of variance (ANOVA) was performed (Table 8.3). There were discernible and highly meaningful differences between the groups for such participant's DPB in different age groups. DPB scores significantly differ among the genders which shows that DPB scores were significantly greater among females rather than males, indicating that women are more engaged in preparedness behaviors than men. DPB scores significantly differ among different occupational, income, and educational groups. There are significant group differences among different income groups in DPB score; those who had high income level had high DPB value.

TABLE 8.3

Comparisons between Groups Using One-Way Analysis of Variance

Variable	Groups	N (sample size)	Mean of Disaster Preparedness Behavior Score	SD	df	F-test	Sig
Age	21–45 years	95	1.51	8.09	1	0.528	0.469
	46–65 years	9	8.20	1.09			
Sex	Male	87	1.39	1.91	1	3.590	0.061
	Female	17	8.41	8.73			
Education	No education	12	1.00	8.21	2	5.453	0.005
	Primary	73	1.91	1.22			
	Secondary	19	0.5758	8.06			
Occupation	Farmer	49	0.55	1.21	4	6.815	0.000
	Labor	29	1.73	1.84			
	Businessman	7	8.84	8.60			
	Service holder	3	1.50	8.12			
	Housewife	16	3.18	3.09			
Marital status	Single	11	3.75	8.12	2	5.513	0.005
	Married	86	1.39	1.98			
	Others	7	0.000				
	>5,000	21					
Income level	5,000–10,000	45	1.0357	1.63137	1	0.803	0.373
	10,000–15,000	28	3.1290	8.43231		0.129	0.722
	15,000–20,000	7	0.000				
	>20,000	3					

8.5 VULNERABILITY ASSESSMENT IN RESPECT TO DISASTER PREPAREDNESS BEHAVIOR TO PROMOTE COMMUNITY CAPACITY FOR DISASTER RISK REDUCTION

8.5.1 COMMUNITY HAZARD RESILIENCE CAPACITY FOR DISASTER PREPAREDNESS

Hazard resilience is measured by the GOAL Toolkit Disaster Resilience Guidance. Resilience is the capacity of individuals, families, and communities to absorb, react, and recover from shocks and pressures in a timely and effective way without suffering long-term negative consequences.

The use of this toolbox for the assessment of response and recovery might be construed in a couple of different ways. It is possible to utilize the toolkit to provide an indicative percentage of resilience based on the evaluation of the critical elements of persistence. The checklist may also be used to assess the degrees of resilience that an individual has.

Table 8.4 represents the community hazard resilience capacity for disaster preparedness. The community resilience score is 40%, representing low resilience.

8.6 HAZARD RANKING AND RISK VALUATION

One focus group discussion was conducted in this area among respected and educated people to evaluate this risk zone on the basis of the previously presented characteristics.

8.6.1 ZONE 1: 800–1,000 SCORES: VERY HIGH-RISK ZONE

Something that happens often and has the potential to cause significant harm. The majority of the community considers it crucial. Requires immediate mitigation and planning for emergency response. Now it can be said that Borunpara is a high-risk zone, and for river erosion this village is highly vulnerable.

TABLE 8.4
Hazard Resilience Capacity for Disaster Preparedness

Percentage (%)	Level	Category	Description
0–20	1	Inadequate resistance	Few people are even aware of the problem, much alone are motivated to do something about it. Actions are restricted to those required by the emergency.
21–40	2	Insufficient resistance to stress	Knowledge and desire to handle the condition(s). Capability to operate (learning and abilities, physical, financial, as well as other sources) remains restricted. Most efforts are one-offs that don't add up to much over time.
41–60	3	Moderate durability	Invention of fixes and their actual use. Effectiveness and capability have both increased significantly. Several, extended approaches are being used.
61–80	4	Resilient	Integrity and consistency. The interventions are comprehensive, touching on all the major facets of the issue, and are interconnected within a logical long-term plan.
81–100	5	High resilience	All parties have adopted a "culture of security" in which DRR is intrinsic to all appropriate strategy, preparation, exercise, approach, and conduct.
The community resilience score		40%	
Category		Medium resilience	

Source: From GOAL Toolkit (2015).

8.7 DISASTER VULNERABILITY SCORE

Disaster vulnerability score is calculated by combining water logging, cyclone, flood, river erosion, and storm surge and salinity intrusion score using the equation: DVS = ([0.1 × Water logging score] + [0.15 × Cyclone score] + [0.225 × Flooding score] + [0.3 × River erosion score] + [0.1 × Storm surge] + [0.125 × Salinity score])/6.

Ratios of geography assigned to each hazard score category are shown in Table 8.6. The research area has a vulnerability score between 0.90 and 1.14. If the value is low, the risk is low, and if it is large, the risk is high. Around a third of the research region (33.70%) has a fire score between 0.25 and 0.5, indicating it is only moderately sensitive to catastrophe. Areas having disaster scores less than 0.25 make up 28.47% of the region, which are low vulnerability areas; areas with scores less than 0.75 make up 28.08% of the region, which are moderately high vulnerability areas. Areas in the highly vulnerable category of disaster make up 15.73% of the region.

8.8 DEMOGRAPHIC ANALYSIS

Table 8.7 shows community practices were compared to demographic factors such as gender, age, occupation, education, and income level. Males made up 83.6% of the total respondents, while females made up 16.34%. Only 70.2% of those polled had completed primary school. A quarter of the respondents had a medium level of income. The majority of those who took part were on a tight budget.

8.9 DISCUSSION

The original study knowledge score indicates that higher levels of knowledge provide many possibilities for warning systems to undergo training. Most responders from the community of Borunpara

TABLE 8.5
Risk Zoning of the Study Area

Serial	Event	Hazard Ranking	Vulnerability Indexing	Social Cost Prioritization	Risk Valuation
1	River erosion	9.5	9	8.5	726.75
2	Cyclone	6.5	6	7	273
3	Salinity	6	6	5	180
5	Storm surge	4	5	4	80
Total:					1,260

TABLE 8.6
Distribution of Areas According to Disaster Vulnerability Score Along with FEMA (2006) Score

Vulnerability Score	FEMA Score	Vulnerability Category	Frequency	Percentage (%)
0.91–0.96	0.25	Low vulnerability	20	28.47
0.97–1.02	0.5	Moderately low vulnerability	30	33.70
1.03–1.08	0.75	Moderate vulnerability	25	28.08
1.09–1.14	1	High vulnerability	14	15.73
Total				100

TABLE 8.7

Participant Characteristics of Demographic Factors

Variable	Groups	Frequency	Percentage (%)
Age	21–45 years	95	91.3
	46–65 years	9	8.6
Sex	Male	87	83.6
	Female	17	16.34
Education	No education	12	11.54
	Primary	73	70.2
	Secondary	19	18.27
Occupation	Farmer	49	45.79
	Labor	29	28
	Businessman	7	6.7
	Service holder	3	8.88
	Housewife	16	15.38
Marital	Single	11	10.57
status	Married	86	88.7
	Others (widowed, divorced)	7	6.7
Income	<5,000	21	20.2
level	5,000–10,000	45	43.27
	10,000–15,000	28	27
	15,000–20,000	7	6.7
	<20,000	3	8.88

lacked sufficient understanding across all aspects with regards to catastrophe. Determinants of DPB's demographic variables are affected by a community's human capital. This was very helpful to the community right from the outset, when it came to disseminating indigenous knowledge and skills for threat minimization. Only 10% of people in these trials had a DPB score of five or above, the threshold for being classified as prepared (Najafi et al., 2017), whereas our findings show that 90% of those who filled out the survey had a DPB score of five or below. More than half of the respondents (58.5%) also revealed that they had taken no measures to prepare for future calamities. Effects of demographic determinants on DPB were not properly found in abundance in the study area because of culture, values, and living in isolation. After a period during a disaster situation, demographic factors on preparedness behavior decrease in activity and, in extreme cases, break down as a result of rivalry and strife for the availability of external resources for rehabilitation. Thus, the role of human capital decreases at the disaster situation among the villagers or neighborhood. Due to this, adequacy of factors is not observed for disaster response, recovery, relief, and preparedness. From this statistical analysis, it is shown that the most effective demographic factors are sex, income, and education.

8.10 CONCLUSION

Past research has shown that higher-income families are more inclined to live in disaster-resistant neighborhoods and to own qualifying houses. The research found that those with lower socio-economic status had a lower likelihood of mitigating the negative effects of risks because they felt powerless to alter the results. In addition to examining DPB's relationship to singleness, the research looked at it in relation to marriage. People who are married are more likely to talk to one another about their security concerns. When compared to their single counterparts, married people tend to have clearer eyesight and more proactive thought processes. One definition of risk perception is an individual's estimation of the threat posed by a disaster or other hazard.

In this analysis, educational attainment was shown to be associated with DPB. As people's ability to think and learn, as well as their access to information, are all enhanced by their formal education, this may be one reason why it helps them be more prepared for catastrophes. Results showed that in the coastal area located to the south-west of Bangladesh, specifically in Borunpara village under Khulna district, marital status, income, and education are significant factors. Government authorities, emergency services, community leaders, and educators must prioritize low- and middle-income individuals, niche communities, and education to increase disaster preparedness.

REFERENCES

Afjal Hossain, M., Imran Reza, M., Rahman, S., & Kayes, I. (2012). Climate change and its impacts on the livelihoods of the vulnerable people in the southwestern coastal zone in Bangladesh. *Climate Change and the Sustainable Use of Water Resources*, 237–259.

Akter, S., & Rahman, M. M. (2023). Demographic determinants of disaster preparedness behavior among the inhabitants of Southwest Coastal Bangladesh. *Coastal Disaster Risk Management in Bangladesh: Vulnerability and Resilience*.

Cutter, S. L., & Finch, C. (2008). Temporal and spatial changes in social vulnerability to natural hazards. *Proceedings of the National Academy of Sciences*, *105*(7), 2301–2306.

Gencer, E. A., & Gencer, E. A. (2013). Natural disasters, urban vulnerability, and risk management: A theoretical overview. *The Interplay between Urban Development, Vulnerability, and Risk Management: A Case Study of the Istanbul Metropolitan Area*, 7–43.

Karim, M. F., & Mimura, N. (2008). Impacts of climate change and sea-level rise on cyclonic storm surge floods in Bangladesh. *Global Environmental Change*, *18*(3), 490–500.

Najafi, M., Ardalan, A., Akbarisari, A., Noorbala, A. A., & Elmi, H. (2017). The theory of planned behavior and disaster preparedness. *PLoS Currents*, *9*.

Najafi, M., Ardalan, A., Akbarisari, A., Noorbala, A. A., & Jabbari, H. (2015). Demographic determinants of disaster preparedness behaviors amongst Tehran inhabitants, Iran. *PLoS Currents*, *7*.

Primer, A. U. (n.d.). *United Nations Development Programme Capacity Development: A UNDP Primer. Public Awareness and Public Education for Disaster Risk Reduction: A Guide*. (n.d.).

Saha, C. K. (2015). Dynamics of disaster-induced risk in southwestern coastal Bangladesh: An analysis on tropical Cyclone Aila 2009. *Natural Hazards*, *75*(1), 727–754.

Sanderson, D. (2000). Cities, disasters and livelihoods. *Environment and Urbanization*, *12*(2), 93–108.

9 Evaluating Agricultural Drought in the Purulia District, West Bengal (India)
An Earth Observation Perspective

Somenath Goswami, Sunil Mahato, Santoshi Mahato, Arnab Kundu, Brijmohan Bairwa, K. K. Chattoraj, and Azizur Rahman Siddiqui

9.1 INTRODUCTION

According to Hagman (1984), drought is a complex and poorly understood natural hazard that can have significant impacts on the environment, society, and the economy. The frequency and resulting losses associated with droughts have increased globally in recent decades, leading to a growing interest in understanding them (Wilhelmi and Wilhite, 2002; Keyantash and Dracup, 2004; Zhou et al., 2013). To define drought, Wilhite and Glantz (1985) proposed four fundamental approaches: meteorological, hydrological, agricultural, and socio-economic. The first three approaches focus on quantifying drought as a physical phenomenon. While drought refers to a deviation from normal conditions, a water deficit is an absolute term used to describe insufficient water availability (Obi Reddy et al., 2001; Alshaikh, 2015). According to Mishra and Singh (2010), a drought is defined as a prolonged absence of precipitation that causes a water shortage that considerably harms crops and other living and non-living objects. In plain English, it is the prolonged lack of water in a location that is deemed "not normal" in comparison to its typical conditions (Mishra and Singh, 2011). There are various drought indices that utilize different data sources. Some indices rely on ground-based data, while others incorporate energy-balance models or a combination of satellite data and energy-balance models (Heim, 2002; Nagarajan, 2010). These indices serve the purpose of assessing the meteorological, agricultural, hydrological, and socio-economic aspects of drought. Drought is familiar as a significant natural disaster that impacts the environment, economy, and global sustainability efforts (Peters et al., 2002). In contrast, quick and inexpensive drought monitoring techniques come from remote sensing from space (Alshaikh, 2015). Agriculture is a main activity that is influenced by environmental elements like soil (edaphic factor), height, slope, and aspect of the landform, as well as temperature, rainfall, and humidity (climatic factors). Due to the region's unfavorable climatic conditions, which typically satisfy the criteria of so-called low rainfall, drought susceptibility, extreme weather and climate, and unfavorable climate for agriculture, the Puruliya district in the south-western part of West Bengal is regarded as the state's most underdeveloped and backward region. The evidence of backwardness is evident everywhere when the average per capita income, the percentage of people living in poverty, crop yield, and cropping intensity are considered (Mishra et al., 2012). The final one tracks the effects of water scarcity as it propagates through socio-economic systems and addresses drought in terms of supply and demand (Wilhite and Glantz, 1985). An agricultural drought is characterized by a lack of sufficient water for plant growth, whereas a meteorological drought is brought on by a drop in precipitation (Patel et al., 2012; Patel and Yadav, 2015). A hydrological drought is characterized by a

DOI: 10.1201/9781003377825-11

shortage of surface and subterranean water supplies. A socio-economic drought is defined as one in which there is insufficient water to meet the demands of particular economic products, frequently in conjunction with another type of drought (Zhang and Jia, 2013; Kundu et al., 2021). One of the world's nations most susceptible to drought is India (Mishra and Singh, 2010). About 70% to 90% of India's annual mean precipitation falls during the south-west monsoon, which lasts from June to September (Kumar et al., 2013). Droughts may develop from a shortage of water due to monsoon rainfall failure. India has experienced droughts at least every three years for the past 50 years (from 1960 to 2020) (Mishra and Singh, 2010). Droughts during the monsoon have a serious impact on the country's economy, water supplies, and agricultural productivity. As an example, the 2002 drought led to a decline of approximately 1% in the gross domestic product of the country, while the 1987 and 1988 droughts in India resulted in a loss of 36 million tonnes of food production (Gadgil et al., 2003). Analyzing the geographical and temporal variations of droughts in detailed and precise temporal and spatial resolutions is essential for effective and sustainable drought mitigation and water resource management. Due to little rainfall and unfavorable soil conditions, West Bengal's lateritic Paschim Medinipur, Bankura, Puruliya, and Birbhum regions are regularly impacted by droughts. About 11,594 km^2 in these four West Bengal districts, in 36 community blocks, have been designated as drought-prone zones, with daily rainfall intensities ranging from 23.51 mm to less than 21.51 mm (West Bengal State Action Plan on Climate Change, 2010). According to Banik et al. (2002), the south-west monsoon's fluctuating rainfall from June to September is the primary factor affecting agricultural production in the Puruliya district. Despite having substantial yearly rainfall (1,331 mm), the western and south-western portions of the Puruliya upland experience drought (Ghosh and Jana, 2018). In years of below average rainfall, this district impacted by the drought lacks a sufficient supply of water, even drinkable water. In the summer, the majority of tanks, streams, and other surface water sources dry up, leaving only groundwater as a source of water supply, which is frequently discovered to contain fluoride (Bhattacharya and Chakrabarti, 2011). Most of the region lacks adequate irrigation facilities, making it difficult to grow more than one crop (Nag, 1998). The current study has been designed in this context to analyze different types of droughts in the district with improved temporal and spatial precision. A working theory has been created that predicts the sequential emergence of drought in the region, starting with meteorological drought and progressing through hydrological drought, agricultural drought, and socio-economic drought. The Vegetation Condition Index (VCI), a well-known drought estimate, was used to test the hypothesis and was repeatedly proved to be helpful for tracking agricultural drought (Gitelson et al., 1998; Kogan, 1997; Unganai and Kogan, 1998; Dutta et al., 2013, 2015). Agricultural drought was researched by Shukla et al. in 2014. The use of the Temperature Condition Index (TCI) assumes that a drought event will decrease soil moisture, causing thermal stress on the land surface, and that the land surface temperature (LST) will be higher in the drought year than it will be in the same month of normal years. According to Singh et al. (2003), when crops are growing, high LST over the course of the growth season indicate adverse weather, such as drought. Further study on crop productivity in a region like Puruliya, where the majority of the population relies on rain-fed subsistence agriculture, is important. Therefore, the goal of this study is to characterize the district's general drought and the agricultural drought in particular using a number of indicators (precipitation, vegetation, and groundwater level) assembled by multi-satellite remote sensing.

9.2   STUDY AREA

In the Indian state of West Bengal, there is a district known as Puruliya. It is located in the western region of the state and shares a western boundary with Jharkhand. The district of Puruliya is located roughly at 23.34°N latitude and 86.36°E longitude. Approximately 6,259 km^2 (2,416 mi^2) make up the district. Hills, woodlands, and plains make up its varied landscape, which is wellknown. The Ayodhya and Dalma hill ranges, which cut through Puruliya, dominate the region's topography. The district is bounded by the Subarnarekha River and its tributaries, including Kangsabati and Kumari, which provide water resources for irrigation and other uses. Puruliya's climate is often characterized by warm summers, cold winters, and moderate monsoon rainfall. The majority of the people in the Puruliya district live in rural areas, where agriculture is their main occupation.

The area has abundant mineral resources, including coal, limestone, and granite reserves, which support the regional economy. Overall, the geographical region around Puruliya offers an attractive environment, a variety of natural aspects, and a distinctive fusion of cultural heritage, making it a desirable location for travelers and researchers alike. Puruliya has a tropical savanna climate (Aw) according to the Köppen classification, and the monsoon season is when it rains the most heavily there. The average summer temperature is 40°C (104°F), and the average winter temperature is 10°C (50°F). The district's climate is characterized by rainfall. The district's primary rainfall source is the south-west monsoon. It typically rains between 1,100 and 1,500 mm each year (43 and 59 in). In the monsoon season, the relative humidity is high, ranging from 75% to 85%. But in the sweltering summer, it drops to 20% to 35%.

The district of Puruliya is crossed by numerous rivers. The most significant ones are Kangsabati, Kumari, Silabati (Silai), Dwarakeswar, Subarnarekha, and Damodar. Even though the district is crossed by multiple rivers, the undulating geography causes 50% of the water to run off. A number of smaller dams, including Futiyary, Murguma, Pardi, Burda, and Gopalpur, are also present and are mostly utilized to irrigate agricultural fields. One of Puruliya's well-known and well-liked waterbodies is Saheb Bandh. It is situated in the center of the town of Puruliya. From December to March, it serves as a refuge for migrating birds that arrive from Bangladesh, Burma, Sindh, and Baluchistan. Due to the topography's irregularities, over half of the rainfall is lost as runoff. Most of the district is covered in leftover soil created by bedrock weathering.

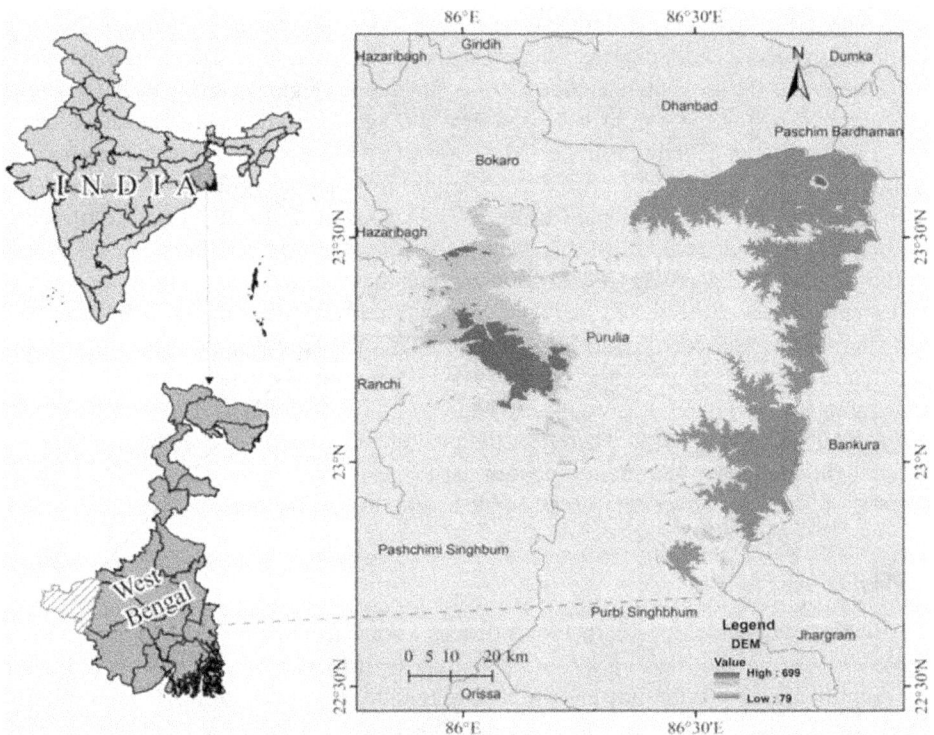

FIGURE 9.1 Location map of the study area.

9.3 BASIS FOR SELECTION OF THE STUDY AREA

Study region Puruliya was selected based on comparison to the rest of the districts identified as 'drought prone' underdeveloped areas in the state. Analyzing factors like per capita income, farming intensity, the percentage of land that is fallow both seasonally and permanently, etc., clearly demonstrates how backward the country is. Its backwardness is typically attributed to the country's diverse physical environment, especially the soil, topography, and weather and climate. The agricultural sector of the district in Puruliya has been hardest hit by the region's severe weather in the state. Therefore, the Puruliya district has been chosen for the study because of the district's agricultural and other minor benefits.

9.4 MATERIALS AND METHODS

9.4.1 Datasets

9.4.2 Shuttle Radar Topography Mission-Digital Elevation Model (DEM)

The Shuttle Radar Topography Mission (SRTM) was a project that collected detailed elevation data from almost all parts of the world using a specially modified radar system onboard the Space Shuttle Endeavour during a mission in February 2000. This data allowed for the creation of a comprehensive and high-resolution digital map of the earth's topography. The heights (elevation) recorded by SRTM were relative to the geoid and were adjusted using geoid separation values to ensure accuracy. The elevation models generated from SRTM data are used in geographic information systems. The United States Geological Survey provides a global digital elevation model with a resolution of 30 m available on their website. In 2014, the United States government announced that the highest-resolution SRTM data would be made accessible to the public, and we used this data to create the location map for our research area.

9.4.3 National Oceanic and Atmospheric Administration's Advanced Very High-Resolution Radiometer

The use of indices based on the National Oceanic and Atmospheric Administration's Advanced Very High-Resolution Radiometer (NOAA-AVHRR), such as the Normalized Difference Vegetation Index (NDVI), VCI, TCI, and Vegetation Health Index (VHI), has been successful in detecting and measuring vegetative drought and assessing its sensitivity to drought (Tarpley et al., 1984; Bhat, 2006; Brown et al., 2008; Barati et al., 2011; Fensholt and Proud, 2012; Do and Kang, 2014; Vrieling et al., 2014; Kundu et al., 2016). In this study, we calculated the VCI using long-term satellite data from NOAA-AVHRR GIMMS (Global Inventory Modelling and Mapping Studies) spanning the years 1984 to 2003. The AVHRR sensor captures data from different channels that are sensitive to various aspects of solar radiation across different spectrums. The GIMMS dataset, which includes NDVI, VCI, TCI, and VHI, has a spatial resolution of 1.1 km at its center and has undergone corrections for factors like cloud cover, sensor degradation, satellite drift, and volcanic aerosols. We trimmed the dataset to match the boundaries of our study region, converted it to image format (.img), and stacked the 15-day composite images from the 20-year period of 1984 to 2004.

9.4.4 Climate Hazards Group InfraRed Precipitation with Station Data

The Climate Hazards Group InfraRed Precipitation with Station (CHIRPS) data is a rainfall dataset that spans over 30 years and covers a large portion of the globe. By combining in-situ station data with satellite imagery at a resolution of 0.05°, CHIRPS produces gridded time series of rainfall (Dinku et al., 2018). This dataset is particularly valuable for trend analysis and monitoring seasonal drought conditions (Kundu et al., 2020).

9.4.5   CROP DATA

Block-wise crop data were brought from the agriculture office of Puruliya district. These data show the production per hectare of rabi and kharif in 2017 and 2018. In this crop data, block-wise production per hectare of about 32 types of crops of two seasons is found.

9.4.6   METHODS

9.4.7   NORMALIZED DIFFERENCE VEGETATION INDEX

For this study, we used a method called NDVI to monitor agricultural drought over a long period of time. NDVI is a valuable tool that helps us track when droughts begin, how they progress, and how widespread they become. Many previous studies have shown the effectiveness of NDVI in various applications, such as understanding vegetation patterns, classifying different types of vegetation, and mapping land cover across large areas. NDVI is particularly useful for monitoring drought, assessing the health of plants, predicting agricultural conditions, and estimating crop yields. The basic idea behind NDVI is that healthy green leaves reflect a lot of near-infrared (NIR) radiation and absorb a significant amount of red visible (R) radiation. This is due to the specific structure of healthy leaves. However, when plants are unhealthy or experiencing water stress, this relationship changes, and they reflect less NIR radiation and absorb more R radiation. We can use this information to measure the health of vegetation using Equation 9.1:

$$NDVI = (NIR-R)/(NIR + R) \qquad (9.1)$$

NDVI is a calculation that compares the amount of Rand NIR light reflected by plants. Its values range from −1 to 1, with different types of landscapes showing different ranges of values. For example, areas with snow, sand, or barren rock have NDVI values below 0.1, while rainforests can have values between 0.6 and 0.8. NDVI has become widely accepted as a measure for monitoring how dry the land is, assessing soil moisture, and determining the health of vegetation. However, there are factors that can introduce errors into NDVI measurements, such as atmospheric interference and changes in satellite performance. It can be challenging to separate the effects of weather from the actual condition of vegetation when using NDVI. Scientists have used NDVI based on NOAA-AVHRR data to classify and quantify vegetative drought, as well as to study how sensitive different areas are to drought.

9.4.8   TEMPERATURE CONDITION INDEX

TCI refers to the shift in thermal condition as determined by brightness temperature from the bands of the NOAA-thermal AVHRR. TCI data may be analyzed to monitor subtle alterations in the vegetation's health brought on, in particular, by thermal stress (Kogan, 1995, 2001, 2002). The TCI can also be expressed as follows (Equation 9.2):

$$TCI = 100 \: [(BT_{max} - BT)/(BT_{max} - BT_{min})] \qquad (9.2)$$

In this study, the seasonal average of weekly brightness temperature (BT), along with its minimum (BT_{min}) and maximum (BT_{max}) values over multiple years, were considered. The NOAA-AVHRR 16-km, seven-day composite TCI for the Puruliya district during 2017 and 2018 was used to assess the thermal conditions as anomalies relative to a 25-year climatology. This climatology was derived based on biophysical and biological principles. The TCI is computed from radiance measurements obtained from the AVHRR sensor in the 10.3–11.3 μm range. These measurements are converted to brightness temperature (BT) using a lookup table. Non-linear adjustments, following the methodology of Weinreb et al. (1990), have been applied to the BT values. Furthermore, measures have been taken to effectively remove high-frequency noise, thus improving the accuracy of the TCI (Kogan, 1997).

9.4.9 Vegetation Condition Index

The VCI is a measure that helps us understand how the condition of vegetation changes over time. It ranges from extremely poor to ideal and is represented as a percentage between 0 and 100. To calculate the VCI, we use the NDVI and the Equation 9.3. The NDVI, $NDVI_{min}$, and $NDVI_{max}$ values are based on the average NDVI over a season, as well as the minimum and maximum values observed over several years. In this study, we used the NOAA-AVHRR dataset with a resolution of 16 km and a seven-day composite to obtain the VCI for the Puruliya district in the years 2017 and 2018. The VCI serves as an indicator of moisture status, showing how the NDVI compares to a 25-year average based on scientific principles related to the environment and ecosystems.

$$VCI = 100 \, [(NDVI - NDVI_{min})/(NDVI_{max} - NDVI_{min})] \tag{9.3}$$

9.4.10 Vegetation Health Index

It is asserted that the VHI accurately depicts the total health of the plants by combining calculations of moisture and heat conditions. Using VCI and TCI, the following formula is calculated (Equation 9.4):

$$VHI = \alpha \, VCI + (1 - \alpha) \, TCI \tag{9.4}$$

where α is a coefficient that establishes how much each of the two indices contributes, and it is usually assumed to be 0.5. Here, we used a substitute for describing vegetative health: the NOAA-AVHRR 16-km, seven-day composite VHI over Puruliya district for the years 2017 and 2018. For the purposes of classifying vegetative drought, Kogan (2002) identified five distinct VHI classes (Table 9.1).

9.4.11 Rainfall Anomaly Index

The Rainfall Anomaly Index (RAI) serves as a valuable tool for identifying the occurrence of meteorological drought in a given year (Van Rooy, 1965). The study revealed a close similarity in the predictions of drought occurrence in 2018 between RAI, a meteorological drought index, and VCI, a remote sensing-based index. Among various other drought monitoring indicators, rainfall anomaly emerges as the most effective and straightforward measure. To calculate rainfall anomalies, the average long-term rainfall for the research area was employed. Annual rainfall data for the Puruliya

TABLE 9.1
Vegetation Health Index Classes

Classes	Vegetation Health Index Value
Extreme drought	<10
Severe drought	≥10 and <20
Moderate drought	≥20 and <30
Mild drought	≥30 and <40
No drought	≥40

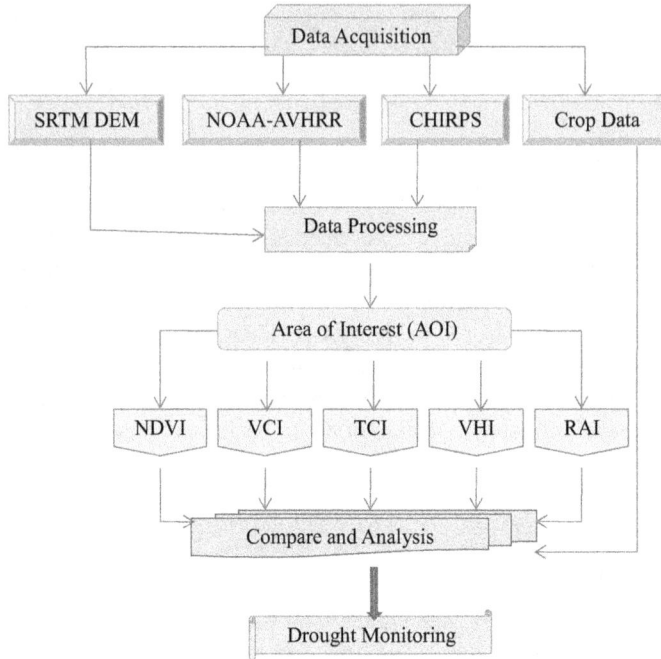

FIGURE 9.2 Flowchart of methodology.

district were collected from 2017 to 2018. Drought years are identified based on low rainfall totals and a negative deviation from the seasonal norm, indicating a departure from the expected precipitation patterns. The following formulas were used to calculate the RAI (Equation 9.5):

$$\text{RAI} = (R - \mu)/\sigma \tag{9.5}$$

where R is the rainfall, μ is long-term average rainfall, and σ is standard deviation.

The following methodology has been employed for this study (Figure 9.2).

9.5 RESULTS AND DISCUSSION

9.5.1 NORMALIZED DIFFERENCE VEGETATION INDEX

The NDVI is a simple way to measure the health and density of vegetation observed in satellite images. It involves calculating the difference between the visible Rand NIR bands, taking advantage of the unique way healthy plants reflect light. This difference is represented as an NDVI value ranging from −1 to 1. The NDVI equation compares the intensity of reflected light in the red and infrared spectrum and divides it by the total intensity to determine the vegetation density (NDVI) at a specific location in the image.

In the most recent study, long-term NDVI for various rain-fed fortnights (June—September) was employed. Despite the fact that many authors have utilized NDVI to analyze drought, several studies have suggested using VCI rather than NDVI alone. The various differences in vegetation health during drought and wet areas were identified using the stacked NDVI layers. The years 2017 and 2018 were picked for their distinct NDVI characteristics in order to evaluate the performance of NOAA-AVHRR generated NDVI across two consecutive years.

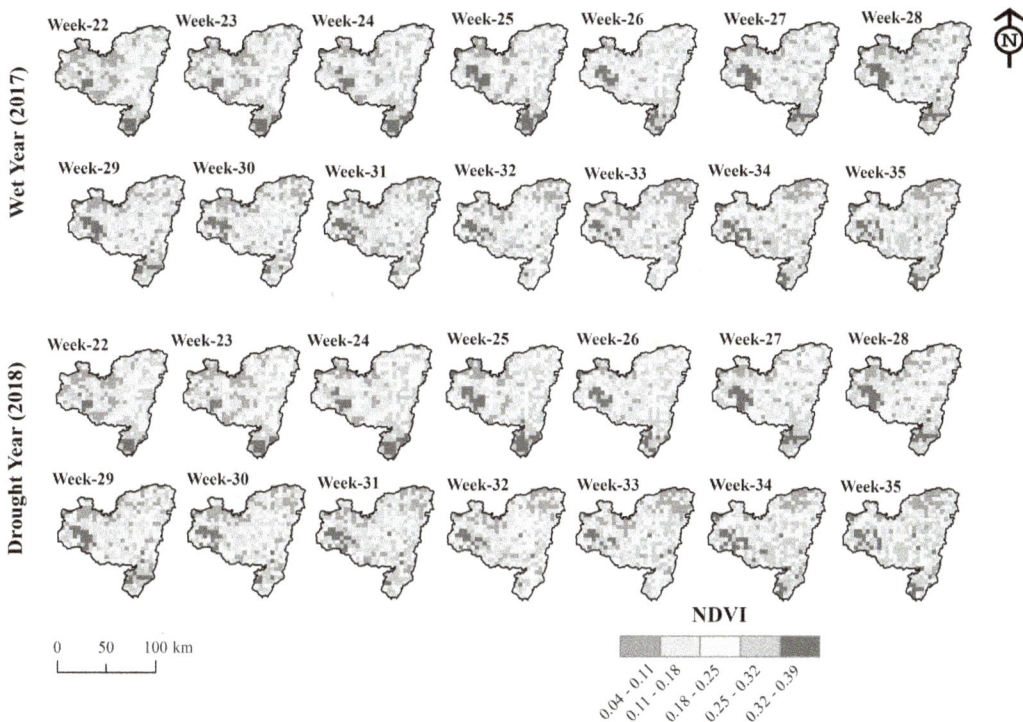

FIGURE 9.3   June to September weekly Normalized Difference Vegetation Index for wet year (2017) and drought year (2018).

When comparing the NDVI of 2017 and 2018, the Puruliya district showed a noticeable improvement. We get to the conclusion that the NDVI value in 2018 is lower than in 2017 (Figure 9.3). As we can see, NDVI values were lower during the first, second, and third weeks of June, third and fourth weeks of July, and first, second, and third weeks of August 2017 compared to the last week of June, the first week of July, the third and fourth weeks of August, and the first week of September. The first, second, and third weeks of June, the entire month of July, and the first, second, and third weeks of August in 2018 all had low NDVI values, while the last week of June, the third and fourth weeks of August, and the first week of September had higher NDVI values. The conclusion that follows is that 2017 was a wet year and 2018 was a dry year (Figure 9.3). The variety in the NDVI reveals regional heterogeneity in the health of the vegetation. The uneven distribution of monsoonal rainfall, which is the main driver of NDVI for this type of diverse geographical variability.

9.5.2   TEMPERATURE CONDITION INDEX

Numerous drought indicators based on remote sensing have been created to show how severe the drought is. One of the most used indexes is the TCI. The long-term TCI for various wet seasons (June to September) is used in the current study. Although many authors have utilized TCI in their studies on drought, many studies still recommend utilizing a TCI in conjunction with NDVI and TCI to assess vegetation in drought situations that have an influence on agriculture. In order to evaluate the efficacy of TCI derived from NOAA-AVHRR thermal bands, we selected two consecutive years, 2017 and 2018, which exhibited distinct TCI patterns (refer to Figure 9.4). TCI is utilized to assess the impact of extreme heat and high moisture on vegetation, providing insights into vegetation stress levels.

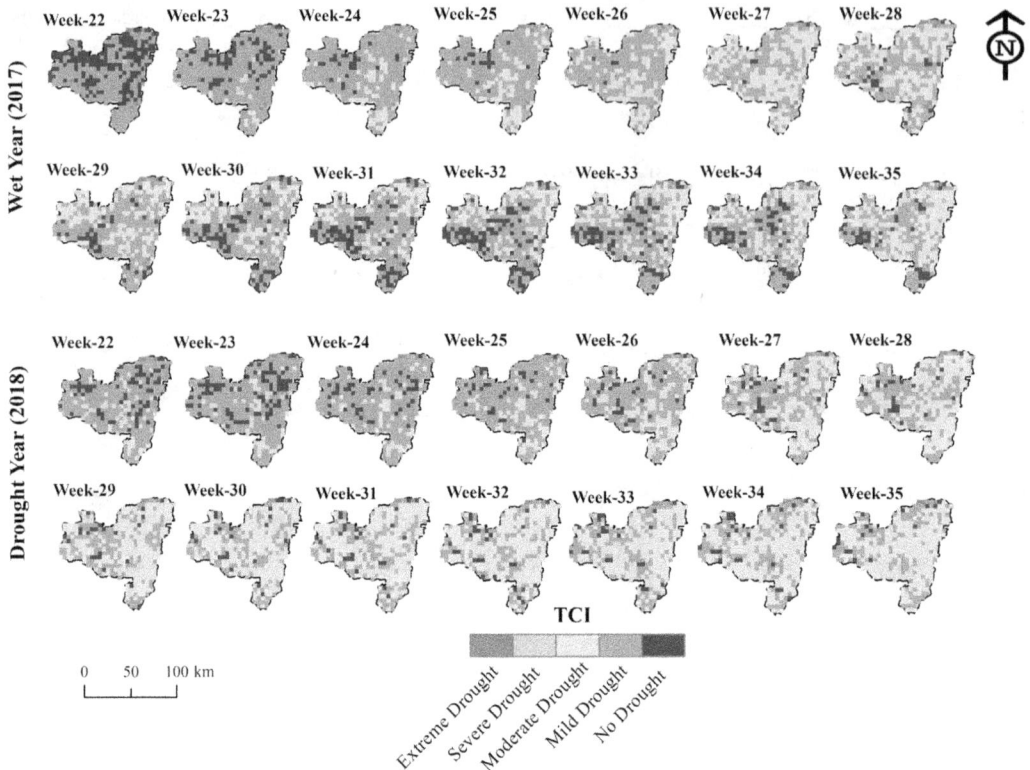

FIGURE 9.4 June to September weekly Temperature Condition Index for wet year (2017) and drought year (2018).

Conditions are assessed with respect to the maximum and minimum temperatures and modified in order to account for the varied vegetation responses to temperature. When compared to all other weeks in TCI 2018, the prevalence of drought is higher in weeks 29, 33, 34, 35, and 36 (Figure 9.4). According to the TCI images from the 2018 drought year, Puruliya's western half is more severely impacted by the drought than its eastern region. When compared to 2017, the TCI image for 2018 is found to be lower, indicating that 2017 was a wet year and 2018 is a dry year.

9.5.3 VEGETATION CONDITION INDEX

Several studies recommend using a VHI rather than a long-term VCI, even though many authors have used long-term VCI for studies on drought. In the current study, the long-term VCI is used for different aspects of the rainy season (June to September) along with NDVI and TCI to assess vegetation in drought scenarios that have an influence on agriculture. VCI layers were presented in Figure 9.5 to show how vegetation health changed in drought and wet years. The years 2017 and 2018 were chosen because of their unique VCI properties in order to assess the performance of NOAA-AVHRR generated VCI. The weeks 29, 30, 31, 32, 33, and 34 observe the highest degrees of drought, according to the VCI 2018 statistics.

Overall the VCI image of the 2018 dry year reveals that Puruliya's western section is more affected by drought than its eastern region. The year 2018 is dry, whereas 2017 is a wet year since there is less VCI discovered in 2018 than there was in 2017. That is caused by monsoon uncertainty in the majority of areas.

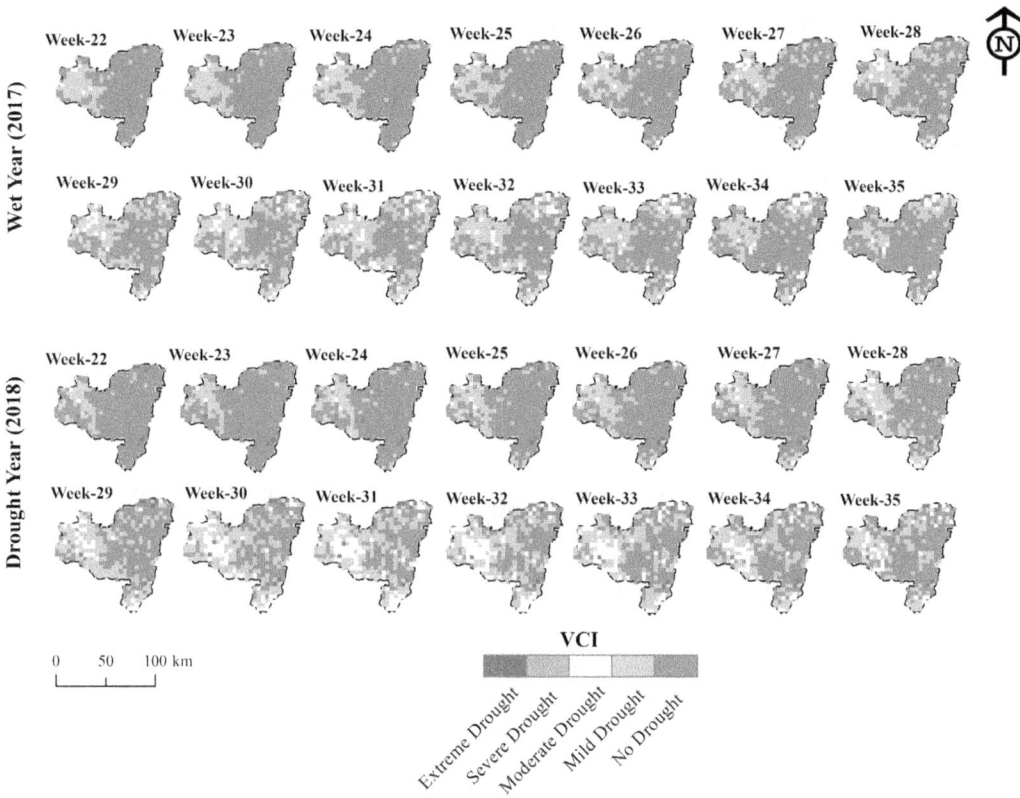

FIGURE 9.5   June to September weekly Vegetation Condition Index for wet year (2017) and drought year (2018).

9.5.4 VEGETATION HEALTH INDEX

In this study, we used a remote sensing indicator called the VHI to examine drought conditions. The VHI combines two other indicators, the NDVI and the LST. Using a formula with a constant value of 0.5, we calculated the VHI by combining the VCI and the TCI. We conducted tests to see how the VHI values are influenced by the LST and NDVI. Based on the VHI values, we categorized the severity of drought into four levels: extreme, severe, moderate, and mild, as well as a no drought category. The VHI gave us a comprehensive understanding of drought by considering both the stress on vegetation and the temperature. We used it to monitor the condition of vegetation throughout the study (Figure 9.6).

9.5.5 RAINFALL ANOMALY INDEX

In contrast to the typical rainfall of 408.44 mm and 311.46 mm, the Puruliya district experienced rainfall of 375.92 mm and 300.08 mm from June to September of 2017 and 2018, respectively. The July rains were helpful in hastening the kharif sowing. The pattern of rainfall over time shows that the first week of July and September, as well as the middle of June and August, see very substantial precipitation. On the other hand, June, July, the first week of August, and the latter week of August saw the lowest rainfall totals (Figures 9.7 and 9.8). Early June, the first week of August, and the final week of August 2018 also showed the highest rainfall totals. The least amount of rainfall occurs during the first two weeks of July, mid-August, and the first week of September (Table 9.2).

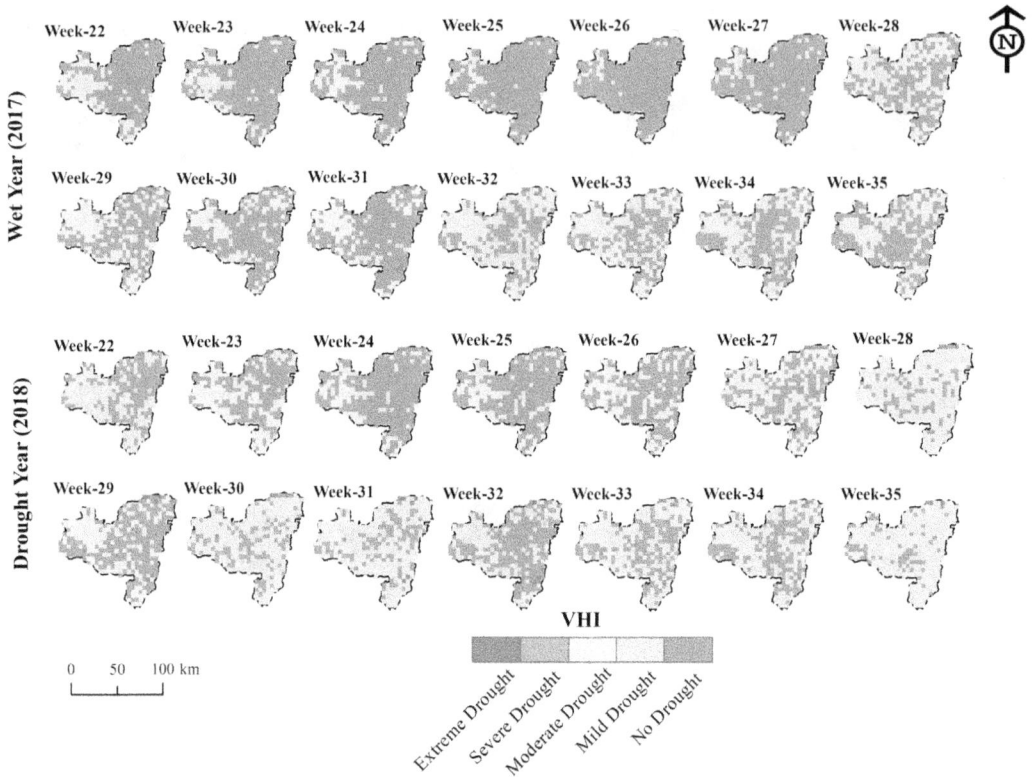

FIGURE 9.6   June to September weekly Vegetation Health Index for wet year (2017) and drought year (2018).

TABLE 9.2
Actual and Normal Rainfall (2017, 2018)

	2017			2018	
Weeks	Actual Rainfall (mm)	Normal Rainfall (mm)	Weeks	Actual Rainfall (mm)	Normal Rainfall (mm)
22	1.8	5.6	22	7.69	3.4
23	3.51	7.12	23	8.36	7.79
24	16.85	6.07	24	11.09	7.62
25	6.83	8.69	25	12.67	11.65
26	9.47	10.32	26	13.28	16.63
27	10	10.42	27	13.4	5.05
28	8.4	9.72	28	11.8	6.11
29	60.22	7.1	29	12.26	17.21
30	10.11	10.82	30	12.77	11.95
31	5.81	10.88	31	10.95	9.26
32	18.31	10.02	32	11	5.67
33	5.01	8.34	33	11.04	9.03
34	7.97	10.41	34	10.91	7.95
35	2.42	9.9	35	10.5	9.09

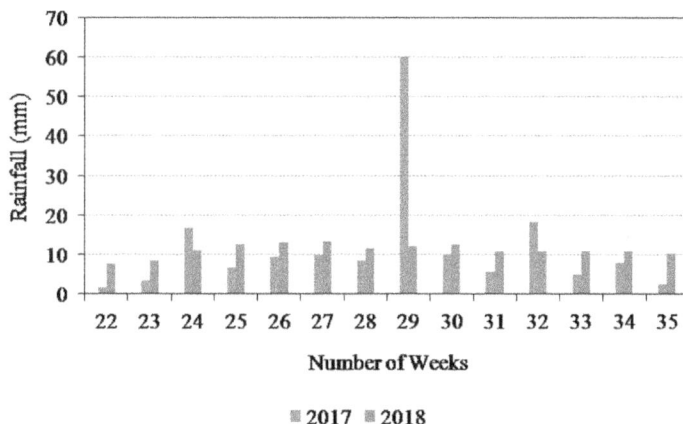

FIGURE 9.7 June to September weekly rainfall in Puruliya

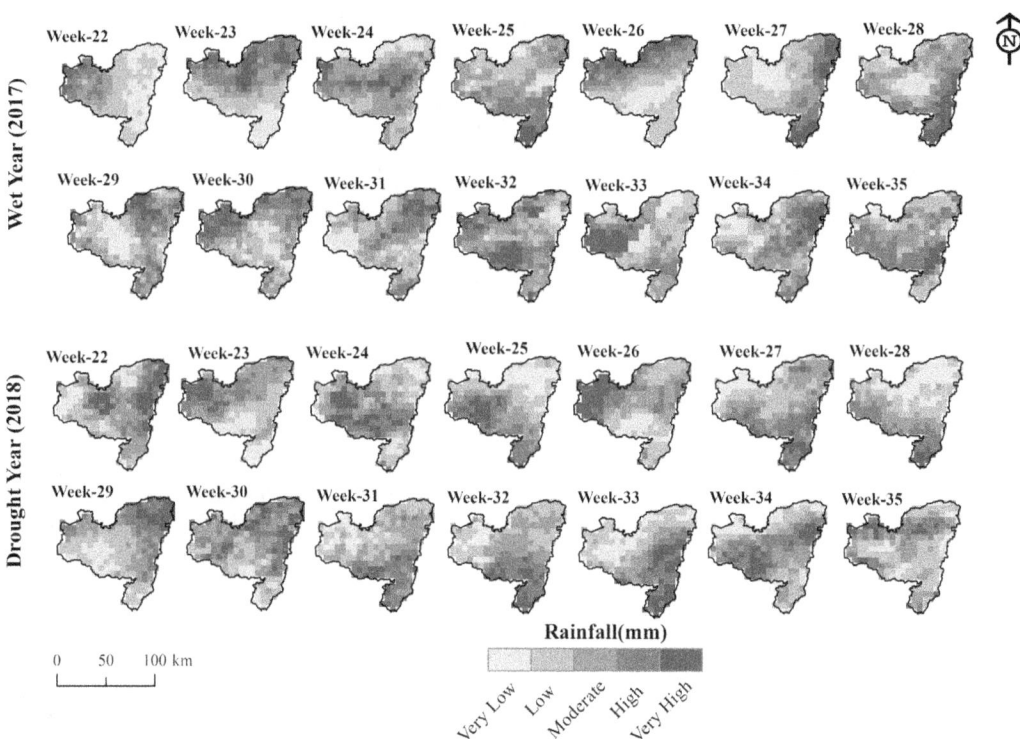

FIGURE 9.8 June to September weekly Rainfall Anomaly Index for wet year (2017) and drought year (2018).

The excessive rain that was recorded in Puruliya during June helped with the kharif sowing process and the growth of early-sown kharif crops. Lack of rain in the months of July and August affects the sustenance of crops. Crop species affect stress levels in different ways.

9.5.6 CROP DATA

Data on seasonal crops have been obtained from the Puruliya Agriculture Office. We noticed from this agricultural data that during 2017 there were 312,437 hectares of kharif crops production,

TABLE 9.3

Coverage of Kharif and Rabi Crops Production in Puruliya (2017 and 2018)

Year	Rabi (tonnes)	Kharif (tonnes)
2017	50,338	312,437
2018	30,130	214,497

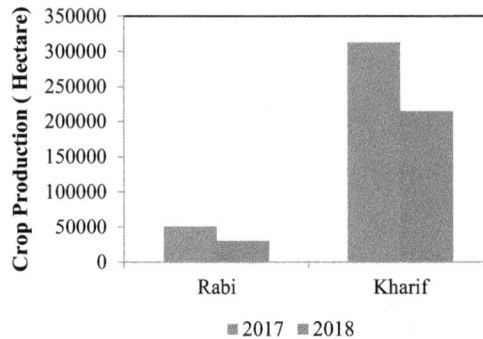

FIGURE 9.9   Crop production (rabi and kharif) for wet and dry years.

50,338 hectares in rabi, and 214,497 hectares of kharif crops production, 50,338 hectares in rabi in 2018. When rainfall and crop production are compared, it is evident that 2017's high rainfall also resulted in high crop production, while 2018's low rainfall caused low crop production (Figure 9.9, Table 9.3). As a result, whereas it did not in 2017, people's financial situation deteriorated in 2018. This indicates that 2017 is a wet year, and 2018 is a drought year.

9.6   CONCLUSION

The identification and mapping of droughts primarily rely on remote sensing–based indicators of vegetation health, such as NDVI, TCI, VCI, and VHI. These indicators provide valuable insights into estimating and predicting crop conditions and potential output across a region. In our study, we employed these techniques to monitor agricultural drought in the Puruliya district of West Bengal, India, specifically during the months of August and September in the kharif season of two different monsoon years: 2017 and 2018. Our findings revealed significantly higher TCI values in the wet monsoon year of 2017 compared to the drought year of 2018. In other words, when compared to 2017, the year 2018 exhibits significantly higher thermal stress. By comparing the VCI between 2018 and 2017, it is seen that the VCI value of 2018 is lower than that of 2017, i.e. the vegetation condition is very bad in Puruliya in 2018 as compared to 2017. Vegetation health in 2017 is very good compared to 2018 in Puruliya. Vegetation health in 2017 is very good due to rainfall in 2017. Comparing the NDVI maps of 2017 and 2018 shows that the vegetation growth of 2017 is very good compared to 2018. Vegetation growth in 2018 is very poor, due to low 2018 rainfall. The rainfall data from CHIRPS show that the rainfall of 2017 is much higher than that of 2018. The maximum rainfall of 2017 is 60.22 mm, and the maximum rainfall of 2018 is 13.4 mm weekly. By looking at seasonal crops data for kharif and rabi seasons from the agriculture office of Puruliya and comparing between 2018 and 2017, it is understood that because of higher rainfall in 2017, crops production is higher; in 2018 seasonal crops production is lower due to less rainfall. So people's financial condition worsened in 2018. From this we can say that 2017 is a wet year, and 2018 is a drought year.

REFERENCES

Alshaikh, A.Y. 2015. Space applications for drought assessment in Wadi-Dama (West Tabouk), KSA. The Egyptian Journal of Remote Sensing and Space Sciences 18, S43–S53.

Banik, P., Mandal, A., Rahman, M.S. 2002. Markov chain analysis of weekly rainfall data in determining drought-proneness. Discrete Dynamics in Nature and Society 7, 231–239.

Barati, S., Rayegani, B., Saati, M., Sharifi, A., Nasri, M. 2011. Comparison the accuracies of different spectral indices for estimation of vegetation cover fraction in sparse vegetated areas. The Egyptian Journal of Remote Sensing and Space Sciences 14, 49–56.

Bhat, G.S. 2006. The Indian drought of 2002: a sub-seasonal phenomenon? Quarterly Journal of the Royal Meteorological Society 132, 2583–2602.

Bhattacharya, H.N., Chakrabarti, S. 2011. Incidence offluoride in the groundwater of Purulia District, West Bengal: a geo-environmental appraisal. Current Science 101(2), 152–155.

Brown, J.F., Wardlow, B.D., Tadesse, T., Hayes, M.J., Reed, B.C. 2008. The vegetation drought response index (VegDRI): a new integrated approach for monitoring drought stress in vegetation. GIScience and Remote Sensing 45, 16–46.

Dinku, T., Funk, C., Peterson, P., Maidment, R., Tadesse, T., Gadain, H., et al. 2018. Validation of the CHIRPS satellite rainfall estimates over eastern Africa. Quarterly Journal of the Royal Meteorological Society 144, 292–312.

Do, N., Kang, S. 2014. Assessing drought vulnerability using soil moisture-based water use efficiency measurements obtained from multi-sensor satellite data in Northeast Asia dryland regions. Journal of Arid Environments 105, 22–32.

Dutta, D., Kundu, A., Patel, N.R. 2013. Predicting agricultural drought in eastern Rajasthan of India using NDVI and standardized precipitation index. Geocarto International 28, 192–209.

Dutta, D., Kundu, A., Patel, N.R., Saha, S.K., Siddiqui, A.R. 2015. Assessment of agricultural drought in Rajasthan (India) using remote sensing derived vegetation condition index (VCI) and standardized precipitation index (SPI). The Egyptian Journal of Remote Sensing and Space Sciences 18, 53–63.

Fensholt, R., Proud, S.R. 2012. Evaluation of earth observation based global long-term vegetation trends—comparing GIMMS and MODIS global NDVI time series. Remote Sensing of Environment 119, 131–147.

Gadgil, S., Vinayachandran, P.N., Francis, P.A. 2003. Droughts of Indian summer monsoon: role of clouds over the Indian Ocean. Current Science 85, 1713–1719.

Ghosh, P.K., Jana, N.C. 2018. Groundwater potentiality of the Kumari River Basin in drought-prone Purulia upland, Eastern India: a combined approach using quantitative geomorphology and GIS. Sustainable Water Resources Management 4, 583–599.

Gitelson, A.A., Kogan, F., Zakarin, E., Spivak, L., Lebed, L. 1998. Using AVHRR data for quantitative estimation of vegetation conditions: calibration and validation. Advances in Space Research 22, 673–676.

Hagman, G. 1984. Prevention better than cure: report on human and natural disasters in the third world. Swedish Red Cross, Stockholm.

Heim, R.R. 2002. A review of twentieth-century drought indices used in the United States. Bulletin of the American Meteorological Society 83, 1149–1166.

Keyantash, J.A., Dracup, J.A. 2004. An aggregate drought index: assessing drought severity based on fluctuations in the hydrologic cycle and surface water storage. Water Resources Research 40, 1–14.

Kogan, F.N. 1995. Application of vegetation index and brightness temperature for drought detection. Advances in Space Research 15, 91–100.

Kogan, F.N. 1997. Global drought watch from space. Bulletin of the American Meteorological Society 78, 621–636.

Kogan, F.N. 2001. Operational space technology for global vegetation assessment. Bulletin of the American Meteorological Society 82, 1949–1964.

Kogan, F.N. 2002. World droughts in the new millennium from AVHRR-based vegetation health indices. Eos, Transactions, American Geophysical Union 83, 562–563.

Kumar, K.N., Rajeevan, M., Pai, D.S., Srivastava, A.K., Preethi, B. 2013. On the observed variability of monsoon droughts over India. Weather and Climate Extremes 1, 42–50.

Kundu, A., Dutta, D., Patel, N.R., Denis, D.M., Chattoraj, K.K. 2021. Evaluation of socio-economic drought risk over Bundelkhand region of India using analytic hierarchy process (AHP) and geo-spatial techniques. Journal of the Indian Society of Remote Sensing 49, 1365–1377.

Kundu, A., Dwivedi, S., Dutta, D. 2016. Monitoring the vegetation health over India during contrasting monsoon years using satellite remote sensing indices. Arabian Journal of Geoscience 9, 144. https://doi.org/10.1007/s12517-015-2185-9.

Kundu, A., Patel, N.R., Denis, D.M., Dutta, D. 2020. An estimation of hydrometeorological drought stress over the central part of India using geo-information technology. Journal of the Indian Society of Remote Sensing 48, 1–9.

Mishra, A.K., Singh, V.P. 2010. A review of drought concepts. Journal of Hydrology 391, 202–216.

Mishra, A.K., Singh, V.P. 2011. Drought modeling—a review. Journal of Hydrology 403, 157–175.

Mishra, V., Smoliak, B.V., Lettenmaier, D.P., Wallace, J.M. 2012. A prominent pattern of year-to-year variability in Indian summer monsoon rainfall. The Proceedings of the National Academy of Sciences USA 109, 7213–7217.

Nag, S.K. 1998. Morphometric analysis using remote sensing techniques in the Chaka sub-basin, Purulia district, West Bengal. Journal of the Indian Society of Remote Sensing 26, 69–76.

Nagarajan, R. 2010. Drought assessment (p. 429). Springer, Dordrecht.

Obi Reddy, B.P., Maji, A.K., Srinivas, C.V., Kamble, K.H., Velayutham, M. 2001. GIS-based basin morphometric information system for terrain and resources analysis. In Proceedings of first National conference on agro-informatics, Dharwad, India (pp. 37–42).

Patel, N.R., Parida, B.R., Venus, V., Saha, S.K., Dadhwal, V.K. 2012. Analysis of agricultural drought using vegetation temperature condition index (VTCI) from terra/MODIS satellite data. Environmental Monitoring and Assessment 184, 7153–7163.

Patel, N.R., Yadav, K. 2015. Monitoring spatio-temporal pattern of drought stress using integrated drought index over Bundelkhand region, India. Natural Hazards 77, 663–677.

Peters, A.J., Walter-Shea, E.A., Ji, L., Vina, A., Hayes, M., Svoboda, M.D. 2002. Drought monitoring with NDVI-based standardized vegetation index. Photogrammetric Engineering and Remote Sensing 68, 71–75.

Shukla, S., McNally, A., Husak, G., Funk, C. 2014. A seasonal agricultural drought forecast system for food-insecure regions of East Africa. Hydrology and Earth System Sciences 18, 3907–3921.

Singh, R.P., Roy, S., Kogan, F. 2003. Vegetation and temperature condition indices from NOAA AVHRR data for drought monitoring over India. International Journal of Remote Sensing 24, 4393–4402.

Tarpley, J.D., Schnieder, S.R., Money, R.L. 1984. Global vegetation indices from NOAA-7 meteorological satellite. Journal of Applied Meteorology and Climatology 23, 4491–4503.

Unganai, L.S., Kogan, F.N. 1998. Southern Africa's recent droughts from space. Advances in Space Research 21, 507–511.

Van Rooy, M.P. 1965. A rainfall anomaly index (RAI) independent of time and space. Notos 14, 43–48.

Vrieling, A., Meroni, M., Shee, A., Mudec, A.J., Woodard, J., (Kees) de Bie, C.A.J.M., et al. 2014. Historical extension of operational NDVI products for livestock insurance in Kenya. International Journal of Applied Earth Observation and Geoinformation 28, 238–251.

Weinreb, M.P., Hamilton, G., Brown, S., Koczor, R.J. 1990. Nonlinearity corrections in calibration of advanced very high-resolution radiometer infrared channels. Journal of Geophysical Research: Oceans 95, 7381–7388.

West Bengal State Action Plan on Climate Change. 2010. Government of West Bengal, Government of India Saraswaty press limited, SAPCCWB, Open access (pp. 1–191).

Wilhelmi, O.V., Wilhite, D.A. 2002. Assessing vulnerability to agricultural drought: a nebraska case study. Natural Hazards 25, 37–58.

Wilhite, D.A., Glantz, M.H. 1985. Understanding the drought phenomenon: the role of definitions. Water International 10, 111–120.

Zhang, A., Jia, G. 2013. Monitoring meteorological drought in semiarid regions using multi-sensor microwave remote sensing data. Remote Sensing of Environment 134, 12–23.

Zhou, S., Duursma, R.A., Medlyn, B.E., Kelly, J.W., Prentice, I.C. 2013. How should we model plant responses to drought? An analysis of stomatal and non-stomatal responses to water stress. Agricultural and Forest Meteorology 182, 204–214.

10 Sectorial Vulnerability of Climate Change in Ziro Valley of Arunachal Pradesh, India

Pritom Saikia, Kesar Chand, Jagdish Chandra Kuniyal, Suraj Kumar Singh, and Shruti Kanga

10.1 INTRODUCTION

The effects of climate change on society can be described in terms of exposure, sensitivity, and adaptive capacity (Füssel and Klein, 2006 Füssel, 2007). A better understanding of communities' vulnerability and adaptive practices is critical to develop well-targeted adaptation policies (Adger et al., 2005; Smit and Wandel, 2006). Adaptation practises are site-specific phenomena that require deeper local-level analysis for better understanding the underlying variables and for better targeting adaptation strategies by local governments, non-governmental organisations, and other policymakers (Boko et al., 2007; Mano et al., 2007; Smit and Wandel, 2006). Previous research indicates the need for a vulnerability assessment method that is both precise enough to capture local characteristics and transferrable to other sites (Vincent, 2007; Below et al., 2012).

Climate change has occurred naturally since earth's formation (Carter and Mäkinen, 2011; Pandey and Bardsley, 2015), but since the twentieth and twenty-first centuries its frequency and intensity are increasing mainly due to human activities (IPCC, 2013). Climate change is impacting human lives in many ways such as socio-economic, extreme events, decreased food production, diseases, etc. along with impacts in the ecological systems. The mountain ecosystems are one of the world's most fragile ecosystems, and the climate change impacts are severe in the mountains (Bhatta et al., 2015). Climate change has already impacted the Himalayas with losses in agrodiversity and changes in the cropping pattern (Gerlitz et al., 2017) and an aggregate decrease in food production (Sinha, 2007). Despite this, the mountain communities are settled in remote places at higher altitudes making them most vulnerable. The populations in the Indian Himalayan region rely heavily on natural resources for their livelihoods and sustenance, making them susceptible to climate-induced changes in local ecosystems (Pandey, 2009).

Climate change vulnerability is the extent to which geophysical, biological, and socio-economic systems are susceptible to and unable to cope with the adverse effects of climate change (Carter et al., 2011). Vulnerability assessment reflects the adaptive capacity of the systems while considering exposure and sensitivity of the systems (Gupta et al., 2019). Climate change is already impacting the agricultural system and natural resources on which the communities are directly dependent (Pandey et al., 2017). Therefore, in the present study the vulnerability of three sectors viz. forest, agriculture, and water resources has been assessed.

Vulnerability assessment has attained much attention in recent times and is attracting significant attention from policymakers. In the present study, vulnerability is viewed as a propensity to be measured based on the three variables of exposure, sensitivity, and adaptive capability, following the Intergovernmental Panel on Climate Change framework. Exposure is the presence of individuals, assets, or ecosystems that may be negatively impacted. Sensitivity refers to the factors that influence a system, whereas adaptive capacity is the ability of natural and human systems to deal with or adapt to a climate that is changing.

DOI: 10.1201/9781003377825-12

The present study sought to identify sector-specific indicators for the assessment of climate change vulnerability in the Ziro valley of Arunachal Pradesh using an entropy weighing method. The four indicators centred mostly on climate change were used to determine exposure (Table 10.1). Ten indicators focused on forest, agriculture, and water resources were utilised to determine the sensitivity. (Table 10.1). The adaptive capacity was assessed based on 13 indicators in terms of conservation practices, knowledge on climate change, and support to coping with the stresses of climate change (Table 10.1).

Table 10.1 also displays the association between each indication and vulnerability. Based on these indicators, a questionnaire was formulated to collect data from households. The purpose of a household-level vulnerability assessment is mainly to develop knowledge that can reduce exposure and sensitivity of the communities (Below et al., 2012; Saroar and Routray, 2012; Pandey et al., 2016). The prime objective of the present research is (1) to assess the overall climate vulnerability of the study region and identify the major drivers of vulnerability and (2) to calculate the Vulnerability Indices (VIs) for the three different categories of settlements.

TABLE 10.1
omponents, Dimensions, Indicators and Their Relationship with Vulnerability

Component	Dimension	Indicator	Relationship with Vulnerability
Exposure	Climate change	Temperature	Positive
		Precipitation	Positive
		Winter duration	Positive
		Summer duration	Positive
Sensitivity	Water resources	Water availability	Positive
		Water availability in agriculture	Positive
		Number of drying springs	Positive
		Water pollution	Positive
		Water conflict	Positive
	Forest resources	Deforestation	Positive
		Shifting cultivation	Positive
		Forest fires	Positive
	Agriculture	Agricultural yield	Positive
		Crop damage	Positive
Adaptive capacity	Water resources	Water conservation practices	Negative
		NGO's role in water conservation	Negative
		Government's role in water conservation	Negative
	Forest resources	Afforestation activities	Negative
		Community participation in forest conservation	Negative
		NGO's role in forest conservation	Negative
		Government's participation in forest conservation	Negative
		Customary laws for forest conservation	Negative
	Agriculture	Cash crop percentage	Negative
		Crop insurance	Negative
		Government's participation in improving agriculture	Negative
		Fertiliser usage	Negative
		Any other adaptive practices	Negative

Note: NGO, non-governmental organisation.

10.2 STUDY AREA

The Ziro valley is situated in north-east India between 93.76°E and 94.00°E and between 27.42°N and 27.65°N of the Lower Subansiri district in the western centre of Arunachal Pradesh. The Ziro valley, also known as the Apatani valley, is located between 1,524 m and 2,738 m above sea level. The valley has an area more than 1,058 km², of which 33 km² is cultivated land while the rest is under forest, plantations, and settlement. The valley lies between the Panior and Kamla (Kuru) Rivers bounded by the surrounding hills. The Kley River, flowing longitudinally, forms the trunk drainage of the valley and tributaries joining it from east and west. The valley itself has a very gentle gradient towards the south with an average elevation of 1,570 to 1,580 m above mean sea level. The surrounding hill ranges have variable elevations between 1,700 and 1,800 m above mean sea level. The soils of the valley are humid black and reddish in color, developed from gneiss and schist overlaid on a wide area with older alluvial deposits (Kala et al., 2008).

The climate of the valley is humid subtropical to temperate, and it receives 235 cm of annual precipitation. Summer temperatures range from 6.3°C to 28.1°C, while winter temperatures range from 1.9°C to 18.4°C. The minimum temperatures in 2011, 2012, and 2013 were −3°C, −4°C, and −9°C, respectively.

The Ziro valley is renowned for its distinctive land-use pattern, natural resource management, and resource conservation. The valley is also known as "Arunachal Pradesh's rice bowl." The Apatani tribe in the valley is renowned for their traditional cultivation of paddy cum fish. People create temporary aquatic systems for the production of rice and fish as part of a very distinctive water management system.

As of the 2011 India census, Ziro had a population of 12,806. Males constitute 50% of the population and females 50%. The average literacy rate of Ziro is about 66%; male literacy rate is 72% and female literacy rate is 60%.

10.3 METHODOLOGY

In the Ziro valley, Arunachal Pradesh, questionnaire-based interviews were undertaken to collect information on the susceptibility of villages to climate change. Random sampling was used to select 105 households from 21 villages. The questionnaire included questions based on the selected vulnerability indicators for several dimensions. The number of samples was kept equal in all the villages as the size of the villages was more or less similar. The interview was briefed and mainly conducted in their local *Apatani* language.

The overall methodology used to assess vulnerability can be categorised into three main steps: (1) selection of indicators and questionnaire formulation associated with the vulnerability; (2) conducting of village-level questionnaire surveys; and (3) quantifying, normalisation, and allocation of weights and VI estimation.

10.3.1 IDENTIFICATION AND SELECTION OF INDICATORS

The selection of suitable indicators in each of the three categories, i.e. exposure, sensitivity and adaptive capacity to determine the vulnerability of a community is the most crucial task. Relevant indicators for the vulnerability assessment were selected based on peer-reviewed literature and contact with various stakeholders in the study area of interest. These indicators represented the conditions of forest resources, water resources, and agriculture. The data were collected by developing a village-level questionnaire based on the identified indicators. The questionnaire contained four distinct components. The first component consisted of generic questions about the responder, whereas the remaining three sections addressed the three dimensions of vulnerability, namely exposure, sensitivity, and adaptive capacity, separately.

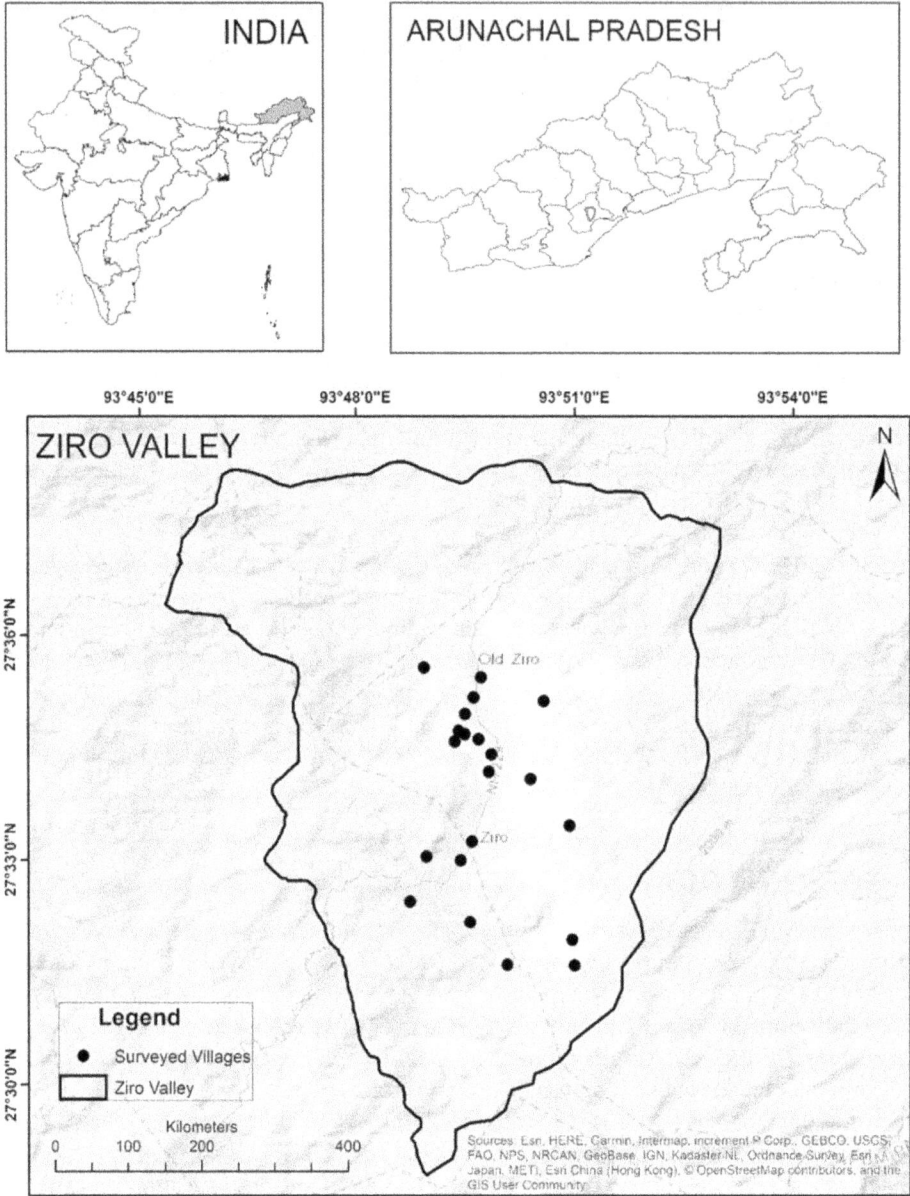

FIGURE 10.1   Location map of the study area.

10.3.2   Data Collection

After the formulation of the questionnaire, it was pre-tested and some redundant questions were eliminated. We adopted a random sampling approach to select 105 households for the survey. Respondents were interviewed in their native Apatani dialect after being briefed on the topic and given informed permission. The present study conducted village-level survey in three categories of settlements for vulnerability assessment. The three different categories of settlements were divided based on their distance from the main township location. The settlements were categorised as town, semi-town, and rural. Each category of settlements consists of seven villages or localities, where we randomly conducted surveys for vulnerability assessment. Each interview lasted 30 minutes on average. The acquired data were filtered and organised prior to being fed into computers for further processing.

10.3.3 QUANTIFYING, NORMALISATION, ALLOCATION OF WEIGHTS, AND VULNERABILITY INDICES ESTIMATION

The qualitative data collected from the questionnaire survey were initially converted to quantitative data using standard coding method. Each indicator was normalised for further processing and for computing the index (Hahn et al., 2009; Gupta et al., 2019). For indicators (I) impacted vulnerability positively i.e., increase vulnerability (Equation 10.1) and for indicators (I) impacted vulnerability negatively i.e., decrease vulnerability (Equation 10.2),

$$x = \frac{(Xi - Min)}{Max - Min}, \tag{10.1}$$

$$x = \frac{(Max - Xi)}{Max - Min}, \tag{10.2}$$

where
Xi is the value of indicator X for ith household.
Max represents the maximum of all the values of X indicator.
Min represents the minimum value of all the values of X indicator.
After normalisation of each indicator, the components were estimated by averaging all associated indicators, using Equation 10.3 to calculate the value of each index:

$$Mv = \frac{\left(\sum_{i=1}^{n} Index\right)}{n} \tag{10.3}$$

where M_v is the index of component v, $Index$ is the value of indicator i, and n is the number of indicators in the index.

The entropy weightage method (EWM) was used to evaluate the weights of the three dimensions (exposure, sensitivity, and adaptive capacity). The entropy method is an objective, quantitative weight assignment method with decent accuracy (Peng et al., 2018). Entropy is used to determine the dispersion degree of a certain index, where the smaller the entropy, the larger is the dispersion degree (Zhang et al., 2014; Zhao et al., 2018). EWM was first used in thermodynamics and is now widely used in engineering, economy, finance etc. (Tian and Du, 2004; Cheng and Zhang, 2003; Guo, 2001; Zhao and Song, 2001; Fang et al., 2004; Li et al., 2004; Zhou, 2003; Xu et al., 2004; Tang et al., 2000; Lin et al., 2003) The entropy value (e_j) of each indicator is calculated using Equation 10.4:

$$e_j = -h\sum_{i=1}^{m} X_{ij} \ln r_{ij}, j = 1, 2, \ldots\ldots, n \tag{10.4}$$

where m is the number of indicators:

$$h = \frac{1}{\ln(m)}$$

Next we calculate the weight vector (w_j) using Equation 10.5:

$$w_j = \frac{1-e_j}{\sum_{j=1}^{n} 1-e_j}, j = 1, 2, \ldots\ldots, n \tag{10.5}$$

where $1 - e_j$ is known as the degree of diversification.

The values obtained for each indicator from Equation 10.5 are the final weights for vulnerability assessment. These weights are multiplied by the value of each dimension in the VI computation.

The final VIs have been calculated using Equation 10.6 (Carbon Tracker and CEEW, 2021):

$$Vulnerability\,(V) = \frac{Exposure\,(E)*Sensitivity\,(S)}{AdaptiveCapacity\,(AC)} \tag{10.6}$$

The present study also evaluated the vulnerability of the indigenous paddy-cum-fish cultivation practiced by the *Apatani* community. A few questions were designed to understand the vulnerability of the paddy-cum-fish cultivation.

10.4   RESULTS

The general discussion with villagers, whose subsistence primarily revolves around agriculture and extraction of fuelwood, fodder, and many other resources from forests, confirms the view that there has been a variation in climate and specifically changes in temperature and rainfall, with a significant increase in summer duration. Eight six percent of the villagers reported changes in climate, especially increase of temperature and changes in rainfall pattern. These changes are imposing negative impacts on the water and forest resources as well as agriculture.

10.4.1   VULNERABILITY OF WATER RESOURCES

The villagers in the region mainly depend on spring water from the surrounding hills for domestic use. Each household has water connections provided by the local government. The water from the nearby springs is diverted and collected in large tanks in the uphill regions, from where the water is distributed to each household. And almost 100% of the villagers depend on this spring water for daily use. An analysis of the surveyed data showed that 58% of the villagers have sufficient availability throughout the year, and 37.5% of the villagers are sensitive to water scarcity, while rest of the villagers reported seasonal variation in water availability. It is to be noted that 13% villagers responded that a significant number of springs in the surrounding hills are drying from the past decade. The analysis of the data also shows that 85% of the population are exposed to polluted water especially during the monsoon season. The people of the villages in the region mainly depend on rain-fed agriculture. The analysis of surveyed data shows that almost 46% of the villagers responded that the amount of water in the agricultural land is decreasing from the past decade. A small portion, i.e. 4% of the villagers, responded that they are facing acute water shortage in the agricultural fields. Moreover, the region has only one river flowing through it which signifies that the region needs water conservation practices. The surveyed data showed that only 20% of the population are conserving water at household levels. Apart from this, the construction of Siikhe lake water conservation project by the government has proved to be beneficial as reported by the nearby villagers.

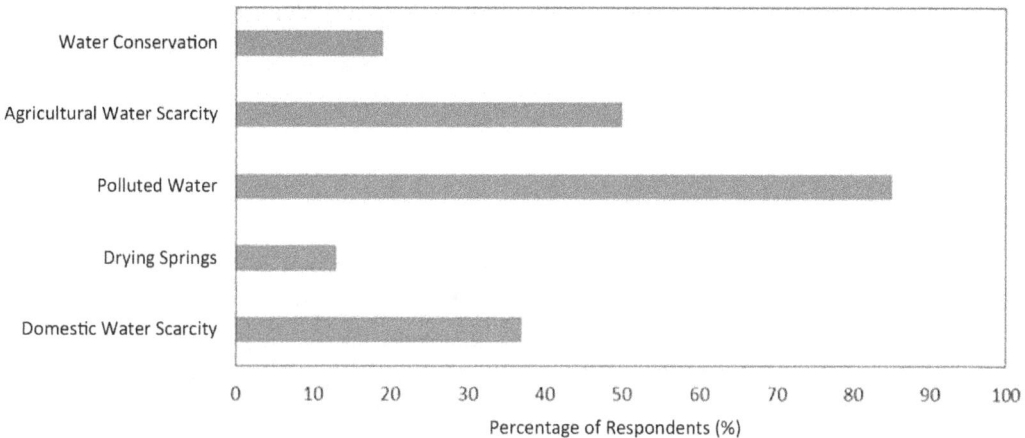

FIGURE 10.2   The percentage of respondents on *x*-axis and surveyed questions on *y*-axis.

10.4.2   Vulnerability of Forest Resources

The villagers in the region extract fuelwood for cooking energy, fodder for livestock, timber, edible items (*salyo, samper, holla, diirang, sankhang, baching, semo*, etc.), and creepers, climbers are also collected which are used as ropes. Nearly all households surveyed extract fuelwood and other resources from the community forests. In the study area, almost all the villagers have sufficient knowledge about forest resources and forest conservation, and they have been practicing it for generations. Nearly 90% of the respondents grow traditional pine forest trees along with bamboo in their farmlands and community forest-lands. But, over the past few years the respondents have witnessed forest degradation and forest clearing for commercial purposes. Clearing of forest for kiwi cultivation is also becoming a major cause of deforestation. The analysis of the surveyed data showed that 88% of the respondent have seen clearing of forest. The respondents have seen the forests are being cleared for construction of roads, kiwi cultivation, quarrying, and human settlements. The analysis also reveals that the respondents of the region do not practice shifting cultivation and have been able to eradicate forest fires completely. It was observed that nearly 44% of the respondents are frequently practicing afforestation activities, and 47% of the respondents practice afforestation activities occasionally. It was also observed that the local *Apatani* community of the region have the practice of community management of their forest-land. They have the practice of plantation, boundary demarcation, and strict monitoring of their community forest-lands. Any illegal activities such as cutting of excess trees is strictly prohibited in the community forests. The local government bodies and non-governmental organisations are also playing an active role in forest conservation in the region.

10.4.3   Vulnerability of Agriculture

Agriculture is the principal source of income for the majority of the population in the study area. The primary crops of the region are paddy, millets, and seasonal vegetables along with few cash crops. The kiwi cultivation has recently gained attention in the study area. This is because of the favorable climatic conditions and economic benefits of the farmers. Apart from these, the farmers of the region also practice the unique and sustainable paddy-cum-fish cultivation. The indigenous

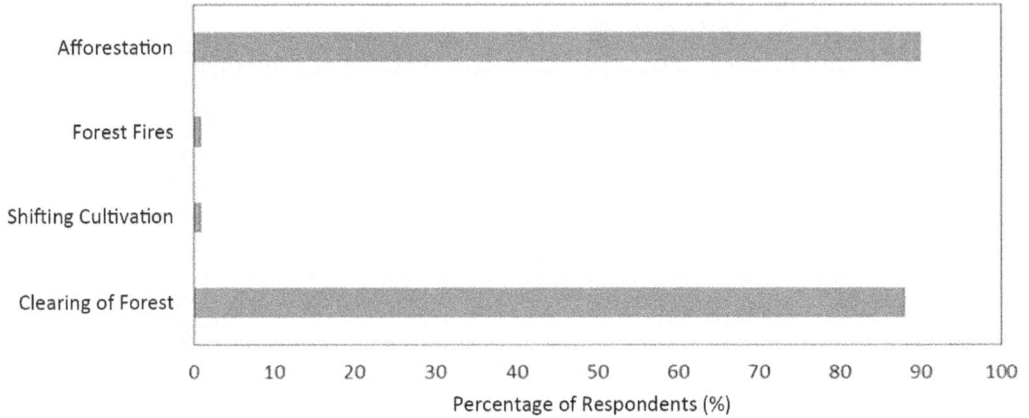

FIGURE 10.3   The percentage of respondents on *x*-axis and surveyed questions on *y*-axis.

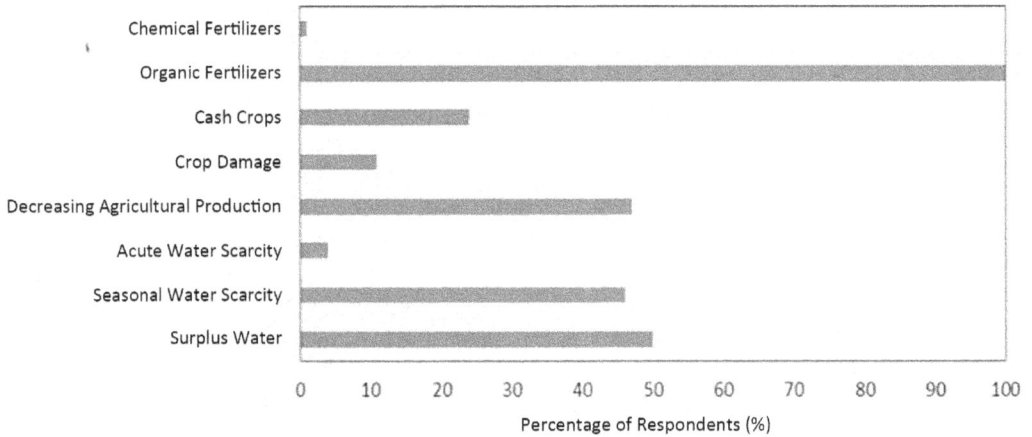

FIGURE 10.4   Graph showing the percentage of respondents on *x*-axis and surveyed questions on *y*-axis.

integration of paddy and fish is a distinct characteristic of the Apatani community which is unique in itself (Saikia et al., 2021). All these agricultural practices in the region are dependent on rainwater. In the present study, the analysis revealed that 50% of the respondents have surplus water for agriculture, whereas 46% of the respondents face scarcity of water seasonally and 4% of the farmers are facing acute shortage of water for agriculture. It is to be noted that 47% of the respondents have also witnessed a decrease in their agricultural production from the past decade. A few number of respondents (11%) have also faced damage of crops mainly due to excess rain or deficit of rain. A few questions on adaptive capacity have also revealed that most of the respondents are aware of changing climate and have adapted a few techniques to cope with it. Almost 24% of the respondents are growing cash crops as an alternative source of income. The most unique and sustainable of the people in the region is the use of organic fertilisers. Almost all respondents are using organic fertilisers in their crops and have never used any chemical fertilisers. Only a few respondents are using herbicides.

10.4.4 Vulnerability of Paddy-cum-Fish Cultivation

The analysis shows that 85% of the respondents have seen a decrease in paddy-cum-fish cultivation. The reason behind this decrease is the hard labor required in paddy-cum-fish, jobs, free rice provided by the government, and kiwi cultivation. It is also seen that according to 21% of the respondents, paddy-cum-fish is no more economically beneficial. Along with it, almost more than 45% of the respondents say that paddy-cum-fish may vanish in the near future. It has been seen that 28% of the respondents may shift to other agricultural practices instead of paddy-cum-fish cultivation.

10.4.5 Vulnerability Indices

A set of 27 indicators of vulnerability was used in the assessment capturing exposure, sensitivity, and adaptive capacity of the villages with respect to current climate risks. Since the VI's can theoretically lie between 0 and 1, with 0 the lowest possible VI and 1 the highest, this indicates all three categories of villages are vulnerable. The current vulnerability assessment shows that the rural villages are most vulnerable, semi-town is moderately vulnerable, and town is least vulnerable.

The rural villages in the region are highly vulnerable to climate change with VI = 0.967, semi-town is moderately vulnerable with VI = 0.860, and town is least vulnerable with VI = 0.835. Figure 10.7 shows the distribution of the villages according to their vulnerability ranking.

10.5 DISCUSSION AND CONCLUSION

Arunachal Pradesh has developed a long-term climate change strategy on climate resilience and response (i.e., Pakke Declaration 2021) to address the development needs of society while building climate resilience. It plans to implement a range of adaptation options and policies to be implemented.

To ensure a climate-resilient community, a targeted local-level policy intervention is required for the vulnerable communities. As an interim intervention, initially a block level vulnerability assessment for entire Arunachal Pradesh is to be done to identify the most vulnerable communities and subsequently identify the factors of vulnerability. In the present study we identified that vulnerability varies in local scale; therefore, adaptation strategies and interventional policies need to consider

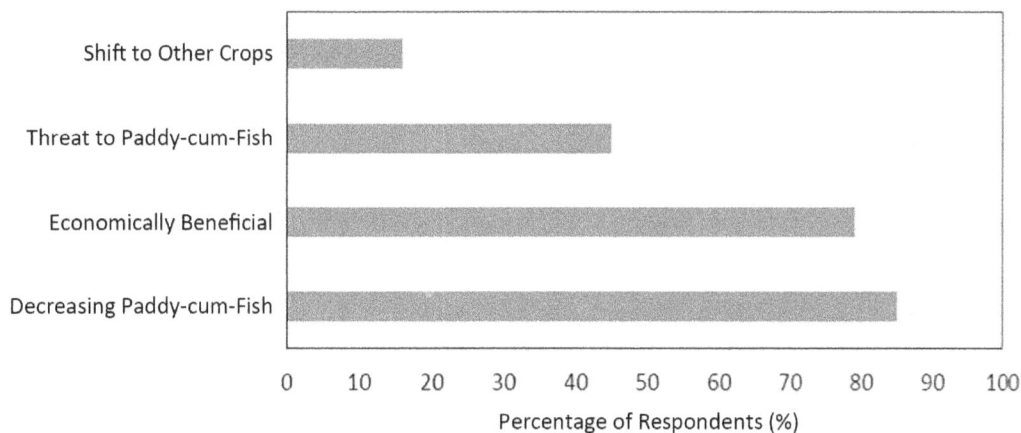

FIGURE 10.5 Graph showing the percentage of respondents on *x*-axis and surveyed questions on *y*-axis.

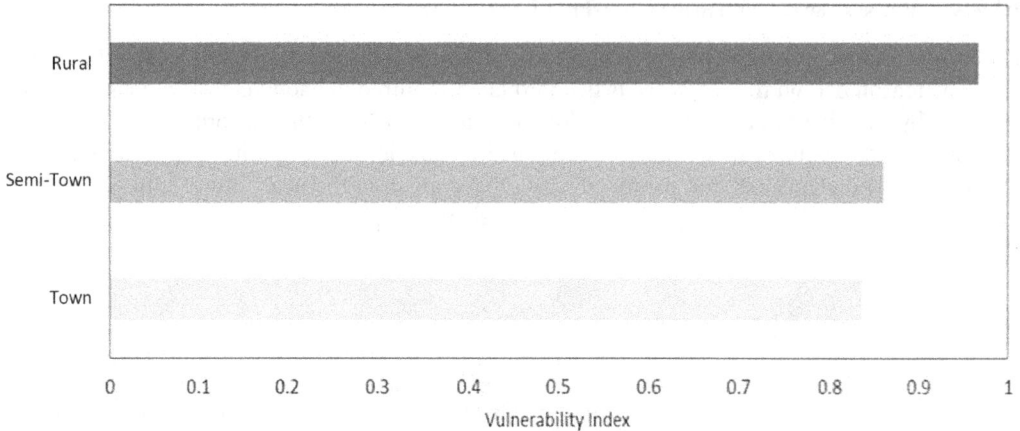

FIGURE 10.6 Vulnerability Indices of the villages.

local circumstances. In the long-term development of improved communication, transportation, access to markets, education, health care, and other services are also important in supporting existing and alternative livelihoods for the communities (Alam et al., 2017).

The farmers must be made aware of the government policies such as access to credit and crop insurance. This will assist the farmers in obtaining resources and appropriate technologies which are needed for adaptation. Local-level policies should focus on developing new crop varieties, high-value crop varieties, and sustainable kiwi cultivation practices in the study region.

Human-induced as well as natural factors influence the change in climate and its associated vulnerability. Once there is a clear understanding of the associated factors identified as indicators of vulnerability, the vulnerability can be mapped using these indicators. In the present study, we assessed the vulnerability of the three categories of villages, viz. rural, semi-town, and town, in the Ziro circle of Arunachal Pradesh. We assessed and calculated the VIs of the three categories of settlements using a set of 27 indicators. Weights were assigned each indicator/group of similar indicators using the entropy method. The extent of vulnerability largely depends upon the types of selected indicators and the allocation of weights. Therefore, in the present study the authors had carefully selected the indicators through literature and consultation with stakeholders, and weights had been allocated with the widely accepted entropy method.

This research took a holistic approach for assessing the vulnerability of the communities with regards to climate change and its impacts on agricultural, forest, and water resources along with the adaptive practices. The study found that the inhabitants of the rural settlements are found to be highly vulnerable, whereas semi-town is moderately vulnerable, and town areas are least vulnerable. The study found that the polluted water quality during monsoon is one of the major factors of vulnerability in the region. Apart from it, the region also experiences water scarcity at times and only a small percentage of the respondents are conserving water. In case of forest resources, deforestation is the major driver of vulnerability. Deforestation due to kiwi cultivation and construction of roads and human settlements has increased considerably in the region. However a majority of the respondents are also involved in plantations of pine and bamboo in their community forest-lands. The major drivers of agricultural vulnerability are erratic rainfall. The respondents reported damage of crops due to excess rainfall and sometimes due to deficit rainfall.

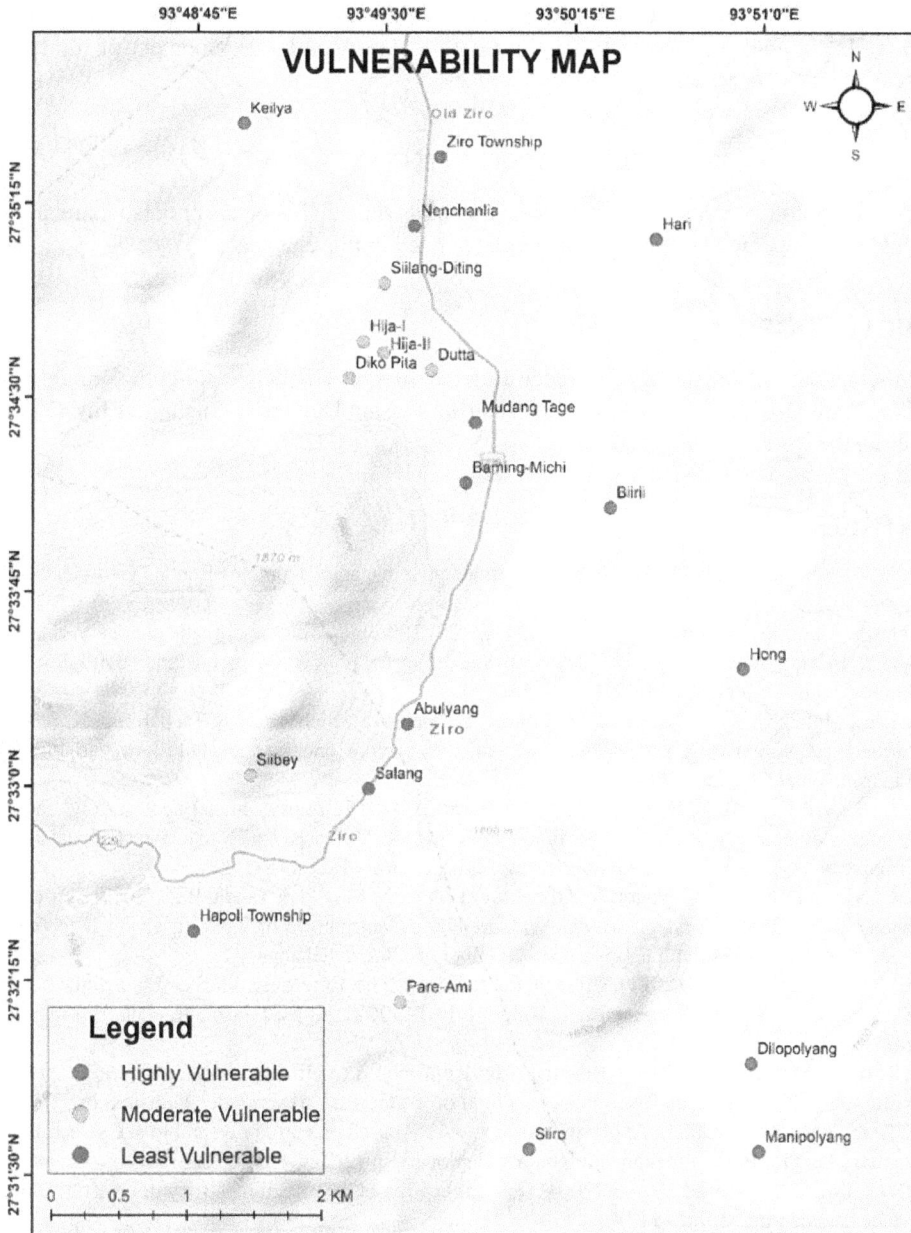

FIGURE 10.7 Map showing the categories of vulnerable villages.

Source: Esri, HERE, Garmin, intermap, increment P Corp., GEBCO, USGS, FAO, NPS, NRCAN, GeoBase, IGN, Kadaster NL, Ordnance Survey, Esri Japan, METI, Esri China (Hong Kong), © OpenStreetMap contributors, and the GIS User Community

Above all, this study has shown that overall the vulnerability index does not differ much in a small area. However, from the present study it can be concluded that there is a certain degree of variation in VIs among the different categories of settlements in a small area like Ziro circle. New policy interventions should focus on improving forest conservation, land-use planning, sustainable kiwi cultivation, water conservation, better irrigation, and alternative sources of livelihood. In order to boost adaptive capacity, effort needs to be made in supporting the socio-demographic profiles and social network of the communities and diversification of livelihood activities. The existing and alternative livelihood activities in the regions can be enhanced by development of transport and communication and access to markets and services.

10.6 DECLARATION OF COMPETING INTEREST

The authors declare that they have no known competing financial interests or personal relationships that could have appeared to influence the work reported in this chapter.

ACKNOWLEDGEMENT

The authors acknowledge the support received from director GBPNIHE for completing this work. The authors are also thankful to Animekh Hazarika, Assam University, Silchar for his guidance in completing the work.

REFERENCES

Adger, W. N., & Vincent, K. (2005). Uncertainty in adaptive capacity. Comptes Rendus Geoscience, 337(4), 399–410.

Alam, G. M., Alam, K., Mushtaq, S., & Clarke, M. L. (2017). Vulnerability to climatic change in riparian char and river-bank households in Bangladesh: Implication for policy, livelihoods and social development. Ecological Indicators, 72, 23–32.

Below, T. B., Mutabazi, K. D., Kirschke, D., Franke, C., Sieber, S., Siebert, R., & Tscherning, K. (2012). Can farmers' adaptation to climate change be explained by socio-economic household-level variables? Global Environmental Change, 22(1), 223–235

Bhatta, L. D., van Oort, B. E. H., Stork, N. E., & Baral, H. (2015). Ecosystem services and livelihoods in a changing climate: Understanding local adaptations in the Upper Koshi, Nepal. International Journal of Biodiversity Science, Ecosystem Services & Management, 11(2), 145–155.

Boko, M., Niang, I., Nyong, A., Vogel, A., Githeko, A., Medany, M., . . . & Yanda, P. Z. (2007). Africa climate change 2007: Impacts, adaptation and vulnerability: Contribution of working group II to the Fourth Assessment Report of the Intergovernmental Panel on Climate Change.

Carbon Tracker and CEEW. (2021). "Reach for the Sun: The Emerging Market Electricity Leapfrog." Council on Energy, Environment and Water, July 14, 2021, https:// carbontracker.org/reports/reach-for-thesun/.

Carter, T. R., & Mäkinen, K. (2011). Revised title: Approaches to climate change impact, adaptation and vulnerability assessment: Towards a classification framework to serve decision-making.

Cheng, T., & Zhang, C. X. (2003). Application of fuzzy AHP based on entropy weight to site selection of solid sanitary landfill. Environmental and Sanitary Engineering, 12(2), 64–67.

Fang, D. C., Liu, G. L., Bang, L. P. rt d. (2004). The application of information entropy in investment decision. Value Engineering, 2, 115–117.

Füssel, H. M. (2007). Vulnerability: A generally applicable conceptual framework for climate change research. Global Environmental Change, 17(2), 155–167.

Füssel, H. M., & Klein, R. J. (2006). Climate change vulnerability assessments: An evolution of conceptual thinking. Climatic Change, 75(3), 301–329.

Gerlitz, J. Y., Macchi, M., Brooks, N., Pandey, R., Banerjee, S., & Jha, S. K. (2017). The Multidimensional Livelihood Vulnerability Index—An instrument to measure livelihood vulnerability to change in the Hindu Kush Himalayas. Climate and Development, 9(2), 124–140.

Guo, C. Z. (2001). Study on the evaluating method of entropy coefficient for stock investment value. Nankai Economics Studies, 5, 65–67.

Gupta, A. K., Negi, M., Nandy, S., Alatalo, J. M., Singh, V., & Pandey, R. (2019). Assessing the vulnerability of socio-environmental systems to climate change along an altitude gradient in the Indian Himalayas. Ecological Indicators, 106, 105512.

Hahn, M. B., Riederer, A. M., & Foster, S. O. (2009). The Livelihood Vulnerability Index: A pragmatic approach to assessing risks from climate variability and change—A case study in Mozambique. Global Environmental Change, 19(1), 74–88.

IPCC. (2013). *Summary for policymakers*. In Climate Change 2013: The Physical Science Basis. Contribution of Working Group. Allen, J. Doschung, A. Nauels, Y. Xia, V. Bex, and P.M. Midgley, Eds., Cambridge University Press, pp. 3–29, doi:10.1017/CBO97 81107415324.004

Kala, C. P., Dollo, M., Farooquee, N. A., & Choudhury, D. C. (2008). Land-use management and wet-rice cultivation (Jebi Aji) by the Apatani people in Arunachal Pradesh, India: Traditional knowledge and practices. Outlook on Agriculture, 37(2), 125–129.

Li, X. H., Li, Y. M., & Gu, Z. H. er d. (2004). Competitive situation analysis of regional logistics development based on AHP and entropy weight. Journal of Southeast University, 34(3), 398–40 I.

Lin, Y., Men, B., & Jia, W. (2003). Application of entropy coefficient method to evaluating on alimentative type of water. Northwest Water Resources & Water Engineering, 13(3), 27–28.

Mano, R., & Nhemachena, C. (2007). Assessment of the economic impacts of climate change on agriculture in Zimbabwe: A Ricardian approach. World Bank Policy Research Working Paper (4292).

Pandey, R. (2009). Forest resource utilization by tribal community of Jaunsar, Uttarakhand. Indian Forester, 135(5), 655.

Pandey, R., & Bardsley, D. K. (2015). Social-ecological vulnerability to climate change in the Nepali Himalaya. Applied Geography, 64, 74–86.

Pandey, R., Jha, S. K., Alatalo, J. M., Archie, K. M., & Gupta, A. K. (2017). Sustainable livelihood framework-based indicators for assessing climate change vulnerability and adaptation for Himalayan communities. Ecological Indicators, 79, 338–346.

Pandey, R., Maithani, N., Aretano, R., Zurlini, G., Archie, K. M., Gupta, A. K., & Pandey, V. P. (2016). Empirical assessment of adaptation to climate change impacts of mountain households: Development and application of an Adaptation Capability Index. Journal of Mountain Science, 13(8), 1503–1514.

Peng, L., Xu, D., & Wang, X. (2018). Vulnerability of rural household livelihood to climate variability and adaptive strategies in landslide-threatened western mountainous regions of the Three Gorges Reservoir Area. China. Clim. Dev. 1–16. https://doi.org/10.1080/17565529.2018.1445613.

Saikia, P., Chand, K., Kuniyal, J. C., & Lodhi, M. S. (2021). Vulnerability assessment of traditional fish-cum paddy cultivation of the Apatani tribe of Arunachal Pradesh, India. Environ Waste Management Recycling, 15(7), 1–7.

Saroar, M. M., & Routray, J. K. (2012). Impacts of climatic disasters in coastal Bangladesh: Why does private adaptive capacity differ? Regional Environmental Change, 12(1), 169–190.

Sinha, S. (2007). Impact of climate change in the highland agroecological region of India, Sahara Time Magazine.

Smit, B., & Wandel, J. (2006). Adaptation, adaptive capacity and vulnerability. Global Environmental Change, 16(3), 282–292.

Tang, R., Guo, C., & Dong, X. (2000). An optimization model with entropic coefficients for management in irrigation water resources. Journal of Hohai University, 28(1), 18–21.

Tian, Q., & Du, Y. X. (2004). Study of performance evaluation for mechanical products based on entropy fuzzy comprehensive review. China Manufacturing Information, 33(3), 97–99.

Vincent, K. (2007). Uncertainty in adaptive capacity and the importance of scale. Global Environmental Change, 17(1), 12–24.

Xu, S. Q., Hu, Z. G., Liu, Q., HUANG, H., & PU, J. P. (2004). Multi-objective decision analysis of diversion standards based on entropy. China Rural Water and Hydropower, 8, 45–47.

Zhang, X., Wang, C., Li, E., & Xu, C. (2014). Assessment model of ecoenvironmental vulnerability based on improved entropy weight method. Scientific World Journal, https://doi.org/10.1155/2014/797814.

Zhao, D. Y., & Song, H. (2001). A method of ameliorative multi-objective synthetic evaluation based on entropy weight and its application. Journal of Ordnance Engineering College, 13(3), 47–51.

Zhao, J., Ji, G., Tian, Y., Chen, Y., & Wang, Z. (2018). Environmental vulnerability assessment for mainland China based on entropy method. Ecological Indicators, 91, 410–422. https://doi.org/10.1016/j.ecolind.2018.04.016

Zhou, M. H. (2003). The research about method of sustainable consumption system measure. Systems Engineering-Theory & Practice 12, 25–31.

11 A Comparative Analysis of Social Capital for Reducing the Effect of Disaster on Social Vulnerabilities in South-West Coastal Bangladesh
A Social Exchange Theory Approach

*Md. Bariul Musabbir, Shahana Akter,
and Md. Mujibor Rahman*

11.1 INTRODUCTION

Because of its vulnerability to common and serious environmental phenomena like cyclones and the related storm surge, Bangladesh has been named as one of the most susceptible nations (Shahid, 2012). Rapid urbanization, commercialization, global warming, and growth all contribute to increase in recurrence as well as rigidity of natural catastrophes. There is evidence that much global "progress" actually makes people more vulnerable by forcing them to relocate to more dangerous locations, where they may be compelled to further destroy the environment in order to ensure their own survival (Wisner et al., 2004). Throughout current history, natural hazards have been seen more and more as a way to stop people from reaching sustainable development goals (SDGs). People have been trying to avoid disasters and reduce risks since the beginning of time (Akter and Rahman, 2023). Various disaster risk reduction (DRR) schemes worldwide have made efforts to reduce vulnerability. In addition to the physical destruction of homes and businesses, lives are lost, and the economy is negatively impacted whenever a natural catastrophe occurs (Sanyal and Routray, 2016). At the community level, where this intangible resource may be mobilized for the benefit of the entire community, social capital has shown significant acts in several steps toward the risk of hazard management stages, both before and after a catastrophe has occurred (Sanyal and Routray, 2016). There are two aspects to social capital: one's mindset and another's actions. Based on the findings of this study, community members' propensity to share their expertise is most influenced by their shared values and norms (Gubbins et al., 2021). Sharing of knowledge is extremely necessary to reduce the likelihood of disasters. The social integration of human relationship networks is especially important for the community. Community knowledge can aid in disaster risk mitigation, disaster preparedness, and disaster recovery (Cadag and Gaillard, 2012). Indigenous peoples have developed resilient skills through generations, and these abilities and understandings have been tested and shown to be useful in preventing and mitigating the effects of catastrophes and other dangers (Magni, 2017).

The knowledge sharing behavior and Indigenous knowledge that cultures have accumulated over time must be well known in order to suggest suitable strategies for their improvement. The subjects of

DOI: 10.1201/9781003377825-13

this research were the key conception of social vulnerability and social capital with respect to knowledge exchange mechanism in the disaster management and reduction of risk among the inhabitants of the south-west coastal region of Bangladesh. The goals of this study are as follows: (1) to evaluate the prevalence of local hazards; (2) to evaluate the degree to which local residents are vulnerable to those hazards; and (3) to determine relationships between knowledge sharing and social exchange variables in the context of social exchange theory. Information sharing behavior and socialization in the south-west coastal area of Bangladesh for DRR have not been studied, however, according to any of the theories mentioned earlier, although most of the considerations have been addressed in the related literature of information sharing utilizing theories of social exchange from diverse angles. Furthermore, philosophical or analytical writings provide sources for potential variables and should be investigated in the sense of social exchange behavior. This chapter looks at the possibility of identifying possible variables.

11.2 STUDY AREA

11.2.1 SELECTION OF THE STUDY AREA

The aims of this study are taken into consideration while selecting the design of the study. The south-western coast of Bangladesh is very hazardous. The three districts of Khulna, Bagerhat, and Satkhira in south-western Bangladesh are among the most disaster-prone in the world. Holdibuniya, Henchi, and Barunpara villages are chosen from the three districts. Because of the prevalence of threats including river erosion, salt water intrusion, and cyclones, these regions were chosen (Figure 11.1).

11.2.2 LOCATION OF BARUNPARA VILLAGE

The distance between the hamlet of Barunpara and the Khulna District is around 22 km. This little community is part of the ninth ward of the Gangarampur Union. This district is home to the four villages of Kathamaree, Moshiardanga, Baronpara, and Maitvanga. This settlement may be found between the longitudes of 22°37′59″ and 22°38′59″ N, and the latitudes of 89°30′40″ and 89°31′23″ E. The land around the settlement of Barunpara was 10 km. The eastern bank of the Pashur River is home to this quaint little town. The erosion caused by the river has caused the settlement to split into two halves, which are now known as Barunpara and Maitvanga. Dip Barunpara is the name that the locals call Maitvanga.

11.2.3 LOCATION OF HOLDIBUNIYA VILLAGE

The Holdibuniya community may be found in the Chila Union of the Mongla Upazila in the Bagerhat district. The upazila known as Mongla is the most populous one in the Bagerhat district. The fact that the Chandpai range of the Sundarbans Forest region is located inside the Mongla Upazila gives it a unique importance. Additionally, it is the country's second seaport.

The location of the settlement known as Holdibuniya may be found between the coordinates of 22°27′17″ to 22°27′37″ N latitude and 89°37′25″ and 89°38′10″ E longitude, almost 4 km distant toward the household section of historic Mongla port. There are nine different wards that make up the Chila union. The 1, 2, and 3 number wards are each included inside their own hamlet. On the northern part of this hamlet is where you will find the Chandpai Union. The Sundarbans may be found in the south-western portion of this hamlet.

11.2.4 LOCATION OF HENCHI VILLAGE

This study area is situated at 22°18′57″ to 22°18′49″ N longitude and 89°11′64″ to 89°11′61″ E latitude in the south-west coastal region in Bangladesh, which may be found at Shyamnagar Upazilla, Atulia Union, Satkhira District, Bangladesh. Burigoalini Union and the Chuna River are located in the southern part of the Henchi village, which is bordered to the east by the Kholpatua River and Barakupat

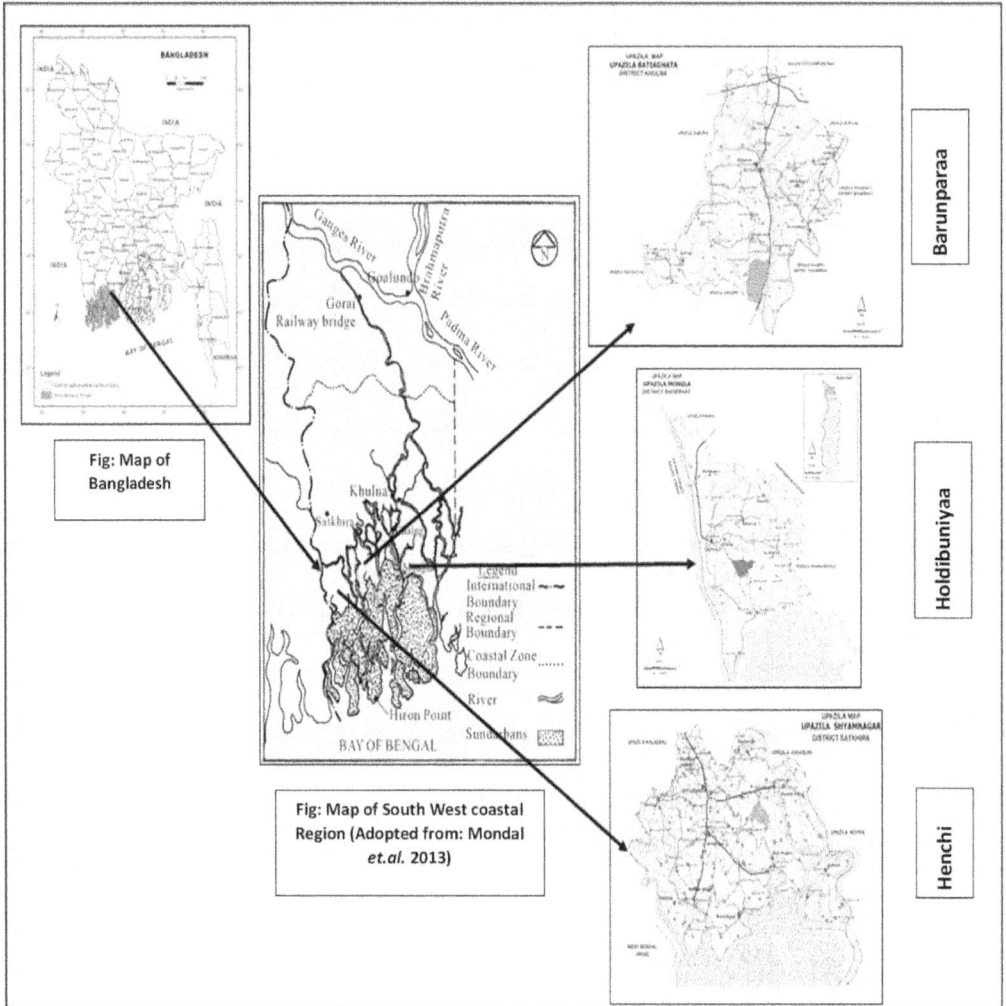

FIGURE 11.1 Map of the study area.

village, to the west by the Iswaripur Union, to the north by the Kashimari Union, and to the south by the Kholpatua River. The Sundarbans mangrove forest is located in close proximity to the neighborhood.

The cyclone Sidr, which struck on 15 November 2007, and the cyclone Aila, which struck on 25 May 2009, both had a profound impact on this community. It is roughly 78 km from the district headquarters to the Atulia Union, which is situated on the outskirts of the Sundarbans, which include the biggest continuous tract of mangrove forest in the world. The Sundarbans are separated from the Atulia Union by the river Malancha.

11.3 METHODOLOGY

11.3.1 Sample Size Determination

The sample size was determined using the following four procedures: (1) constructing a Pearson's correlation coefficient matrix between socioeconomic characteristics, (2) analyzing socioeconomic factors, (3) figuring out the right size of a sample, and (4) sampling using a stratified random design. For sample size determination, Equation 11.1 is applied:

$$n = \frac{z^2 pqN}{e^2 (N-1) + z^2 pq} \tag{11.1}$$

where N is the total number of people in the population, n is the number of people in the sample, p is the sample percentage, q is the difference between p and q, and Z is the standard variate at the specified level of confidence 95% and error (0.05). Additionally, the population of Barunpara, N, is estimated to be 200. Using the procedure and equation provided, we can determine that there should be 89 people sampled from Barunpara village. The N for Holdibuniya village was 307 once again. Holdibuniya village requires a sample size of around 102 if one follows the procedures and uses the calculation provided.

Furthermore, the sample size for Henchi village was 156. The formula and procedures lead to a calculated sample size of 102 for the settlement of Henchi.

11.3.2 Social Vulnerability and Community Hazard Risk Assessment

For indexing vulnerability in the study area at the community level, vulnerability was framed as a subset of risk, because vulnerability and risk of any community under stressed environmental conditions are interlinked. Several factors, each measured on a scale from 1 to 10, add together to form a person's overall level of risk. Community disaster assessment methods, as outlined by Ferrier and Haque (2003), consist of the following four phases: (1) disaster recognition and regular analysis, (2) vulnerability estimation and effect analysis, (3) community implications evaluation, and (4) risk zoning.

Step 1: Disaster Recognition and Regular Analysis

Occurrence Characteristics	Score
In my nation, it is not uncommon for extreme events to occur (i.e. ones that exceed some criterion).	10
Occurs at least once per month in my community	9
Occurs at least once per year in my community	8
Has occurred once in the past decade in my country	7
Has occurred at least once in the past in my country	6
Occurrence is commonplace in my province/territory	5
Occurs at least once per year in my province/territory	4
Has occurred at least once in the past in my province/territory	3
Has occurred at least once in the past in Canada	2
Has occurred somewhere in the world, has some potential for occurrence	1

Step 2: Vulnerability Estimation and Effect Analysis

Impact Characteristics	Score
Causes a lot of people to be hurt or die. Generates economic harm from which my community would not be able to recoup financially.	10
As a consequence, at least ten people perish and others are critically harmed. Private residences and public infrastructure destroyed. Government discredit. Community-wide disruption. Emergency proclamation.	9
Causes severe injuries to many people and the deaths of five to ten. There was damage to both private and public buildings. Large costs for the hospital setting. Situations of normalcy are being disrupted just in a certain area.	8

(Continued)

(Continued)

Impact Characteristics	Score
Causes fewer than five deaths and significant injuries. Individual property damage. State infrastructure decay. Medical system losses exceed service expenses.	7
Causes one fatality and many serious injuries. Adversely affected by the transfer to private hands. An essential municipal service or utility has been severely disrupted. There are a lot of private residences that are unsafe to live in.	6
Causes serious harm to a large number of people. Certain homes and businesses sustained significant harm. Disruption of governmental amenities and services at a regional level. There are certain houses in each of these communities that are not suited for human occupancy.	5
It is possible to find communities where some of the houses are uninhabitable. Loss of utility or civic access in one specific area.	4
Results in widespread minor injuries, no major injuries. The majority of individual residences have suffered severely but are still habitable.	3
Causes a few people to sustain light damages. Localized environmental destruction.	2
The outcome is that nobody gets hurt. No material loss.	1

Step 3:   Community Implications Evaluation

Distinctive Features in Social Assessments	Score
It is possible that residents will not give their stamp of approval to a costly reaction, or would rather put off preventative measures.	10
The possibility for harm is minor, thus this is not a high community priority.	9
The projected damage is low, and the priority of response efforts should be evaluated in light of other societal requirements.	8
The social fallout should be minor, but a reaction is preferable.	7
It is reasonable to anticipate some social fallout, thus efforts should be made to prepare for them and lessen their impact if possible.	6
There is a general consensus among residents that the threat is only mild to moderate, thus they would rather take their time implementing preventative measures.	5
The community sees the risk of harm as low, therefore they would rather take a variety of quick and slow measures to prepare for it.	4
The public sees this as a threat and would rather take preventative measures right now than wait and hope for the best.	3
The community is aware of the possible negative and adverse effects.	2
The community understands this to be the most severe and perhaps catastrophic result.	1

Step 4:   Risk Zoning

1000	900	800	700	600	500	400	300	200	100
900	810	720	630	540	450	360	270	180	90
800	720	640	560	480	400	320	240	160	80
700	630	560	490	420	350	280	210	140	70
600	540	480	420	360	300	240	180	120	60
500	450	400	350	300	250	200	150	100	50
400	360	320	280	240	200	160	120	80	40
300	270	240	210	180	150	120	90	60	30
200	180	160	140	120	100	80	60	40	20
100	90	80	70	60	50	40	30	20	10

Note: 800–1,000 scores: very high-risk zone; 500–799 scores: high-risk zone; 200–499 scores: moderate risk zone; 1–199 scores: low-risk zone.

11.3.3   THE STATE OF THE COMMUNITY'S RESILIENCY

Documented in 2013, GOAL's "GOAL and Resilience: a Guidance Note" emphasizes development with adaptability training and reaction toward rising incidence as well as effect for crisis situations impacting overwhelmingly the globe's most vulnerable and impoverished communities. Community resilience status of each village is indicated through a community resilience status chart. Dual interpretations can be made of measuring the group resilience by using this toolkit. The toolbox will contain a predicted level of resilience based on an analysis of the key elements of resilience. There is also the option of using the toolkit to evaluate one's degree of resilience.

11.3.4   PUBLIC DISASTER PREPARATION INDEX

The US Department of Homeland Security developed public readiness indexing as a method to assess disaster preparedness in the year 2005. The Public Readiness Index (PRI) is focused on emergency preparedness knowledge elements and is measured on a scale of 0 to 3.

Knowledge Elements

1. Is there a catastrophe or contingency plan in place for your area?
2. Can you tune in to the radio's emergency programming station if a disaster strikes?
3. Have you seen any communication in the last 30 days that encourages residents to make their homes and neighborhoods more resilient to natural disasters?

11.3.5   HYPOTHESIZED RELATIONSHIPS BETWEEN KNOWLEDGE SHARING AND SOCIAL EXCHANGE VARIABLES

In this study pattern, this analysis proposes seven hypotheses. The hypothesis depicts the effect of sharing desire, reciprocity, altruism, trust, and sharing behavior on group information sharing behavior and attitudes.

H1. Members' propensity to share has a favorable effect on their habits of sharing information and expertise.

H2. As a rule, the level of trust between members has a constructive effect on the propensity to share valuable information.

H3. There is a favorable correlation between members' levels of trust and their willingness to provide information.

H4. As members help one another out, it increases the likelihood that they will share what they know.

H5. Knowing that other members will return the favor encourages them to share what they know.

H6. Individuals' propensity for charity affects their openness to share information.

H7. Participants' willingness to help one another out has a beneficial effect on the rate of information exchange inside the group.

11.4   DATA COLLECTION

Some sequential orders are followed in order to obtain the primary data. In the study area, a preliminary survey is carried out. The preliminary survey is then used to create a questionnaire. Finally, a household survey is carried out using a pretested questionnaire. The information is gathered through a personal interview in this case. The information required for the analysis is collected using a series of questionnaires. The method is also called structured interview. Specifically, there will be a random selection, a focus group discussion (FGD), and an in-depth interview. The data presented in Table 11.1 displays the participant's eligibility requirements.

TABLE 11.1
Methods Used to Include Participants

Method	Eligibility Requirements
Interview	• Leader of family in excess of 18 years of age
Focus group discussion	• With local honored persons (eight to ten persons)
	• With different occupational groups
	• With both male and female groups
Key Informant Interview	• With local civil society (teacher, religious leader, village mathabar)
	• During a meeting with officials from the community
	• Supported by medical staff
	• Along with volunteers from non-governmental organizations

Secondary data, such as statistical data, surveys, and maps, were gathered from a variety of government and nongovernment sources. Demographic data are gathered from nongovernmental organizations (NGOs) and union parishes in those areas. Disaster- and health-related data were gathered from the upazila health complex, referred national and foreign papers, published booklets, articles and author presentations, various government and nongovernment policy documents, seminar library of environmental science discipline at Khulna University, Bangladesh Bureau of Statistics (BBS), and meteorological data from the Bangladesh Meteorological Department.

Following the data collection from secondary and primary sources, attempts are made to process the information. Following sorting, data and information are classified and interpreted in accordance with the goals, and analysis is performed using various computer programs such as MS Office 2013, SPSS 21, Minitab, and others.

11.5   RESULTS AND DISCUSSION

11.5.1   DEMOGRAPHY

The population distribution shows that there are 200 families in Barunpara, 156 families in Henchi, and 307 families in Holdibuniya. The number of households in each square mile inside the community of Barunpara is 50, while it is 41 inside the community of Henchi and 55.81 in the community of Holdibuniya. As a whole, this region has a density of 50 households per square kilometer. The combined populations of Barunpara, Henchi, and Holdibuniya are 1,950 people. Barunpara has 485 men living there, Henchi has 178, and Holdibuniya has 743. There are 465 women living in Barunpara, 200 in Henchi, and 541 in Holdibuniya altogether. Between the villages of Barunpara, Henchi, and Holdibuniya, there are a total of 247 children aged 0–18 years. Barunpara village has an 89% literate population, Henchi village has a 65.80% literate population, and Holdibuniya village has a 56% literate population (secondary data from Upazila Parishad, 2019).

11.5.2   OCCURRENCE OF NATURAL DISASTERS

Natural disasters of many kinds are shown in Figure 11.2. As can be seen in Figure 11.2, river erosion occurs 32% more often in Barunpara than any other natural catastrophe. Henchi shows signs of having experienced the consequences of saline intrusion (35%). Holdibuniya community has

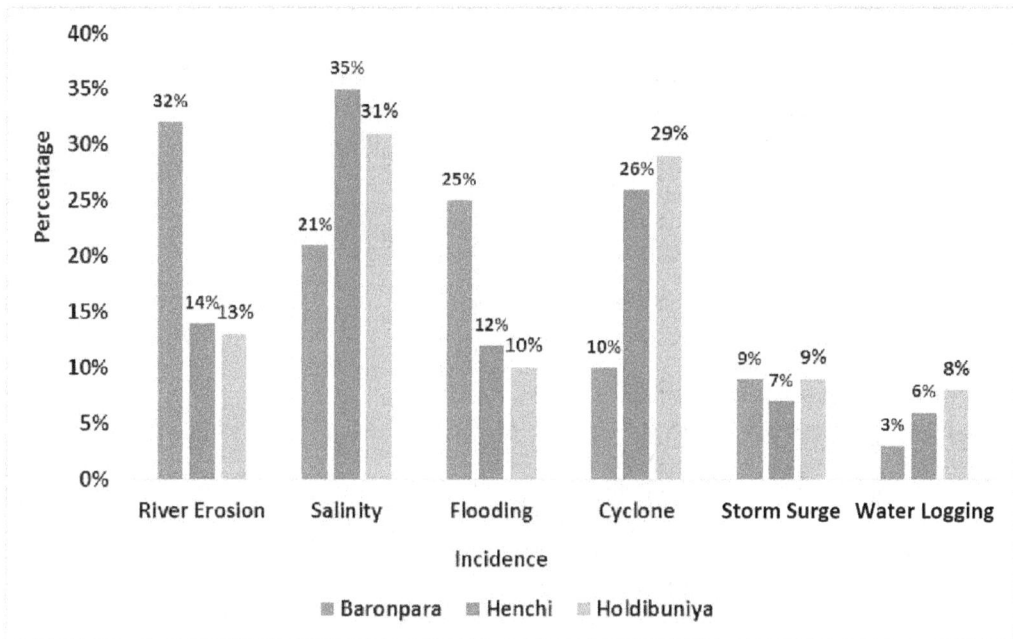

FIGURE 11.2 Incidence of natural hazards.

Source: Field Survey (2019).

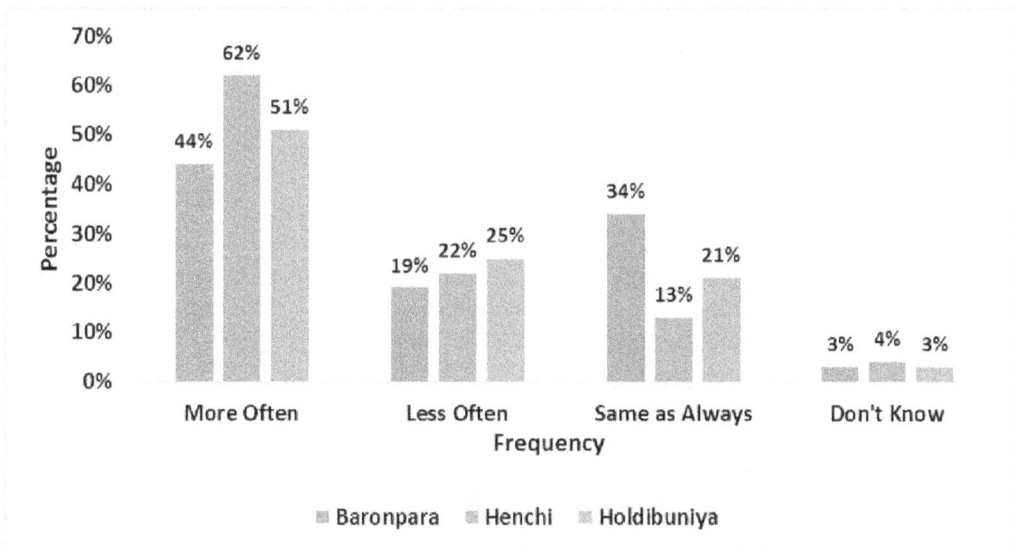

FIGURE 11.3 Frequency of natural hazards.

Source: Field Survey (2019).

seen a 29% increase in the frequency of cyclones and their aftermath. The areas of Barunpara and Henchi that have been hit most by the floods are 25% and 12%, correspondingly. The consequences of flooding and storm surges are insignificant in comparison to those of other natural disasters. The human toll from such natural disasters cannot be measured here.

There are about 5.2% as many natural disasters as there were a decade ago. The results suggest that 62% of people in Henchi village, 44% in Barunpara village, and 51% in Holdibuniya village all agree that they commonly encounter various natural disasters. Although 34% of people in Barunpara, 13% of people in Henchi, and 21% of people in Holdibuniya share this view, the frequency of natural disasters is widely regarded to be constant. It is worth noting that the vast majority of residents in certain areas believe that natural disasters occur with more regularity there.

11.5.3 Social Vulnerability and Community Hazard Risk Assessment

After a method of risk evaluation and risk rating, Table 11.1 reveals the threats to and vulnerabilities of the local population. Discussion on the current circumstances in these three settlements has been included.

Barunpara: A frequent event that may have severe and detrimental effects. Considered important by the populace at large. If you are looking at risk scores, this is in the 800–1,000 range, which is quite dangerous.

Henchi: The cumulative effect might be severe if it seldom occurs, or less severe if it occurs often. People tend to give it an average rating. The range from 500 to 799 is this section's territory. This is a really dangerous area.

Holdibuniya: A frequent event that may have severe and detrimental effects. Considered important by the populace at large. If you are looking at risk scores, this is in the 800–1,000 range, which is quite dangerous (Table 11.2).

11.5.4 Community Resilience Status

Table 11.3 indicates the results for the three villages of Barunpara, Henchi, and Holdibuniya in terms of their resilience. At Barunpara, the community has a resilience rating of 60%. The resiliency of this community is somewhere in the middle. This is more evidence of Henchi village's strong resiliency. With a resilience score of 57%, this community is in the middle of the resilience spectrum. Also, the table demonstrates Holdibuniya village's resilience level. With a score of 63%, the community is quite resilient. The village's level of resilience is somewhere in the middle.

11.5.5 Knowledge Scoring for Disaster Preparedness

The PRI is a tool for gauging the degree to which residents of a certain area have taken precautions to protect themselves and their family in the event of a catastrophe. The PRI assesses individual and family readiness by measuring their preparation and providing a practical "score."

Barunpara

Figure 11.4 shows the lowest knowledge score is now in first place. There are 38.20% of people who have a poor knowledge score. Out of all families, 29% fall into the "medium awareness" category, while 34% have a "high knowledge" score. A very high level of knowledge is notable.

Holdibuniya

Figure 11.5 shows the maximum possible score in this group is a mediocre 49.02% on the knowledge scale. A total of 37% of the population has a poor knowledge score, and another 13% also has poor knowledge score.

Henchi

According to the data shown in Figure 11.6, the residents of Henchi village have a knowledge of disasters that ranges from moderate (33%), to low (56%), to high (11%).

TABLE 11.2
Risk Scoring

Serial	Event	Barunpara				Henchi				Holdibuniya			
		Hazard Ranking	Vulnerability Indexing	Social Cost Prioritization	Risk Valuation	Hazard Ranking	Vulnerability Indexing	Social Cost Prioritization	Risk Valuation	Hazard Ranking	Vulnerability Indexing	Social Cost Prioritization	Risk Valuation
1	River erosion	8	7	6	331	5	4	5	100	8	7	6	336
2	Salinity	5	6	6	180	8	7	6	336	5	6	6	180
3	Flooding	3	3	4	36	4	4	5	80	3	3	4	36
4	Cyclone	7	6	5	210	7	5	5	175	7	6	5	210
5	Storm Surge	3	4	4	48	4	5	4	80	3	4	4	48
6	Water logging	2	3	2	12	2	3	2	12	2	3	2	12
	Total score				817	Total score			743	Total score			822

Source: Field Survey (2019); adopted from Ferrier and Haque (2003).

TABLE 11.3

Community Social Resilient Status

Percentage (%)	Level	Category	Description
0–20	1	Poor resistance to damage	Both the problem and the need to fix it are mostly unknown. Only emergency measures will be taken.
21–40	2	Little resilience	Knowledge of the problem and a willingness to work to solve it. Ability to operate (learning and capabilities, personal, technical, and other factors) persists. One-time, narrow procedures are typical.
41–60	3	Average resilience	Conceiving problems and working to solve them. Greater and more significant action capacity. There are more interventions, and they tend to last longer.
61–80	4	Resilient	Congruence and full integration. The scale and scope of the interventions are commensurate with the magnitude and complexity of the issue, and they are interconnected within a comprehensive, long-term plan.
81–100	5	High resilience	All parties have adopted a "culture of security in which disaster risk reduction is intrinsic to all effective legislation, organization, operation, approach, and conduct."
Community resilience score		Barunpara	60% (Medium resilience)
		Henchi	57% (Medium resilience)
		Holdibuniya	63% (Resilient)

Source: Data of focus group discussion in 2019 adapted Basu, et al. (2013).

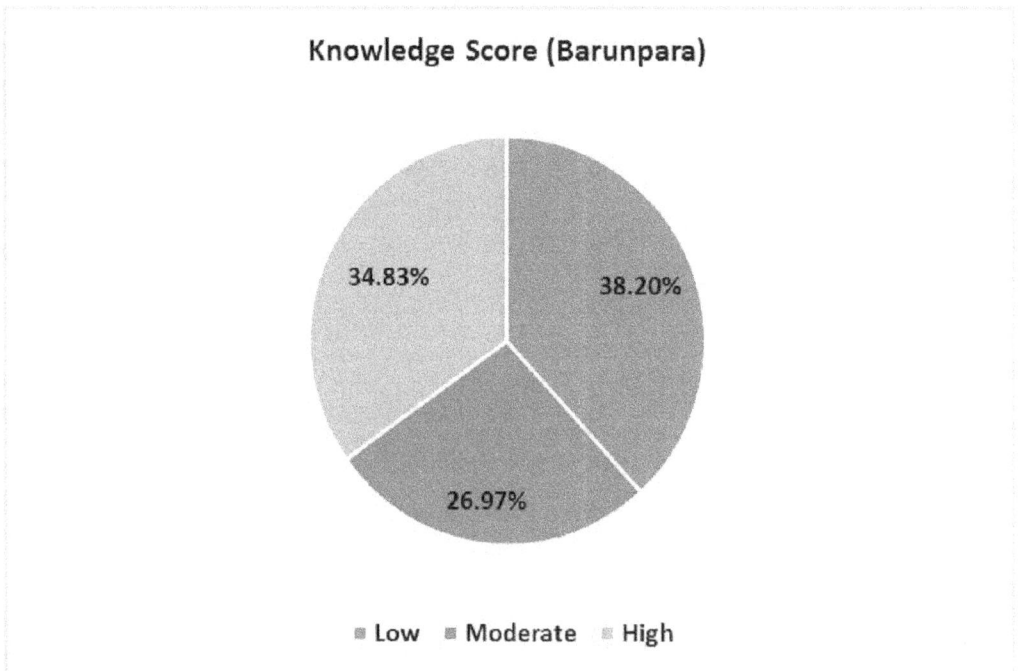

FIGURE 11.4   Knowledge about disaster preparedness in the community of Barunpara village.

Source: Field Survey (2019).

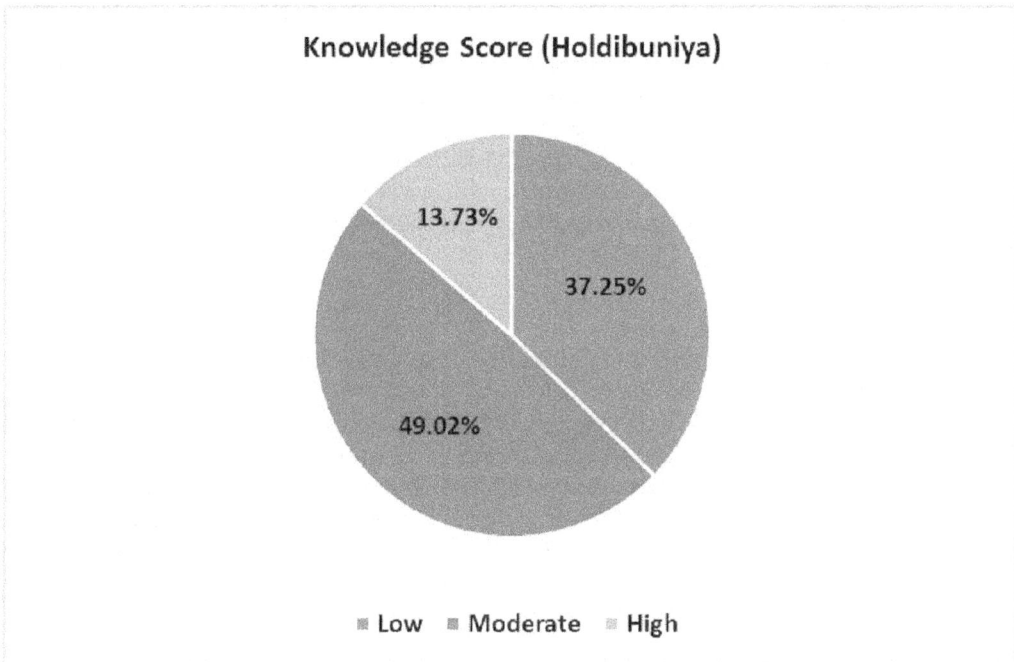

FIGURE 11.5 Knowledge about disaster preparedness in the community Holdibuniya village.
Source: Field Survey (2019).

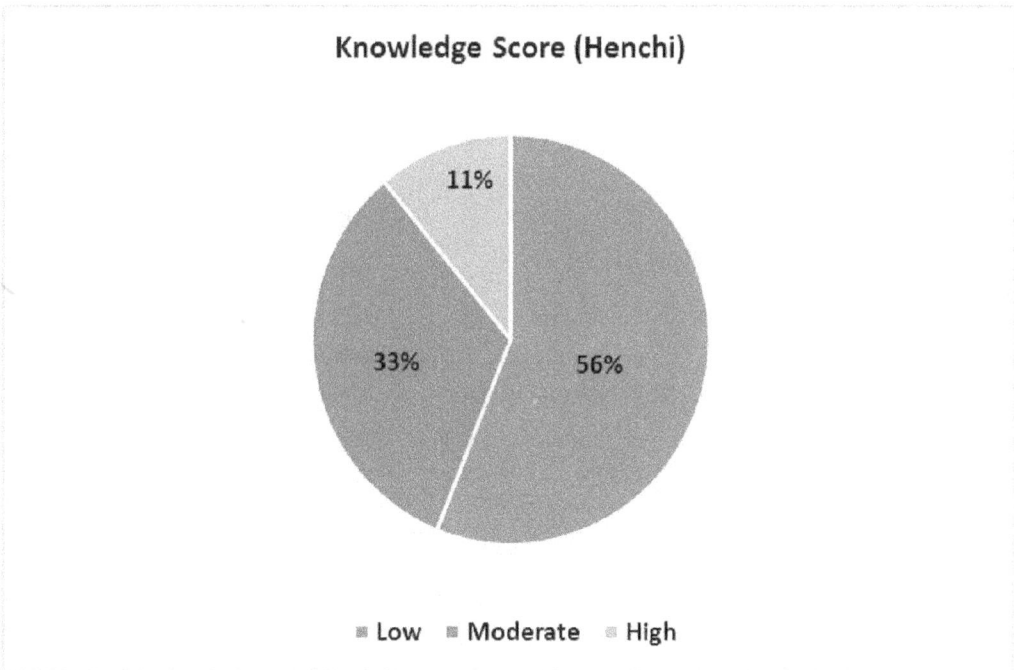

FIGURE 11.6 Knowledge about disaster preparedness in the community Henchi village.
Source: Field Survey (2019).

11.5.6 Influence of Social Capital for Knowledge Sharing Among Community

Here we can see the correlation matrix between the three towns of Barunpara, Henchi, and Holdibuniya. All of the correlations are positive and statistically significant. Matrix of correlation for exogenous variables are shown 0.05 volume of significance. These factors have been formatted into five equal quintiles, which has resulted in their having a normal distribution. Consequently, we utilize Pearson's R to examine how strongly the variables are connected to one another under consideration. A correlation matrix is "square" because its rows and columns include the same information about the variables being correlated. It demonstrates how many social factors are linked to certain individuals.

Barunpara

All correlations in Table 11.4 are meaningful and statistically significant (at the 0.05 level). Despite the fact that just an 0.05 level of significance is required for this study, all of the results are significant. The confidence that individuals have in their neighbors has the weakest link with the advantages that come from people participating. The weakest correlation recommends that members of the community should participate in a variety of get-togethers within the community and should discuss this topic with their neighbors. Despite this, the connection is one that is both beneficial and meaningful. There is conflicting information in the matrix of Table 11.5. Donation is significantly correlated with participation in politics, club meetings attended, and religious organization engagement. There is a significant correlation between most forms of social conduct. All coefficients are less than 0.08; therefore, heteroscedasticity is not an issue.

Henchi

The tale told by the Table 11.6 matrix is complicated. It is safe to trust the vast majority of people on any social topic, and there are strong positive relationships between the advantages of participating in various social organizations. Those in the area who regularly attend social events report feeling closer to one another and more comfortable exchanging information with one another in the event of a crisis.

The correlation matrix shown in Table 11.7 shows strong positive associations between participation in religious and social groups and the acquisition of information relevant to emergency preparation. The notation indicates a correlation between the two variables in question.

Holdibuniya

Both government responsiveness and community gain through social organization membership are positively correlated, as seen in Table 11.8 of the correlation matrix. The notation indicates a connection between the two variables in this expression.

TABLE 11.4
Attitudes and Social Capital: A Correlational Analysis (Barunpara)

	Common Belief	Have Faith in Nearby People	Government Is Responsive	Humans Are Kind and Willing to Help	Achieving Success through Involvement
Common belief	1.000				
Have faith in nearby people	0.104	1.000			
Government is responsive	0.217	0.188	1.000		
Humans are kind and willing to help	0.359	0.408	0.228	1.000	
Achieving success through involvement	0.351	0.342	0.323	0.323	1.000

TABLE 11.5
Analysis of the Correlation Matrix for the Behavioral Aspect of Social Capital (Barunpara)

	Participation in Religious Practice	When It Comes to Politics	Participants	Contribution	Participates at a Group Conference
Participation in religious practice	1.000				
When it comes to politics	0.138	1.000			
Participants	0.232*	0.250*	1.000		
Contribution	0.362**	0.462**	0.258*	1.000	
Participates at a group conference	0.358**	0.400**	0.273*	0.355**	1.000

* It is statistically meaningful that there is a correlation between these two variables (two-tailed) at 0.05 magnitude.
** The significance threshold for the correlation is 0.01 (two-tailed).

TABLE 11.6
Relationship Chart Analyzing the Attitude Component of Social Capital (Henchi)

	Common Belief	Have Faith in Nearby People	Government Is Responsive	Humans Are Kind and Willing to Help	Achieving Success through Involvement
Common belief	1.000				
Have faith in nearby people	0.090	1.000			
Government is responsive	−0.073	0.260*	1.000		
Humans are kind and willing to help	0.475**	−0.233	−0.017	1.000	
Achieving success through involvement	−0.209	0.463**	0.627**	−0.081	1.000

TABLE 11.7
Graphing the Behavioral Aspects of Social Capital's Correlations (Henchi)

	Participation in Religious Practice	When It Comes to Politics	Participants	Contribution	Participates at a Group Conference
Participation in religious practice	1.000				
When it comes to politics	0.516**	1.000			
Participants	−0.261	−0.057	1.000		
Contribution	0.177	0.099	0.636**	1.000	
Participates at a group conference	0.035*	0.183	−0.256	−0.175	1.000

* It is statistically meaningful that there is a correlation between these two variables (two-tailed) at 0.05 magnitude.
** The significance threshold for the correlation is 0.01 (two-tailed).

TABLE 11.8
Attitudes and Social Capital: A Correlational Analysis (Holdibuniya)

	Common Belief	Have faith in Nearby People	Government Is Responsive	Humans Are Kind and Willing to Help	Achieving Success through Involvement
Common belief	1.000				
Have faith in nearby people	0.156	1.000			
Government is responsive	0.238	0.210	1.000		
Humans are kind and willing to help	0.312	0.463	0.285	1.000	
Achieving success through involvement	0.305	0.399	0.197	0.407	1.000

TABLE 11.9
The Holdibuniya Matrix: A Correlation Analysis of the Behavioral Aspect of Social Value

	Participation in Religious Practice	When It Comes to Politics	Participants	Contribution	Participates at a Group Conference
Participation in religious practice	1.000				
When it comes to politics	0.130	1.000			
Participants	0.242*	0.207*	1.000		
Contribution	0.339**	0.463**	0.270**	1.000	
Participates at a group conference	0.310**	0.393**	0.196*	0.388**	1.000

* It is statistically meaningful that there is a correlation between these two variables (two-tailed) at 0.05 magnitude.

** The significance threshold for the correlation is 0.01 (two-tailed).

According to the correlation matrix shown in Table 11.9, the level of civic engagement and charitable giving is the most strongly correlated of all the variables studied. It is computed with a 95% degree of confidence, and a 0.05 margin of error is allowed.

11.5.7 HYPOTHESIZED RELATIONSHIPS BETWEEN KNOWLEDGE SHARING AND SOCIAL EXCHANGE VARIABLES

Barunpara

Table 11.10's findings point to the importance of trust (0.791), reciprocity (0.681), openness to sharing (0.673), and readiness to share (0.606) in influencing the rate at which local communities share information with one another. Knowledge sharing behavior, on the other hand, may be predicted, at least in part, by altruism.

Henchi

Compassion and a desire to share are the most influential elements among communities for information sharing in catastrophe risk reduction, as shown in Table 11.11. The results reveal an altruism value of 0.824 and a sharing willingness value of 0.653.

TABLE 11.10
Testing Hypotheses: Findings (Barunpara)

Hypothesis			Estimates	Standard Error	P-Value	Description
Acts of generosity	←	Knowledge sharing	0.606	0.080	<0.05	Support
Belief	←	Knowledge sharing	0.791	0.064	<0.05	Support
Nobility	←	Knowledge sharing	0.479	0.084	<0.05	Support
Cooperation	←	Knowledge sharing	0.681	0.067	<0.05	Support
The spirit of cooperation	←	Knowledge sharing	0.673	0.081	<0.05	Support

TABLE 11.11
Testing Hypotheses: Findings (Henchi)

Hypothesis			Estimates	Standard Error	P-Value	Description
Acts of generosity	←	Knowledge sharing	0.213	0.083	<0.05	Support
Belief	←	Knowledge sharing	0.446	0.083	<0.05	Support
Nobility	←	Knowledge sharing	0.824	0.061	<0.05	Support
Cooperation	←	Knowledge sharing	0.221	0.084	<0.05	Support
The spirit of cooperation	←	Knowledge sharing	0.653	0.064	<0.05	Support

Holdibuniya

Table 11.12 demonstrates the predictive power of a readiness to share (0.673), a desire to give (0.648), and a desire to receive (0.662). It is also a big reason why people confide in one another and share information.

11.6 CONCLUSION

The result of this study shows that people of Barunpara and Holdibuniya face regular occurrences of disaster with potentially serious and damaging impacts having 60% and 63% of social resilience status, respectively. The Henchi people have a social resilience status of 57%, meaning that they can withstand the cumulative consequences of disasters that are either infrequent but severe or frequent but comparatively less severe. There is a lack of what is necessary to reduce catastrophe risk in such communities due to a lack of understanding about disaster preparation behavior. A variety of variables influence the spread of knowledge, and all of these aspects are positively correlated. The results of the test of the thesis based on the hypothesis are significant and illustrate the effects of the experiment. Many social organizations and activities are affected by the extent of people's social capitals. The effects of social capital on one's outlook and actions have been explored here. Several

TABLE 11.12

Testing Hypotheses: Findings (Holdibuniya)

Hypothesis			Estimates	Standard Error	P-Value	Description
Acts of generosity	←	Knowledge sharing	0.568	0.073	<0.05	Support
Belief	←	Knowledge sharing	0.648	0.068	<0.05	Support
Nobility	←	Knowledge sharing	0.437	0.086	<0.05	Support
Cooperation	←	Knowledge sharing	0.673	0.066	<0.05	Support
The spirit of cooperation	←	Knowledge sharing	0.662	0.078	<0.05	Support

analyses of social dimensions have suggested a positive association. A community's ability to generate and share information as well as put that knowledge into practice is posited as a means of lowering its catastrophe risk. The aim of this study was to establish a connection between social capital and community knowledge sharing using social exchange theory to identify community information exchange activity that would help mitigate disaster effects in these areas. Further research is needed to discern people's characteristics and the incentives that motivate different people's actions so that knowledge sharing behaviors can be developed at the group level. The research will look at how people share their knowledge based on socioeconomic and environmental factors.

REFERENCES

Akter, S., & Rahman, M. M. (2023). Demographic Determinants of Disaster Preparedness Behavior among the Inhabitants of Southwest Coastal Bangladesh. *Coastal Disaster Risk Management in Bangladesh: Vulnerability and Resilience.*

Basu, M., Srivastava, N., Mulyasari, F., & Shaw, R. (2013). Making cities and local governments ready for disasters: for disasters: A critical overview of a recent approaches. *Risk, Hazards & Crisis in Public Policy, 4*(4), 250–273.

Cadag, J. R. D., & Gaillard, J. C. (2012). Integrating knowledge and actions in disaster risk reduction: The contribution of participatory mapping. *Area, 44*(1), 100–109.

Ferrier, N., & Haque, C. E. (2003). Hazards risk assessment methodology for emergency managers: A standardized framework for application. *Natural Hazards, 28,* 271–290.

Gubbins, C., & Dooley, L. (2021). Delineating the tacit knowledge-seeking phase of knowledge sharing: The influence of relational social capital components. *Human Resource Development Quarterly, 32*(3), 319–348.

Magni, G. (2017). Indigenous knowledge and implications for the sustainable development agenda. *European Journal of Education, 52*(4), 437–447.

Sanyal, S., & Routray, J. K. (2016). Social capital for disaster risk reduction and management with empirical evidences from Sundarbans of India. *International Journal of Disaster Risk Reduction, 19,* 101–111.

Shahid, S. (2012). Vulnerability of the power sector of Bangladesh to climate change and extreme weather events. *Regional Environmental Change, 12,* 595–606.

Wisner, B., Blaikie, P. M., Blaikie, P., Cannon, T., & Davis, I. (2004). *At Risk: Natural Hazards, People's Vulnerability and Disasters.* Psychology Press.

12 Spatio-Temporal Analysis of Urban Growth on North Guwahati's Forest and Agricultural Land

Sneha Deka

12.1 INTRODUCTION

The world is urbanising rapidly, and spatial change has become more prominent over the past few years. Along with economic and technological development, the transformation of a geographic location from rural to urban influences the area's social and physical environment. The world underwent rapid urban growth in the twentieth century. According to the United Nations (UN), in the 1990s, only 13% of the world's population lived in urban areas ('Percentage of Population at Mid-Year Residing in Urban Areas by Country/Area 2000–2050', n.d.). Nevertheless, urban dwellers have dramatically increased over the past two decades. As of 2022, the world urban population was 2,356,985,000 of the total 8 billion (18th edition of *Demographia World Urban Areas*, Wendell Cox, 2022; 'UN Report', 2023). According to this, by 2050, approximately 68% of the world's population will live in urban areas (UN; 'Most Urbanized Countries 2023', n.d.). This will add up to 2.5 billion people in urban areas worldwide.

There are significant differences in the rate of urbanisation, as it is much faster in developing countries than in developed ones (Soubbotina, 2004; Kiamba, 2012; Aburas et al., 2018; Follmann, 2021; Chatterjee & Roy, 2021; Arshad et al., 2022). In countries like India, the urbanisation rate has accelerated in the contemporary period (Liu et al., 2018; Xu et al., 2020). The urbanisation process is mainly initiated by industrialisation in the developed world, whereas it is not the same in the developing world. In developing countries, the permanent concentration of a larger population in a relatively smaller area triggers the process of urbanisation. The rapid growth of urbanisation will become one of the critical concerns for change in the ecological landscape of these urban areas over the coming years.

India is one of the developing countries that has witnessed fast-paced urban growth in the past few years. Studies have shown that urbanisation has led to a high population influx to the country's major cities. There was an increase of approximately 91 million urban population in 2011 compared to 2001 (Mundhe, N., Jaybaye, R.G; A Spatio-Temporal Analysis of Urbanization in India, 2014). The increasing urban population with high living standards (United Nations Department of Economic and Social Affairs, n.d.; Jedwab et al., 2015; 'Urban Population and Demographic Trends', n.d.; 'Urban Population and Demographic Trends | Urban Indicators Database', n.d.; 'Urbanization', n.d.) puts ecological pressure on the land. Incorporating remote sensing and geographic information systems (GIS) in monitoring urban spaces assists in a balanced development process, securing the sustainability of such areas (Nkeki, 2016; Drswideg, 2018; Murayama et al., 2021; Li et al., 2023). Over time, land-use and land-cover (LULC) change represent insight into urban areas' growth processes and situations. To sustain the urban areas, a planned development with a level of understanding of the spatial patterns will serve as an advantage for a balanced and strategic city.

DOI: 10.1201/9781003377825-14

12.2   LITERATURE REVIEW

The available literature shows that urbanisation, urban growth, urban fringe, urban sprawl, land-cover transition, land use, and other related studies are mainly formed based on the developed world's cities. Over the years, at a global level, scholars have worked on the growth of urbanisation and its impact on its surroundings. Most of these works are based on urban regions of the developed world. Even in India, scholars have presented works on urbanisation based on major cities like Delhi and Kolkata and areas like parts of central India, West Bengal, Jammu and many others.

Mundhe et al. (2014) talked about the spatial patterns of urbanisation at a national level in India from 1991 to 2011. They gave an overall view of the change in population concentration in India by categorising the country into different groups by region based on population density.

In their paper, Saha et al. (2022) talk about the temporal change and possible future spatial scenario of a part of West Bengal. The spatial change detection over time with the help of techniques such as the Markov chain model, Pearson's chi-square test, random group surveys and discussions, kappa coefficient (k), etc., have proved to be effective in drawing a picture of the spatial utilities of urban areas.

Kar et al. (2018) produced spatio-temporal data on the urban sprawl of Nagpur city of Maharashtra by supervising the LULC changes and presented pointers related to maintaining a sustainable urban environment. The research exhibits the possibilities of GIS and temporal satellite data in evaluating the spatio-temporal options of the urban and peri-urban landscape for sustainable management of land resources (Monitoring spatio-temporal dynamics of urban and peri-urban landscape using remote sensing and GIS—A case study from Central India, Kar et al., 2018).

Delhi, the capital city of India, also underwent hasty urban growth and a high population influx in the last few decades. In terms of population, it became the second-largest city after Mumbai (Census tables | Government of India, n.d.; Moghadam & Helbich, 2013). This change acted as a chief indicator for numerous urban, environmental and socio-economic problems, namely deterioration in the air quality, groundwater, surface water, and green spaces (Guttikunda & Calori, 2013; Kumar et al., 2016; Li et al., 2023; Maiti & Agrawal, 2005; Nagar et al., 2017; Ramachandraiah et al., 2004; Sahu et al., 2018; Sharma et al. 2018). The city has become home to many dwellers, giving rise to the rapid growth of urban areas (Chadchan & Shankar, 2012; Jain et al., 2016; Mohan et al., 2011; 'The World Is Becoming Increasingly Urbanized—UN', n.d.; 'Largest Urban Areas Globally by Population 2022', n.d.). Tripathy and Kumar (2019) illustrate the LULC change (1989–2014) for modelling the urban growth of Delhi between 1994 and 2024 from the acquired data of the research work (Monitoring and modelling spatio-temporal urban growth of Delhi using Cellular Automata and geoinformatics).

Saha et al. (2022) talked about the spread of urban sprawl and its influence on the sustainability of a region. They analysed the pattern of urban growth at the zonal level and for a municipality corporation of West Bengal. Through remote sensing and GIS techniques, Saha et al. (2022) portray the growth pattern of Siliguri Municipal Corporation (A spatio-temporal analysis of urban growth and identification of urban sprawling of Siliguri Municipal Corporation, West Bengal, India).

Mumbai is another major city in India, witnessing the most accelerated urban growth. This fast-paced change has had unconventional effects on urban sprawl and population change. Moghadam and Helbich (2015) explore the spatio-temporal evolution of urban development in Mumbai city from 1973 to 2010, presenting the land-use categories most affected by urban expansion and attributing the past urban growth process. The scholars were given a model showing the city's future growth patterns from 2010 to 2030.

With the fast-growing urbanisation, North-East India has also begun to develop welcoming urban footprints and dwellers. Guwahati (Assam) is one such city that has been growing faster over a few years. This city's land-use pattern is changing at an alarming rate, hampering the natural environment. The amount of forest cover, open area, and water bodies declined, whereas built-up and industrial sites have increased from 1972to 2016. With increased industries and technological development, the population influx has also led to pressure over land and resources (Impact of Urban Growth on Landuse, A Case Study Of Guwahati City, Assam, Talukdar, K. K., 2018).

Previous studies and research show that the urban growth process and change over land have influenced the landscape of cities in various ways. However, very few studies focus on the north-eastern part of India. This research tries to illuminate the temporal change in LULC of North Guwahati. This research attempts to cover the gaps available and portray the evolution and impact of urban growth in Assam's much less-known geographic location.

12.3   PROBLEM AND OBJECTIVE

In recent years, the city of Guwahati has undergone various development and redevelopment processes for urban growth. North Guwahati, a neighbour of Guwahati city, has also felt the influence of urban growth due to its geographical location. With the development in the economic sector, the permanent settlement of the population has been expanding over time. The spread of urban growth has hence started influencing the landscape of North Guwahati. There have been a few very noticeable differences in some parts of the study area. This research attempts to analyse the pattern of urban growth over time with the help of GIS techniques. It also throws light on the alteration of LULC pattern through analysis of Landsat satellite imagery and the impact on the forest and agricultural land. It also tries to predict the possible causes of such rapid change.

12.4   DATA SOURCE AND METHODOLOGY

To study the temporal change in the LULC pattern of the study area, satellite images were acquired from Land Viewer, Earth Observation System Data Analytics (EOSDA) and EarthExplorer, United States Geological Survey (USGS). All the acquired images were captured by the spacecraft Landsat 8 with sensor ID OLI_TIRS (Operational Land Imager, Thermal Infrared Sensor). The LULC pattern was analysed from the satellite images by unsupervised classification. GIS platforms such as Erdas Imagine 2014, ArcMap 10.3, and MapInfo 4.0 were used to analyse the images. The data obtained after analysing the images were further processed and represented in graphs and figures using Microsoft Excel.

12.5   STUDY AREA

The area under study for this research is North Guwahati, a satellite town situated on the Brahmaputra's northern banks and to the north of Guwahati city in the north-eastern state of Assam (Figure 12.1). There are about 35 districts in Assam, and the state capital, Dispur, is located in the Kamrup Metropolitan district ('Home | Assam State Portal', 2023).

North Guwahati falls under two different districts, Kamrup Rural and Kamrup Metropolitan (Figure 12.2). The satellite town is inhabited by approximately 54,477 netizens (Census tables | Government of India, n.d.; NEWS, NE NOW 2021), of which 36,405 belong to the Kamrup Rural district and 18,072 belong to the Kamrup Metropolitan district. There are 38 villages and towns in the study area, of which 32 are in Kamrup Rural, and six are in Kamrup Metropolitan (International Res Jour Managt Socio Human, 2022). The study area extends between 26°10′34.39″ N to 26°91′45.84″ N latitude and 91°35′18.2″ E to 91°46′2.75″ E longitude.

12.6   LAND-USE AND LAND-COVER PATTERN

The LULC pattern of North Guwahati has changed over time. About two decades ago, this region was covered with natural vegetation, forests, wetlands, and very few settlements and built-up areas. Around 2007, one of the significant changes that took place was the clearing up

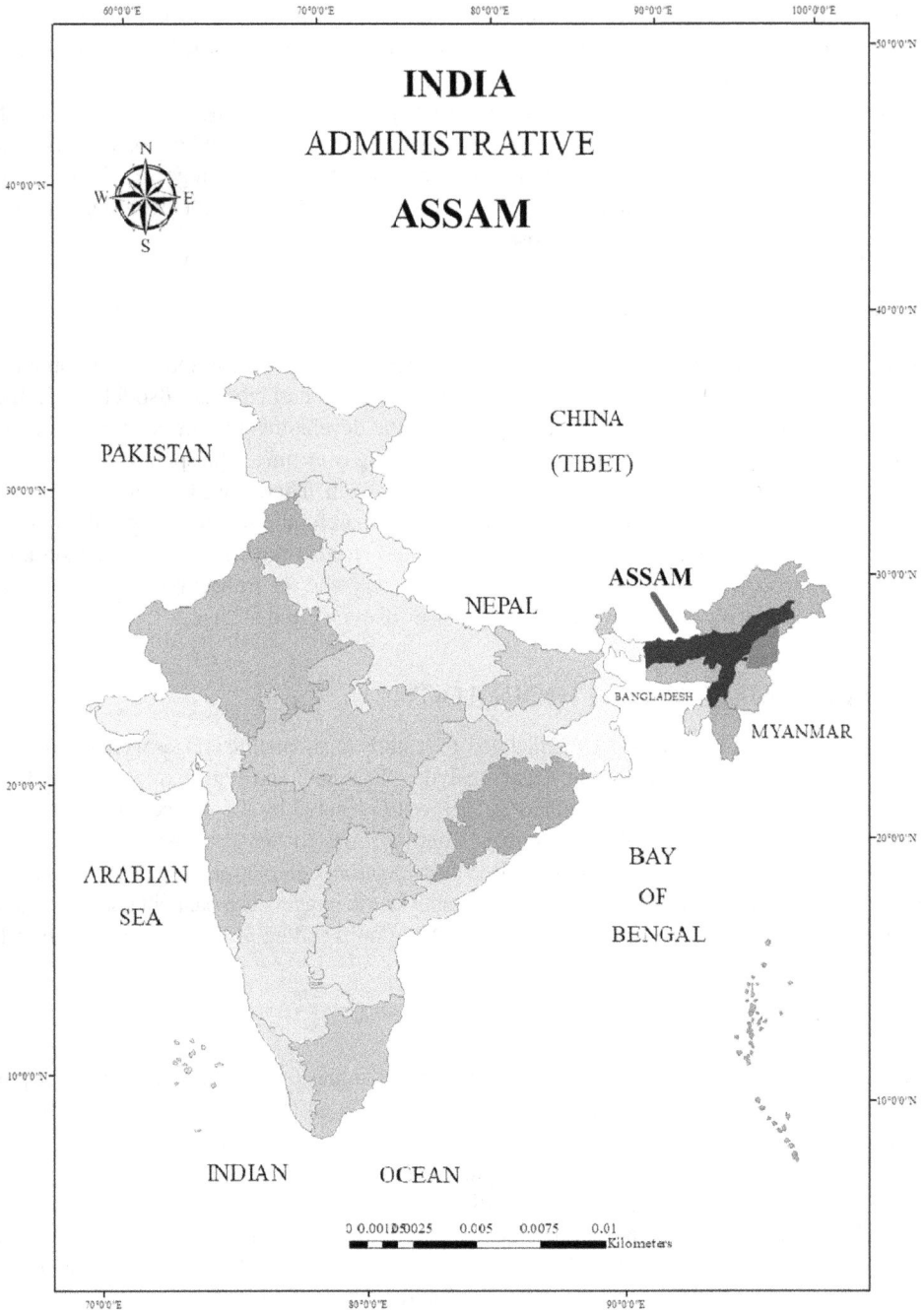

FIGURE 12.1　Map of India showing the location of Assam.

of a large sum of natural vegetation and the filling up of wetlands of Brahmaputra to construct a bridge over the Brahmaputra river. The construction work brought change to the northern and southern banks of the river. The construction lasted for almost a decade, and by 2017, communication from the metropolitan city to the satellite town had become faster and more accessible due to the reduction of congestion. During and after the connectivity improvement with the satellite town, North Guwahati welcomed the establishment of various organisations, such as

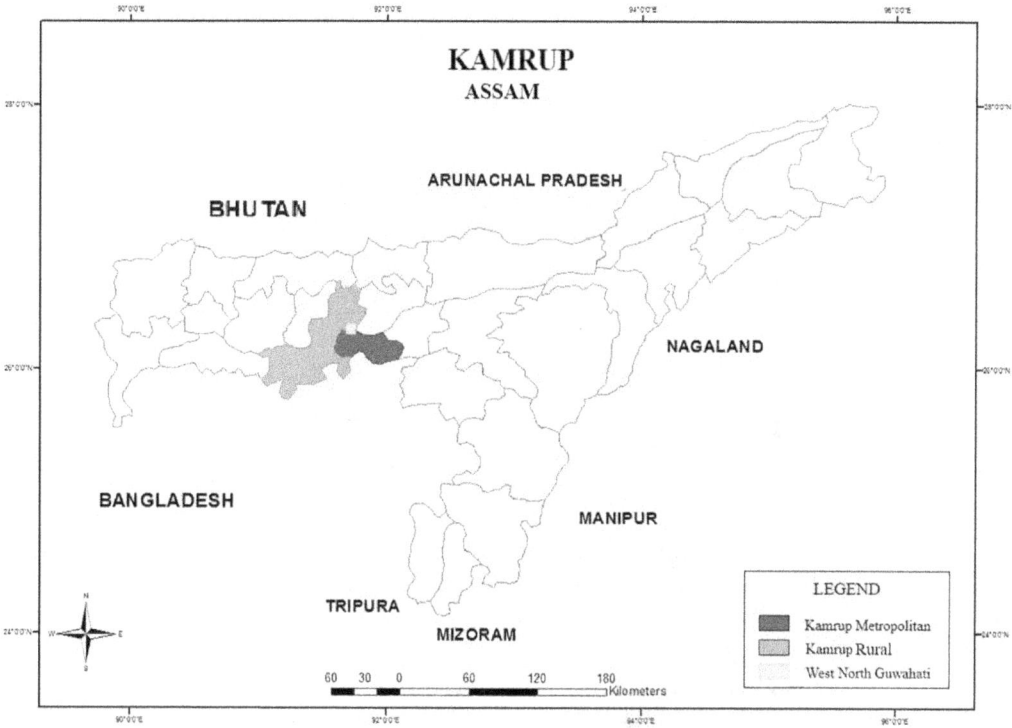

FIGURE 12.2   Map of Assam showing the location of Kamrup Rural and Kamrup Metropolitan districts and the study area, North Guwahati.

educational, medical, industrial, and trade. With time, the concentration of the population also started growing due to the introduction of new possibilities for livelihood. This research work studies the change of LULC in three different years, i.e. 2013, 2016, and 2022, through unsupervised classification.

12.6.1   22 November 2013

In 2013, the influence of urban growth was less felt over other geographical attributes in North Guwahati. The map in Figure 12.3 shows different LULC features covering the study area by unsupervised classification.

12.6.2   Water Body

The Brahmaputra river is the water body running along the southern part of the study area, which separates North Guwahati from the metropolitan city of Guwahati. This river's drainage basin covers an area of 70,634 km^2 alone throughout Assam.

Due to its geographic location near the mighty river, the water table of the study area is high. This is one of the factors leading to the presence of shallow water bodies such as streams and wetlands. The stream in the north-western part of the study area is the Puthimari river, a tributary to the river Brahmaputra. The shallow water bodies observed in the southern part of the study area and to the east of the dense natural vegetation are lakes situated in and around the campus of the Indian Institute of Technology, Guwahati. The ones observed to the west of the dense natural vegetation areas in the southern part of the study area, and the east are wetlands and swamps.

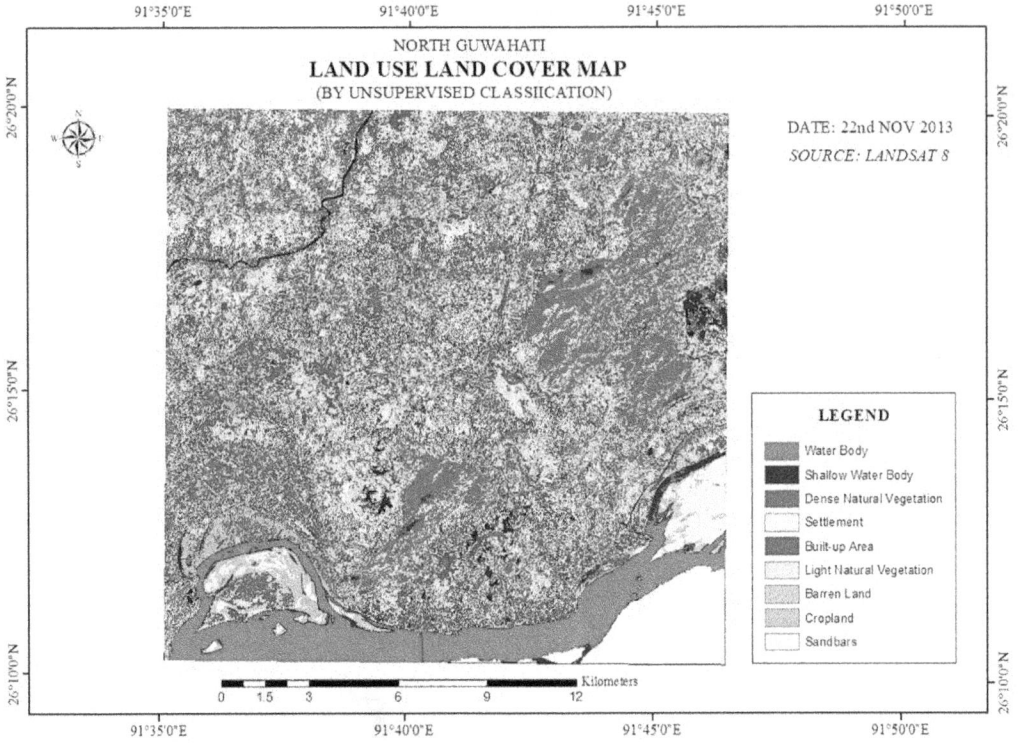

FIGURE 12.3 Land-use/land-cover map of North Guwahati, 2013.

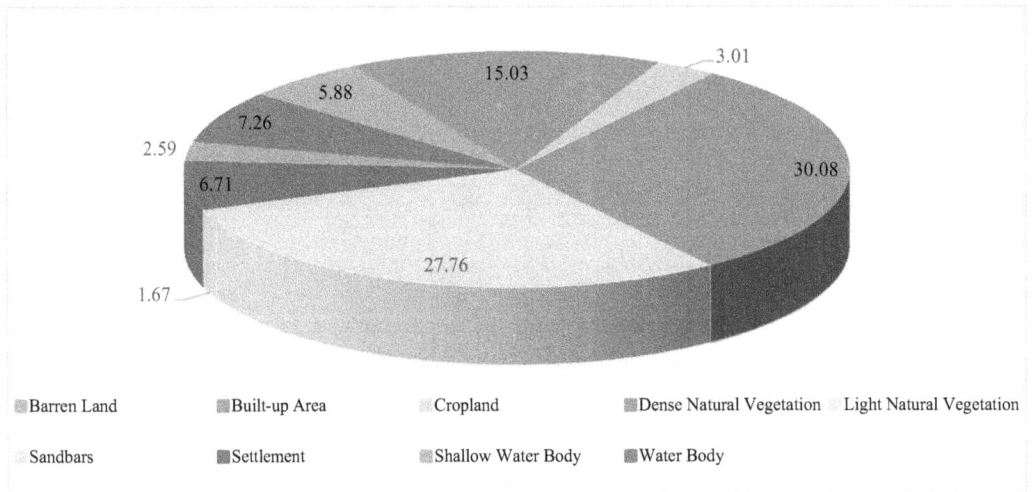

FIGURE 12.4 Graph showing the percentage of area covered by different land-use/land-cover patterns, 2013. The following analysis regarding various attributes can be generated from the map.

12.6.3   Dense Natural Vegetation

The dense natural vegetations to the south and east of the study area are hills with dense forests. The others are mostly forested areas over flat landscapes. It can also be observed along the River Puthimari and the roadway. It can be assessed that most of the portion of the study area was covered by dense natural vegetation.

12.6.4   Settlement

The concentration of population in North Guwahati was scattered throughout the area. Most settlements can be observed near the water bodies and the roadways. Nevertheless, the area covered by the settlement is less than that of natural vegetation.

12.6.5   Built-up Area

After natural vegetation, the built-up area covered most of North Guwahati. The built-up area represented by the unsupervised classification map includes roadways, railway tracks, industries, factories, and clusters of buildings. The Erdas Imagine software did not recognise some routes as they were unmetalled roads or local streets. The long stretch of built-up area running from north to south across the River Brahmaputra is National Highway (NH) 27, and the one running along it is a railway line. The area recognised as built-up over the Brahmaputra is the older Saraighat bridge, and the new bridge was still under construction in 2013. The roadway stretching towards the west from the bridge in the southern part of the study area is NH 427. The built-up area recognised over the sandbars are *char*[1] areas.

12.6.6   Light Natural Vegetation

In the analysed map, light natural vegetation covers the grasslands and less crowded forested areas. As the satellite image collected is from 22 November 2013, a few agricultural lands were cleared out, why they were recognised as light natural vegetation. Most of the light natural vegetation areas to the east of the NH 427 were empty stretches of land.

12.6.7   Barren Land

The barren lands were scattered throughout the study area and were much less than the other geographical attributes. A maximum of it was noticed over the sandbars and towards the south-eastern part of the study area. Some stretches of barren land were observed in and around NH 27 and near the dense natural vegetation.

12.6.8   Cropland

The area analysed as cropland in the south-western part of North Guwahati was agricultural land used for commercial agriculture. Other than this, the areas recognised as cropland were mainly used for subsistence agriculture.

12.6.9   Sandbars

As the satellite image captured was during the winter, the sandbars were visible due to the depletion of the water level of the river Brahmaputra. This river is perennial and deep, so it does not dry up entirely.

12.7   16 DECEMBER 2016

According to the analysed map of 2016, the influence of urban growth in the study area was much more potent compared to 2013.

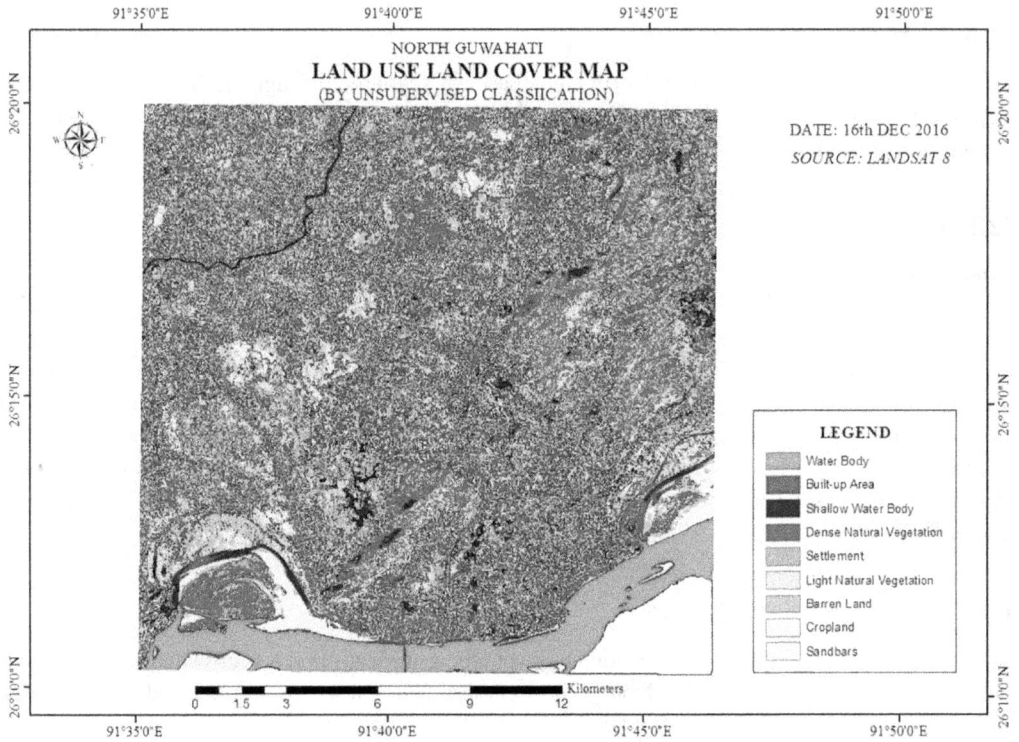

FIGURE 12.5   Land-use/land-cover map of North Guwahati, 2016.

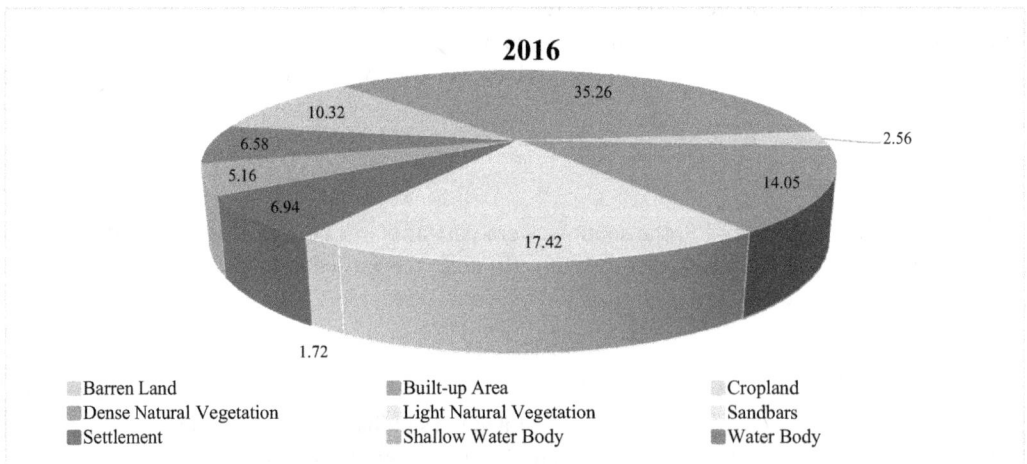

FIGURE 12.6   Graph showing the percentage of area covered by different land-use/land-cover patterns, 2016.

The following analysis can be generated from the map.

12.7.1 WATER BODY

The area covered by the water body, i.e., the Brahmaputra, remained the same as in 2013.

12.7.2 BUILT-UP AREA

The built-up area covered most of the location of all the analysed attributes. Due to its rapid increase, the roadways and railway tracks cannot be distinguished from the others, compared to 2013.

12.7.3 SHALLOW WATER BODY

The noticeable difference in terms of shallow water bodies was the decrease in the area covered by the wetlands towards the east of the northern hills. Another visible change to the south of the study area was that, due to the decline in the river's water level, shallow water bodies were observed towards the edge of the river body.

12.7.4 DENSE NATURAL VEGETATION

The area covered by dense natural vegetation decreased in comparison to 2013. It chiefly covered the hilly regions and the area along the River Puthimari. Some of it can be noticed along the local streets.

12.7.5 SETTLEMENT

The area covered by settlement was also more than in 2013. More concentration of population can be noticed along NH 427 and near the shallow water bodies at the slope of the hills. Some patches of settlement can also be observed in the *char* area.

12.7.6 LIGHT NATURAL VEGETATION

Light natural vegetation also faced a decline and covered only the areas along the river Puthimari and local roads. Some stretches of it can be observed in the hilly areas also. Another patch of light natural vegetation can be noticed near the riverbanks towards the south-east.

12.7.7 BARREN LAND

The area of barren land increased in comparison to 2013. In the south-western part of the study area, where croplands were present in 2013, most regions were analysed as barren land in 2016. Others are scattered throughout the study area, and some patches were observed near the River Puthimari.

12.7.8 CROPLAND

The area covered by cropland was seen to have increased, but there was also a decrease towards the south-western part. The area covered by commercial agricultural land increased compared to 2013.

12.7.9 SANDBARS

The sandbars in the south-western part were covered mainly by built-up areas, which means an increase in the area covered by *char*. On the other hand, new patches were noticed in the south-eastern part. Another important observation was that the sandbars on the south-west were connected to the mainland near the Saraighat bridge, which means that, even if the area covered by it decreased due to built-up areas, there was an increase too.

12.8   15 NOVEMBER 2022

From the prepared map of 2022, it is prominent that urban growth had a noticeable impact on North Guwahati.

The following analysis can be generated from the map.

12.8.1   SANDBARS

The area covered by sandbars decreased significantly compared to the previous years. In 2022 it covered only the south-eastern part of the study area, and the other parts observed in the earlier years were covered up by different attributes.

12.8.2   DENSE NATURAL VEGETATION

The area covered by dense natural vegetation was the least in 2022 compared to the previous years. It can be noticed only in the hilly regions of the study area and towards the extreme south-western corner.

12.8.3   SHALLOW WATER BODY

The area covered by shallow water bodies was seen to have increased over the years. This may be due to the rise in the riverbed of Brahmaputra. The wetlands observed towards the eastern

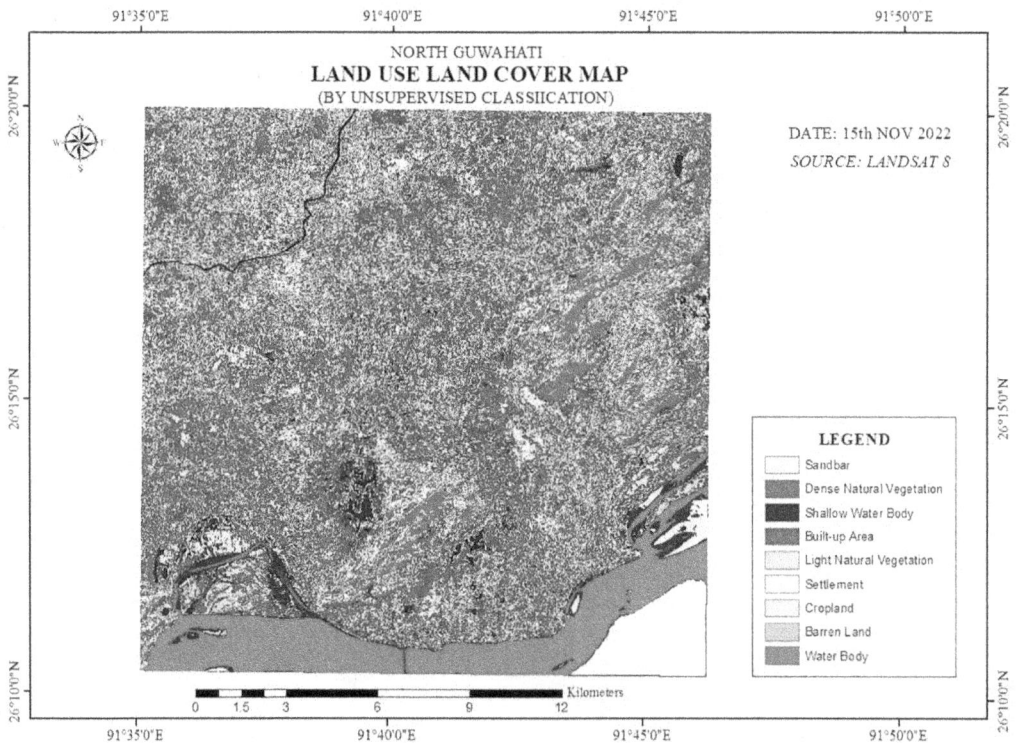

FIGURE 12.7   Land-use/land-cover map of North Guwahati, 2022.

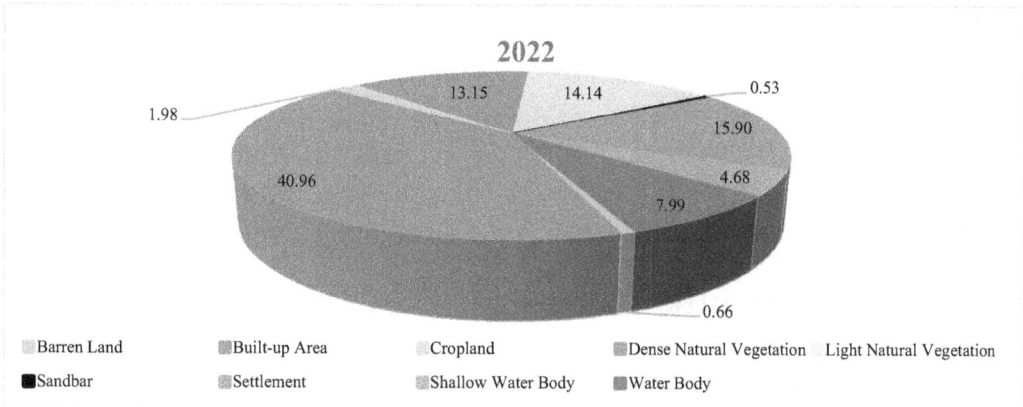

FIGURE 12.8 Graph showing the percentage of area covered by different land-use/land-cover patterns, 2022.

part of the study area in 2013 decreased until 2022. Nevertheless, more shallow water bodies were noticed towards the south-western part of the study area, where cropland was observed in 2013.

12.8.4 BUILT-UP AREA

In 2022, the built-up area covered the maximum of the study area. Most of it was concentrated near NH 27 and NH 427. Some concentration was also observed near the hilly areas and also in the hilly regions. Another important observation was that towards the south-western part of the study area, over River Brahmaputra, spots of the built-up area could be noticed. This is because of the construction work undergoing since 2019 for a new bridge connecting Guwahati with the satellite town. Due to this, the south-western region on the mainland has also been covered by more of a built-up area.

12.8.5 LIGHT NATURAL VEGETATION

The area covered by light natural vegetation is more than that of dense natural vegetation. Most of it can be noticed in the hill areas, along the river Puthimari and local roads and in the north-western part of the study area. Some stretches of it can also be observed in the hilly regions. Another patch of light natural vegetation can be noticed near the riverbanks towards the south-east.

12.8.6 SETTLEMENT

The area covered by settlement was more this year than in the previous years. The concentration of population increased near NH 27 and N 427 and around the areas near River Puthimari. Some patches of settlement can also be observed in the *char* area.

12.8.7 CROPLAND

The area covered by cropland decreased compared to the previous years. As in the earlier years, a part of cropland was observed towards the south-western region, but the site has reduced in comparison. A few small patches were also noticed in the west and south-eastern part of the study area.

12.8.9 BARREN LAND

Barren land was almost absent this year. A small part was noticed towards the north near the hill slopes.

12.8.10 WATER BODY

The area covered by the water body is the only attribute which did not change much over the years.

12.9 ANALYSIS AND INTERPRETATION

The LULC pattern of North Guwahati has changed over the years, and urban growth's influence is undeniable. The percentage of area covered by built-up area kept on increasing over the years, but natural vegetation kept on decreasing. The rate of change of the various geographical attributes cannot be interpreted as sustainable.

Figure 12.9 compares the data collected over three years of North Guwahati. From the prepared clustered bar graph, the following interpretation can be devised:

- Even though 2016 witnessed an increase in barren lands, most of it was utilised by 2022. This can be considered a positive change because the land could serve some purpose for vegetation, settlement, or urban development.
- The area covered by built-up increased over the years, indicating that new industries, factories, and institutions for health and education were established in this area. Another improvement was that the roadways were converted to mettled roads.
- The percentage of cropland was not as high as the other attributes and also faced a decline over the years. Compared to built-up and settlement areas, this region does not seem to have an active primary economic sector.
- The highest negative impact of urban growth was felt in the dense natural vegetation areas, i.e. forested areas. Clearing up forested regions at such a fast rate is not a sustainable way of urbanising. Loss in forest areas leads to an imbalance in the area's ecology, displacing the then-existing flora and fauna.
- Light natural vegetation, i.e. grasslands and less crowded natural vegetation, also decreased over the years. Even though the rate of decline in light natural vegetation was not as rapid as dense natural vegetation, it still cannot be considered a sustainable urbanisation process.

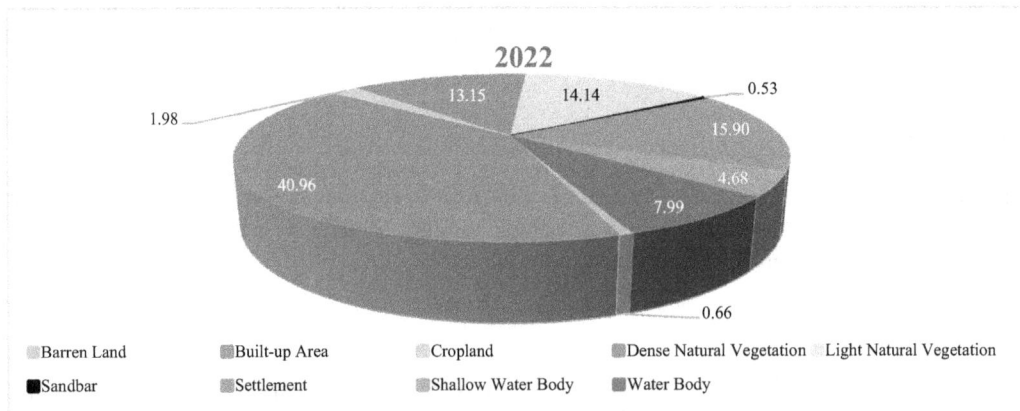

FIGURE 12.9 Graph comparing the area covered by different land-use/land-cover patterns for 2013, 2016, and 2022.

- The change in sandbars was not as drastic because its percentage was already low. In 2022, the difference in its area occurred chiefly due to the increased population in the char regions.
- The increase in settlement in 2022 might indicate that the decreased areas in the barren land, natural vegetation areas, and croplands were probably utilised for settlement.
- The area occupied by shallow water bodies fluctuated over the years. Although there was a decrease in wetlands and swamps in 2022 compared to 2013, due to the rise of the riverbed, there was an increase in shallow water towards the edge of the river Brahmaputra.
- The area occupied by the Brahmaputra did not fluctuate significantly. This may be due to its untamable nature and strong current.

12.10   CONCLUSION

The application of GIS techniques, open-source satellite images, land use, and land-cover analysis shows that North Guwahati has experienced a severe reduction in forest and agricultural land in the last decade. This is due to rapid urban growth and the need for proper sustainable development planning. The primary triggers of rapid urban growth were the shift in demographic dynamics and government policies regarding the new master plan of the Guwahati Metropolitan Development plan, 2050 ('Master Plan Guwahati 2025 Maps | Guwahati Metropolitan Development Authority | Government of Assam, India', n.d.) and the concentration of new socio-economic sectors. The clearance of forest-land, grasslands, wetlands, and croplands may result in total ecosystem degradation of the town shortly if necessary sustainable actions are not taken up by concerned authorities. Local climate change and depletion of soil quality may result due to the lack of forest resources. All the ecologically important regions must be well connected to ensure a sustainable supply of ecosystem services and maintain a balanced urban environment. Therefore, the analysis of the satellite town's LULC patterns produces important data for effective ecological planning of the region. The research also indicates that the current development strategies adopted by the government for urbanisation have further increased the rate of the vulnerability of the agricultural and natural vegetation land to built-up and settlement areas. Therefore, the concerned authorities, planners, and policymakers should conduct research to control and monitor such imbalanced unsustainable development patterns. Another important observation was that the concerned authorities should provide proper high-resolution satellite images and recent data for open access for research. Due to a lack of region-specific high-resolution satellite images and current administrative maps, this study lacks proper area recognition, leaving room for further research about the study area.

NOTE

1. *Char* areas or *Chapori* (Assamese term) are riverine areas (small islands) on the river Brahmaputra, inhabited mainly by migrated population, locally termed as the 'backward' people.

REFERENCES

Aburas, M. M., Y. Ming Ho, M. F. Ramli, and Z. Hanan Ash'aari. 2018. 'Monitoring and Assessment of Urban Growth Patterns Using Spatio-Temporal Built-up Area Analysis'. Environmental Monitoring and Assessment 190 (3): 156. https://doi.org/10.1007/s10661-018-6522-9.

Arshad, S., S. R. Ahmad, S. Abbas, A. Asharf, N. Asad Siddiqui, and Zia ul Islam. 2022. 'Quantifying the Contribution of Diminishing Green Spaces and Urban Sprawl to Urban Heat Island Effect in a Rapidly Urbanizing Metropolitan City of Pakistan'. Land Use Policy 113 (February): 105874. https://doi.org/10.1016/j.landusepol.2021.105874.

Census tables | Government of India. n.d. https://censusindia.gov.in/census.website/data/census-tables.

Chadchan, J., and R. Shankar. 2012. 'An Analysis of Urban Growth Trends in the Post-Economic Reforms Period in India'. International Journal of Sustainable Built Environment 1 (1): 36–49. https://doi.org/10.1016/j.ijsbe.2012.05.001.

Chatterjee, S., and S. Roy. 2021. 'Chapter 27—Land Use Land Cover Dynamics with the Outgrowth of Burdwan Town (India): Problems with Sustainable Solutions'. In Modern Cartography Series, edited by Gouri Sankar Bhunia, Uday Chatterjee, Anil Kashyap, and Pravat Kumar Shit, 10: 603–26. Land Reclamation and Restoration Strategies for Sustainable Development. Academic Press. https://doi.org/10.1016/B978-0-12-823895-0.00032-4.

Drswideg. 2018 .'الجغرافيا و أبحاث دراسات: الجغرافيا. Spatio-Temporal Analysis of Urban and Population Growths in Tripoli Using Remotely Sensed Data and GIS'. الجغرافيا (blog). https://swideg-geography.blogspot.com/2018/04/spatiotemporal-analysis-of-urban-and.html.

Follmann, A. 2021. 'Urbanisation and Agriculture: Farmers Under Urban Growth'. Urbanet. 11 November 2021. www.urbanet.info/urbanisation-and-agriculture/.

Guttikunda, S., and G. Calori. 2013. 'A GIS Based Emissions Inventory at 1 km × 1 km Spatial Resolution for Air Pollution Analysis in Delhi, India'. Atmospheric Environment 67: 101–11. https://doi.org/10.1016/j.atmosenv.2012.10.040.

'Home | Assam State Portal'. n.d. Accessed 5 January 2023. https://assam.gov.in/.

International Res Jour Managt Socio Human. 2022. 'Land Use Change/Land Cover in the Western Fringe Area of Guwahati Under Agglomeration Dr. Daisy Das'. www.academia.edu. https://www.academia.edu/78916539/Land_use_Change_Land_Cover_in_the_Western_Fringe_Area_of_Guwahati_Under_Agglomeration_Dr_Daisy_Das.

Jain, S., P. Aggarwal, P. Sharma, and P. Kumar. 2016. 'Vehicular Exhaust Emissions Under Current and Alternative Future Policy Measures for Megacity Delhi, India'. Journal of Transport and Health 3 (3): 404–12. https://doi.org/10.1016/j.jth.2016.06.005.

Jedwab, R., L. Christiaensen, and M. Gindelsky. 2015. 'Demography, Urbanization and Development: Rural Push, Urban Pull and. . . Urban Push'? The World Bank eBooks. https://doi.org/10.1596/1813-9450-7333.

Kar, R., G. P. O. Reddy, N. T. Kumar, and S. Singh. 2018. Monitoring Spatio-Temporal Dynamics of Urban and Peri-Urban Landscape Using Remote Sensing and GIS—A Case Study From Central India'. The Egyptian Journal of Remote Sensing and Space Science 21 (3): 401–11. https://doi.org/10.1016/j.ejrs.2017.12.006.

Kiamba, A. 2012. The Sustainability of Urban Development in Developing Economies'. Consilience: Journal of Sustainable Development 8 (1): 20–25. https://doi.org/10.7916/d8xd11c8.

Kumar, K., J. A. Kozak, L. S. Hundal, A. Cox, H. Zhang, and T. C. Granato. 2016. 'In-Situ Infiltration Performance of Different Permeable Pavements in a Employee Used Parking Lot—A Four-Year Study'. Journal of Environmental Management 167: 8–14. https://doi.org/10.1016/j.jenvman.2015.11.019.

'Largest Urban Areas Globally by Population 2022'. n.d. Statista. Accessed 3 January 2023. www.statista.com/statistics/912263/population-of-urban-agglomerations-worldwide/.

Li, Yan, Wen Yan, Sai An, Wanlin Gao, Jingdun Jia, Sha Tao, and Wei Wang. 2023. 'A Spatio-Temporal Fusion Framework of UAV and Satellite Imagery for Winter Wheat Growth Monitoring'. Drones 7 (1): 23. https://doi.org/10.3390/drones7010023.

Liu, P. Y., D. Ding, and N. Ravenscroft. 2018. 'The New Urban Agricultural Geography of Shanghai'. Geoforum 90: 74–83. https://doi.org/10.1016/j.geoforum.2018.02.010.

Maiti, S., and P. K. Agrawal. 2005. 'Environmental Degradation in the Context of Growing Urbanization: A Focus on the Metropolitan Cities of India'. Journal of Human Ecology 17 (4): 277–87.

'Master Plan Guwahati 2025 Maps | Guwahati Metropolitan Development Authority | Government of Assam, India'. n.d. Accessed 6 January 2023. https://gmda.assam.gov.in/documents-detail/master-plan-guwahati-2025-maps.

Moghadam, H. S., and M. Helbich. 2013. 'Spatiotemporal Urbanization Processes in the Megacity of Mumbai, India: A Markov Chains-Cellular Automata Urban Growth Model'. Applied Geography 40: 140–49. https://doi.org/10.1016/j.apgeog.2013.01.009.

Mohan, M., S. Tiwari, S. Payra, S. Verma, and D. S. Bisht. 2011. Visibility Degradation During Foggy Period Due to Anthropogenic Urban Aerosol at Delhi, India'. Atmospheric Pollution Research 2 (1): 116–20. https://doi.org/10.5094/apr.2011.014.

'Most Urbanized Countries 2023'. n.d. Accessed 3 January 2023. https://worldpopulationreview.com/country-rankings/most-urbanized-countries.

Mundhe, Nitin, and Ravindra Jaybhaye. 2014. 'A Spatio-Temporal Analysis of Urbanization in India'. Maharashtra Bhugholshastra Sanshodhan Patrika 31 (June): 68–73.

Murayama, Yuji, Matamyo Simwanda, and Manjula Ranagalage. 2021. 'Spatiotemporal Analysis of Urbanization Using GIS and Remote Sensing in Developing Countries'. Sustainability 13 (7): 3681. https://doi.org/10.3390/su13073681.

Nagar, A. K., S. K. Jauhar, and M. Pant. 2017. 'Sustainable Educational Supply Chain Performance Measurement Through DEA and Differential Evolution: A Case on Indian HEI'. Journal of Computational Science 19: 138–52. https://doi.org/10.1016/j.jocs.2016.10.007.

NEWS, NE NOW. 2021. 'Projection of District-Wise Population of Assam for 2021'. NORTHEAST NOW. 21 November 2021. http://nenow.in/north-east-news/assam/projection-population-assam-2021.html.

Nkeki, F. N. 2016. 'Spatio-Temporal Analysis of Land Use Transition and Urban Growth Characterization in Benin Metropolitan Region, Nigeria'. Remote Sensing Applications: Society and Environment 4: 119–37. https://doi.org/10.1016/j.rsase.2016.08.002.

'Percentage of Population at Mid-Year Residing in Urban Areas by Country/Area 2000–2050'. n.d. Percentage of Population at Mid-Year Residing in Urban Areas by Country/Area 2000–2050. https://data.unhabitat.org/datasets/percentage-of-population-at-mid-year-residing-in-urban-areas-by-country-area-2000–2050/explore.

Ramachandraiah, G., M. S. Reddy, S. Basha, V. S. Kumar, and H. V. Joshi. 2004. Distribution, Enrichment and Accumulation of Heavy Metals in Coastal Sediments of Alang–Sosiya Ship Scrapping Yard, India'. Marine Pollution Bulletin 48 (11–12): 1055–59. https://doi.org/10.1016/j.marpolbul.2003.12.011.

Saha, P., R. Mitra, K. Chakraborty, and M. Roy. 2022. 'Application of Multi Layer Perceptron Neural Network Markov Chain Model for LULC Change Detection in the Sub-Himalayan North Bengal'. Remote Sensing Applications: Society and Environment 26: 100730. https://doi.org/10.1016/j.rsase.2022.100730.

Sahu, S. P., S. S. R. Kolluru, and A. K. Patra. 2018. 'A Comparison of Personal Exposure to Air Pollutants in Different Travel Modes on National Highways in India'. Science of the Total Environment 619–620: 155–64. https://doi.org/10.1016/j.scitotenv.2017.11.086.

Shafizadeh-Moghadam, H., and M. Helbich. 2015. 'Spatiotemporal Variability of Urban Growth Factors: A Global and Local Perspective on the Megacity of Mumbai'. International Journal of Applied Earth Observation and Geoinformation 35: 187–98. https://doi.org/10.1016/j.jag.2014.08.013.

Sharma, N., S. Taneja, V. Sagar, and A. Bhatt. 2018. 'Forecasting Air Pollution Load in Delhi Using Data Analysis Tools'. Procedia Computer Science 132: 1077–85. https://doi.org/10.1016/j.procs.2018.05.023.

Soubbotina, T. P. 2004. Beyond Economic Growth: An Introduction to Sustainable Development, Second Edition. © Washington, DC: World Bank. http://hdl.handle.net/10986/14865 License: CC BY 3.0 IGO.

Talukdar, K. K. T. 2018. 'Impact of Urban Growth on Landuse, A Case Study Of Guwahati City, Assam'. International Journal of Engineering Science Invention (IJESI) 7 (8): 26–32. http://www.ijesi.org/papers/Vol(7)i8/Version-2/E0708022632.pdf.

Tripathy, Pratyush, and Amit Kumar. 2019. 'Monitoring and Modelling Spatio-Temporal Urban Growth of Delhi Using Cellular Automata and Geoinformatics'. Cities 90 (July): 52–63. https://doi.org/10.1016/j.cities.2019.01.021.

United Nations Department of Economic and Social Affairs. n.d. 2014 revision of the World Urbanization Prospects | Latest Major Publications. https://www.un.org/en/development/desa/publications/2014-revision-world-urbanization-prospects.html.

'Urbanization'. n.d. Accessed 4 January 2023. https://geography.name/urbanization/.

'Urban Population and Demographic Trends'. n.d. Accessed 3 January 2023. https://data.unhabitat.org/pages/urbanpopulation-and-demographic-trends.

'Urban Population and Demographic Trends | Urban Indicators Database'. n.d. Accessed 3 January 2023. https://data.unhabitat.org/pages/urban-population-and-demographic-trends.

'The World Is Becoming Increasingly Urbanized—UN'. n.d. Weekend Post. www.weekendpost.co.bw/17932/news/the-world-is-becoming-increasingly-urbanized-un/.

Xu, K., X. Han, X. Xia, M. Zhao, and X. Li. 2020. 'Synergistic Effects Between Financial Development and Improvements in New-Type Urbanization: Evidence from China'. Emerging Markets Finance and Trade 56: 2055–72.

13 Assessing Impact of Super Cyclone Amphan on Natural Vegetation in Sundarban Biosphere Reserve, India

Tania Nasrin, Md Nawaj Sarif, Mohd Ramiz,
Mohammad Hashim, Sk Mohibul, Durgesh Dwivedi,
Masood Ahsan Siddiqui, and Lubna Siddiqui

13.1 INTRODUCTION

Natural hazards are physical events that have the potential to cause property damage, loss of life, and disruption of economic and social activities (Sarif et al., 2022). India is one of the countries with the highest risk of natural disasters. The nation is experiencing significant human suffering and financial loss as a result of repeated disasters (Konda et al., 2018). Extreme damaging features of this natural hazard include high winds, heavy rain, towering storm surges, and coastal flooding, all of which pose a threat to human lives, destroy property, and have negative effects on the environment (Shultz et al., 2005). Among these calamities, cyclones are regarded as the most terrible and destructive to human habitation in this country (Subhani et al., 2021). The history of natural disasters along coastlines around the world has served as a stark reminder of how vulnerable coastal areas are (Saravanan et al., 2018). The Indian Ocean has a significant potential for cyclone damage, particularly in the Bay of Bengal and the Arabian Sea (Ghorai et al., 2017). The Sundarban region consists primarily of the northern and southern 24 Pargana districts of West Bengal, located in the northern part of the Bay of Bengal (Sahana & Sajjad, 2019), which are the poorest and most underdeveloped districts in the state; as a result, this region is regarded as highly vulnerable and cyclone prone (Ali et al., 2020). The Sundarban Biosphere Reserve (SBR) is well known for its diverse environment, mangrove forests, and coastline setting. Geographically, this region is affected by strong floods and cyclones that cause a tremendous loss of natural and human life (Sahana et al., 2021). Tropical cyclones and other atmospheric threats, as well as floods, storm surges, salinisation, and erosion, are frequent occurrences in this region (Ali et al., 2020). Tropical cyclones are regarded as one of the most dangerous natural threats in the coastal areas of Sundarban (Hoque et al., 2021). Amphan is the first super cyclone to develop in the Bay of Bengal since the super cyclone that struck Odisha in 1999 (Nasrin et al., 2023). On 20 May 2020, Amphan unleashed devastation across West Bengal, coastal Odisha, and parts of Bangladesh (Sharma et al., 2022). Between 1530 and 1730 hours (IST), it made a catastrophic landfall over the SBR while maintaining the strength of a super cyclonic storm with winds of 155–165 km/h (Cyclone Warning Division, 2021). This super cyclonic storm reported both the greatest sustained 3-minute wind speeds of more than 240 km/h (130 knots) and the largest 1-min wind gusts of up to 260 km/h. (Khan et al., 2021). The cyclone has a substantial influence on several environmental factors in addition to destroying the infrastructure and economy (Hoque et al., 2016).

To determine the best usage and conduct post-disaster analysis, land-use/land-cover (LULC) inventories are essential. To implement effective cyclone impact management strategies, it is

DOI: 10.1201/9781003377825-15

necessary to evaluate the total impact of tropical cyclones (Nasrin et al., 2023). Therefore, it is necessary to develop technologies that allow for rapid evaluation of affected areas. To determine the best operation and conduct post-disaster analysis, LULC assessments are essential. Land use represents what way land is utilised by humans (Abujayyab & Karaş, 2019). Land cover is a fundamental variable that influences the physical material which causes the earth's surface to comprise artificial surfaces built by human activities (Konda et al., 2018). LULC classification is necessary for decision-makers, resource managers, administrators, and scientists to manage the environment and living circumstances (Ramanamurthy & Vijayasaradhi, 2021). Recently, machine learning and remote sensing techniques have been integrated to track and investigate environmental hazards around the world (Ahmed et al., 2023). Machine learning (ML) has become an essential instrument for researchers to analyse their data and infer accurate and reliable outcomes (Khan & Sudheer, 2022). The majority of earlier studies examining the effects of tropical cyclones used a land-cover change detection technique led by on-the-ground impact assessment that focused on particular consequences in the area (Rodgers et al., 2009; Nasrin et al., 2023). Several scholars used LULC in different research fields, for example Talukdar et al. used LULC in the Teesta river basin (Talukdar et al., 2021), Hasan et al. performed ML classification in the Khulna district (Hasan et al., 2022), Mohibul et al. used LULC in Kolkata urban agglomeration (Mohibul et al., 2022a), and Saravanan performed LULC for cyclone vulnerability assessment in Tamil Nadu coast (Saravanan et al., 2018). In studies conducted on LULC classification, the ML models, such as maximum likelihood classification (Kaliraj et al., 2017; Samanta et al., 2021), Random Forest (Behera et al., 2022; Masroor et al., 2022), support vector machine (SVM) (Talukdar et al., 2021; Nasrin et al., 2023), classification and regression tree (Moisen, 2008), decision tree (Talukdar et al., 2021), artificial neural network (Zhang et al., 2023), and Bayesian networks (Mayfield et al., 2017) have been applied in different regions around the world. SVM is the most popular technique for LULC classification and provides greater accuracy than the other conventional techniques (Das et al., 2022).

The Normalized Difference Vegetation Index (NDVI), a measure of vegetation growth and coverage, is frequently used to characterise the spatiotemporal aspects of LULC, particularly the proportion of cover provided by vegetation (Konda et al., 2018). Mangrove forests suffer greatly as a result of tropical cyclones (Konda et al., 2018). Several scholars evaluated the effects of tropical cyclones on forests and other vegetation using the NDVI (Bhowmik & Cabral, 2013; Mitra et al., 2020; Samanta et al., 2021). Several hydrodynamic numerical models have been developed for cyclone impact assessment and can stimulate the changes of LULC and NDVI. However, the studies on the changes in LULC pattern and the effect of coastal vegetation through Sentinel-2 images are very limited.

Therefore, through the distribution of NDVI to evaluate the damage to vegetation in possibly impacted regions and change detection of pre-cyclone and post-cyclone, this study seeks to determine the influence of cyclone Amphan on LULC patterns. However, there are very limited studies on the impact of super cyclone Amphan. The research on the modifications in LULC and NDVI patterns before and after a cyclone can assist local government agencies and emergency management authorities in promptly determining the consequences of the storms. The study brings a fresh perspective because it looks at the effect estimate of Amphan using high-resolution Sentinel-2 images and the ML classification method. This study can be used by future academics to better understand the effects of cyclones and the efforts made by various governmental and non-governmental organisations to create and carry out development strategies in cyclone-affected areas.

13.2 DATA AND METHOD

The SBR region of India stretches from $88°2'27.42''$ to $89°5'46.06''$ E and $21°33'32.62''$ to $22°38'15.66''$ N in the south-western portion of the Ganges-Brahmaputra delta (Bhadra et al., 2018), shown in Figure 13.1. The entire area of the SBR is 9,630 km², of which 5,367 km² is divided into 19 community development blocks (Bhadra et al., 2020). Thirteen blocks are located in the South 24 Parganas

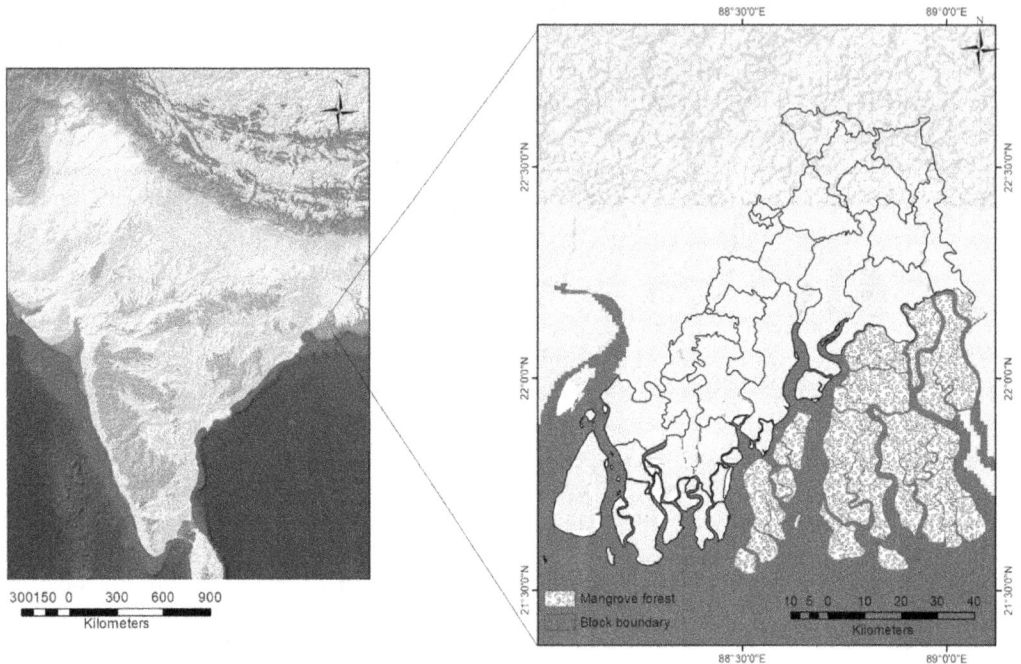

FIGURE 13.1 Location map: Sundarban Biosphere Reserve, India, and community development blocks in Sundarban Biosphere Reserve.

district and six blocks of North 24 Parganas districts of West Bengal in India (Sahana et al., 2016). The SBR is home to the longest continuous mangrove stretch on earth and provides a habitat for a variety of animals and plants (Sahana et al., 2021). The SBR is inhabited by many living things, including the Royal Bengal Tiger, Ganges dolphins, spotted deer, estuary crocodiles, rhinoceroses, and many bird and fish species (Sahana et al., 2021). The Indian Sundarban consists of 102 islands, of which 102 are in equilibrium and make up the forest of the mangrove reserve, and 54 of which are habitable and dispersed throughout the buffer and transition zone of the SBR (Nasrin et al., 2023). The monsoon season brings heavy rainfall, and the reserve endures high relative humidity and temperatures throughout the year. The region experiences 43°C temperature in the summer season and 11°C temperature in the winter season (Sahana & Sajjad, 2019). The average yearly precipitation is between 150 and 200 cm (Sahana et al., 2021). The delta, historically reclaimed from pristine mangrove forests, consists of multiple production land and waterscapes, such as extensive agricultural land, mangrove-protected areas, river and tidal streams, aquaculture ponds, mudflats, and rural and semi-urban settlements (DasGupta et al., 2019). It delights in a tropical climate with copious monsoon rainfall. The average annual precipitation in the SBR is 15 to 20 cm. The region has a tropical climate with brief, arid winters between November and February.

Mangrove habitat, human life, and other infrastructures are all in danger from cyclone landfall and storm surges in this area because of its proximity to the Bay of Bengal (Datta et al., 2012). Over the past four decades, 225 cyclones ranging from mild to severe have affected the Bay of Bengal (Bhargava & Friess, 2022). Since 2007, the intensity and frequency of tropical cyclones in the northern Bay of Bengal have increased (Sahoo & Bhaskaran, 2016). Recent estimates indicate that approximately 200,000 farmers were adversely impacted by seawater intrusion and pluvial inundation resulting from cyclone Amphan in 2020 (Marcinko et al., 2021). The strongest storm of the twenty-first century was the super cyclone Amphan, which made landfall on 20 May 2020 and had a maximum sustained wind speed of 230 km/h (Team, 19 May 2020). Amphan had the largest impact over the past two decades, costing India 14 billion USD in economic losses and 129 lives in India and Bangladesh (Report, 20 April, 2021).

13.3 DATA SOURCES

Satellite data are required for the assessment of damage due to super cyclone Amphan. To conduct this study, we accessed Sentinel-2A satellite data before and after cyclone Amphan. To access and obtain Sentinel-2A data, the Copernicus Open Access Hub was utilised. Sentinel-2A has 13 spectral bands covered by its optical instrument inventory, with 10-m, 20-m, and 60-m spatial resolutions. Super cyclone Amphan passed across the Sundarbans coast on 20 May 2020. Pre-cyclone data were gathered on 14 May 2020. After the cyclone, there were no immediately available cloud-free images. The accuracy of the categorisation process can be significantly reduced by images with clouds (Islam et al., 2018). Because of this, we were unable to choose any of the images from June to October. Post-cyclone images were collected on 11 November 2020. The first damage assessment using supervised classification and NDVI has been done by analysing the spatial data. Land use and cover mapping have been carried out using the SVM technique based on adopted classes such as water bodies, dense forests, waterlogged, open forests, agricultural land, fallow land, settlement, sand and beaches, and swamp. Using the NDVI method, an NDVI image was generated to examine the spatial distribution of the affected vegetation areas.

13.4 METHODOLOGY

13.4.1 IMAGE PRE-PROCESSING

For LULC change detection, a supervised classification method could be more intuitive (Konda et al., 2018). For LULC classification, Sentinel-2A images were used. Methodological flowcharts are shown in Figure 13.2. In this study, different geographic information systems (GIS) and remote sensing tools like ENVI Classic, ArcGIS 10.6, QGIS 3.16, and ERDAS Imagine 2015 were used. After mosaicking with the QGIS 3.16 Semi-Automatic Classification plug-in, all satellite pictures were changed from the Top of Atmosphere format to the Bottom of Atmosphere format (Rai et al., 2021; Sarif et al., 2021). Using ERDAS Imagine 2015, a stacking of the different bands was done. Later, a subset was done on the same platform.

13.4.2 LULC CLASSIFICATION AND ACCURACY ASSESSMENT

The SVM classification method was used for LULC classification and change detection. Based on the return values of satellite images, LULC data are created by classifying raw satellite data into LULC classes (Zain et al., 2021). Following the pre-processing of the images, the study area was classified into nine classes so that various classification procedures could be performed following our study objectives. Several studies have suggested that SVM classification produces superior results compared to other types of supervised classification (Das, 2009; Twisa & Buchroithner, 2019). ML-based classification SVM algorithms were performed for LULC classification. Using ENVI software (Exelis Visual Information Solutions, Herndon, United States), training sets were constructed for these classification methods. The selected polygons were utilised as a region of interest for each class in the training set and for LULC classification. Nine LULC classes were determined in the research area using the Level-I LULC classification technique of the National Remote Sensing Centre (NRSC), India. These included water bodies, dense forests, waterlogged, open forests, agricultural land, fallow land, settlement, sand and beaches, and swamp. Using the SVM classification algorithm, two images with different dates are independently classified. For accurate change detection findings, accurate classifications are

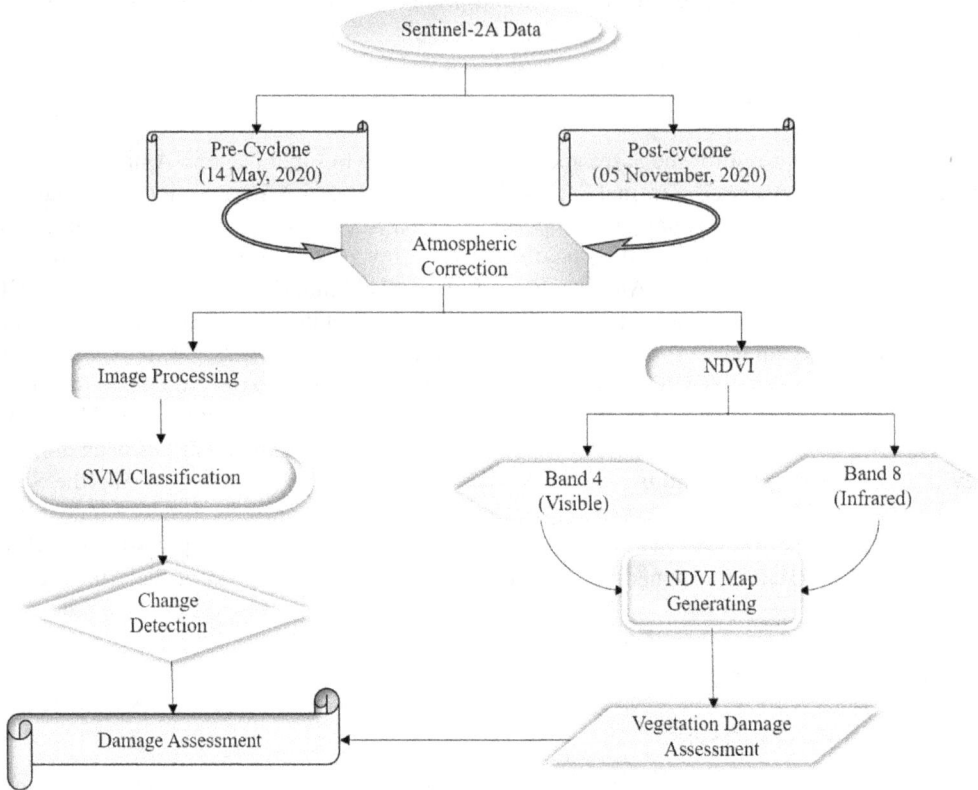

FIGURE 13.2   Methodological flowchart of the study.

essential (Alam et al., 2020, Nasrin et al. 2023). By contrasting it with data from the region of interest or ground truth, accuracy assessment determines how well a classification corresponds to the real world (Kafi, 2014). Using Erdas Imagine software, an accuracy evaluation was done to evaluate the better outcomes of the different classes. If individual classification data are to be used for change detection, accuracy evaluation for that classification is crucial (Twisa & Buchroithner, 2019). With 55 points assigned to each of the nine classes, 495 stratified random samples were selected using the accuracy assessment tool. The actual classes identified from the Google Earth satellite images were compared to each class of categorised LULC. However, when the classification findings are contrasted with real ground circumstances, accuracy evaluation mandates that an appropriate number of samples per map class be acquired (Kafi, 2014). By developing confusion metrics for each classification, the total accuracy and kappa statistics for all applied classification algorithms were calculated. Using the overlay procedure in ArcGIS, the LULC change-transition matrix was computed to quantify the area converted from one LULC category to another during the study period.

13.4.3   Change Detection

Using multiple-temporal images of a location, change detection is a method for examining both short- and long-term changes in LULC (Kumar et al., 2021). Finding the areas on photos that show a change in the feature of interest between two or more image dates is the goal of change detection

(Zain et al., 2021). Change detection enhances resource management and utilisation by helping to better comprehend the relationships and interactions between human and natural events (Lu et al., 2004). This research compares each class before and after the cyclone and the area lost and gained by class changes. The QGIS MOLUSCE plug-in developed these pre- and post-cyclone change detection maps.

13.4.4 NORMALIZED DIFFERENCE VEGETATION INDEX

The NDVI is a computation of the photosynthetic output of a pixel in a satellite picture, based on many spectral bands (Bhowmik & Cabral, 2013; Mohibul et al., 2022b). To examine the geographical extent of the impacted regions of the vegetation, the classic NDVI technique was created. The NDVI calculation is carried out with the images of pre-cyclone and post-cyclone in the study area. The NDVI image was divided into five classes from which the NDVI threshold range was used to separate the NDVI range vegetation. Images from the study region taken before and after the storm are used to calculate the NDVI. Based on the concept that developing green plants significantly absorb light in the visible spectrum (known as photo-synthetically active radiation) while strongly reflecting light in the near-infrared (NIR), NDVI calculations are made. This premise serves as the foundation for the concept of vegetative "spectral signatures (patterns)" (Bhowmik & Cabral, 2013). Due to the limited spectral resolution of the satellite images, only the red image band was modified for this study rather than the entire range of photo-synthetically active radiation. Thus, the NDVI is calculated by Equation 13.1:

$$NDVI = (NIR - Red)/(NIR + Red) \qquad (13.1)$$

The NDVI value has been computed using the raster calculator in ArcGIS 10.6. Band 8 and band 4 were utilised to compute the NDVI because in Sentinel-2A images, band 8 stands for the NIR band and band 4 for the red band. A value between −1 (usually water) and 1 (strongest vegetative growth) is produced by this formula (Sreelekha, 2019). The training data were classified mainly into five major classes which showed water, value from −0.46 to 0.15, bare soil 0.015 to 0.25, followed by sparse vegetation (0.25 to 0.35), moderate vegetation (0.35 to 0.5), and dense vegetation (0.5 to 1). The values of the NDVI threshold are shown in Table 13.1.

13.5 RESULTS AND DISCUSSION

13.5.1 LAND USE/LAND COVER BEFORE AND AFTER THE CYCLONE AMPHAN

Tropical cyclones are the most destructive of the natural hazards that threaten the coastlines of the world every year (Halder et al., 2021). Nine types of LULC were targeted for mapping in the SBR during pre-cyclone and post-cyclone, including water bodies, dense forests, waterlogged open forests, agricultural land, fallow land, settlement, sand and beaches, and swamp. The producer's

TABLE 13.1

Normalized Difference Vegetation Index Classification in the Study Area

Sl No	Description	NDVI Value Threshold	Reference
1	Water	−0.046–0.15	(Akbar et al. 2019)
2	Bare soil	0.015–0.25	(Mehta et al. 2021)
3	Sparse vegetation	0.25–0.35	(Ding and Wang, 2002)
4	Moderate vegetation	0.35–0.0.5	(Mehta et al. 2021)
5	Moderate vegetation	0.5–1	(Dalezios et al. 2001)

Note: NDVI, Normalized Difference Vegetation Index.

accuracy (PA) and user's accuracy (UA) of SVM are shown in Table 13.2 for each class of pre-cyclone and post-cyclone. These results provide a significant platform for future analyses of LULC modifications. Moreover, the spatial representation of LULC types of pre-cyclone and post-cyclone are shown in Figure 13.3a and 13.3b. In pre-cyclone and post-cyclone images, the pattern of LULC

TABLE 13.2

Support Vector Machine Classification and Accuracy Assessment of Pre-cyclone and Post-cyclone

Sl No	Classes	Pre-Cyclone		Pre-Cyclone	
		PA (%)	UA (%)	PA (%)	UA (%)
1	Water logged	58.33	87.5	93.02	100
2	Dense Forest	96.04	95.1	90.91	100
3	Fellow land	95.65	93.62	90.24	92.5
4	Swamp	85.71	85.71	97.3	90
5	Open forest	33.33	40	100	100
6	Settlement	66.67	66.67	85.71	90
7	Waterbody	98.77	95.83	97.5	97.5
8	Agricultural land	86.21	92.59	97.56	100
9	Sand and beaches	66.67	66.67	100	45
10	Overall accuracy	**90.50%**		**91.25%**	
11	Kappa statistics	**0.91**		**0.90**	

Note: PA, producer's accuracy; UA, user's accuracy.

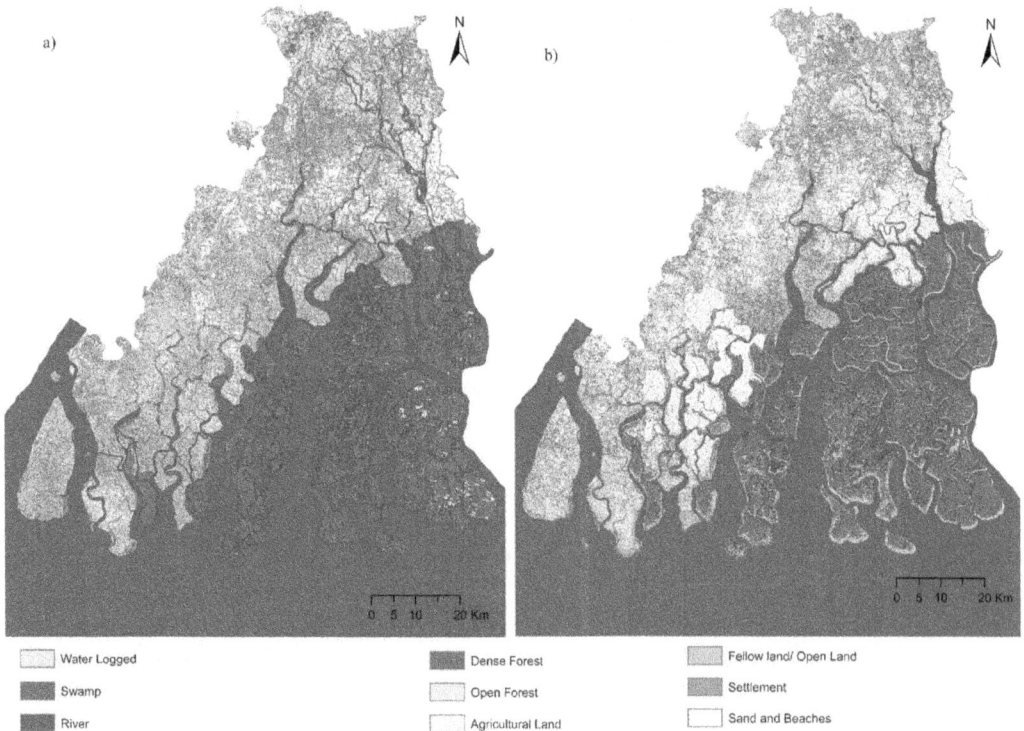

FIGURE 13.3   Land-use/land-cover classification of the study area: (a) pre-cyclone and (b) post-cyclone.

TABLE 13.3

Total Area Before and After the Cyclone and the Rate of Change

LULC Classes	Total Area Before Cyclone (km²)	Total Area After Cyclone (km²)	Rate of Change (km²)
Water bodies	4553.84	4543.46	−0.23
Dense forest	1896.59	1170.63	−38.28
Waterlogged	466.59	336.53	−27.88
Open forest	457.61	444.05	−2.96
Agricultural land	1264.95	1891.30	49.51
Fallow land	998.11	1067.33	6.93
Settlement	523.65	487.69	−6.87
Sand/beaches	52.14	61.21	17.40
Swamp	534.09	754.41	39.57

Note: LULC, land use/land cover.

as the percentage of the total area is shown in Table 13.3. In pre-cyclone SVM classification, the overall accuracy and kappa statistics are 90.50 and 0.91, respectively, and post-cyclone overall accuracy and kappa statistics are 91.25% and 0.90, respectively. According to Anderson's classification scheme, these kappa values for the two classification results are acceptable for the study area because they satisfy the minimum 85% accuracy requirement (Twisa & Buchroithner, 2019).

Before the cyclone, the area under water bodies was 4,553.84 km²; the area under dense forests was 1,896.59 km²; the area under waterlogging was 466.59 km²; the area under open forest was 457.61 km²; the area under agricultural lands was 1,264.95 km²; the area under fallow lands was 998.11 km²; the area under settlements was 523.65 km²; the area under sand and beaches was 52.14 km²; and the area under swamp was 534.09 km². The cyclone caused a shift in the LULC pattern of the study area which now includes 4,543.46 km² of water bodies, followed by dense forest (1,176.63 km²), waterlogging (336.53 km²), open forest (444.53 km²), agricultural land (1,891.30 km²), fallow land (1,067.33 km²), settlements (476.69 km²), sand and beaches (61.21 km²), and swamp (745.41 km²). Dense forests lost 1,896.60 to 1,170.63 km², whereas wasteland increased from 998.12 to 1,087.33 km². Open forest decreased from pre-cyclone to post-cyclone 2.96 km². The water bodies and settlements exhibited the smallest degree of change within the LULC. There were 133.13 km² of open forest converted to agriculture. This conversion mostly serves seasonal farming, from 523.65 km² to 487.69 km² were occupied by settlements. Figure 13.5 displays a few photographs that were captured after the super cyclone arrived from different blocks of Sundarban. Photos show the devastation of the different parts of the study area which was captured by the authors.

13.5.2 CHANGE DETECTION

Changes in LULC patterns have direct spatial and temporal effects on the subsistence of people in coastal areas (Kaliraj et al., 2017). The result of change detection (Figure 13.4) indicates changes in the study area due to super cyclone Amphan in the SBR. The LULC classes of dense forest, open forest, agricultural land, fallow land, sand and beaches, and swampy were mapped with the greatest changes caused by super cyclone Amphan. Approximately 6.93% more fallow or open land was created as a result of the partial destruction of the densely forested area. On the other hand, 100.66 km² of open forest was destroyed and left fallow. Following that, 238.49 km² of dense forests were transformed into open forests. Although the cyclone seriously weakened the settlements, this is not visible in the data since residents quickly renovated their houses following the disaster. Hoque et al., (2016), who conducted object-based classification for the cyclone Sidr in Sundarban, also observed the impact of cyclones on LULC change detection (Hoque et al., 2018). Often, the Sundarbans are

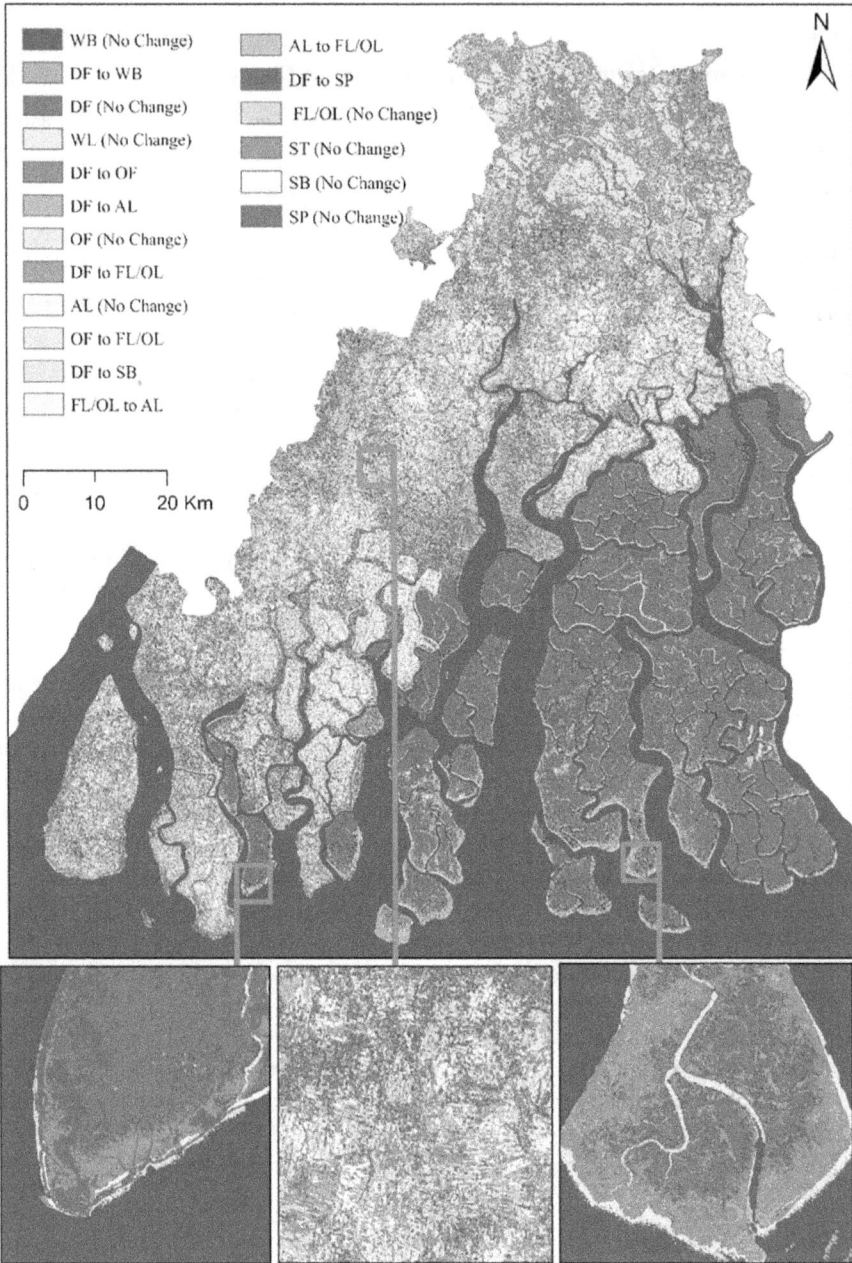

FIGURE 13.4 Land-use/land-cover change detection map of Sundarban Biosphere Reserve pre-cyclone and post-cyclone.

hit by powerful cyclones and storm surges, destroying the littoral and making it impossible for Sundarbans to migrate to the sea (Islam et al., 2018; Rai et al., 2021).

13.5.3 Normalized Difference Vegetation Index

Using the spectral vegetation index, the NDVI images were generated. The regions that are vegetated on the NDVI map are bright, whereas the parts that are not vegetated (such as rivers, ponds,

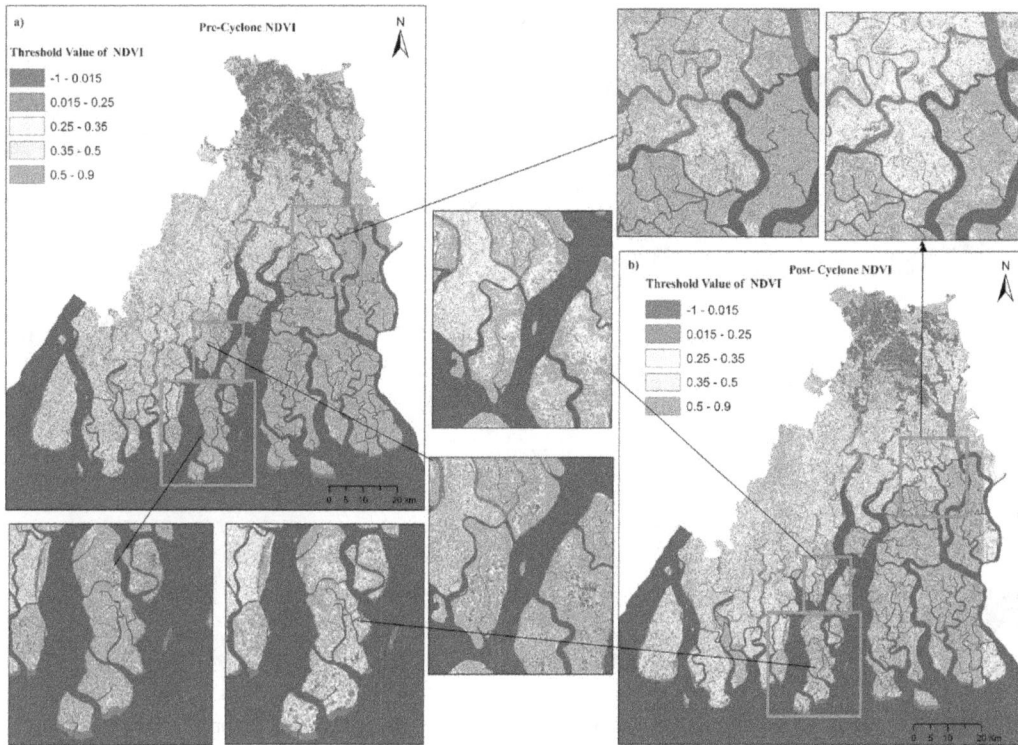

FIGURE 13.5 Images showing the destruction caused by the super cyclone Amphan in different parts of Sundarban: (a) location, (b) Kakdwip, (c) Patharpratima, (d) Mathurapur II, (e) Jaynagar II, (f) G-Plot.

sandy areas, and buildings) are often dark. Images of the study area have been classified into five categories for the identification of vegetation using NDVI threshold value. Threshold values from −1 to 0.015 are showing water bodies and swamps (Akbar et al., 2019). Threshold values from 0.015 to 0.25 are showing bare soil, sandy areas, road, settlements, etc. (Mehta et al., 2021), followed by sparse vegetation (0.25–0.35) (Ding & Wang, 2002), moderate vegetation (0.35–0.5) (Mehta et al., 2021), and dense vegetation (0.5–1) (Dalezios et al., 2001). Healthy plants have a high NDVI value as a result of their high infrared reflectance and low red reflectance. In the study area, the NDVI value varies from −1 to 1. Figure 13.6 shows pre-cyclone and post-cyclone NDVI maps where pre-cyclone images show more dark green, which has a higher NDVI value, than post-cyclone images. In the post-cyclone NDVI map, the spectral reflectance was relatively low, and this is primarily due to the destruction. Therefore, it can be concluded that pre-cyclone NDVI values are slightly higher in the study area than post-cyclone NDVI values. Konda et al. (2018) calculated NDVI for vegetation damage assessment for the Hudhud cyclone. Comparing the results of our analysis to those of other studies reveals that Sundarban mangroves and natural vegetation have sustained substantial losses (Bhowmik & Cabral, 2013; Konda et al., 2018). Based on analyses of the cyclone's effect on mangroves (Behera et al., 2022; Khatun et al., 2022), it is evident that cyclone Amphan was a key contributing factor in the majority of mangrove plants losing their leaves. Some of them are uprooted, but the government has taken new planning to recover the damage within 6 months (Figure 13.5d).

13.6 CONCLUSION

Super cyclone Amphan caused catastrophic damage as it made landfall in Sundarban with winds of 230 km/h (Mishra et al., 2021). Cyclone Amphan was a key contributing factor in the majority

FIGURE 13.6   Normalized Difference Vegetation Index map of the study area: (a) pre-cyclone map and (b) post-cyclone map.

of mangrove forest destruction. The study additionally discovered the difference between NDVI results in the study area before and after a cyclone. The dense forest, particularly in the southern and south-eastern regions of the SBR, had the most modification, with 38.28% of it being lost throughout the pre-Amphan and post-Amphan phases. The disaster management authority and other government agencies may thus use the study's findings to improve their disaster management strategies and reduce the effects of coastal hazards like cyclones and storm surges. For this, extra attention could be paid to the places where Amphan has caused the most changes in LULC patterns as these are the most vulnerable places. Additionally, this study contains certain flaws that could skew the findings. This study used Sentinel-2 satellite data, which have a spatial resolution of 10 m and could not accurately represent destructions. To verify the results, we visited the site. For the examination of the effects of natural disasters like cyclones and storm surges, high- and extremely high-resolution satellite photos from the French satellite SPOT-5 or the Indian Remote Sensing Satellite LISS-IV may be used in future research. If immediate post-cyclone field data were available, the study would also produce higher-quality results. This research offers a basis for the management authority to create a disaster management strategy by providing the effect of the cyclone that can be specifically studied to determine implications in terms of social, physical, economic, and environmental aspects. As a consequence, this research will unquestionably assist the appropriate authorities in taking the required and immediate action, and it will also inspire other researchers to do similar activities in opposition to cyclone damage assessment.

ACKNOWLEDGEMENT

The authors are grateful to the United States Geological Survey for providing them with complimentary access to the satellite data. The authors are thankful to the Department of Geography, Jamia Millia Islamia, New Delhi, for providing the laboratory space for this research.

REFERENCES

Abujayyab, S. K. M., & Karaş, İ. R. (2019). Employing neural networks machine learning algorithm. *Baltic Journal of Modern Computing, 8*(2), 370–378. https://doi.org/10.22364/bjmc.2020.8.2.12

Ahmed, I. A., Talukdar, S., Naikoo, M. W., Parvez, A., Pal, S., Ahmed, S., & Mosavi, A. (2023). A new framework to identify most suitable priority areas for soil-water conservation using coupling mechanism in Guwahati urban watershed, India, with future insight. *Journal of Cleaner Production, 382,* 135363.

Akbar, T. A., Hassan, Q. K., Ishaq, S., Batool, M., Butt, H. J., & Jabbar, H. (2019). Investigative spatial distribution and modelling of existing and future urban land changes and its impact on urbanization and economy. *Remote Sensing, 11*(2). https://doi.org/10.3390/rs11020105

Alam, A., Bhat, M. S., & Maheen, M. (2020). Using Landsat satellite data for assessing the land use and land cover change in Kashmir valley. *GeoJournal, 85*(6), 1529–1543. https://doi.org/10.1007/s10708-019-10037-x

Ali, S. A., Khatun, R., Ahmad, A., & Ahmad, S. N. (2020). Assessment of cyclone vulnerability, hazard evaluation and mitigation capacity for analyzing cyclone risk using GIS technique: A study on sundarban biosphere reserve, India. *Earth Systems and Environment, 4*(1), 71–92. https://doi.org/10.1007/s41748-019-00140-x

Behera, M. D., Prakash, J., Paramanik, S., Mudi, S., Dash, J., Varghese, R., Roy, P. S., Abhilash, P. C., Gupta, A. K., & Srivastava, P. K. (2022). Assessment of tropical cyclone Amphan affected inundation areas using Sentinel-1 satellite data. *Tropical Ecology, 63*(1), 9–19. https://doi.org/10.1007/s42965-021-00187-w

Bhadra, T., Das, S., Hazra, S., & Barman, B. C. (2018). Assessing the demand, availability and accessibility of potable water in Indian Sundarban biosphere reserve area. *International Journal of Recent Scientific Research, 9*(3), 25437–25443.

Bhadra, T., Hazra, S., Sinha Ray, S. P., & Barman, B. C. (2020). Assessing the groundwater quality of the coastal aquifers of a vulnerable delta: A case study of the Sundarban Biosphere reserve, India. *Groundwater for Sustainable Development, 11,* 100438. https://doi.org/10.1016/j.gsd.2020.100438

Bhargava, R., & Friess, D. A. (2022). Previous shoreline dynamics determine future susceptibility to cyclone impact in the Sundarban mangrove forest. *Frontiers in Marine Science, 9*(March), 1–12. https://doi.org/10.3389/fmars.2022.814577

Bhowmik, A. K., & Cabral, P. (2013). Cyclone Sidr impacts on the Sundarbans floristic diversity. *Earth Science Research, 2*(2). https://doi.org/10.5539/esr.v2n2p62

Cyclone Warning Division. (2021). Super cyclonic storm Amphan over southeast bay of Bengal (16 May-21 May 2020). *India Meteorological Department,* (January), 1–73. https://rsmcnewdelhi.imd.gov.in/uploads/report/26/26_936e63_amphan.pdf

Dalezios, N. R., Domenikiotis, C., Loukas, A., Tzortzios, S. T., & Kalaitzidis, C. (2001). Cotton yield estimation based on NOAA/AVHRR produced NDVI. *Physics and Chemistry of the Earth, Part B: Hydrology, Oceans and Atmosphere, 26*(3), 247–251.

Das, T. (2009). Land use/land cover change detection: An object oriented approach, Münster, Germany Thesis. *Thesis, March,* 70.

Das, T., Shahfahad, Naikoo, M. W., Talukdar, S., Parvez, A., Rahman, A., Pal, S., Asgher, M. S., Islam, A. R. M. T., & Mosavi, A. (2022). Analysing process and probability of built-up expansion using machine learning and fuzzy logic in English bazar, West Bengal. *Remote Sensing, 14*(10), 2349. https://doi.org/10.3390/rs14102349

DasGupta, R., Hashimoto, S., Okuro, T., & Basu, M. (2019). Scenario-based land change modelling in the Indian Sundarban delta: An exploratory analysis of plausible alternative regional futures. *Sustainability Science, 14*(1), 221–240. https://doi.org/10.1007/s11625-018-0642-6

Datta, D., Chattopadhyay, R. N., & Guha, P. (2012). Community based mangrove management: A review on status and sustainability. *Journal of Environmental Management, 107,* 84–95.

Ding, H., & Wang, X. K. (2002). Research on algorithm of decision tree induction. *Proceedings of 2002 International Conference on Machine Learning and Cybernetics, 2*(November), 1062–1065. https://doi.org/10.1109/icmlc.2002.1174546

Ghorai, D., Devulapalli, S., & Paul, A. K. (2017). Cyclone vulnerability assessment of Tamil Nadu coast, India using remote sensing and GIS techniques. *Journal of Remote Sensing Technology, 5*(1), 32–43. https://doi.org/10.18005/jrst0501004

Halder, N. K., Merchant, A., Misbahuzzaman, K., Wagner, S., & Mukul, S. A. (2021). Why some trees are more vulnerable during catastrophic cyclone events in the Sundarbans mangrove forest of Bangladesh? *Forest Ecology and Management, 490.* https://doi.org/10.1016/j.foreco.2021.119117

Hasan, M. Z., Leya, R. S., & Islam, K. S. (2022). Comparative assessment of machine learning algorithms for land use and land cover classification using multispectral remote sensing image. *Khulna University Studies*, (October), 33–46. https://doi.org/10.53808/kus.2022.icstem4ir.0124-se

Hoque, M. A. A., Phinn, S., Roelfsema, C., & Childs, I. (2016). Assessing tropical cyclone impacts using object-based moderate spatial resolution image analysis: A case study in Bangladesh. *International Journal of Remote Sensing*, *37*(22), 5320–5343. https://doi.org/10.1080/01431161.2016.1239286

Hoque, M. A. A., Phinn, S., Roelfsema, C., & Childs, I. (2018). Modelling tropical cyclone risks for present and future climate change scenarios using geospatial techniques. *International Journal of Digital Earth*, *11*(3), 246–263. https://doi.org/10.1080/17538947.2017.1320595

Hoque, M. A. A., Pradhan, B., Ahmed, N., Ahmed, B., & Alamri, A. M. (2021). Cyclone vulnerability assessment of the western coast of Bangladesh. *Geomatics, Natural Hazards and Risk*, *12*(1), 198–221. https://doi.org/10.1080/19475705.2020.1867652

Islam, K., Jashimuddin, M., Nath, B., & Nath, T. K. (2018). Land use classification and change detection by using multi-temporal remotely sensed imagery: The case of Chunati wildlife sanctuary, Bangladesh. *Egyptian Journal of Remote Sensing and Space Science*, *21*(1), 37–47. https://doi.org/10.1016/j.ejrs.2016.12.005

Kafi, K. M. (2014). *An analysis of LULC change detection using remotely sensed data: A case study of Bauchi city an analysis of LULC change detection using remotely sensed data : A case study of Bauchi city. January 2015.* https://doi.org/10.1088/1755-1315/20/1/012056

Kaliraj, S., Chandrasekar, N., Ramachandran, K. K., Srinivas, Y., & Saravanan, S. (2017). Coastal land use and land cover change and transformations of Kanyakumari coast, India using remote sensing and GIS. *Egyptian Journal of Remote Sensing and Space Science*, *20*(2), 169–185. https://doi.org/10.1016/j.ejrs.2017.04.003

Khan, A., & Sudheer, M. (2022). Machine learning-based monitoring and modeling for spatio-temporal urban growth of Islamabad. *Egyptian Journal of Remote Sensing and Space Science*, *25*(2), 541–550. https://doi.org/10.1016/j.ejrs.2022.03.012

Khan, M. J. U., Durand, F., Bertin, X., Testut, L., Krien, Y., Islam, A. K. M. S., Pezerat, M., & Hossain, S. (2021). Towards an efficient storm surge and inundation forecasting system over the Bengal delta: Chasing the supercyclone Amphan. *Natural Hazards and Earth System Sciences*, *21*(8), 2523–2541. https://doi.org/10.5194/nhess-21-2523-2021

Khatun, M., Rahaman, S. K., Garai, S., Ranjan, A., Ghosh, B. G., Kumar, A., & Tiwari, S. (2022). Assessing the impact of super cyclone Amphan on Indian Sundarban biosphere reserve. *Indian Journal of Ecology*, *49*(6), 2236–2242.

Konda, V. G. R. K., Chejarla, V. R., Mandla, V. R., Voleti, V., & Chokkavarapu, N. (2018). Vegetation damage assessment due to hudhud cyclone based on NDVI using Landsat-8 satellite imagery. *Arabian Journal of Geosciences*, *11*(2). https://doi.org/10.1007/s12517-017-3371-8

Kumar, M., Kalra, N., Singh, H., Sharma, S., Singh Rawat, P., Kumar Singh, R., Kumar Gupta, A., Kumar, P., & Ravindranath, N. H. (2021). Indicator-based vulnerability assessment of forest ecosystem in the Indian Western Himalayas: An analytical hierarchy process integrated approach. *Ecological Indicators*, *125*(December 2020), 107568. https://doi.org/10.1016/j.ecolind.2021.107568

Kumar, R., Rai, A., Mishra, V., Diwate, P., & Arya, V. (2021). Performance evaluation of supervised classifiers for land use and land cover mapping using Sentinel-2 MSI image. *Journal of Geosciences Research, 6,* 231, 241.

Lu, D., Mausel, P., Batistella, M., & Moran, E. (2004). Comparison of land-cover classification methods in the Brazilian Amazon basin. *Photogrammetric Engineering and Remote Sensing*, *70*(6), 723–731. https://doi.org/10.14358/PERS.70.6.723

Marcinko, C. L. J., Nicholls, R. J., Daw, T. M., Hazra, S., Hutton, C. W., Hill, C. T., Clarke, D., Harfoot, A., Basu, O., Das, I., Giri, S., Pal, S., & Mondal, P. P. (2021). The development of a framework for the integrated assessment of SDG trade-offs in the Sundarban Biosphere Reserve. *Water (Switzerland)*, *13*(4). https://doi.org/10.3390/w13040528

Masroor, M., Avtar, R., Sajjad, H., Choudhari, P., Kulimushi, L. C., Khedher, K. M., Komolafe, A. A., Yunus, A. P., & Sahu, N. (2022). Assessing the influence of land use/land cover alteration on climate variability: An analysis in the Aurangabad district of Maharashtra State, India. *Sustainability (Switzerland)*, *14*(2). https://doi.org/10.3390/su14020642

Mayfield, H., Smith, C., Gallagher, M., & Hockings, M. (2017). Use of freely available datasets and machine learning methods in predicting deforestation. *Environmental Modelling and Software*, *87*, 17–28. https://doi.org/10.1016/j.envsoft.2016.10.006

Mehta, A., Shukla, S., & Rakholia, S. (2021). Vegetation change analysis using normalized difference vegetation index and land surface temperature in greater GIR landscape. *Journal of Scientific Research*, 65(03), 01–06. https://doi.org/10.37398/jsr.2021.650301

Mishra, M., Acharyya, T., Santos, C. A. G., Silva, R. M. da, Kar, D., Mustafa Kamal, A. H., & Raulo, S. (2021). Geo-ecological impact assessment of severe cyclonic storm Amphan on Sundarban mangrove forest using geospatial technology. *Estuarine, Coastal and Shelf Science*, 260(June). https://doi.org/10.1016/j.ecss.2021.107486.

Mitra, A., Dutta, J., Mitra, A., & Thakur, T. (2020). Amphan super cyclone: A death knell for Indian Sundarbans. *eJournal of Applied Forest Ecology (eJAFE)*, 8(1), 41–48.

Mohibul, S., Sarif, N., Parveen, N., Khanam, N., Siddiqui, M. A., Naqvi, H. R., Nasrin, T., & Siddiqui, L. (2022a). *Wetland Health Assessment Using DPSI Framework: A Case Study in Kolkata Metropolitan Area*. 1–27.

Mohibul, S., Siddiqui, L., Siddiqui, M. A., Sarif, M. N., Parveen, N., Islam, M. S., & Nasrin, T. (2022b). Spatio-temporal analysis of land use/land cover change using STAR method in Kolkata urban agglomeration. In: Haroon S., Lubna S., Atiqur R., Mary T., Masood A.S. (eds) *Challenges of Disasters in Asia: Vulnerability, Adaptation and Resilience* (pp. 187–207). Singapore: Springer Nature Singapore.

Moisen, G. G. (2008). Classification and Regression Trees. *Encyclopedia of Ecology, Five-Volume Set, February*, 582–588. https://doi.org/10.1016/B978-008045405-4.00149-X

Nasrin, T., Ramiz, M., Sarif, M. N., Hashim, M., Siddiqui, M. A., Siddiqui, L., . . . & Mankotia, S. (2023). Modeling of impact assessment of super cyclone Amphan with machine learning algorithms in Sundarban Biosphere Reserve, India. *Natural Hazards*, 117(2), 1945–1968.

Ramanamurthy, B. V., & Vijayasaradhi, B. (2021). Change detection analysis in LULC of the upstream Thandava reservoir using RS and GIS applications. *IOP Conference Series: Materials Science and Engineering*, 1025(1). https://doi.org/10.1088/1757-899X/1025/1/012034

Rodgers, J. C., Murrah, A. W., & Cooke, W. H. (2009). The impact of Hurricane Katrina on the coastal vegetation of the Weeks Bay Reserve, Alabama from NDVI data. *Estuaries and Coasts*, 32(3), 496–507. https://doi.org/10.1007/s12237-009-9138-z

Sahana, M., Ahmed, R., & Sajjad, H. (2016). Analyzing land surface temperature distribution in response to land use/land cover change using split window algorithm and spectral radiance model in Sundarban Biosphere Reserve, India. *Modeling Earth Systems and Environment*, 2(2). https://doi.org/10.1007/s40808-016-0135-5

Sahana, M., Rehman, S., Ahmed, R., & Sajjad, H. (2021). Analyzing climate variability and its effects in Sundarban Biosphere Reserve, India: Reaffirmation from local communities. *Environment, Development and Sustainability*, 23(2), 2465–2492. https://doi.org/10.1007/s10668-020-00682-5

Sahana, M., Rehman, S., Paul, A. K., & Sajjad, H. (2021). Assessing socio-economic vulnerability to climate change-induced disasters: Evidence from Sundarban Biosphere Reserve, India. *Geology, Ecology, and Landscapes*, 5(1), 40–52. https://doi.org/10.1080/24749508.2019.1700670

Sahana, M., & Sajjad, H. (2019). Vulnerability to storm surge flood using remote sensing and GIS techniques: A study on Sundarban Biosphere Reserve, India. *Remote Sensing Applications: Society and Environment*, 13(November), 106–120. https://doi.org/10.1016/j.rsase.2018.10.008

Sahoo, B., & Bhaskaran, P. K. (2016). Assessment on historical cyclone tracks in the bay of Bengal, east coast of India. *International Journal of Climatology*, 36(1), 95–109.

Samanta, S., Hazra, S., Mondal, P. P., Chanda, A., Giri, S., French, J. R., & Nicholls, R. J. (2021). Assessment and attribution of mangrove forest changes in the Indian Sundarbans from 2000 to 2020. *Remote Sensing*, 13(24). https://doi.org/10.3390/rs13244957

Saravanan, S., Jennifer, J., Singh, L., & Abijith, D. (2018). Cyclone vulnerability assessment of Cuddalore coast in Tamil Nadu, India using remote sensing, and GIS. *MATEC Web of Conferences*, 229. https://doi.org/10.1051/matecconf/201822902022

Sarif, N., Siddiqui, L., Parveen, N., & Saha, M. (2021). Evolution of river course and morphometric features of the river Ganga : A case study of up and downstream of Farakka Barrage. *International Soil and Water Conservation Research*, 9(4), 578–590. https://doi.org/10.1016/j.iswcr.2021.01.006

Sarif, N., Siddiqui, L., Siddiqui, M. A., Parveen, N., Islam, S., Khan, S., Khanam, N., Mohibul, S., Shariq, M., & Nasrin, T. (2022). *Household-Based Approach to Assess the Impact of River Bank Erosion on the Socio-Economic Condition of People: A Case Study of Lower Ganga Plain.*

Sharma, S., Suwa, R., Ray, R., & Mandal, M. S. H. (2022). Successive cyclones attacked the world's largest mangrove forest located in the bay of Bengal under pandemic. *Sustainability (Switzerland)*, 14(9). https://doi.org/10.3390/su14095130

Shultz, J. M., Russell, J., & Espinel, Z. (2005). Epidemiology of tropical cyclones: The dynamics of disaster, disease, and development. *Epidemiologic Reviews*, *27*(1), 21–35.

Sreelekha, M. (2019). Accuracy assessment of supervised and unsupervised classification using NOAA data in NDHRA Pradesh region. *International Journal of Engineering Research And*, *V8*(12), 60–64. https://doi.org/10.17577/ijertv8is120065

Subhani, R., Saqib, S. E., Rahman, M. A., Ahmad, M. M., & Pradit, S. (2021). Impact of cyclone yaas 2021 aggravated by Covid-19 pandemic in the southwest coastal zone of Bangladesh. *Sustainability (Switzerland)*, *13*(23), 1–13. https://doi.org/10.3390/su132313324

Talukdar, S., Eibek, K. U., Akhter, S., Ziaul, S., Towfiqul Islam, A. R. M., & Mallick, J. (2021). Modeling fragmentation probability of land-use and land-cover using the bagging, random forest and random subspace in the Teesta river basin, Bangladesh. *Ecological Indicators*, *126*(March), 107612. https://doi.org/10.1016/j.ecolind.2021.107612

Twisa, S., & Buchroithner, M. F. (2019). Land-Use and Land-Cover (LULC) change detection in Wami river basin, Tanzania. *Land*, *8*(9). https://doi.org/10.3390/land8090136

Zain, W. M., Idris, S. R. A., Ramsi, M. F. M., & Nordin, S. K. (2021). Analysing land-use land cover (LULC) and development change in nearby university campuses' area: A case of Universiti Teknologi MARA Negeri Sembilan, Malaysia. *Journal of Science and Technology*, *13*(2), 25–37.

Zhang, M., Kafy, A. Al, Xiao, P., Han, S., Zou, S., Saha, M., Zhang, C., & Tan, S. (2023). Impact of urban expansion on land surface temperature and carbon emissions using machine learning algorithms in Wuhan, China. *Urban Climate*, *47*(July 2022), 101347. https://doi.org/10.1016/j.uclim.2022.101347

14 Estimation of Surface Runoff by Soil Conservation Services– Curve Number Method in the Sanjai River Basin, Jharkhand

Arunashis Chandra and Swati Mondal

14.1 INTRODUCTION

Water is one of the most needed and extremely precious natural resources. In India, where the demand for water is always increasing as this is the world's largest populated country, the government emphasizes increasing crop production for the fulfillment of the higher rate of sustainability in agriculture. Thus, sustainable water management practices should be considered as one of the main themes to ensure requisite water resources for agricultural land. To fulfill this purpose, water resource management and sustainable development modeling techniques or surface runoff estimation methods are very important.

Surface runoff is a hydrological incident in the hydrological cycle. Surface runoff estimation is fundamental for the evaluation of potential water output in a watershed, planning for water conservation, groundwater, artificial recharge structure, flood, etc. According to Sarangi et al. (2005), "most of the agricultural watersheds are ungauged without having past records of measured runoff and accurate runoff data are hardly available in India". The scenario is common not only in India but also in other countries. To overcome this problem, several methods and models have been introduced, such as Top Model (Beven 1982), Soil Conservation Services–Curve Number (SCS-CN) (SCS, National Engineering Handbook 1985), Kineros (Woolhiser et al. 1990), Hydrological Engineering Centre Model 1 and Hydrologic Modeling System 1990–2001, Geomorphological Instantaneous Unit *Hydrograph* (Kumar et al. 2007), etc.

In the present study, runoff modeling has been applied through the geographic information systems (GIS) platform. ArcGIS10.1 and Erdas Imagine 14 software have been used to estimate the surface runoff volume of the Sanjai River basin by the SCS-CN model. This model was created by the United States Department of Agriculture (USDA 1972a, 1972b). "This model has a long, fruitful application history and is generally referred to as 'blue collar' hydrology" (Hawkins et al. 2009; Hawkins 1973, 1993).

14.2 STUDY AREA

The present study area for the SCS-CN model is the Sanjai River basin. The river Sanjai is a subtributary of the Subarnarekha River. The study area extends from 22°25′50″ N to 22°26′12″ N latitude and 85°17′33″ E to 86°05′18″ E longitude. The basin is located mostly in Paschimi Singhbhum and Saraikela district and a very negligible part of Ranchi and Khunti district (Figure 14.1).

14.3 METHODS AND MATERIALS

The methodology for the SCS-CN runoff model involves various steps. These steps are described as follows.

DOI: 10.1201/9781003377825-16

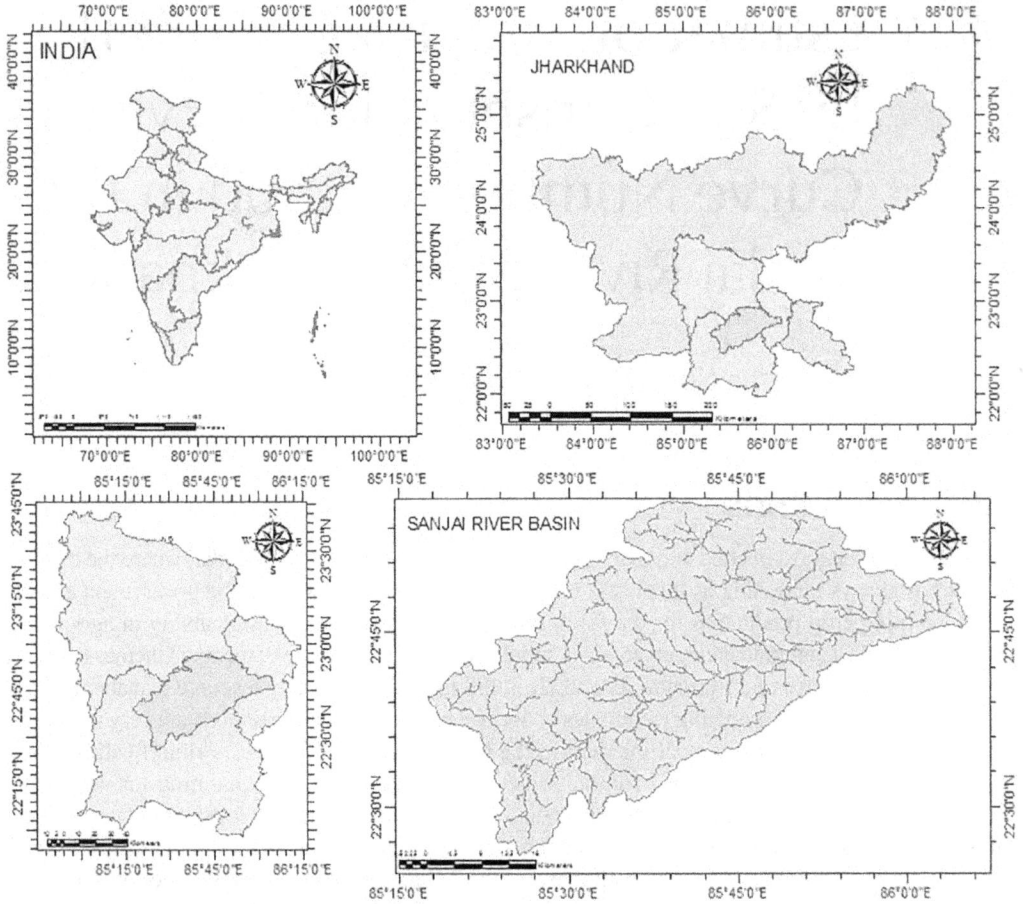

FIGURE 14.1 Location map of the study area.

The land-use/land-cover (LULC) map is extracted from Landsat 8 Operational Land Imager (OLI) with a 30-m resolution, which was acquired in April 2021. The soil map was collected from the National Bureau of Soil Science and Land Use Planning. Daily rainfall data were taken from Indian Meteorological Department, Ranchi. The watershed map was extracted from the Digital Elevation Model (DEM).

At first, the LULC map was determined through Erdas Imagine 14 software. Then the soil map was digitized from the published soil map by the National Bureau of Soil Science and Land Use Planning. The watershed map was created in ArcGIS 10.1 software. After that soil map and LULC map were superimposed on each other. Then the superimposed LULC and soil map was dissected in multiple micro-watersheds. There are nine micro-watersheds in my research area. The curve number is assigned based on different LULC and hydrological soil group. The weighted CN is calculated on the basis of each LULC area and each CN:

Weighted CN = (Area of dense vegetation × CN) + (Area of cropland × CN) + (Area of settlement × CN) . . . . (Area of canal × CN)/(Total area of the sub-watershed)

The soil retention rate is calculated by converting the weighted CN with different antecedent moisture conditions (AMC).

Finally, the runoff volume of each sub-watershed is calculated by considering the soil retention rate and antecedent rainfall volume.

14.3.1 SOIL CONSERVATION SERVICES–CURVE NUMBER MODEL

The SCS-CN method was created by the National Engineering Handbook section of Hydrology of Soil Conservation Services, USDA. This method is based on the water balance equation and two hypotheses (USDA; sec 4 Hydrology):

> The first suggestion states about the ratio of the real quantity of direct runoff to the maximum possible runoff is equal to the ratio of the amount of real infiltration to the quantity of the potential maximum retention.
>   The second hypothesis states that the amount of early abstraction is some fraction of the probable maximum retention.

Satheeshkumar et al. 2017

The SCS-CN method is used to estimate direct runoff of the study area. "SCS assumed that the ratio of actual retention to potential maximum retention was equal to the ratio of actual runoff to

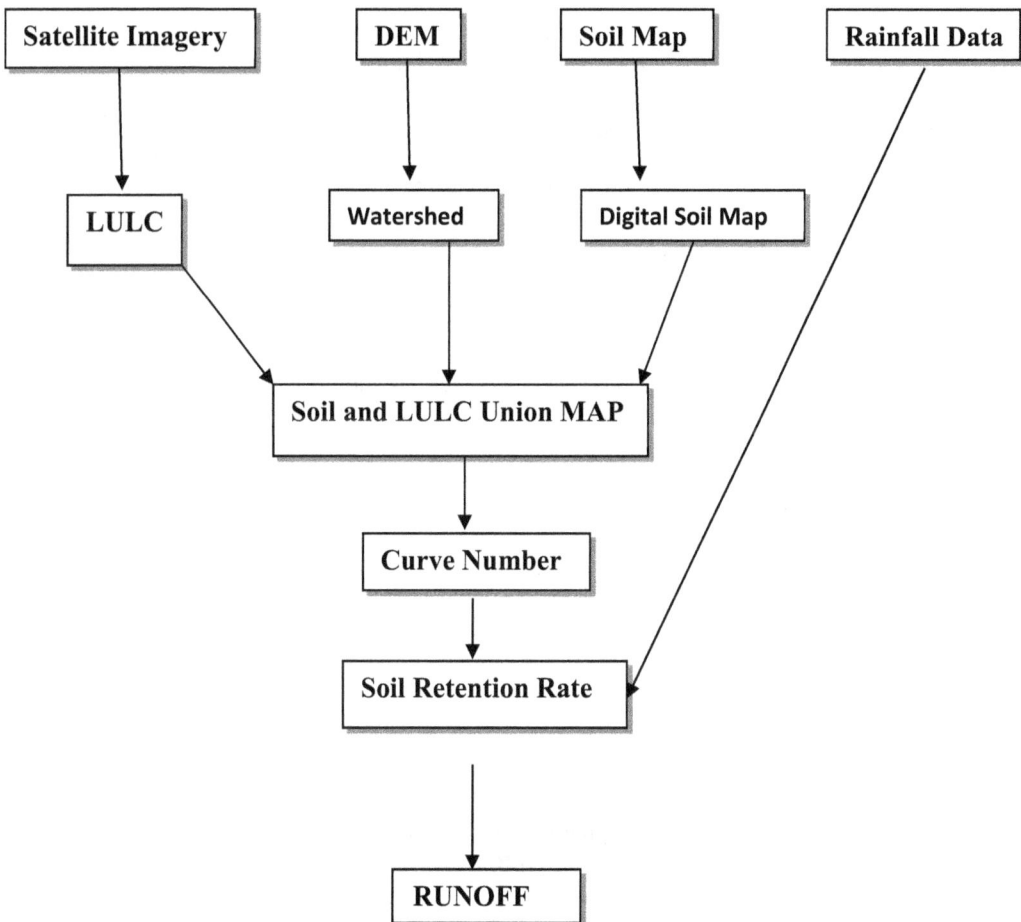

FIGURE 14.2 Flow diagram of the method used to estimate runoff.

potential maximum runoff the latter being rainfall minus initial abstraction" (USDA; NEH sec 4. Hydrology):

$$F/S = Q/(P - Ia) \tag{14.1}$$

where F is actual retention (in millimeters), S is potential maximum retention (in millimeters), Q is the accumulated runoff depth (in millimeters), P is the accumulated rainfall depth (in millimeters), and Ia is initial abstraction (in millimeters).

After the start of rainfall, all the additional rainfall turns into either actual retention or runoff:

$$F = P - Ia - Q \tag{14.2}$$

After combining all these equations,

$$Q = (P - Ia)^2/P - Ia + S \tag{14.3}$$

where Q is total runoff (in millimeters); P is the depth of rainfall (in millimeters); Ia stands for initial abstraction (in millimeters), S surface storage. An empirical relation was obtained for the first abstraction after taking into account interception, infiltration, and the start of the watershed runoff. The relationship is

$$Ia = 0.3 \text{ (for Indian condition)}$$

S is the potential maximum retention, and it is shown in Equation 14.4:

$$S = (25400/CN) - 254 \tag{14.4}$$

CN ranges between 0 and 100, where $S = \infty$ and $S = 0$, respectively:

$$Q = (P - 0.3S)^2/P - 0.7S \tag{14.5}$$

where Q is the surface runoff volume, S is potential water retention, and P is the precipitation.

This method represents the efficiency of soil for infiltration of water in respect of LULC and AMC.

"As per US Soil Conservation Services soil is divided into 4 hydrologic soil groups A, B, C, and D with respect to the rate of potentiality and rate of final infiltration" (Satheeshkumar et al. 2017).

14.3.2 ANTECEDENT MOISTURE CONDITION

On the basis of CN, there are three types of humid conditions (AMC) on land:

- Dry condition (CN I) (wilting point not yet reached)
- Average condition (CN II)
- Saturated condition (CN III) (Table 14.1)

AMC I and III conditions are easily computed through AMC II on the basis of CN. "The following equations are used to calculate the conditions, AMC-I and AMC- III" (Chow et al. 1988):

$$CN (I) = CN (II)/(2.281 - 0.0128 CN (II)) \tag{14.6}$$
$$CN (III) = CN (II)/0.427 + 0.00573 CN (II) \tag{14.7}$$

where
    CN I = Curve number for dry conditions
    CN II = Curve number for average conditions
    CN III = Curve number for saturated conditions

14.3.3 Hydrological Soil Group

A soil classification has been developed by SCS on the basis of infiltration rate, which involves four types of soil:

- A—Indicates the potentiality of low runoff and high infiltration rate.
- B—Indicates moderate infiltration rate and "moderately drained to well-drained soil with moderately fine to moderately deep texture" (USDA, NEH section 4, Hydrology).
- C—It consists of slow rate of infiltration when continuously wetted and the water transmission rate is low.
- D—It has a potentiality of high runoff, and the infiltration rate is very low when thoroughly wetted.

Now the weighted CN will be calculated as in the following equation:

$$\text{Weighted Curve Number (WCN)} = (CN*Area)/Total\ Area \qquad (14.8)$$

For the calculation of the depth of surface runoff, Equations 14.4 and 14.5 will be applied.

TABLE 14.1
Antecedent Moisture Conditions

		Five-Day Antecedent Rainfall (mm)	
AMC Class	Soil Condition	Dormant Season	Growing Season
I	Dry	<13	<36
II	Average	13–28	36–53
III	Saturated	>28	>53

Source: After Soil Conservation Service (1972).

Note: AMC, antecedent moisture conditions.

TABLE 14.2
Hydrological Soil Group

Hydrological Soil Group	Soil Texture	Runoff Potentiality	Water Transmission	Final Infiltration
A	Deep, well-drained, sands and gravels	Low	High rate	>7.5
B	Moderately deep to deep, moderately well to well drained	Moderate	Moderate rate	3.8–7.5
C	Clay loams, shallow sandy loam, soils with moderate to fine texture	Moderate	Moderate rate	1.3–3.8
D	Clay soils that swell significantly when wet	High	Low rate	<1.3

Source: Klingebiel and Montgomery (1961).

14.4   RESULTS AND DISCUSSION

Soil: The hydrological soil group developed four types of soil on the basis of infiltration rate, namely, A, B, C, and D. The A soil group consists of 823 km² of the study area; B soil group has an area of 415 km²; C soil group has an area of 289 km²; and the extension of D soil group is 743 km².

LULC: The land use/land cover classification map has been developed by supervised classification. Eleven classes have been created through this process, as follows:

- Dense vegetation includes 668.97 km² which is 29.47% of the total area.
- Fairly dense vegetation has 315.79 km² and 13.97% of the total area.
- Cropland extends over 1,137.95 km², which is 50.13% of the total area.
- Settlement extends of 53.44 km² (2.35% of total area).
- Inland water body includes lakes, ponds, reservoirs, etc. It consists of 41.59 km² of area and 1.83% of the total area.
- The river area includes 8.86 km² and 0.39% of the total area.
- Wetland has an area of 1.12 km² and 0.05% of the total area.
- The study area is characterized by a number of hills and upland. This rocky outcrop involves over 32 km² area and 1.42% of the total area.
- Fallow land covers 0.68 km² and 0.03% of the total area.
- Built-up area is developed by anthropogenic activities. Nowadays this type of area is extending. In my study area, it consists of 9.07 km² and 0.40% of the total area.
- Canal is extending over 0.35 km² area and 0.02% of the total area.

The soil and LULC maps are combined through the Union tool in ArcGIS 10.1. Then CN values for hydrologic soil cover complexes were assigned based on different LULC and soil category. The watershed was then divided into nine sub-watershed and a combined soil and LULC map was also extracted on the basis of these nine sub-watersheds.

The total volume of the surface runoff is 981.84607 mm. As the basin is characterized with secondary porosity, the runoff value is quite high with respect to annual rainfall, that is, 1,054.6 mm.

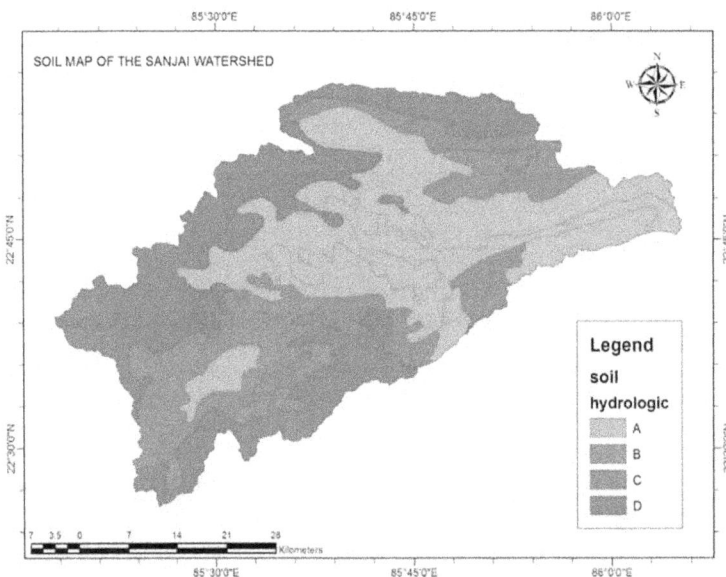

FIGURE 14.3   Hydrologic soil map of study area 3.

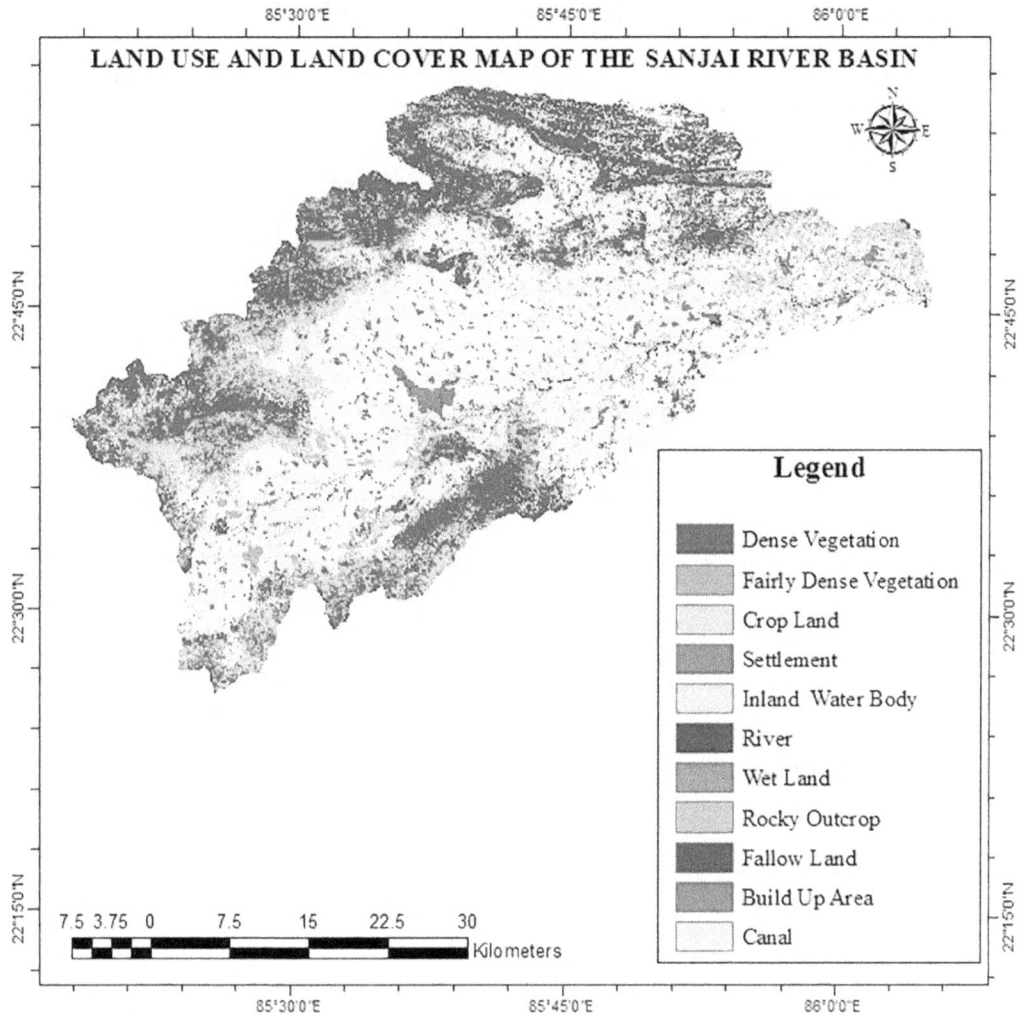

FIGURE 14.4   Land-use and land-cover map of the study area.

From the earlier discussion, it is clear that the runoff characteristics are structurally controlled. The runoff volume is higher in hilly sub-watersheds and lower in plain ones.

14.5   CONCLUSION

The SCS-CN method is helpful for the estimation of surface runoff. This method is also applicable in watershed management. The amount of runoff varies with the change of LULC and soil conditions in the Sanjai River basin. As the basin is divided into sub-watersheds, the highest amount of runoff is observed in watershed 1 as this is a hilly and mountainous area; the lowest amount of runoff is observed in watershed 9 as this is a plain land. The LULC map indicates that the area of cropland is highest among all classes in the basin. The soil map shows that the soil category A is the predominant soil in the basin. Finally, with the help of the SCS-CN technique, artificial recharge points can be determined.

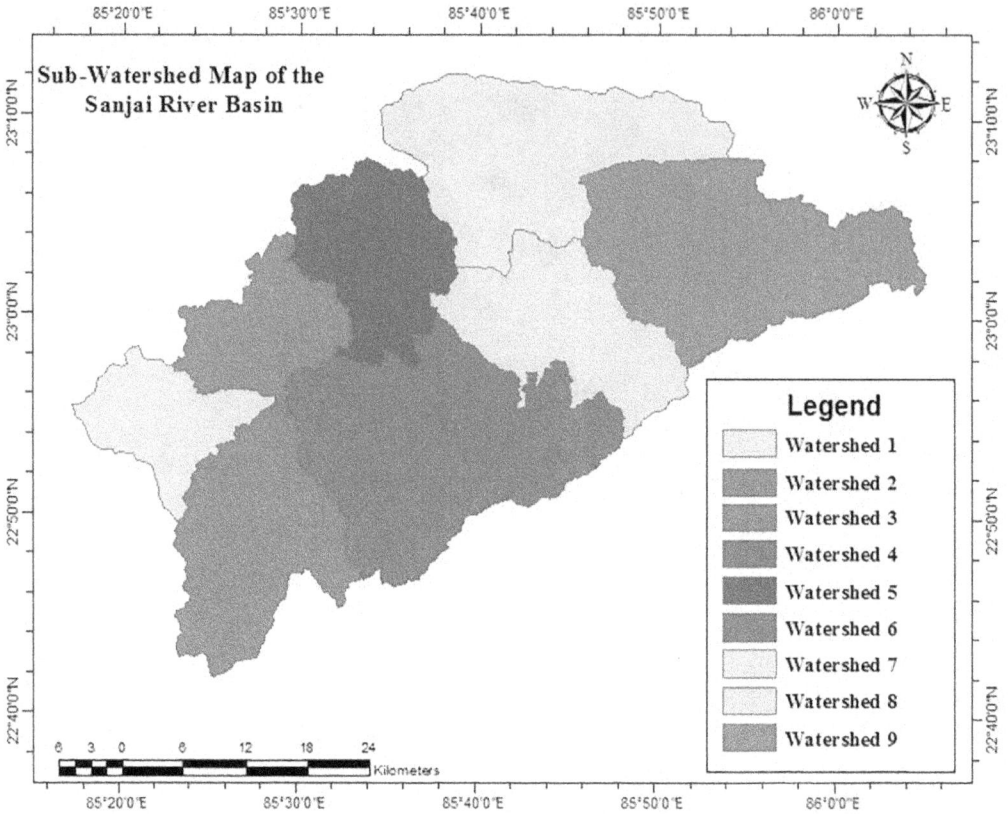

FIGURE 14.5   Sub-watershed map of the study area.

TABLE 14.3
Runoff Volume

Sub-Watershed No.	Area of Sub-Watershed (km²)	Runoff Volume (mm)
1	334	151.69
2	385	127.88
3	202	116.84
4	148	127.88
5	231	96.88
6	195	125.99
7	245	80.86
8	135	80.86
9	259	72.97
Total Runoff = 981.85 mm		

REFERENCES

Beven, K. J. (1982). On subsurface stormflow, an analysis of response times. Hydrological Sciences Journal, 27, 505–521

Chow, V. T., Maidment, D. R., Mays, L. W. (1988). Applied Hydrology. McGraw-Hill, New York

Hawkins, R. H. (1973). Improved prediction of storm runoff in mountain watersheds. Journal of Irrigation and Drainage Engineering, 99, 519–523

Hawkins, R. H. (1993). Asymptotic determination of runoff curve numbers from data. Journal of Irrigation and Drainage Engineering, 119, 334e345

Hawkins, R. H., Ward, T. J., Woodward, D. E., Van Mullem, J. A. (2009). Review of Curve number hydrology-state of practice by R. H. McCuen In: SCE, Reston, VA, Journal of Hydraulic Engineering 14(9), 20191-4400; ISBN 978-0-7844-1004-2.

Klingebiel, A. A. Montgomery, P. H. (1961). Land-capability classification. In: Michael Aide. (ed) Agricultural Handbook 210, Soil Conservation Service. U.S. Govt. Printing Office, Washington, DC, 21 p

Kumar, V., Singh, P., Singh, V. (2007). Snow and glacier melt contribution in the Beas River at Pandoh Dam, Himachal Pradesh, India. Hydrological Sciences Journal, 52(2), 376–388

Patel, J., Singh, N. P., Indra, P., Khalid, M. (2017). Surface runoff estimation using SCS-CN method—a case study on Bhadar watershed, Gujrat, India. Imperial Journal of Interdisciplinary Research, 3, 5

Sarangi, A., Bhattacharya, A. K. (2005). Comparison of artificial neural network and regression models for sediment loss prediction from Banha watershed in India. Agricultural Water Management, 78, 195–208

Satheeshkumar, S., Venkateswaran, S., Kannan, R. (2017). Rainfall-runoff estimation using SCS-CN and GIS approach in the Pappiredipatti watershed of the Vaniyar sub-basin, South India. Modeling Earth Systems and Environment, 3, 24

SCS, National Engineering Handbook (1985). Section 4: Hydrology, Soil Conservation Service. USDA, Washington, DC. https://www.scirp.org

Shadeed, S., Mohammad, A. (2010). Application of GIS-based SCS-CN method in West Bank catchments, Palestine. Water Science and Engineering, 3(1), 1–13

Soil Conservation Service (SCS). (1972). National Engineering Handbook, Section 4: Hydrology. Department of Agriculture, Washington, DC, 762 p

United States Department of Agriculture (USDA). (1972a). National Engineering Handbook. Section 4. Hydrology. Soil Conservation Service, US Government Printing Office, Washington, DC

United States of Department of Agriculture (USDA). (1972b). National Engineering Handbook of Hydrology. Handbook. Section 4. Chapter 21

Woolhiser, D. A., Smith, R. E. Goodrich, D. C. (1990). KINEROS, A kinematic runoff and erosion model: Documentation and User Manual. U.S. Department of Agriculture, Agricultural Research Service, ARS-77, 130 p

15 Impact of Land-Use Change on Rural Development
A Study of Population Dynamics in Canning-I Block, South 24 Parganas, West Bengal, India

Amiya Kumar Sarkar

15.1 INTRODUCTION

As the population continues to grow, more space is required to accommodate the growing population, and this gave rise to the development of functional zones, particularly in rural settlements. Population growth has been high in recent decades, but the per capita income is exorbitantly low, which requires a land utilization survey with a view to extracting the utmost possible produce from the land. The ultimate goal of land-use study is to suggest planning for better utilization of available land to society. Rural development is the process of improving the quality of life and economic well-being of people living in rural areas, often relatively isolated and sparsely populated areas. Rural development has traditionally centered on the exploitation of land-intensive natural resources such as agriculture and forestry. However, changes in global production networks and increased urbanization have altered the character of rural areas. Increasingly, tourism, niche manufactures, and recreation have replaced resource extraction and agriculture as dominant economic drivers. The need for rural communities to approach development from a wider perspective has created more focus on a broad range of development goals rather than merely creation incentives for agriculture or resource-based businesses. Education, entrepreneurship, physical infrastructure, and social infrastructure all play an important role in developing rural regions. Thus, land use is influenced by economic, cultural, political, historical, and land-tenure factors at multiple scales. Land cover, on the other hand, is one of the many biophysical attributes of the land that affect how ecosystems function (Meyer and Turner, 1992). The objectives of this chapter are to analyze different categories of land use and their changing patterns and to identify the changes in response to population increase and rural development of Canning-I block.

15.2 LOCATION OF CANNING-I BLOCK

Canning-I block (Figure 15.1) is situated in the western part of the Canning subdivision and eastern part of south 24 Parganas (West Bengal, India), between 22°08′13″ N to 22°24′16″ N latitude and 88°30′53″ E to 88°41′53″ E longitude and encompasses an area of 187.86 km². The block is divided into ten gram panchayats (Bansra, Daria, Dighirpar, Gopalpur, Hatpukuria, Itkhola, Matla-I, Matla-II, Nikarighata, and Taldi) and 60 revenue mouzas or villages.

DOI: 10.1201/9781003377825-17

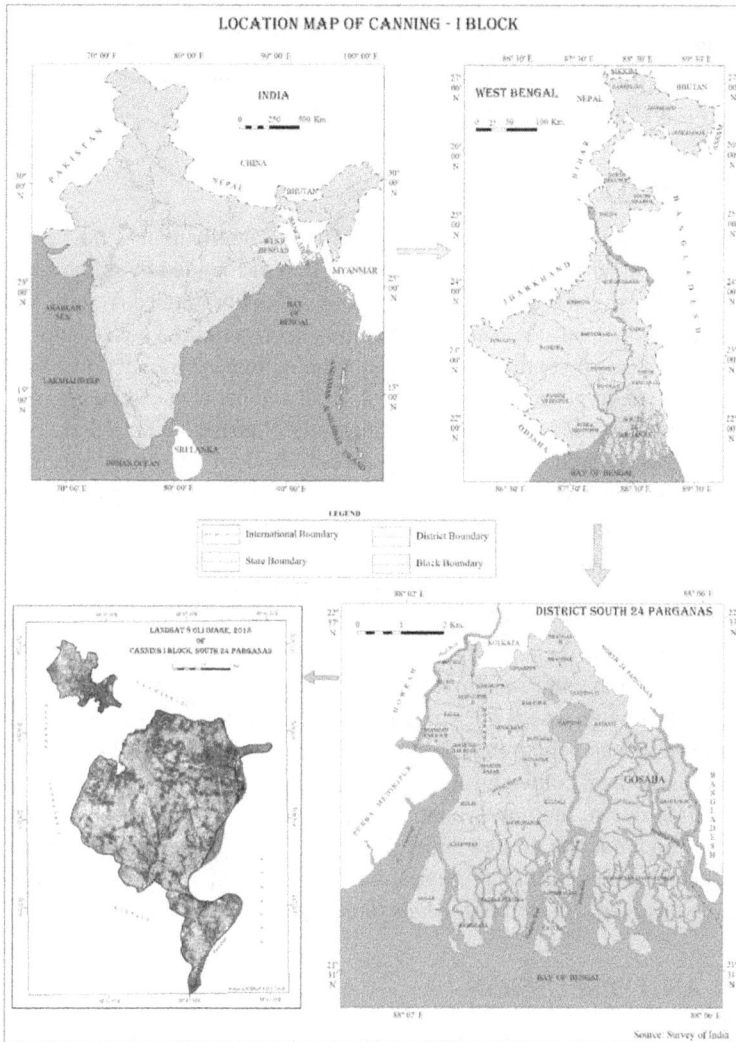

FIGURE 15.1   Location map of Canning-I block.

15.3   DATABASE AND RESEARCH METHODOLOGY

The study has been done with the help of relevant primary and secondary data, and those were represented by suitable cartographic techniques. The collection of data was made from different sources, such as for the preparation of the land-use/land-cover (LULC) map in 1968 topo-sheets are used, and for preparation of the LULC map in the year 1993 US Geological Survey (USGS) Landsat 5 Thematic Mapper (TM) image was used and for the preparation of LULC map of 2018 USGS Landsat 8 Operational Land Imager (OLI) image was used. Other secondary data were collected from different government offices and non-governmental organizations. Such data were collected from various sources like demographic data from the Census office (Census of India; year: 1961–2011) and land-use data from the District Statistical Hand Book (Govt. of W.B.), Agricultural data from the agricultural office (Govt. of W.B.), block maps, district maps from the National Atlas and Thematic Mapping Organisation, and other statistical data from different government offices of West Bengal were obtained. The results were analyzed with the help of different statistical, cartographic, and graphical representations and geographic information systems (GIS) techniques. For graphical representation and statistical analysis, MS Excel 2016 was used, and mapping was done with the help of different GIS software. After analyzing the data and information, they have been represented through different types of thematic and non-thematic maps.

15.4 PHYSICAL SETUP

Canning-1 block located in the north-eastern part of the Sunderbans is a vast new alluvial plain intersected by a large number of tidal rivers and creeks. It is surrounded by saline and brackish water and has soil, water, air, flora, and fauna as its components of a vibrating ecosystem. The study area is located in a miogeosynclinal furrow, which means that the origin of the area cannot be explained by the single infilling of a shallow continental shelf with the sediments brought down by the Ganga and/ or the Brahmaputra rivers. Synthesizing all the geological and geophysical pieces of evidence, it has been postulated that the ferralitic-lateritic landforms in the Pleistocene period are only a part of the south Bengal basin, now covered by recent to sub- recent alluvium. The study area is a typical deltaic plain with elevation ranging between 5 and 10 ft above mean sea level. It is a poorly drained land in between levees where the soil is fine and unconsolidated. The area is basically drained by a network of large tidal creeks in this low-lying coastal deltaic plain. The orientation, geomorphic setting, and materials of the river deposits are the resultant products of the complex interaction of several factors like the tectonic framework of the Bengal basin, geological, climatological, physical, chemical, and biological processes. The area under study is cyclone prone. Soils of the study area are locally called 'Nonamati'. Ganga alluvium and its salinized part are considered as parent materials. It is normally salt-free where calcite or magnesite is present as rich divalent in this soil. A large percentage of prawn seed collectors are women and children. Traditionally, the area under study is characterized by small farmers and paddy and prawn cultivators. The mono-crop pattern of agriculture is still in practice in the Sunderbans region due to the non-availability of irrigation waters. Lack of water for irrigation causes crop failure and reduction of area under multiple cropping. The land is used in the vast areas of the Canning-I block for agricultural purposes. A few winter crops and vegetables are cultivated by the farmers where rainwaters are stored in Khals (creek) and bills (tanks).

15.5 RESULT AND DISCUSSION

15.5.1 Areal and Temporal Variation in Land-Use Pattern and the Impact on Rural Development Changing Land-Use Pattern of Canning-I Block Since 1968

15.5.1.1 Extent of Land-Use Categories in Canning-I Block 1968

Canning-I block occupies 17.02% of Canning sub-division area and 1.89% of District 24 Parganas South. The distribution of different LULC categories for the study year is presented in Figure 15.2 and Table 15.1. The source of the LULC map of 1968 is the Survey of India topographical map with scale 1:50,000 (Figure 15.2). When we take the block as a whole into consideration, it is observed that the most extensive land-use category of Canning-I block in 1968 was agricultural land, which covered an area of 141.68 km² and occupied 75.49% of the area under study (Figure 15.3).

The second most extensive land-use category was rural settlement, which covered 30.32 km² and occupied 16.14% of the total geographical area of Canning-I block. Other extensive LULC categories were a foot track that occupied 0.49 km² (0.26%), unmetalled road that occupied 0.35 km² (0.18%), metal road that occupied 0.14 km² (0.07%), railway that occupied 0.26 km² (0.14%), embankment that occupied 2.23 km² (1.19%), urban settlement that occupied 0.45 km² (0.24%), sand deposition that occupied 1.12 km² (0.60%), mud flat that occupied 1.87 km² (0.99%), creak that occupied 0.76 km² (0.40%), canal that occupied 0.28 km² (0.15%), river that occupied 6.51 km² (3.47%), and pond that occupied 1.4 km² (0.75%) of the total geographical area under study.

FIGURE 15.2   Canning-I block, 1968.

15.5.1.2   Extent of Land-Use Categories in Canning-I Block 1993

The distribution of different LULC categories for the year 1993 has been presented in Figure 15.5 and Table 15.1. The source of the LULC map of 1993 is Landsat 5 TM with resolution 30 m (Figure 15.4 and Table 15.1). In 1993, the most extensive cover was agricultural land which covers 135.53 km^2 and 72.53% area of the total geographical area under study. The second most extensive cover is the rural settlement which covers 36.27 km^2 and occupies 19.33% of the land under study area. Other extensive categories in 1993 were unmetalled road that occupies 0.52 km^2 (0.28%), metalled road that occupies 0.19 km^2 (0.1%), railway property area that occupies 0.26 km^2 (0.14%), embankment that occupies 2.89 km^2 (1.54%), urban settlement that occupies 0.52 km^2 (0.28%), aquacultural land that occupies 3.69 km^2 (1.97%), and water bodies like creek, canal, river, and pond area that occupy 0.47 km^2 (0.25%), 0.28 km^2 (0.15%), 5.92 km^2 (3.15%) and 0.82 km^2 (0.43%), respectively, of the total geographical area under study.

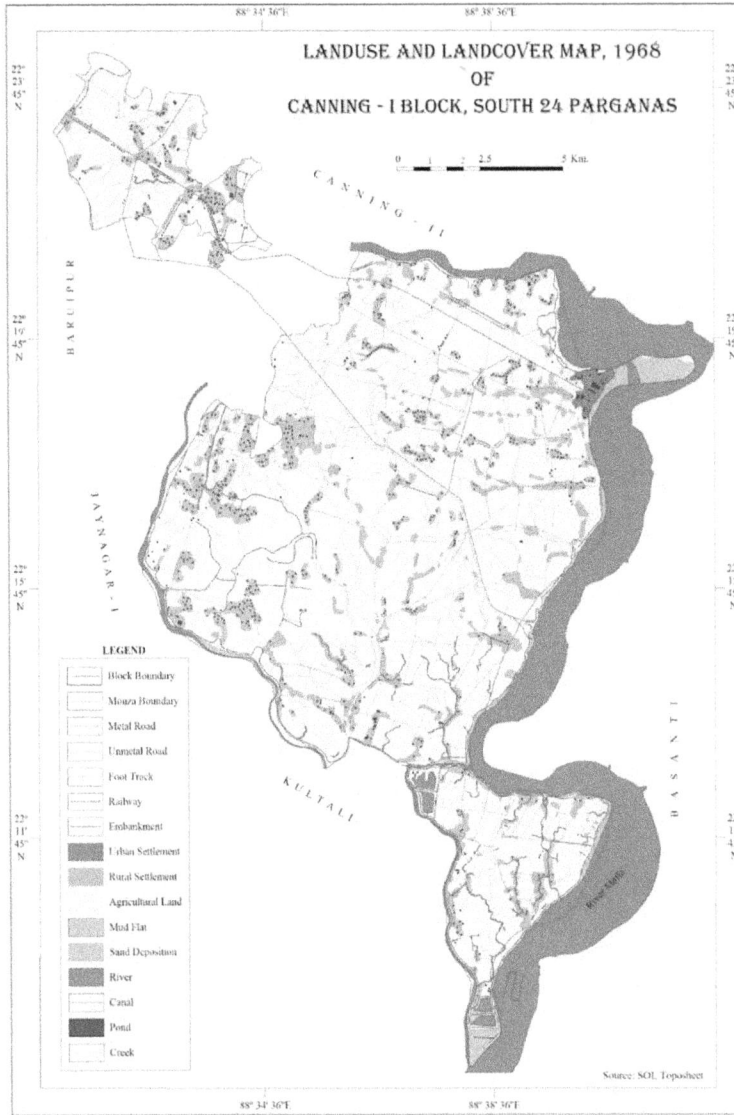

FIGURE 15.3   Canning-I, land-use/land-cover map, 1968.

TABLE 15.1
Land-Use/Land-Cover Types of Canning-I Block, Years 1968, 1993, 2018

Type	1968	Percentage (%)	1993	Percentage (%)	2018	Percentage (%)	Changes (%)
Foot track	0.49	0.26	0.32	0.17	0	0	−0.26
Unmetal road	0.35	0.18	0.52	0.28	0.12	0.06	−0.12
Metal road	0.14	0.07	0.19	0.1	1.77	0.94	0.87
Railway	0.26	0.14	0.26	0.14	0.26	0.14	0
Embankment	2.23	1.18	2.89	1.54	2.89	1.54	0.36
Brick kiln	0	0	0	0	0.63	0.34	0.34
Rural settlement	30.32	16.14	36.27	19.33	59.83	31.85	15.71

TABLE 15.1 (*Continued*)
Land-Use/Land-Cover Types of Canning-I Block, Years 1968, 1993, 2018

Type	1968	Percentage (%)	1993	Percentage (%)	2018	Percentage (%)	Changes (%)
Urban settlement	0.45	0.24	0.52	0.28	0.78	0.42	0.18
Agricultural land	141.68	75.43	135.53	72.21	102.07	54.33	−21.1
Aquacultural land	0	0	3.69	1.97	8.26	4.4	4.4
Plantation	0	0	0	0	2.65	1.41	1.41
Sand deposition	1.12	0.6	0	0	0	0	−0.6
Mud flat	1.87	0.99	0	0	0	0	−0.99
Marshy land	0	0	0	0	2.6	1.38	1.38
Creak	0.76	0.4	0.47	0.25	0.35	0.18	−0.22
Canal	0.28	0.15	0.28	0.15	1.11	0.59	0.59
River	6.51	3.47	5.92	3.15	4.15	2.21	−1.26
Pond	1.4	0.75	0.82	0.43	0.39	0.21	−0.54
Total area	**187.86**	**100**	187.86	100	**187.68**	**100**	—

Source: Computed by author (area in km²).

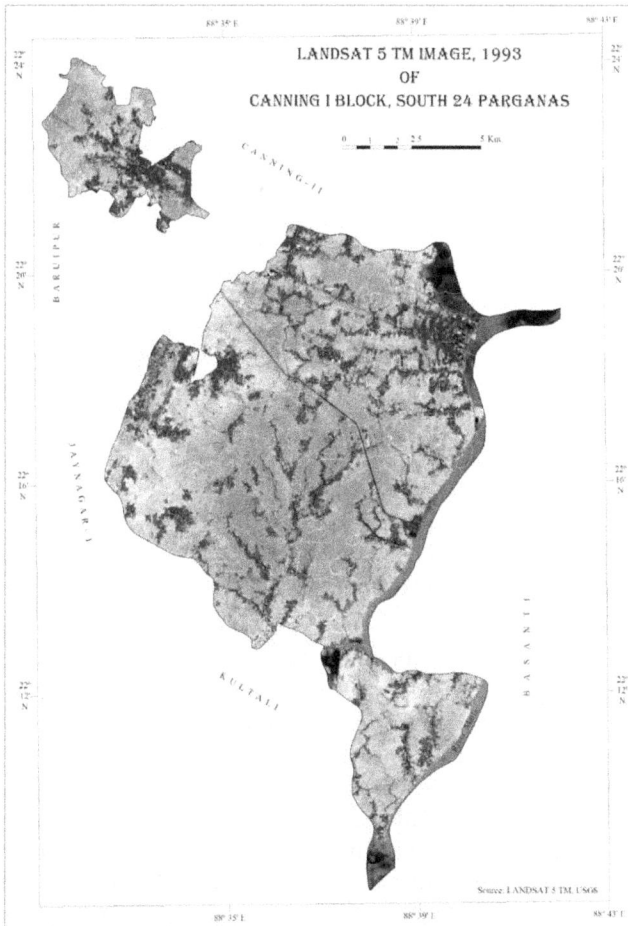

FIGURE 15.4   Landsat Thematic Mapper image, land use 1993.

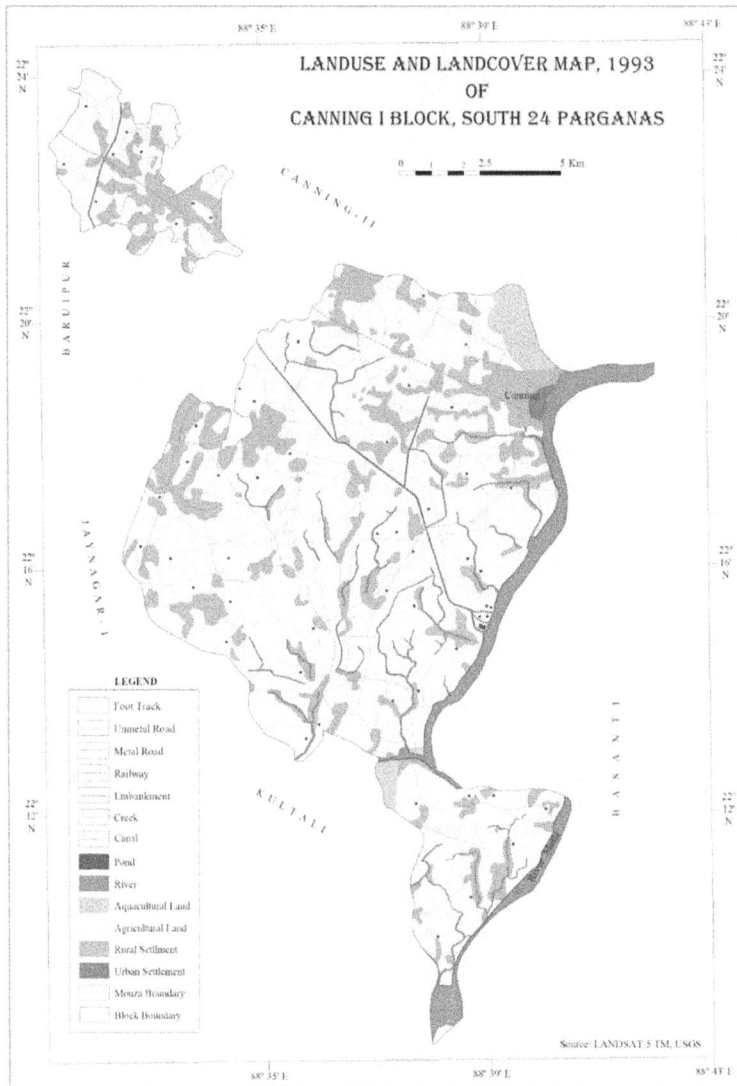

FIGURE 15.5   Land use/land cover 1993.

15.5.1.3   Extent of Land-Use Categories in Canning-I Block 2018

The distribution of different LULC categories for the study year 2018 is presented in Figure 15.7 and Table 15.1. The source of the LULC map of 2018 is Landsat 8 OLI with resolution 28 m (Figure 15.6 and Table 15.1). In 2018, the most extensive cover is agricultural land which covers 102.07 km^2 and 54.33% of the total geographical area of Canning-I block.

The second most extensive cover is the rural settlement which covers 59.83 km^2 and occupies 31.84% of land under study area. Other extensive categories in the year 2018 are unmetalled road that occupies 0.12 km^2 (0.06%), metalled road that occupies 1.77 km^2 (0.94%), railway property area that occupies 0.26 km^2 (0.14%), embankment that occupies 2.89 km^2 (1.54%), brick kiln land that occupies 0.63 km^2 (0.34%), urban settlement that occupies 0.78 km^2 (0.42%), aquacultural land that occupies 8.26 km^2 (4.40%), plantation that occupies 2.65 km^2 (1.41%), marshy land that occupies 2.60 km^2(1.38%), and water bodies like canal, river, pond area that occupy 0.59 km^2 (0.59%), 4.15 km^2 (2.21%), and 0.39 km^2 (0.21%), respectively, of the total geographical area under study.

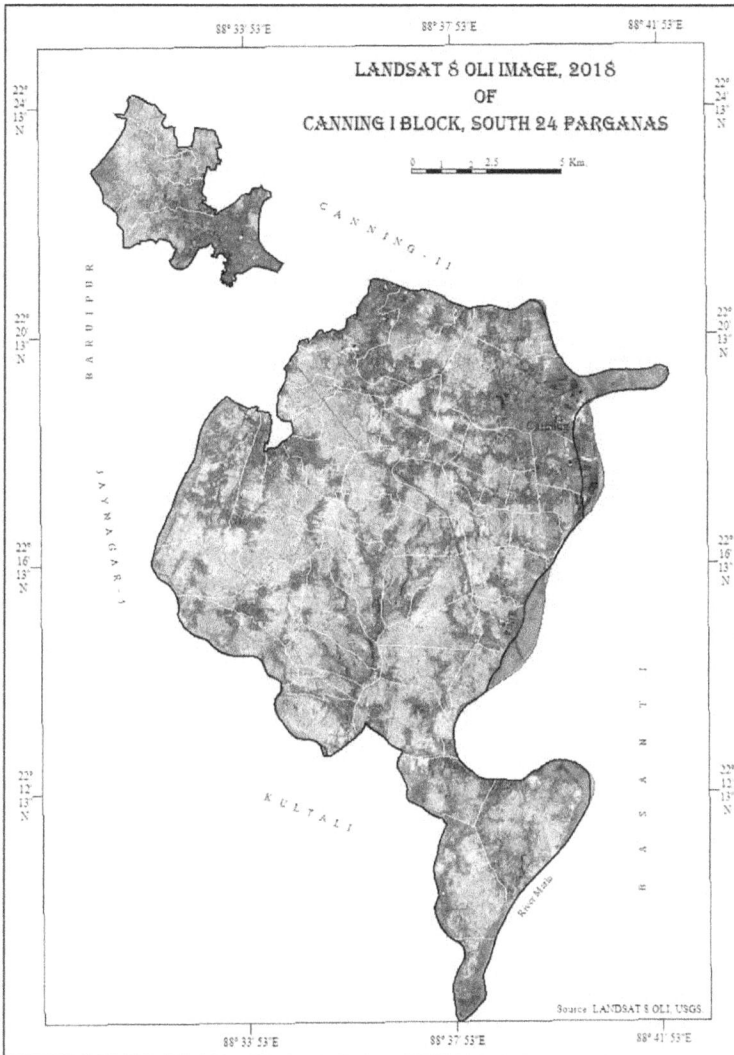

FIGURE 15.6 Landsat Operational Land Imager image 2018.

15.6 TOTAL VOLUME OF CHANGE IN LAND USE OF CANNING-I BLOCK 1968–2018 AND IMPACT ON RURAL DEVELOPMENT

The major positive total volume of changes in LULC categories are metal road (1164%), embankment (30%) and rural settlement (97%), urban settlement (73%), and canal (296%) of Canning-I block. On the other hand, the major negative total volume of changes in LULC categories are unmetalled road (−66%), agricultural land (−28%), creek (−54%), river (−36%), and pond (−72%) of Canning-I block (Figure 15.8).

15.7 PERCENTAGE CHANGE OF TOTAL AREA IN LAND USE 1968–2018 OF CANNING-I BLOCK AND IMPACT ON RURAL DEVELOPMENT

There are some positive and negative percentage changes in the total area in Canning-I block of the area under study. The positive percentage change of total area took place in the metalled road

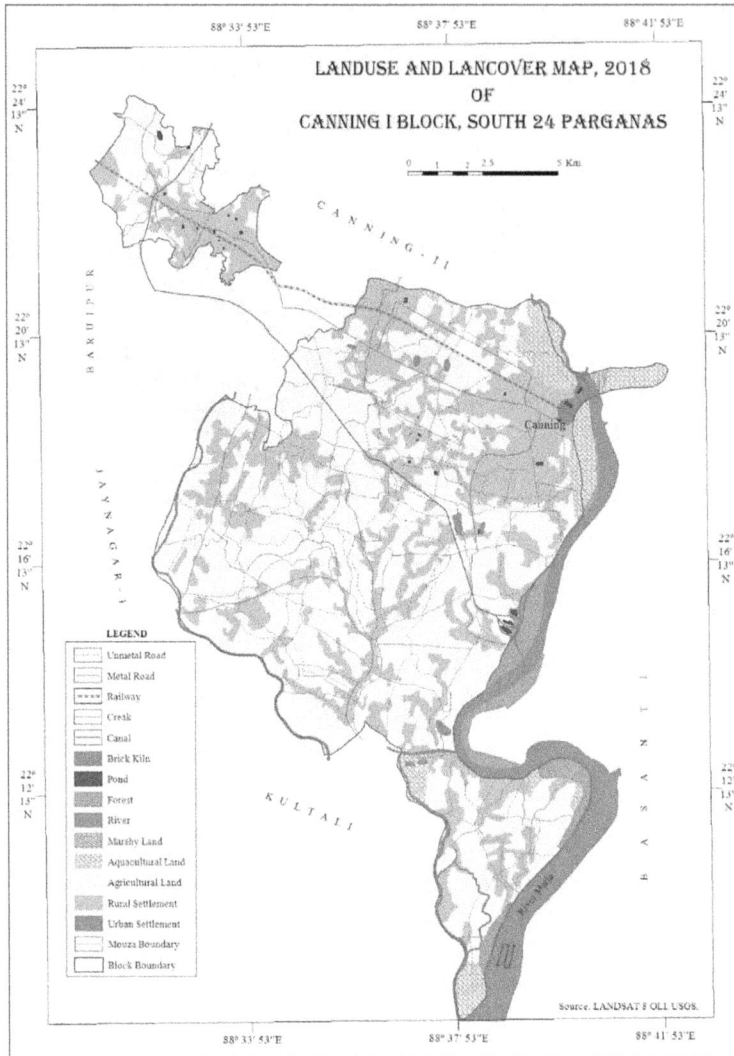

FIGURE 15.7   Land-use/land-cover map 2018.

(0.87%), embankment (0.36%), brick kiln (0.34%), rural settlement (15.71%), urban settlement (0.18%), aquacultural land (4.40%), plantation (1.41%), marshy land (1.38%), and canal (0.44%) areas under Canning-I block. And the negative percentage change of total area took place in foot track (−0.26%), unmetalled road (−0.12%), agricultural land (−21.1%), sand deposition (−0.6%), mudflat (−0.99%), creek (−0.22%), river (−1.26%), and pond (−0.54%) (Figure 15.9).

15.8   LAND UTILIZATION AND RURAL DEVELOPMENT IN CANNING-I BLOCK

15.8.1   CHANGE OF LAND PER HUNDRED POPULATION OF CANNING-I BLOCK 1968–2018

Land per hundred population 1968: In the year 1968, the amount of land per hundred population was 0.327 km^2, in 1971 it was 0.302 km^2 of Canning-I block, and land per capita was 0.003 km^2. In the year 2001, it was 0.159 km^2 per hundred population, and land per person was 0.0015 km^2 (Table 15.2).

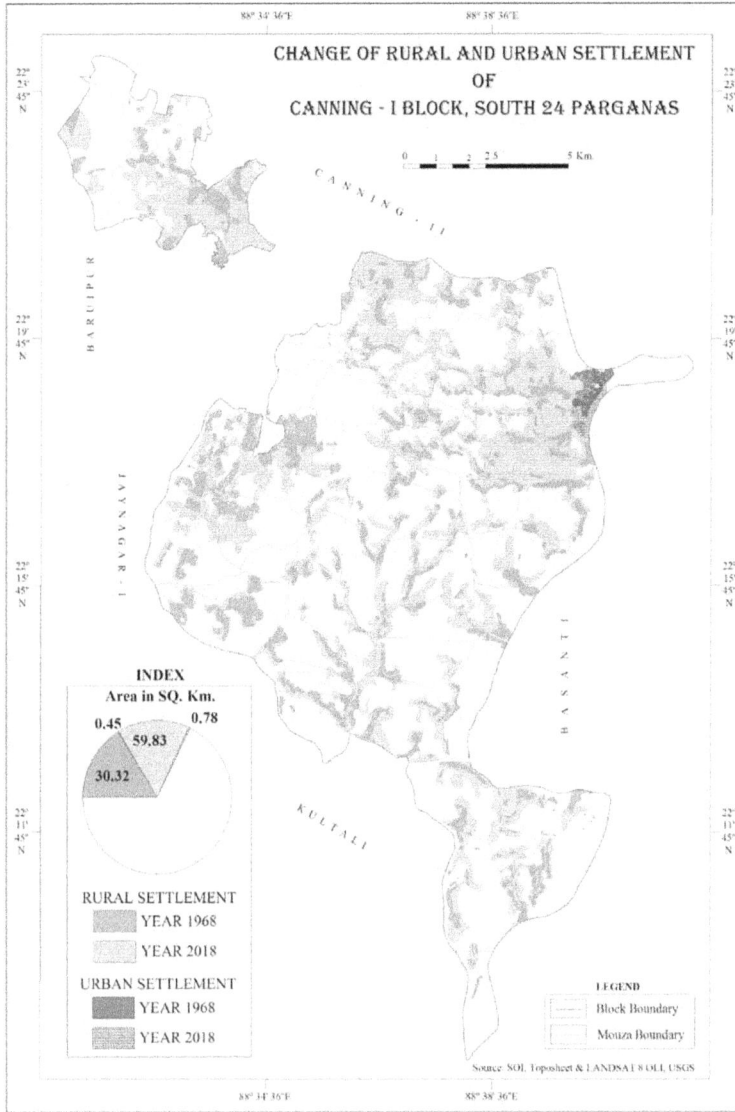

FIGURE 15.8 Change of rural urban settlement.

In the year 2011, the average maximum land per hundred population in Canning-I block was 0.120 km² and the minimum was 0.0012 km². The total volume of changes in land per hundred population from 1968 to 2018 of Canning-I block was −67% (Figure 15.10).

Land per hundred population 1971: To identify the spatial distribution of land per hundred population of Canning-I block in 1971, the area has been divided into six zones (Figure 15.11a). The classes are <0.8, 0.8–1.6, 1.6–2.4, 2.4–3.2, 3.2–4.0, and >4.0 km² of land per hundred population. Sixty villages had below 0.8 km² land per hundred population, one village in the north-western part of the block with 1.6–2.4 km² of land per hundred population, and another village in this part had >4.0 km² of land per hundred population.

Land per hundred population 2011: To show the distribution of land per hundred population, the area has again been divided into six classes such as <0.32, 0.32–0.64, 0.64–0.96, 0.96–1.28, 1.28–1.60, and >1.60 km² of land per hundred population (Figure 15.11b).

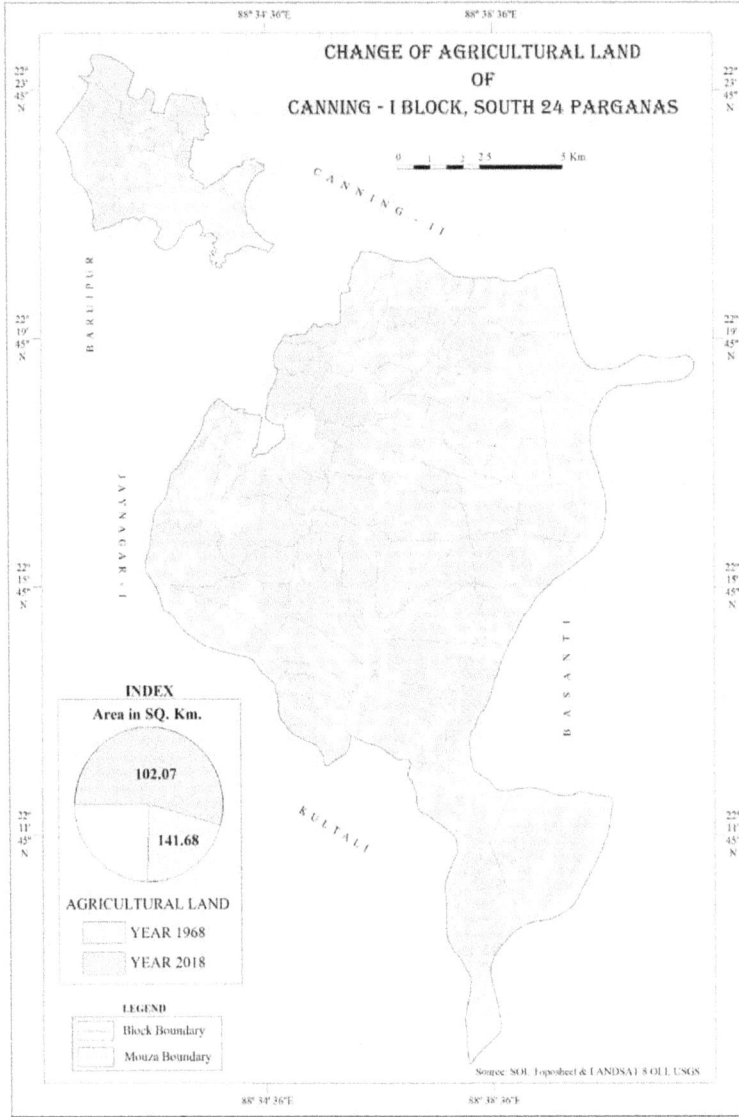

FIGURE 15.9   Change of agricultural land.

TABLE 15.2

Changes of Land per Hundred Population of Canning-I Block from 1968 to 2018

Year	1968	1971	1981	1991	2001	2011	2018
Land per hundred population (km²)	0.327	0.302	0.338	0.196	0.159	0.120	0.108
Volume of change		−67%					

Source: Computed by author.

CHANGING LAND / 00' POPULATION OF CANNING - I
BLOCK FROM 1968 - 2018

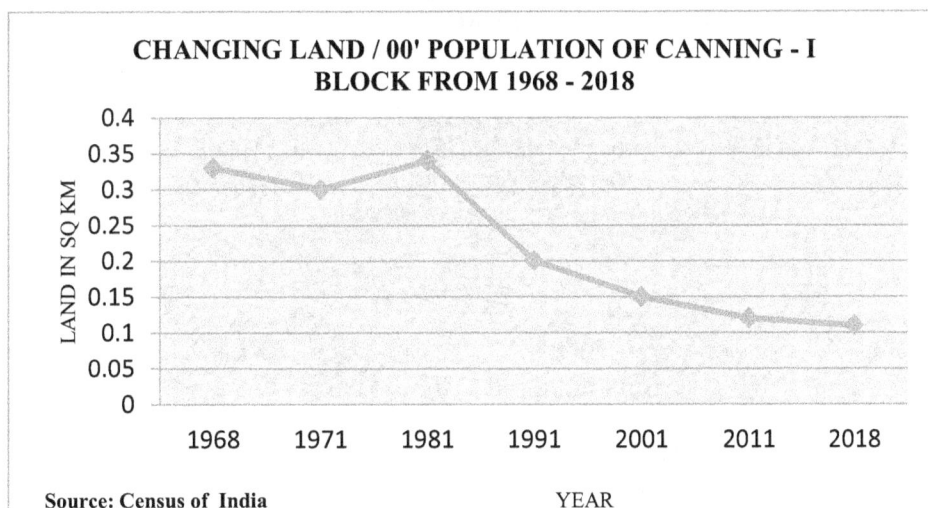

FIGURE 15.10   Changing land per hundred population 1968–2018.

There were about 58 villages with <0.32 km^2 of land per hundred population, and in the north and north-western parts, there were two villages with 0.32–0.64 km^2 of land per hundred population. There was only one village in the north-western part with more than 1.60 km^2 of land per hundred population.

15.9   CHANGE OF LITERATES PER HUNDRED POPULATION OF CANNING-I BLOCK 1968–2018

In 1971, it was identified that in the Canning-I block, there were 22 literates per hundred population, and in the year 2011, the figure increased to 56 literates per hundred population.

In this block, the total number of change of literates per hundred population of Canning I block is 186% (Figure 15.12).

To show the distribution of literates per hundred population, the Canning-I block (Figure 15.13) has been divided into six regions with the following cases: <12, 12–20, 20–28, 28–36, 36–44, and >44. The above 44 literates per hundred population have been recorded by the villages of Dighirpar (176) and Poramura (172).

In the year 2011, the distribution of literate persons per hundred population is shown in Figure 15.13. The six classes of distribution are <13, 13–26, 26–39, 39–52, 52–65, and >65 literates/hundred population. There are 14 villages with >65 literates per hundred population; these are mainly concentrated in the east and north-western part of this block.

15.10   CHANGE OF DIFFERENT VILLAGE AMENITIES AVAILABLE IN CANNING-I BLOCK 1971–2011

15.10.1   CHANGES OF VILLAGE AMENITIES OF CANNING-1 BLOCK

Changes of educational facilities: In 1971, the Canning-1 block had the following facilities of education, viz., 64 primary schools, three middle schools, six secondary schools, and one college (Table 15.4, Figure 15.14). And in 2011, it increased to 122 pre-primary schools, 114 primary schools, five middle schools, 15 secondary schools, eight senior secondary schools, and one college.

Changes of medical facilities: In 1971, there were only seven dispensaries; in 2011, there were 35 primary health centers, three maternity and child welfare centers, one tuberculosis clinic, 11 dispensaries, and two family welfare centers (Table 15.4).

Changes of power supply facilities: In this block, only Taldi village (J.L. no. -78) had a power supply facility in 1971, and in 2011 it thoroughly changed. For example, 53 villages had power supply for domestic use, eight villages had power supply for agricultural use, 11 villages had power supply for commercial use, and six villages had power supply for all uses.

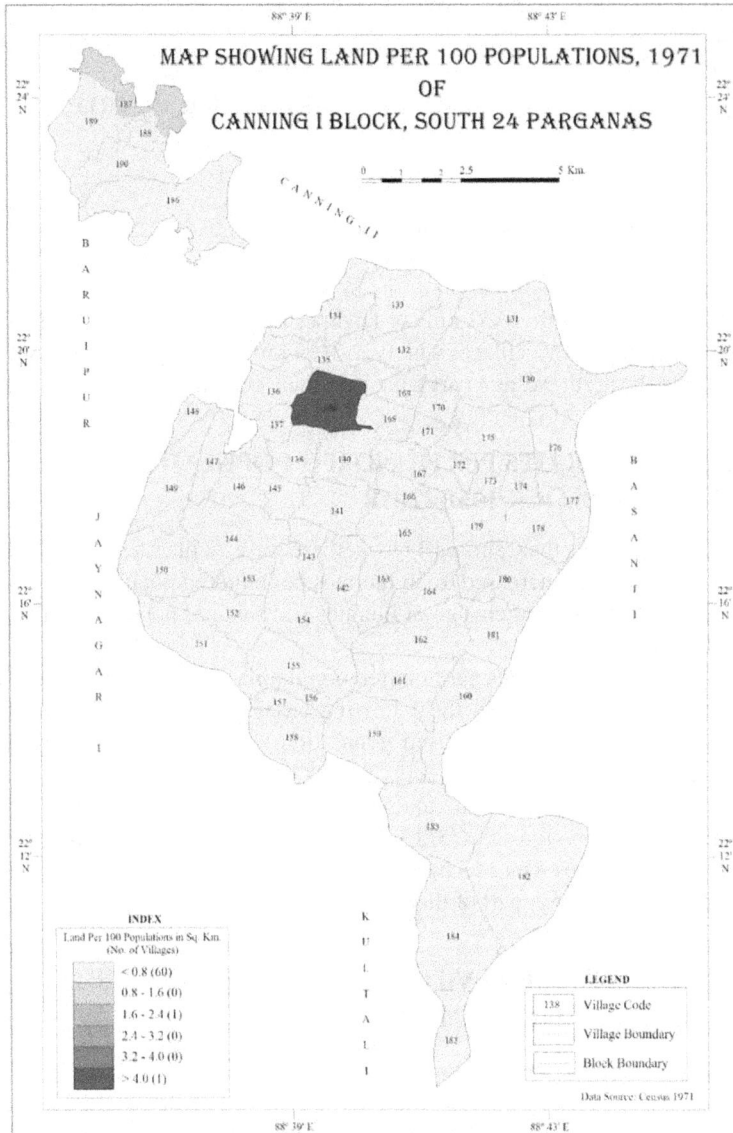

FIGURE 15.11   Map showing land per hundred population (a) 1971 and (b) 2011.

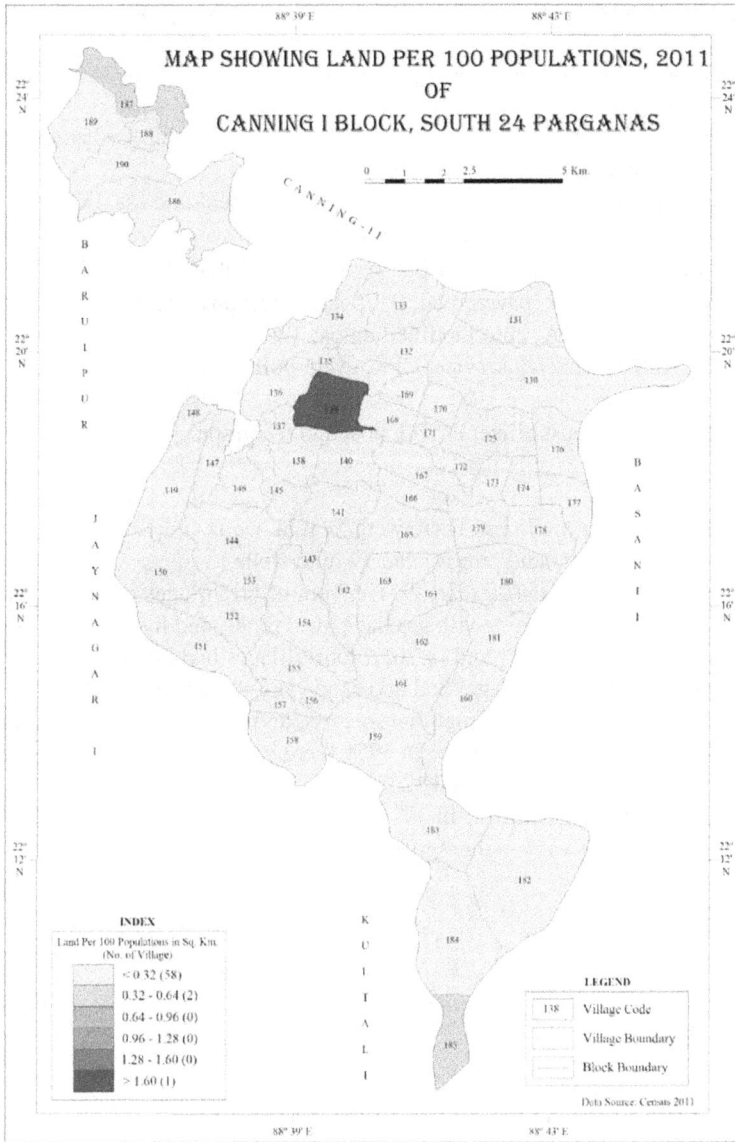

FIGURE 15.11 (Continued)

TABLE 15.3
Change of Literates per Hundred Population of Canning-I Block from 1968 to 2018

Year	1968	1971	1981	1991	2001	2011	2018
Literate per hundred population	22	22	27	31	45	56	63
Volume of change		186%					

Source: Computed by author.

15.11   BREAKDOWN OF VILLAGE AMENITIES AND NOMENCLATURE

PP, pre-primary school; M, middle school; P, primary school; SS, senior secondary school;
S, secondary school; C, college

D, dispensary; MCW, maternity and child welfare centre; PHS, primary health centre; FWC,
family welfare centre; TBC, tuberculosis clinic

ED, power supply for domestic use; EAG, power supply for agricultural use; EC, power supply for
commercial use; EA, power supply for all uses

TW, tap water/tube-well water; HP, hand pump; TK, tank water; WW, well water

KR, kacha road/mud road; PR, paved road; PT, post, telegraph and telephone; BS, bus service
(public or private); AMA, auto/modified autos; T&V, taxis and vans; CDA, cart driven
by animals; RFS, river ferry service; CPR, cycle-pulled rickshaws (manual and machine
driven)

PO, post office; SPO, sub-post office; P&TO, post and telegraph

RM, regular market; WH, weekly market

Changes in drinking water facilities: There were 59 tube-wells and 19 tanks in 1971; in 2011,
there were two wells, 49 hand pumps, and 19 tube-wells.

Changes of communication facilities: In 1971, 57 out of 60 villages of the Canning-1 block
had kacha road or mud road, nine villages had paved roads, and five villages had post, telegraph, and telephone facilities. And in 2011, four villages had bus services, eight villages
had auto/modified autos, 15 villages had taxis and vans, 43 had paddled rickshaws (manual
and machine-driven), two had animal-driven carts, and three had river ferry service.

Changes of post and telegraph facilities: In 1971 there were 11 post offices in this block; in
2011 there were three post offices and 12 sub-post offices.

Changes of market or haat facilities: In 1971, there was one regular market and 16 weekly
haat. And in 2011, there were eight regular markets and 25 weekly haat.

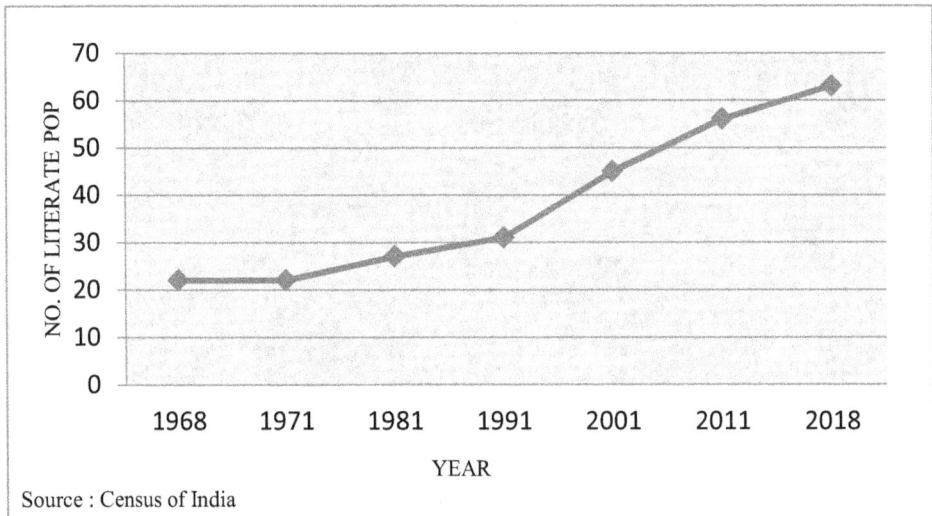

FIGURE 15.12   Change in number of literature persons per hundred population of Canning-1 block.

Source: Census of India.

FIGURE 15.13 Literacy per hundred population in (a) 1971 and (b) 2011.

Source: Census of India.

15.12 CONCLUSION

Land utilization of the Canning-I block is influenced by economic, cultural, political, historical, and land-tenure factors at multiple scales. Land cover, on the other hand, is one of the many biophysical attributes of the land that affect the ecosystems' function. This area is only a part of the South Bengal basin which is now covered by recent to sub-recent alluvium. Mangrove swamps, mudflats, sand depositions, forests, constructed river embankments, reclaimed swamps, and marshy areas are found in the study area. The study area lies typically in the deltaic plains at 5–10 feet above mean sea level. The characteristic feature is the poorly drained land in between levees where the soil is fine and unconsolidated. Traditional shrimp culture and the entire value chain associated with it is a major

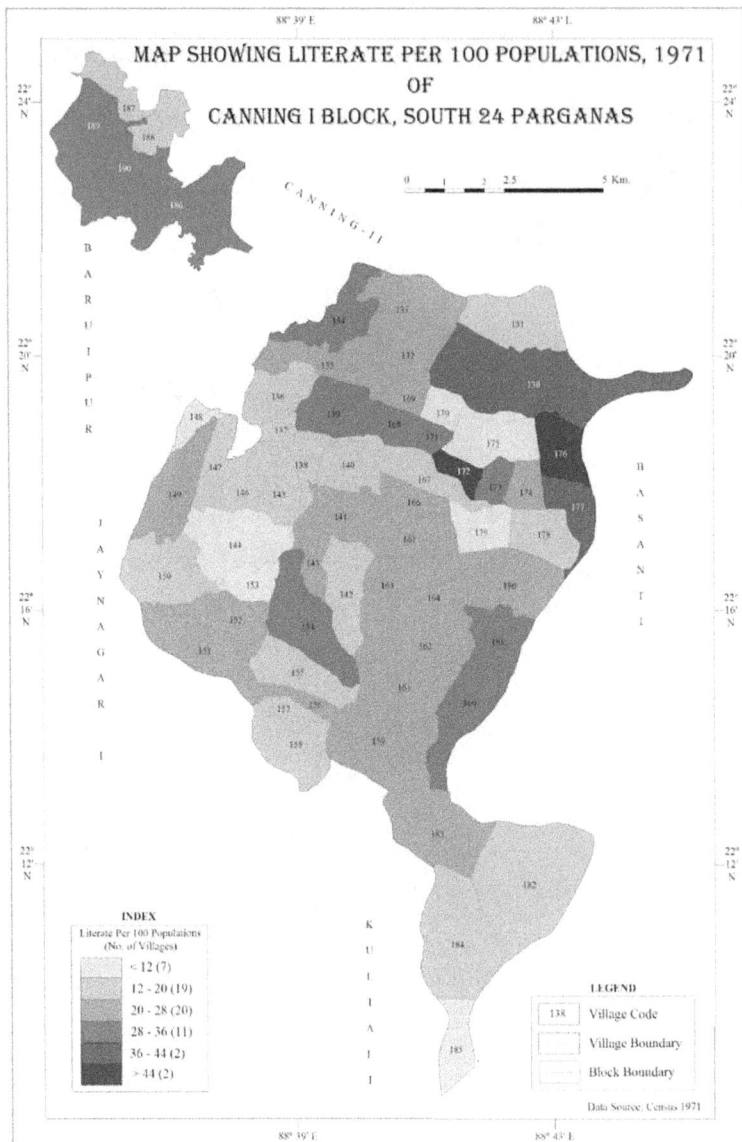

FIGURE 15.13   Continued

TABLE 15.4

Change in Village Amenities of Canning-I Block from 1971 to 2011

Village Amenities	1971	2011
Education	P64, M3, S6, C1	PP122, P114, M25, S15, SS8, C1
Medical	D7	PHS35, MCW3, TBC1, D11, FWC2
Power supply	E1	ED53, EAG8, EC11, EA6
Drinking water	TW59, TK19	WW2, HP49, TW19
Communication	KR57, PR9, PT5	BS4, AMA8, T&V15, CPR43, CDA2, RFS3
Post and telegraph	PO11	PO3, SPO12
Market/Haat	RM1, WH16	RM8, WH25

Source: Computed by author.

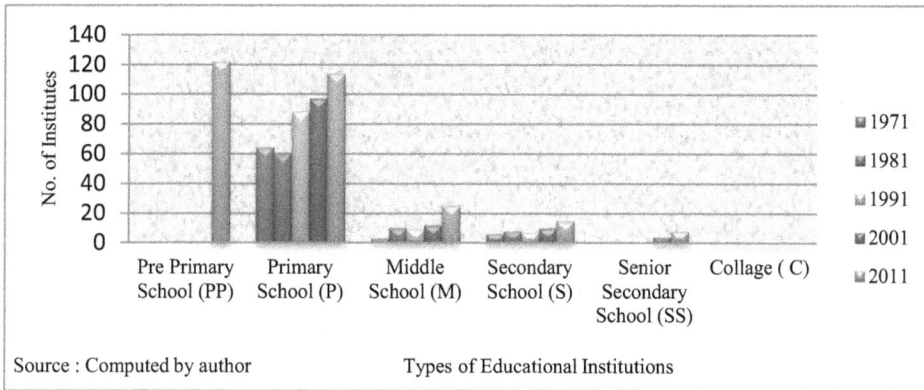

FIGURE 15.14 Educational Institutes in Canning-I Block from 1971 to 2011.

source of livelihood for people in the area, but unfortunately, the Indian open market economic policies of the 1990s influenced the entire value chain of the aquaculture sector of this study area. Paddy cultivation and fishing are the mainstay of the economy of these areas. Sustainable aquaculture as well as shrimp–paddy culture has the potential to help the affected cultivators to not only adapt to climate change but also improve their economic condition. The positive percentage change of total area under study took place in the metalled road (0.87%), embankment (0.36%), brick kiln (0.34%), rural settlement (15.71%), urban settlement (0.18%), aquacultural land (4.40%), plantation (1.41%), marshy land (1.38%), and canal (0.44%) areas under Canning-I block. And the negative percentage change of total area took place in foot track (−0.26%), unmetalled road (−0.12%), agricultural land (−21.1%), sand deposition (−0.6%), mudflat (−0.99%), creek (−0.22%), river (−1.26%), and pond (−0.54%). The total volume of changes in land per hundred population from 1968 to 2018 of Canning-I block was −67%. And the total volume of changes of literates per hundred population of Canning-I block was 186%. In 2011, the educational facility increased to 122 pre-primary schools, 114 primary schools, 25 middle schools, 15 secondary schools, eight senior secondary schools, and one college. And access to medical facilities also improved: there were 35 primary health centers, three maternity and child welfare centers, one tuberculosis clinic, 11 dispensaries, and two family welfare centers. All other village amenities were improving towards development.

REFERENCE

Meyer, B.W., Turner, II, B.L. (1992) Human Population Growth and Global Land-Use/Cover Change, Clark University, MA.

Part III

Application of Remote Sensing and
GIS-Based Approach in Agricultural
Practice and Climatic Issues

16 Application of Fuzzy Machine Learning Model for Mapping Diseased Sugarcane Ratoon Fields

Shruti Pancholi and Anil Kumar

16.1 INTRODUCTION

Sugarcane (*Saccharum officinarum*) is a significant economic crop in the world (Chen et al., 2021; Sumesh et al., 2021). Among the major crops in India, sugarcane is the most important one for agricultural income in the country. More than 80 countries grow it, with India being among the top five. Tropical and subtropical climates are ideal for cultivating it. Mostly grown for the production of sugar and bio-ethanol, it is a highly energy-efficient crop (Varma et al., 2018). Complex biological interactions are affected by accelerated climate change differently, causing changes that are often difficult to predict. In addition to climate change directly affecting crop yields and quality, diseases will change and remain important despite climate change (Newton et al., 2011). Subsistence farmers in developing countries suffer the greatest losses due to crop diseases, which account for 16% of global yield losses (Mahmuti et al., 2011). The concept of food security encompasses several factors, including production, nutritional value, stability, and access (physical and economic) and is not solely reliant on production. There are many components to food security that are concerned by plant diseases (Savary & Willocquet, 2020).

Sugarcane crops suffer serious yield and quality losses due to top borer, depending on the level of infestation (*Sugarcane-Insect Management*, n.d.). Insect pests that are most prevalent are sugarcane borer pests that threaten farmers' productivity. Damage caused by the top shoot borer includes bore holes at the top of the shoot, bunchy tops, shot holes in emerging leaves, and dead hearts. It is estimated that farmers lose 20% to 30% of their yield due to top shoot borer (*Top Three Sugarcane Pests of Economic Importance—AGRIVI*, n.d.). A variety of curative measures are needed to address this problem, where disease detection and mapping are crucial. It is necessary to determine the extent and location of affected crops before applying chemicals for disease control (Apan et al., 2010).

Remote sensing technology plays an important role in crop disease detection. In addition to identifying crop conditions, such as diseases, remote sensing data can provide useful information for agricultural management (Liu et al., 2014; Weiss et al., 2020). Mapping crop diseases has also been done with satellite imagery. Wheat infestations of the take-all disease were mapped using Landsat imagery with 30-m spatial resolution (Badage, 2018). Franke et al., in their research, examined the potential of multispectral remote sensing for analyzing crop diseases across time and space. They exploited a decision tree and Normalized Difference Vegetation Index (NDVI) data for wheat crops (Franke & Menz, 2007). Amarasingam et al. detected white leaf disease in the sugarcane plantation from the Unmanned Aerial Vehicle data using deep learning methods (Amarasingam et al., 2022). Apan et al. detected sugarcane 'orange rust' disease affected areas from Hyperion-1 hyperspectral satellite imagery using 40 spectral indices which focus on the leaf internal structure, leaf water content, and leaf pigments (Apan et al., 2010). Temporal remote sensing data are widely being exploited

DOI: 10.1201/9781003377825-19

for the study of changing phenomenon like agriculture and land-use/land-cover monitoring. Many studies have utilized the temporal satellite data for crop stages monitoring and mapping. Masialeti et al. mapped crops for different years using a temporal NDVI reference curve library over a single growing season for crops. They aimed to study the role of temporal vegetation index data for distinguishing a crop type (Masialeti et al., 2013). Li et al. proposed a rice-cropping area detection based on phenology of the crop derived from temporal MODIS NDVI product (Li et al., 2022).

Indian state Uttar Pradesh grows sugarcane twice a year, from February to March and April to May. Usually, the crop consists of two planting varieties: plant and ratoon. Freshly sown seeds are considered plants with a 12-month life cycle. Ratoon is the second crop from the harvested sugarcane plant, and it has an 8- to 9-month growing season (Misra et al., 2014).

This study aims to find the diseased fields of sugarcane ratoon utilizing the fuzzy machine learning model of modified possibilistic c-means (MPCM) with individual sample as the mean training method in order to handle the heterogeneity within the target class. A kernel-based approach is also used in this study to eliminate the problem of non-linear separation of spectrally similar classes.

16.2 MATHEMATICAL CONCEPTS

16.2.1 TEMPORAL INDICES

Sugarcane crop phenology was studied using time-series data. The inclusion of temporal information is expected to increase the dimensionality of the input data, which in turn increases the processing complexity of the data. To overcome this problem and enhance the vegetation-based features, a spectral index–based approach is used in this study. The choice of spectral index was dependent on the target crop, i.e. sugarcane. In the initial stages of sugarcane ratoon sprouting, there is considerable amount of soil exposure, which affects the spectral signature of the crop itself. Modified Soil-Adjusted Vegetation Index (MSAVI) is a spectral index developed to highlight the vegetation in the regions where the soil is significantly more than the crop. This index overcomes the issues faced while using other vegetation indices like NDVI and Normalized Difference Red Edge which do not provide accurate data in such situations. MSAVI2 is an upgrade to MSAVI as it is parameter independent. MSAVI2 can be computed using Equation 16.1:

$$MSAVI2 = \frac{(2 \times NIR + 1 - \sqrt{2 \times (NIR + 1)^2 - 8 \times (NIR - R)})}{2} \qquad (16.1)$$

16.2.2 MACHINE LEARNING MODEL

In remote sensing technology, machine learning models are widely used for prediction and classification. Each pixel in an image is assigned to a defined class in conventional image classification. It is common to encounter mixed pixels when classifying satellite images in the real world. A pixel of the satellite image does not essentially signify the identical land-cover class on the ground. The concept of image classification was then introduced with fuzzy logic to overcome this problem (Bezdek et al., 1984; Zadeh, 1965). Soft classification can handle mixed pixels using a fuzzy model. Every pixel is classified into distinct classes based on its membership in each class. Members are classified according to distance from cluster centers to individual pixels (Bezdek et al., 1984). In 1993, Krishnapuram et al. introduced possibilistic c-means (PCM), an improvement on the fuzzy c-means algorithm that can handle noise as well (Krishnapuram & Keller, 1993). PCM was further modified to create the MPCM algorithm that overcomes PCM's drawbacks, i.e. sensitivity to good initialization (Li et al., 2003). The MPCM algorithm was not proved to handle the spectral overlap between similar classes as well as non-linear separation between classes. To resolve this problem, kernel concept was brought into the picture which was to map the spectrally similar classes on higher dimensional feature space where they are linearly separable (Wu & Zhou, 2008). The objective function of the model used for MPCM is given in Equations 16.2 through 16.5.

$$J_{MPCM}(U,V) = \sum_{i=1}^{N}\sum_{j=1}^{c}(u_{ji})^{m}\left\|x_{i}-v_{j}\right\|^{2} + \sum_{i=1}^{N}\eta_{i}\sum_{J=1}^{c}(\lambda_{i}-u_{ji})^{m} \tag{16.2}$$

where $\lambda_i > 0$

$$\lambda_{i} = u_{j}loglog\left(u_{ji}\right) \tag{16.3}$$

$$v_{j} = \frac{\sum_{i=1}^{N}u_{ij}x_{i}}{\sum_{i=1}^{N}u_{ij}}, \text{where } v_{j} \text{ is the cluster center} \tag{16.4}$$

$$u_{ij} = e^{\left(\frac{-d_{ij}^{2}}{\eta_{i}}\right)}, \forall i, j \tag{16.5}$$

16.2.3 TRAINING APPROACH

Traditional classifiers make use of the statistical measures to represent a cluster or class. This leads to incorrect representation of the class when a high level of heterogeneity is present in the training data. To take into consideration the individual effect of pixels, an individual sample as a mean training method has been designed. This approach reduces the misclassification and truly represents the class. When incorporated with MPCM classifier, the objective function has been modified as in Equation 16.6:

$$J_{MPCM}(U,V) = \sum_{i=1}^{N}\sum_{j=1}^{c}(u_{ji})^{m}\left\|x_{i}-v_{j}\right\|^{2} + \sum_{i=1}^{N}\eta_{i}\sum_{J=1}^{c}(\lambda_{i}-u_{ji})^{m} \tag{16.6}$$

where v_j is the individual sample value instead of mean of the cluster as cluster centre in Equation 16.2. The objective function remains the same as the MPCM objective function with individual sample taken in place of the mean.

16.2.4 KERNEL CONCEPT

The PCM classifier is sensitive to a good initialization and tends to generate co-incident clusters. As a result, more local minimums are produced which decrease the overall accuracy of the classification. To overcome these limitations of PCM, MPCM was introduced.

MPCM alone was not proven to provide a solution to the spectrally overlapping classes, hence the kernel function was combined in the MPCM classifier. The kernel function maps the input space to a higher dimensional feature space where the overlapping classes separate linearly. The expression for kernel function is given in Equation 16.7:

$$X = (x1,\ldots\ldots\ldots,xM) \rightarrow \varphi(X) = (\varphi(X1),\ldots\ldots,\varphi(XN)) \tag{16.7}$$

Then $\left\|x_i - v_j\right\|$ is mapped into space φ as given in Equation 16.8:

$$\left\|x_{i}-v_{j}\right\| \rightarrow \left\|\phi(x_{i})-\phi(vj)\right\| \tag{16.8}$$

Using Equation 16.8, the objective function of MPCM given in Equation 16.2 is transformed as in Equation 16.9:

$$J_m = \sum_{i=1}^{N}\sum_{j=1}^{C}(u_{ji})^m \left\| \phi(x_i) - \phi(v_j) \right\|^2 + \sum_{i=1}^{N}\eta_i\sum_{J=1}^{C}(\lambda i - u_{ji})^m \qquad (16.9)$$

In this study, we used the Kernel with Moderate Decreasing (KMOD) as it handles highly correlated sparse data (Ayat et al., 2001), as in the case of healthy and diseased sugarcane ratoon fields. The KMOD function can be expressed as given in Equation 16.10:

$$K(x_i, v_k) = \left[exp\left(\frac{\gamma}{\sigma^2 + \left\| x_i - v_k \right\|^2} \right) \right] - 1 \qquad (16.10)$$

where σ and $\gamma > 0$ are the controlling parameters.

16.3   STUDY AREA AND DATASET

16.3.1   Study Area

Muzaffarnagar, Uttar Pradesh, was identified as the study area for the research, which covered an area of 93 km². This region was chosen because it is abundantly cultivated with sugarcane crops. This region is located at 29.33° N latitude and 77.85° E longitude (Figure 16.1).

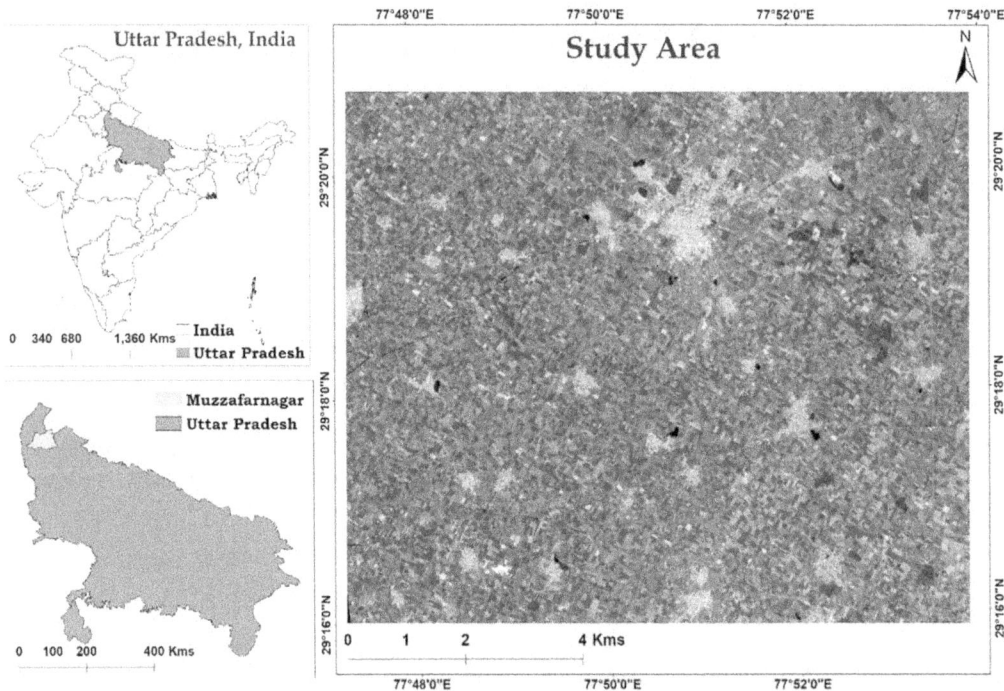

FIGURE 16.1   Study area.

16.3.2 Dataset

Considering its high spatial and temporal resolutions, PlanetScope was used in this study. Featuring daily global coverage at 3-m resolution, it is a suitable product available for crop-type mapping in the Indian agricultural sector. An Indian village consists of many small farms owned by many families of farmers. In the Indian agricultural scenario, farms are usually sized between 800 and 1600 m² or between 0.20 and 0.40 acres, which makes mapping crop types challenging. Remote monitoring of small-sized crop fields was possible using PlanetScope's DOVE temporal dataset. PlanetScope DOVE sensor characteristics are shown in Table 16.1.

The temporal data were acquired for the period of 9 months, i.e. from February to October 2022. The time period was considered covering the stages of phenology of the sugarcane ratoon fields. The sugarcane ratoon fields grown from February 2022 were kept as the target class. To reduce the dimensionality of the input dataset, the MSAVI2 vegetation index was derived from the multispectral temporal satellite images. The dates for which the satellite data was acquired are mentioned in Table 16.2.

16.4 METHODS

The methodology followed for this study is depicted in Figure 16.2.

TABLE 16.1
DOVE Sensor Characteristics

Characteristic	Specifications
Spectral resolution	Four bands: red, green, blue, near infrared
Spatial resolution	3 m
Revisit period	1 day
Swath	25 m

TABLE 16.2
Satellite Data Acquisition Dates

Image	Dates
Ratoon	06/02/2022, 11/02/2022, 19/02/2022, 27/02/2022, 09/03/2022, 14/03/2022, 21/03/2022, 28/03/2022, 06/04/2022, 14/04/2022, 23/04/2022, 28/04/2022, 07/05/2022, 14/05/2022, 25/05/2022, 29/05/2022, 12/062022, 25/06/2022, 08/07/2022, 08/08/2022, 05/09/2022, 11/09/2022

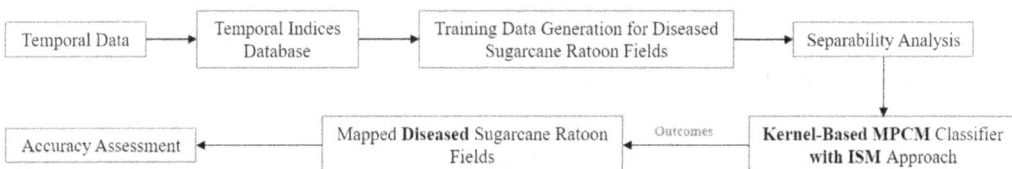

FIGURE 16.2 Methodology adopted.

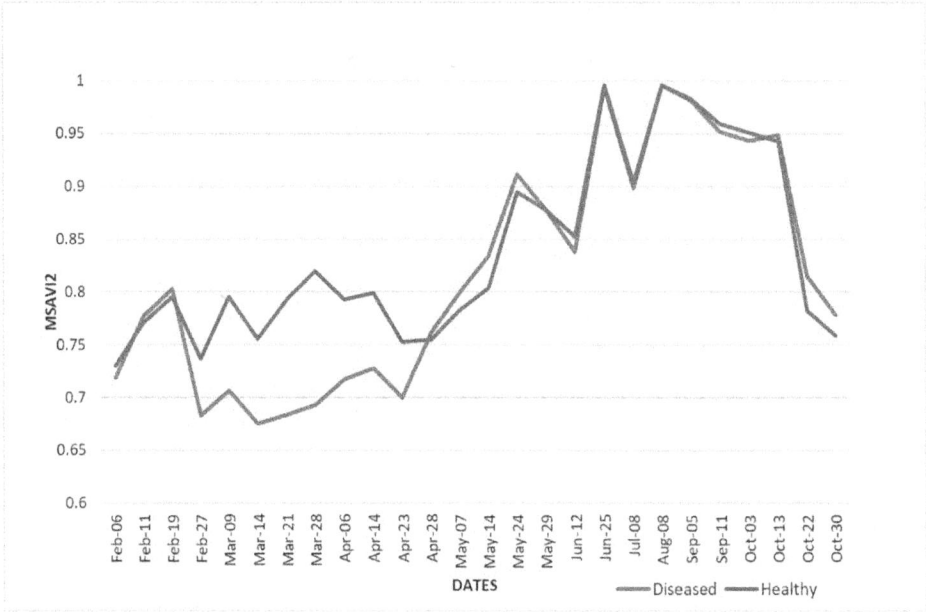

FIGURE 16.3   Temporal MSAVI2 curve plotted for diseased and healthy ratoon fields.

16.4.1   TEMPORAL INDICES DATABASE GENERATION

The time-series data enables unique identification of the diseased and healthy ratoon fields with the help of temporal index data. The phenology curve of healthy and diseased sugarcane ratoon fields sown in February 2022 was plotted against the MSAVI2 index values (Figure 16.3). This curve highlights the major disease period which is from February to April, about 3 months. When compared, this period contrasts with the tillering stage of the sugarcane ratoon life cycle.

For reduction in data dimensionality, an optimum combination of dates was selected through separability analysis. In this stage, the distance between non-target classes and target class was observed with a different combination of considered dates' data. The combination of dated images with highest separation between the non-target and target classes was identified and used for further study. The optimum combination of dates after separability analysis is given in Table 16.3.

The temporal indices graph for the optimized dates dataset is shown in Figure 16.4.

TABLE 16.3
Optimized Temporal Dates Combination

Target class	Optimized dates
Ratoon	19/02/22, 27/02/22, 14/03/22, 28/03/22, 23/04/22, 28/04/22, 14/05/22, 12/06/22, 22/10/22, 30/10/22

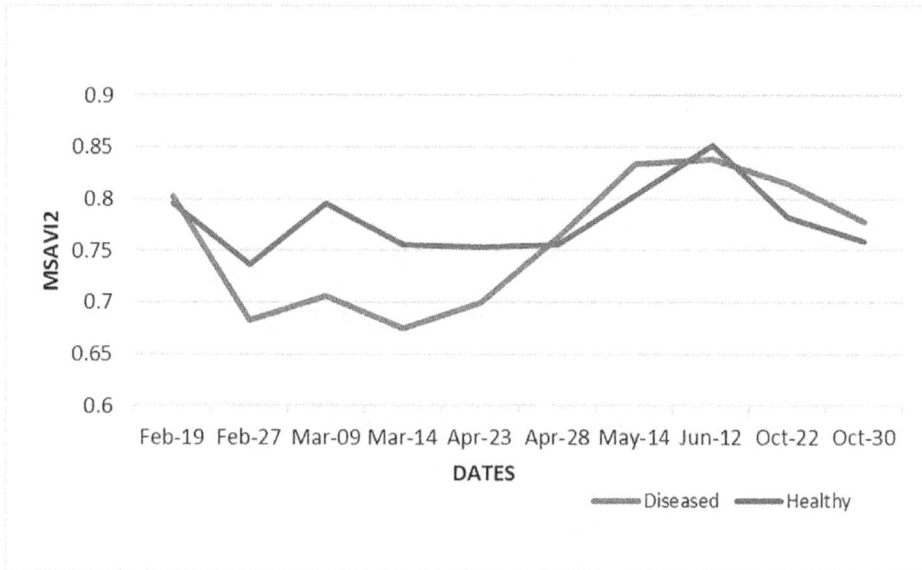

FIGURE 16.4 Optimized temporal MSAVI2.

16.4.2 CLASSIFICATION

The training fields information was collected during a field survey conducted on 21 September 2022. The geo-tagged training fields were used as training data for the MPCM model. The variability within the disease-affected fields of sugarcane ratoon was significant since the spread of the disease was not found to be uniform throughout the field during the field visit. To handle this variability, the concept of individual sample as mean (ISM) was used while training the model.

16.4.3 ACCURACY ASSESSMENT

Each pixel had a membership value associated with only one target class (e.g., plant or ratoon). The mean membership difference (MMD) was used to evaluate algorithm output accuracy (Devinda & Kumar, 2020). MMD and variance parameters were used to assess the accuracy of the mapped output. In training and testing fields, MMD is a measure of closeness between membership values. For the identical classes, the difference between training and testing fields should be near zero, and for the classes not belonging to the target class, it should be near one. Variability within the target classes was measured using the variance parameter. To measure heterogeneity within the testing field, variance values were computed for the membership values of the test field (Anam Sabir Indian Institute of Remote Sensing, 2022). The proposed ISM training approach should result in minimum variance within the same class fields.

16.5 RESULTS AND CONCLUSION

It was shown in the study that one of the major pests to infect the sugarcane cultivation in the Western Uttar Pradesh region is top shoot borer. This pest attacks the sugarcane ratoon at the stage of tillering, i.e. a month after it starts growing. The pesticides are applied during the ripening stage, i.e. 4 months after the growing period. The classification results are shown in Figure 16.5.

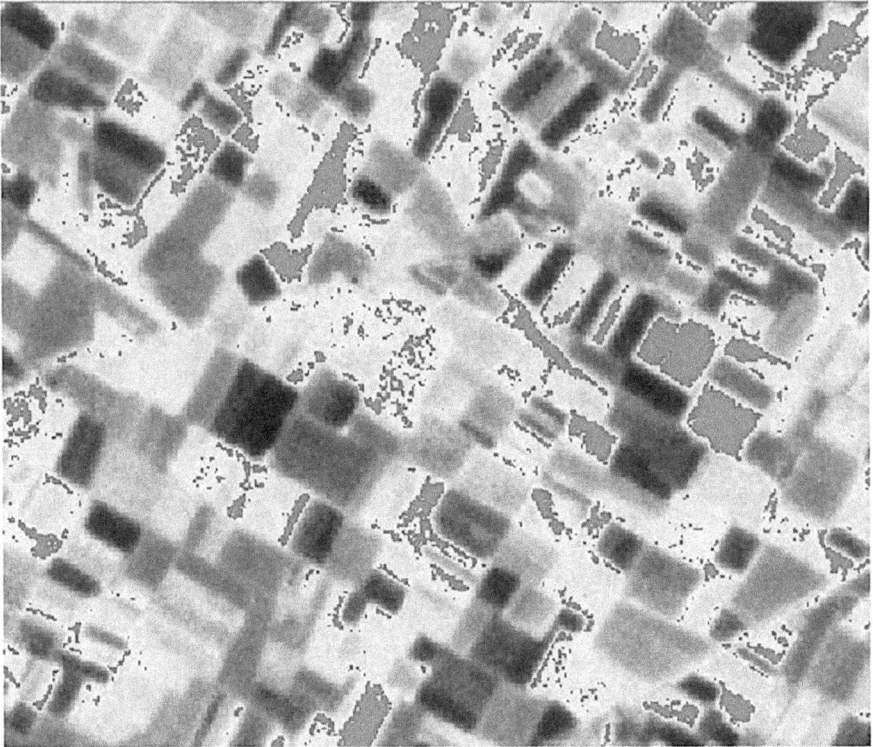

(a)

(b)

FIGURE 16.5 Mapped fields of sugarcane ratoon: (a) healthy and (b) diseased.

The mapped healthy and diseased sugarcane ratoon fields were validated using MMD and variance parameters. The MMD values were 0.026144 and 0.047059 for healthy and diseased fields, respectively. The variance values were 0.000465 and 0.000644 for healthy and diseased, respectively. This study indicates application in the area of crop-type mapping and disease extent mapping and monitoring. The proposed model can reduce the need for intensive manual surveying and save time and resources. The aspect of food security can also be addressed through this study in order to prevent the spread of diseases in crops. It is important to study its extent and trend.

REFERENCES

Amarasingam, N., Gonzalez, F., Salgadoe, A. S. A., Sandino, J., & Powell, K. (2022). Detection of white leaf disease in sugarcane crops using UAV-derived RGB imagery with existing deep learning models. *Remote Sensing*, *14*(23), 6137. https://doi.org/10.3390/RS14236137

Anam Sabir Indian Institute of Remote Sensing. (2022). *CNN Based Deep Learning vs Fuzzy Machine Learning Models for Medicinal Plant Mapping Using Dual-Sensor Temporal Remote Sensing Data*. Andhra University.

Apan, A., Held, A., Phinn, S., & Markley, J. (2010). Detecting sugarcane 'orange rust' disease using EO-1 Hyperion hyperspectral imagery. *International Journal of Remote Sensing*, *25*(2), 489–498. https://doi.org/10.1080/01431160310001618031

Ayat, N. E., Cheriet, M., Remaki, L., & Suen, C. Y. (2001). KMOD: A new support vector machine kernel with moderate decreasing for pattern recognition: Application to digit image recognition. *Proceedings of the International Conference on Document Analysis and Recognition, ICDAR, January*, 1215–1219. https://doi.org/10.1109/ICDAR.2001.953976

Badage, A. (2018). Crop disease detection using machine learning : Indian agriculture Anuradha. *5*(9), 866–869. www.irjet.net/archives/V5/i9/IRJET-V5I9158.pdf

Bezdek, J. C., Ehrlich, R., & Full, W. (1984). FCM: The fuzzy c-means clustering algorithm. *Computers & Geosciences*, *10*(2–3), 191–203. https://doi.org/10.1016/0098-3004(84)90020-7

Chen, J., Wu, J., Qiang, H., Zhou, B., Xu, G., & Wang, Z. (2021). Sugarcane nodes identification algorithm based on sum of local pixel of minimum points of vertical projection function. *Computers and Electronics in Agriculture*, *182*, 105994. https://doi.org/10.1016/J.COMPAG.2021.105994

Devinda, C. S., & Kumar, A. (2020). Application of fuzzy machine learning algorithm in agro-geography. *Khoj: An International Peer Reviewed Journal of Geography*, *7*(1), 30–46. https://doi.org/10.5958/2455-6963.2020.00004.1

Franke, J., & Menz, G. (2007). Multi-temporal wheat disease detection by multi-spectral remote sensing. *Precision Agriculture*, *8*(3), 161–172. https://doi.org/10.1007/S11119-007-9036-Y/TABLES/1

Krishnapuram, R., & Keller, J. M. (1993). A possibilistic approach to clustering. *IEEE Transactions on Fuzzy Systems*, *1*(2), 98–110. https://doi.org/10.1109/91.227387

Li, B., Peng, S., Shen, R., Yang, Z. L., Yan, X., Li, X., Li, R., LiI, C., & Zhang, G. (2022). Development of a new index for automated mapping of ratoon rice areas using time-series normalized difference vegetation index imagery. *Pedosphere*, *32*(4), 576–587. https://doi.org/10.1016/S1002-0160(21)60053-X

Li, K., Huang, H. K., & Li, K. L. (2003). A modified PCM clustering algorithm. *International Conference on Machine Learning and Cybernetics*, *2*, 1174–1179. https://doi.org/10.1109/ICMLC.2003.1259663

Liu, J., Miller, J. R., Haboudane, D., Pattey, E., & Nolin, M. C. (2014). Variability of seasonal CASI image data products and potential application for management zone delineation for precision agriculture. *Canadian Journal of Remote Sensing*, *31*(5), 400–411. https://doi.org/10.5589/M05-023

Mahmuti, M., West, J. S., Watts, J., Gladders, P., & Fitt, B. D. L. (2011). Controlling crop disease contributes to both food security and climate change mitigation. *International Journal of Agricultural Sustainability*, *7*(3), 189–202. https://doi.org/10.3763/IJAS.2009.0476

Masialeti, I., Egbert, S., & Wardlow, B. (2013). A comparative analysis of phenological curves for major crops in Kansas. *GIScience & Remote Sensing*, *47*(2), 241–259. https://doi.org/10.2747/1548-1603.47.2.241

Misra, G., Kumar, A., Patel, N. R., & Zurita-Milla, R. (2014). Mapping a specific crop—A temporal approach for sugarcane ratoon. *Journal of the Indian Society of Remote Sensing*, *42*(2), 325–334. https://doi.org/10.1007/S12524-012-0252-1/TABLES/6

Newton, A. C., Johnson, S. N., & Gregory, P. J. (2011). Implications of climate change for diseases, crop yields and food security. *Euphytica, 179*(1), 3–18. https://doi.org/10.1007/S10681-011-0359-4/METRICS

Savary, S., & Willocquet, L. (2020). Modeling the impact of crop diseases on global food security. *Annual Review of Phytopathology, 58,* 313–341. https://doi.org/10.1146/ANNUREV-PHYTO-010820-012856

Sugarcane-Insect Management. (n.d.). Ikisan. Retrieved May 11, 2023, from www.ikisan.com/tn-sugarcane-insect-management.html

Sumesh, K. C., Ninsawat, S., & Som-ard, J. (2021). Integration of RGB-based vegetation index, crop surface model and object-based image analysis approach for sugarcane yield estimation using unmanned aerial vehicle. *Computers and Electronics in Agriculture, 180,* 105903. https://doi.org/10.1016/J.COMPAG.2020.105903

Top Three Sugarcane Pests of Economic Importance-AGRIVI. (n.d.). Retrieved May 11, 2023, from www.agrivi.com/blog/top-three-sugarcane-pests-of-economic-importance/

Varma, K., Madugula, S., Chandra Sekhar, V., Kishore Varma, P., Kumar, K. V. K., Suresh, M., Kumar, N. R., & Sekhar, V. C. (2018). Potentiality of native *Pseudomonas* spp. in promoting sugarcane seedling growth and red rot (*Colletotrichum falcatum* Went) management. *International Journal of Current Microbiology and Applied Sciences, 7*(2), 2855–2863. https://doi.org/10.20546/ijcmas.2018.702.348

Weiss, M., Jacob, F., & Duveiller, G. (2020). Remote sensing for agricultural applications: A meta-review. *Remote Sensing of Environment, 236,* 111402. https://doi.org/10.1016/J.RSE.2019.111402

Wu, X. H., & Zhou, J. J. (2008). Modified possibilistic clustering model based on kernel methods. *Journal of Shanghai University (English Edition), 12*(2), 136–140. https://doi.org/10.1007/S11741-008-0210-2

Zadeh, L. A. (1965). Fuzzy sets. *Information and Control, 8*(3), 338–353. https://doi.org/10.1016/S0019-9958(65)90241-X

17 Comparative Assessment of Landslide Risk Modelling by Bivariate Model in East Sikkim Himalaya, India

Sk Asraful Alam, Ramkrishna Maiti, and Sujit Mandal

17.1 INTRODUCTION

The mountainous region's vulnerability to landslides has significantly increased over the past few decades as a result of climate change, urbanization, and industrialization practices that have become extremely effective. According to Chang et al., developing nations suffer the most from landslides, which account for 95% of all landslide disasters and cost 0.5% of gross domestic product annually (1995). Intense monsoonal rains, urbanization, road development, and deforestation have rendered the Sikkim Himalaya increasingly vulnerable and one of the most dangerous areas in the world due to its young mountain range's steep slopes and delicate geological formations. Additionally, geo-dynamic activity is higher in the Himalayan region. The earth's surface has unexpectedly changed as a result of landslides, which are more likely to trigger natural disasters in mountainous areas. Landslides in the Himalayan region are estimated to kill more than 200 people and cause property damage in excess of US$1 billion each year (Naithani, 1999). Landslides are becoming more frequent mainly because of their socio-economic impact and growing awareness of the pressures of urbanization on mountains (Aleotti and Chowdhury, 1999).

Landslide risk assessment (LRA) is one of the crucial elements for the advancement and management of the alpine environment. The probability of undesirable consequences, loss, suffering, or damage to the human population and the assets these people value as a result of landslide events is referred to as "landslide risk" by Lee and Zones (2004). The significant increase in landslides over the past few decades has increased the significance of assessing the landslide susceptibility zonation (LSZ) and the LRA maps for management and planning. LSZ maps can be produced through both qualitative and quantitative techniques (Soeters and van Westen, 1996; Aleotti and Chowdhury, 1999). The Rorachu watershed in the Sikkim Himalaya can be managed and planned using this knowledge. The vulnerability of the element-at-risk and the spatial probability of landslide occurrences have been considered in the current environment. The likelihood that a landslide would occur in this area and the vulnerability of the element at risk are the two elements that determine the LRA.

The Sikkim Himalaya is one of the area's most susceptible to landslide disasters. Therefore, the compilation of a LSZ map and LRA is very beneficial for planning and development. A range of qualitative and quantitative models can be used to map the LSZ and the landslide risk. Akgun et al. (2008), Wan and Lei (2009), Wan (2009), Akgun and Needet (2010), Pradhan and Youssef (2010), Pradhan et al. (2012), and Bednarik et al. (2012) completed several studies on the use of geographic information systems (GIS) for LSZ. A wide variety of statistical techniques are used to map landslide hazards and potential danger zones. In this study, three statistical models (Index of Entropy [IOE], weight of evidence [WOE], and Bivariate Statistical Index [BSI]) were utilised to produce landslide susceptibility

DOI: 10.1201/9781003377825-20

and risk maps for the Rorachu watershed in the South Sikkim Himalaya. The BSI model is used by many academics, such as Van Westen et al. (1997), Rautela and Lakhera (2000), Cevik and Topal (2003), and Regmi et al. (2014). The WOE method refers to the landslide susceptibility and landslide risk zonation mapping methodologies employed by Oh and Lee (2011), Dahal et al. (2008b), Singh et al. (2005), Lee and Choi (2004), and Ma et al. (2019). The IOE model, based on bivariate statistical analysis, has been successfully applied by Constantin et al. (2011), Devkota et al. (2012), Jaafari et al. (2014), Youssef et al. (2014a), and Pourghasemi et al. (2013b) for landslide hazard zonation mapping. In the current work, LSZ and LRA of the Rorachu watershed in the east Sikkim Himalaya have been carried out using BSI, IOE, and WOE models. This study's goal is to assess their likelihood and vulnerability. The investigation's main goals were to identify at-risk communities and routes. The Receiver Operating Characteristics (ROC) curve, landslide density (LD), and success rate curve (SRC) approaches were used to evaluate the performance of each model using conventional accuracy metrics.

17.2   THE STUDY AREA

The study area is in the east Sikkim district of the Sikkim Himalaya. The Rorachu watershed is bounded by 27°17′14.67″ N to 27°23′48.50″ N latitudes and 88°35′51.40″ E to 88°43′11.98″ E longitudes, covering an area of around 69.125 km² (Figure 17.1). High relative relief and steep slopes, along with immensely rugged surfaces, are significant physiographic characteristics of the East Sikkim Himalaya. The elevation range of the Rorachu watershed is 4,100 m and 816 m, respectively. The elevation of this watershed differs significantly from its south-western (Ranipool) to its north-western (Pandramaile) edges. The Rorachu watershed enjoys a good climate all year round with an average

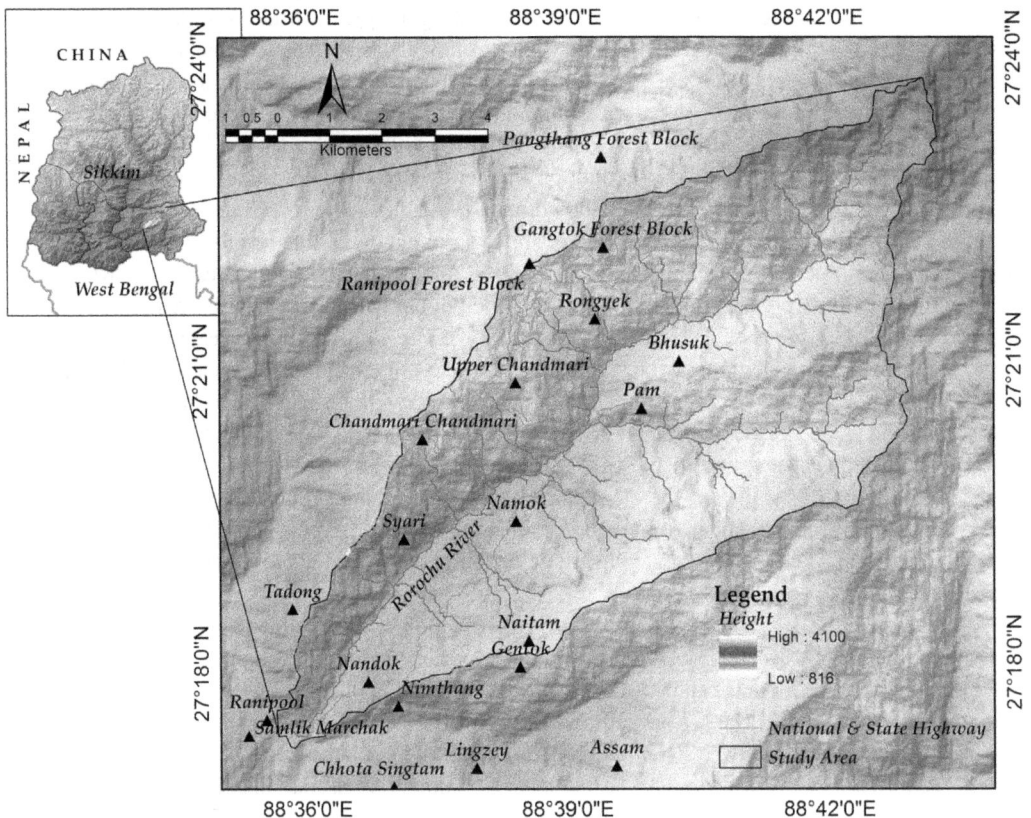

FIGURE 17.1   Rorachu watershed located in the north Sikkim Himalaya.

maximum temperature of 21°C in the summer and an average minimum temperature of 1°C in the winter. Slopes can have an angle as low as 00 or as high as 710. Samdong, Ranipool, and Deorali are the significant minor markets in this research region, whilst Gangtok and Tadong are the major towns. The region is rapidly growing as a result of neighbouring urbanisation in the cities of Gangtok and Ranipool. Numerous local roads, notably NH 31A, pass through these watershed areas, and more roads will be constructed. The Rorachu watershed is mostly made up of farmed land, villages, rocky, arid terrain, and wooded areas. Slope instability is being accelerated by the varied physiography and climate of the Rorachu watershed as well as fast development. In the Rorachu watershed, climate change, variable rainfall, and urbanisation all increase the likelihood and danger of landslides.

17.3 MATERIALS AND METHODS

In the current study, 13 landslide causative factors were considered when modelling the Landslide Susceptibility Index (LSI) and LRA. These factors included elevation, geology, slope, soil, drainage density (DD), road density (RD), rainfall, Normalized Difference Vegetation Index (NDVI), and aspect. In order to use geospatial approaches to run the landslide susceptibility model and landslide risk model, it is crucial to compile a digital database. For modelling landslide susceptibility and risk, the geographic database has been well planned and implemented (Table 17.1 and Figure 17.3). ArcGIS 10.2, SPSS 23, and R software were utilized in the study to model landslides using both categorical and continuous data.

17.3.1 Preparation of Landslide Inventory

According to Guzzetti et al. (1999), the core component of landslide vulnerability assessments is the mapping of the landslide inventories. The crucial information the landslide inventory map

---

TABLE 17.1

Sources of Information on Different Landslide Causative Factors

Feature Layer	Source	Thematic Data Layer	Resolution
Topographical map	Survey of India, Kolkata. Map no. 78A/11	Drainage density	1:50,000
Google Earth image	www.earth.google.com	Road density	30 × 30 m
Geological map	Geological survey of India (GSI)	Geological map	1:250,000
Soil map	NBSS & LUP Regional Centre, Kolkata	Soil map	1:400,000
Landsat 8 OLI	www.earthexplorer.usgs.gov	Land-use/land-cover (LULC) map	30 × 30 m
		NDVI map	30 × 30 m
Rainfall data	www.worldclim.org	Rainfall Distribution map	1 × 1 km
ASTER GDEM	www.earthexplorer.usgs.gov	Elevation map	30 × 30 m
		Slope map	30 × 30 m
		Topographic Wetness Index (TWI) map	30 × 30 m
		Topographic Position Index (TPI) map	30 × 30 m
		Stream Power Index (SPI) map	30 × 30m
Topographical map, Google Earth image, Satellite data and GPS survey	Field study using GPS and internet	Landslide inventory map	Not to scale (after surveying, resampling method was used to match other layers)

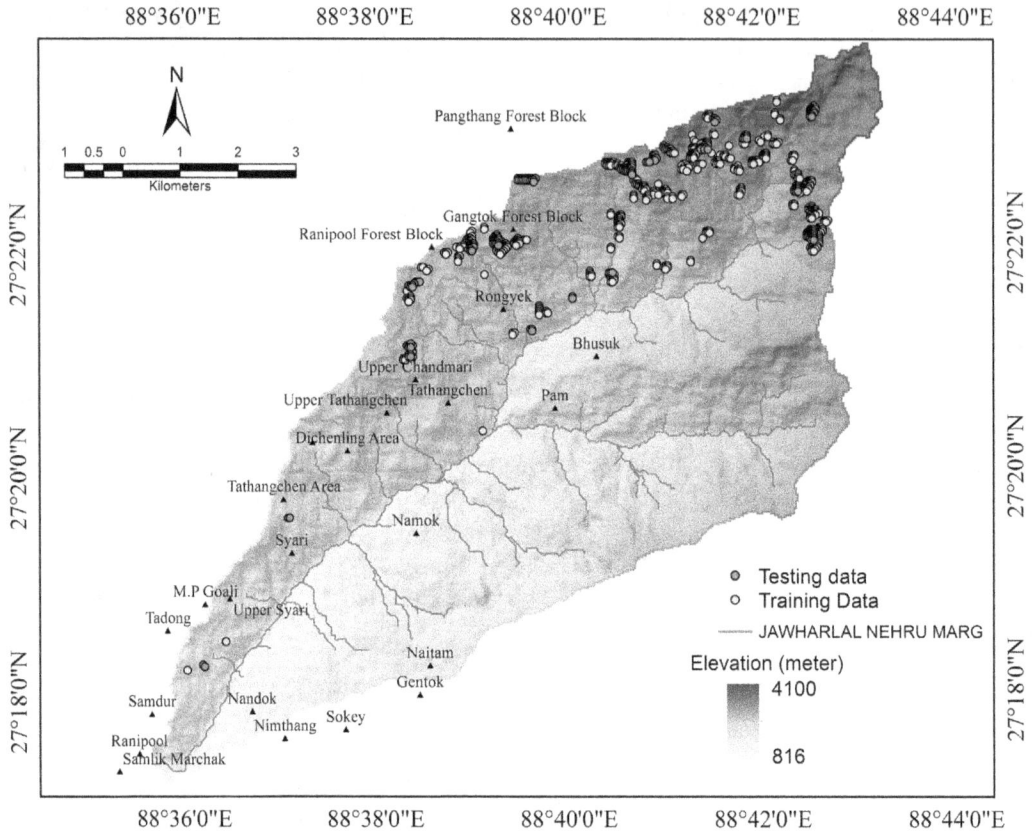

FIGURE 17.2   Landslide inventory map of Rorachu watershed.

provides is helpful in assessing the danger and risk of landslides. Accurate detection and identification of landslides are essential for probabilistic studies of landslide susceptibility and risk
(Rawat and Joshi, 2016; Mandal and Mandal, 2017). The landslide inventory map was produced
using aerial photographs, the Landsat 8 Operational Land Imager (OLI) (30-m) picture, Google
Earth (Quickbird image, 0.60 m), and a GPS survey. Finally, all the data were vectorised using
ArcGIS 10.3 software. The Rorachu watershed has a total of 153 large and small landslides, covering an area of 0.644 km² (Figure 17.2). For the sake of estimating landslide susceptibility and
risk, all landslide data have been transformed from vector to raster format. Out of the 153 total
landslides, about 123 (80%) were randomly chosen for model training, and the other 30 (20%)
were used for model validation. In the Rorachu watershed, rock slides, debris slides, and earth
slides account for most landslides.

17.3.2 Landslide Causative Factors and Their Descriptions

According to Ayalew and Yamagishi (2005), there is no such criterion for choosing causal elements
for the examination of landslide vulnerability. Elevation, geology, slope, soil, DD, RD, rainfall,
NDVI, and slope aspect, as well as Topographic Position Index (TPI), Stream Power Index (SPI),
Topographic Wetness Index (TWI), and land use/land cover (LULC) are the variables that control
slope instability modelling and risk modelling (Table 17.4 and Figure 17.3). Researchers from all
over the world are using these criteria (Wu et al. 2017).

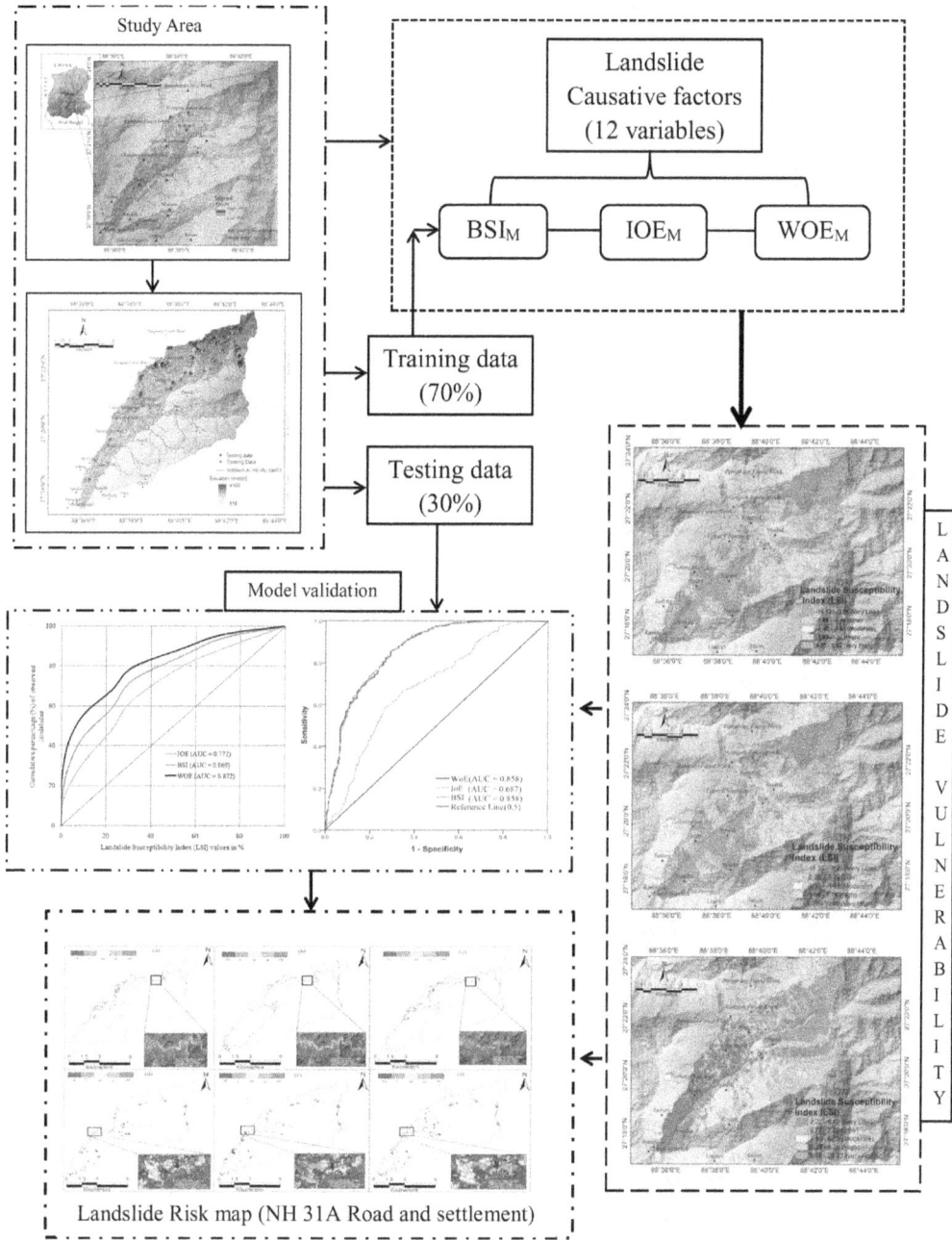

FIGURE 17.3 Methodological flowchart for identifying landslide risk zone in Rorachu watershed.

17.3.2.1 Geology, Elevation, Slope, Soil, and Drainage Density

The Rorachu watershed's lithological and structural changes frequently result in a difference in the strength of the soil and rocks, which has a substantial impact on the incidence of slope instability (Pradhan and Lee, 2010a). Five lithological units, namely the Basic Intrusive, Chungthang Formation, Gorubathan Formation, Lingtse Gnesis, Kanchenjunga Gnesis, and Darjeeling Gnesis (undifferentiated), are present in the research region, which is characterized by its geology (Figure 17.4 and Table 17.2). The district resource map of east Sikkim, which was gathered by the

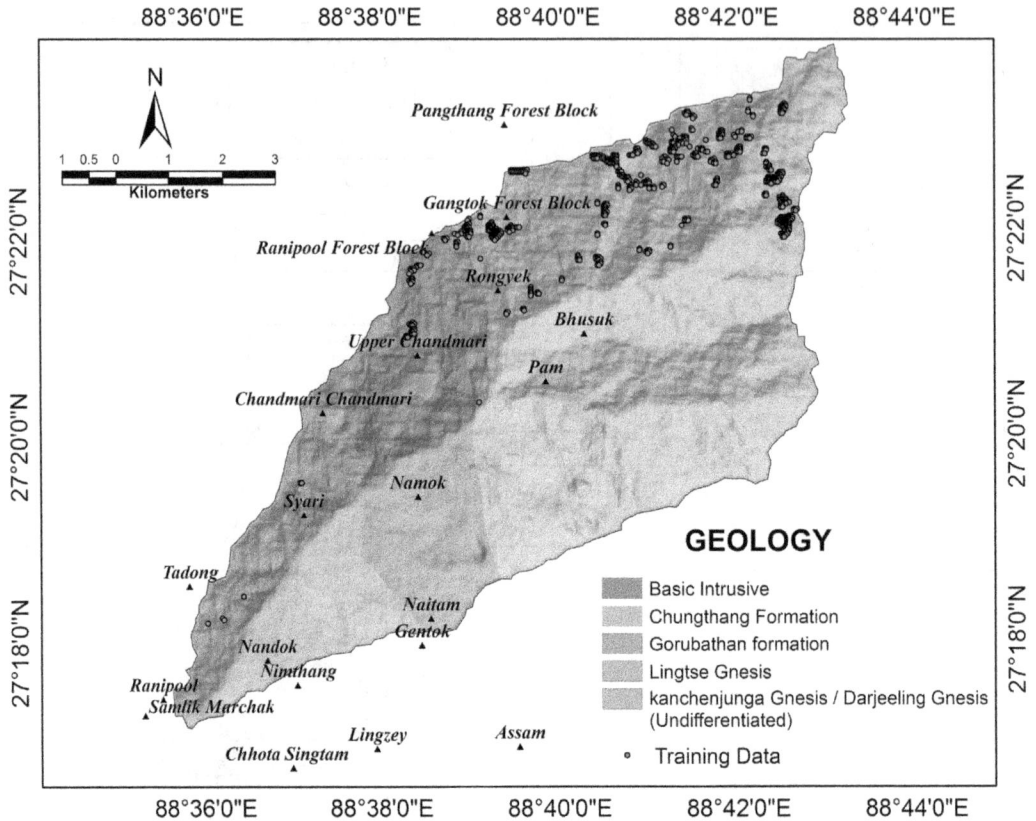

FIGURE 17.4   Geological map of the Rorachu watershed.

Geological Survey of India (GSI), Kolkata, was used to create the geology map of the Rorachu watershed. The Darjeeling Gnesis or the Kanchenjunga Gnesis covers a significant portion of this watershed. The fundamental intrusive is ranked first in terms of size (43.02%), followed by the basic lithological unit of the basic Kanchenjunga Gnesis (21.10%), the Chungthang formation (18.48%), the Gorubathan formation (12.12%), and the Lingtse Gnesis (5.35%). Landslide activities are frequent in the Sikkim Himalaya due to various structural disturbances, including fractures, faults, cracks, and joints. The stability of rocks and the likelihood of landslides are influenced by the lithological properties.

One of the important factors that has been extensively employed in landslide susceptibility and risk modelling is elevation, often known as altitude. In each geographic location, elevation determines the landslide occurrence parameter. Various geological and geomorphological processes regulate it (Ayalew et al. 2005; Pourghasemi, 2008). The elevation in the current study area varies from 816 to 4100 m (Figure 17.5a). The five categories on the elevation map each have a resolution of 30 × 30 m. Most of the landslides have been found to occur in the Rorachu watershed's medium and high elevation zones.

According to Lee and Min (2001), slope gradient is yet another important causal component in the evaluation of slope stability. The interaction of the slope and the slope's material qualities (cohesion, porosity, permeability, and bonding) results in the stability of the slope. Due to decreased shear stress, gentle slopes are less likely to experience slope instability (Dai et al. 2001). In contrast, the shear stresses increase as the slope gradient increases and vice versa. Using the natural break approach in ArcGIS 10.3, slope maps are divided into five groups in the current study. In this

TABLE 17.2
Description of the Geological Parameters of the Rorachu Watershed

Era	Formation	Characteristics	Lithology
Meso-Proterozoic	Lingtse gneiss	The gneisses are sheet-like bodies of coarse- to medium-grained, foliated to strongly lineated granite mylonite. These are streaky, banded, augen gneisses or porphyroblastic gneisses and are traversed by concordant and discordant pegmatite veins. Amphibolite intrusives with sharp contacts are also recorded within gneisses. The most characteristic feature of the Lingtse granite is the presence of a stretching lineation.	Granite gneiss (mylonite)
Proterozoic (Undifferentiated	Basic intrusive	Basic intrusive rocks are characterized by large crystal sizes, and as the individual crystals are visible, the rock is called phaneritic. This is formed as the magma cools underground and while cooling may be fast or slow; cooling is slower than on the surface, so larger crystals grow.	Tourmaline/biotite leuco granite, schroll rock/ pegmatite, aplite (Undifferentiated)
	Gorubathan formation	The formation consists of mappable, monotonous sequence of inter-banded chlorite sericite schist/ phyllite, quartzite, meta greywacke, pyritiferrous black slate/ carbon phylllite, basic meta volcanics. Chlorite phyllite is dark green to light green, whereas the quartz chlorite phyllite is only light green in color.	Inter-banded chlorite-sericite schist/phyllite and quartzite, meta-greywacke (quartzo feldspathic greywacke), pyritiferous black slate, biotite phyllite/mica schist, biotite quartzite, mica schist with garnet, with/ without staurolite, chlorite quartzite
	Kanchenjunga gneiss/ Darjeeling gneiss	The gneisses, dominantly comprising quartz, feldspar, and biotite (with minor amounts of other minerals) have been classified into three types, i.e. (1) banded/streaky gneisses/migmatites, (2) augen bearing biotite gneiss with/without garnet, kyanite, sillimanite, and (3) sillimanite granite gneisses. Mapping of these rocks as individual units is very difficult because they are characterized by frequent interchanging and gradational features among themselves.	Banded/streaky migmatite, augen bearing (garnet) biotite gneiss with/without kyanite, sillimanlte with palaeosomes of staurolite, kyanite, mica schist, biotite gneiss, sillimanite granite gneiss
	Chungthang formation	The main rock types of this formation are quartzites, garnet-kyanite-staurolite bearing biotite schist, calc silicate rock, graphitic schist, and amphibolite.	Quartzite 2. Garnet kyanite sillimanite biotite schist/Garnetiferous mica schist Chungthang 3. Calc-silicate, carbonaceous schist formation

Rorachu watershed, there are more than 30% of places with a slope angle between 35° and 70°, with a range from 0° to 70° (Figure 17.5b).

The intensity, duration, and amount of precipitation in this region as well as the physical properties of the soil, such as its texture, structure, porosity, permeability, compactness, etc., affect how saturated the soil is. In the Rorachu watershed, hills make up more than 90% of the landscape. There are six different types of soil found in the Rorachu watershed (Figure 17.5c, Table 17.3), including coarse loamy humic dystrudepts, coarse loamy humic lithic dystrudepts, coarse loamy typic hapludolls, fine loamy fluventic eutrudepts, fine skeletal cumuli hapludolls, and loamy skeletal entic hapludolls. All soil types in the Rorachu watershed have been transformed to vector polygons, which have subsequently been converted to raster format (30 × 30 m grid).

According to Horton (1932, 1945) and Strahler (1952), drainage density is calculated as the sum of all the streams and rivers in a grid divided by the grid's total area (Equation 17.1). The term DD describes the degree to which a river basin is drained by stream channels. DD is influenced by an area's physical and climatic environments. The DD aids in determining how much a mountain slope's shear strength is reduced, which affects slope instability. ArcGIS 10.3's Euclidean distance approach was used to calculate the drainage density of the Rorachu river basin into a 30 × 30 m grid (Figure 17.5d), and it was then divided into five categories using the natural breaking method.

$$D_d = (L_t/A_{basin}) \tag{17.1}$$

where, D_d represents drainage density, L_t represents total length of the streams in that grid, and A_{basin} represents total length of the grid area.

TABLE 17.3
Description of Soil Parameters in the Rorachu Watershed

Mapping Unit	Soil Name	Soil Code	Characteristics
Inceptisols	Coarse loamy humic dystrudepts	S001	Very deep, well-drained, moderately rapid permeable coarse loamy soil is found in structural benches and foot slope of mountain associated with moderately shallow to deep, little stony, excessively drained coarse loamy soil with moderate erosion.
	Coarse loamy humic Pachic dystrudepts	S002	Moderately rapid permeability occurs in upland slopes associated with moderately deep, well-drained coarse loamy soil with medium runoff, little stony, excessively drained fine loamy soils with moderate erosion.
	Coarse loamy typic hapludolls	S003	Excessively drained, deep coarse loamy soil having little stoniness and slight to moderate erosion is found mainly in the ridges associated with moderate deep to deep coarse loamy soil with little stoniness and moderate erosion.
	Fine-loamy fluventic eutrudepts	S004	Moderate permeability with moderately shallow to deep, well-drained fine loamy soil is found in steep slope, moderately high saturated hydraulic conductivity and moderate erosion associated with very deep, well-drained fine loamy upland soils.
Mollisols	Fine-skeletal cumilic hapludolls	S005	Moderately deep to very deep, excessively drained soils with gravelly surface, little stoniness and moderate erosion is found in very steep slope associated with moderately shallow to deep, slight stoniness, excessively drained, moderately erosion-prone coarse loamy soil.
	Loamy skeletal entic hapludolls	S006	Excessively drained, gravelly loamy soil mainly found in very steep hillside with small stoniness and moderate erosion associated with moderately shallow to deep, slight stoniness, moderately deep to deep, excessively drained, moderately erosion-prone gravelly loamy soil.

Source: According to Mandal and Mandal (2017a).

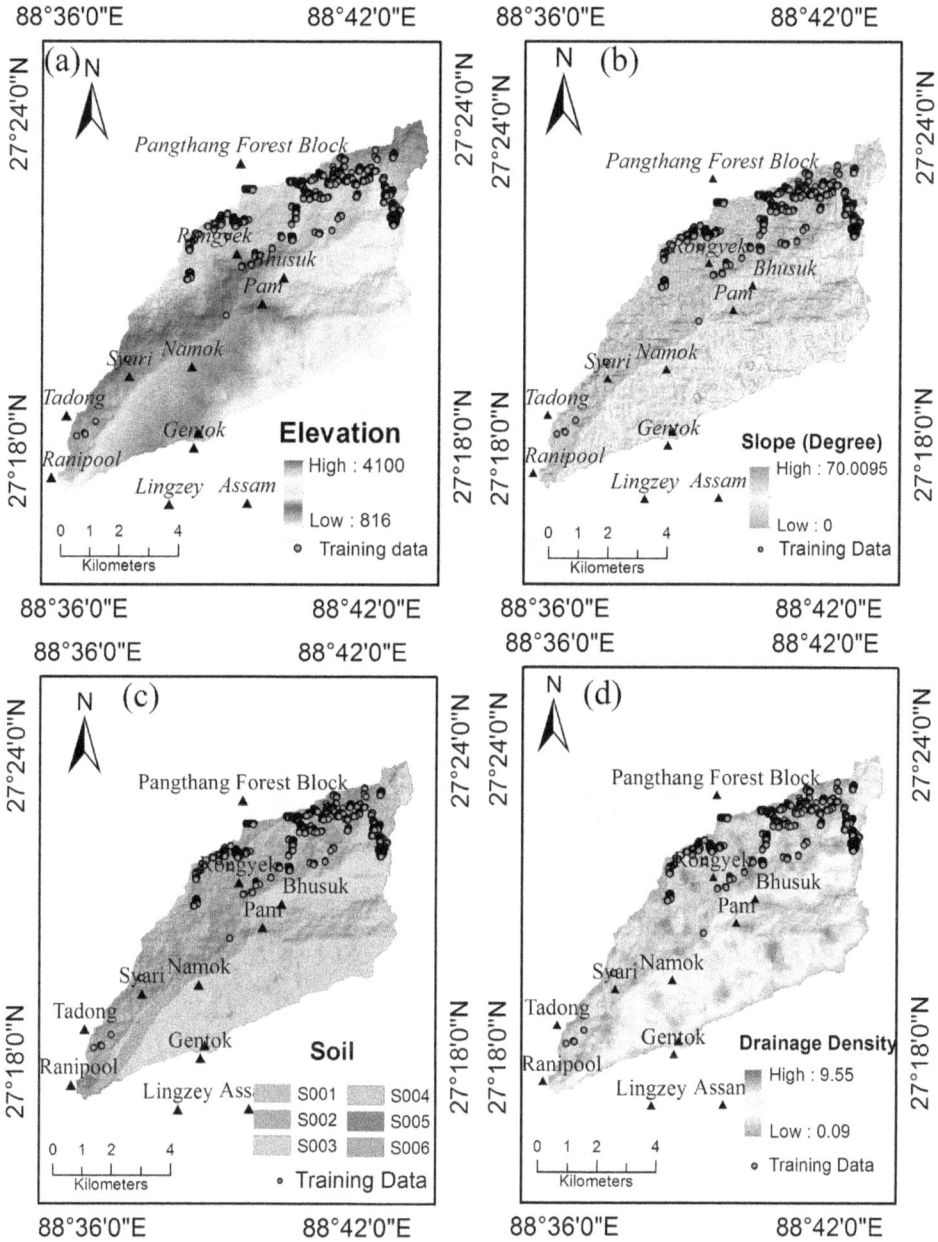

FIGURE 17.5   Landslide conditioning factors: (a) elevation, (b) slope, (c) soil, and (d) drainage density.

17.3.2.2   Road Density, Normalized Difference Vegetation Index, Slope Aspect, Topographic Position Index

Landslides are encouraged by the high road density, which weakens the slope and soil. The development and expansion of road networks, which are all aspects of human activity, are to blame for slope instability. Roads alter the slope's natural gradient and obstruct the flow of surface water (Marcini, 2010). The route map was made using topographical maps and Google Earth. ArcGIS 10.3 created 30 × 30 m grid cells (Figure 17.6e) for this Rorachu watershed area's road density, which was then divided into five groups.

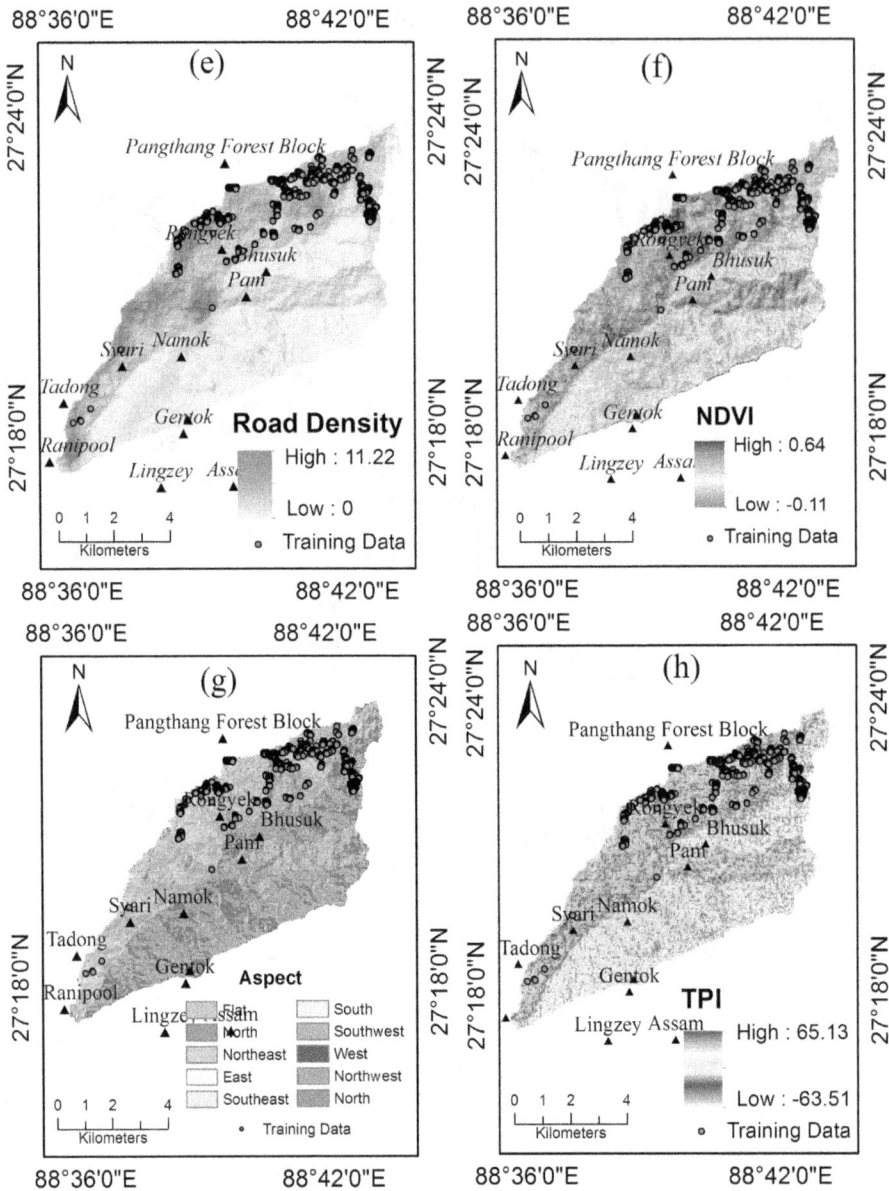

FIGURE 17.6 Landslide conditioning factors: (e) road density, (f) Normalized Difference Vegetation Index, (g) aspect, and (h) Topographic Position Index.

The NDVI is a numerical indicator that is used for the vegetation conditions of the surface. Using the equation NDVI = (NIR − R)/(NIR + R), where NIR stands for the near-infrared band and R for the red band of the satellite picture, the NDVI has been calculated. In the Rorachu watershed, where NDVI values vary from −0.11 to 0.64, NDVI was evaluated by combining Landsat 8 OLI picture in ERDAS 9.2 image processing software (Figure 17.6f). Positive numbers represent a healthy vegetative cover, which improves slope stability and soil cohesiveness. In the sections of the Rorachu watershed, which are more susceptible to soil erosion and slope failure, a negative NDVI score implies no plant cover. The slope aspect of the land is the compass direction of the steepest slope. Physical and biological elements that are connected to landslide threats might be impacted by the slope faces. Temperature and vegetation cover are both significantly influenced by slope aspect. The digital elevation model has been used to prepare the slope aspect map. According to Ercanoglu and Gokceoglu (2002), the slope aspect is connected to physiographic tendencies and the primary direction of precipitation. The slope aspect in the current study is broken down into ten categories: flat, north, north-east, east, south-east, south, south-west, west, north-west, and north.

The method known as the TPI is increasingly utilized to calculate the topographic slope positions and automate the classification of landforms. According to Guisan et al. (1999), TPI designates the higher, middle, and lower portions of the landscape. Locations with positive TPI values are higher than their surroundings (ridges). Locations with negative TPI values are lower than their surroundings (valleys). Values near zero on the TPI denote either flat terrain or slopes that never change. TPI was estimated in the research region using SAGA GIS software, and its value ranged from −63.51 to 65.13. For the evaluation of landslide sensitivity and landslide risk, the TPI is crucial (Figure 17.6h).

17.3.2.3 Stream Power Index, Topographic Wetness Index, Land Use/Land Cover, and Rainfall

SPI is a measure of the erosive power of the flowing water. The SPI has been estimated based on slope and specific catchment area (SCA). The SPI is defined after Moore and Grayson (1991), E (Equation 17.2):

$$SPI = (As \times \tan\beta) \tag{17.2}$$

where As is the SCA, and β is the local slope gradient measured in degrees, respectively. In the Rorachu watershed, SPI values vary from 0 to 145.37, and the study area is classified into five classes (Figure 17.7i) after Hengl et al. (2003) (Equation 17.3):

$$As = (Am \times P^2 / \sum Li) \tag{17.3}$$

In this equation, P is the pixel size, Am is the cumulative drainage fraction from m neighbors, and $\sum Li$ is derived as the sum of lengths for drainage pixels.

The TWI is a significant factor for landslide susceptibility and risk modeling. It is frequently employed to quantify processes that control topography or the flow of water. TWI refers to the accumulation of water at a particular point of time in any grid cell. For shallow landslide modeling, TWI has been used by various researchers (Gokceoglu et al. 2005; Yilmaz, 2009a, 2009b). In this study, the TWI map was prepared by SAGA GIS software after Beven and Kirkby (1979). The TWI map was classified into five categories (Figure 17.7j). TWI is defined as follows:

$$TWI = \ln\left(\frac{a}{\tan\beta}\right) \tag{17.4}$$

where a is the cumulative upslope area draining through a point (per unit contour length) and $\tan\beta$ is the slope angle at the point, which is used to replace approximately the hydraulic gradient under steady-state conditions (Poudyal et al. 2010).

The LULC map was created using supervised classification techniques in the ERDAS 9.2 program and was validated using Google Earth image and field data and Landsat 8 OLI satellite image data of 2019. Forested land encourages infiltration and drainage, which lowers slope failure. Due to the saturation of the covered soil, cultivated land influences the stability of the slope (Devkota et al. 2012). Different types of land cover, including step cultivation, open forest, settlements, bare soil, landslide areas, rivers, and dense forests, can be seen in the studied region. In the Rorachu

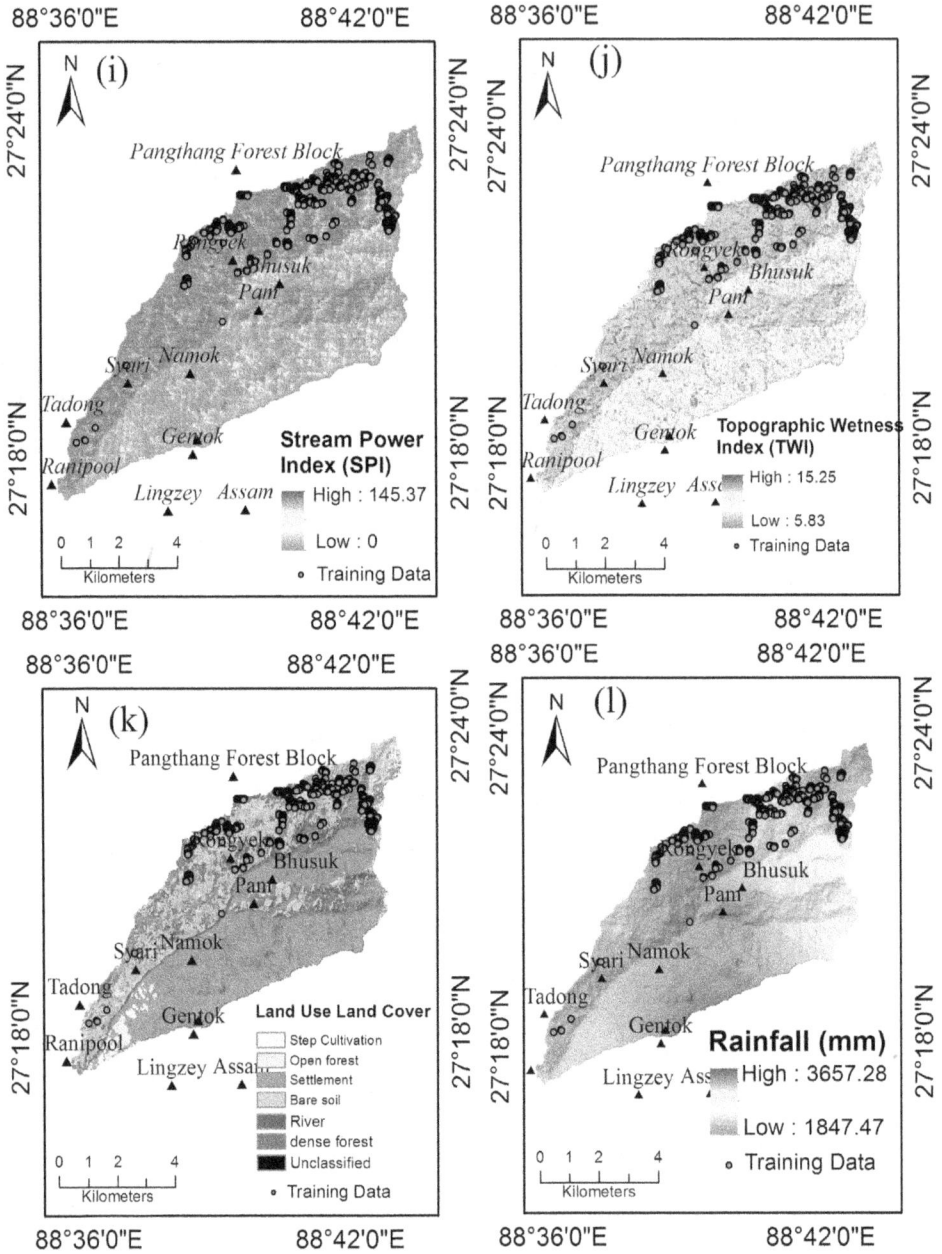

FIGURE 17.7 Landslide conditioning factors: (i) Stream Power Index, (j) Topographic Position Index, (k) land use/land cover, and (l) rainfall.

watershed, the forest (open and dense, 59%) takes up most of the land. Settlements (3.47%) and bare land (3.23%) are the next largest land uses (Figure 17.7k, Table 17.4).

Rainfall is one of the most important triggering factors for landslide events in the Rorachu watershed. The rainfall map was prepared using world climatic data and applying inverse distance weighted (IDW) modeling and then was classified into five categories. Rainfall in Rorachu watershed ranges between 1,847 mm and 3,657 mm (according to www.geog.ucsb.edu/~bodo/TRMM/#tif). Maximum rainfall occurs between June and August (according to IMD data, Table 17.5, and Figure 17.7).

TABLE 17.4
Landslide Causative Factors and Their Sub-Classes for Landslide Susceptibility and Risk Mapping

Factors	Sub-Class
Elevation	816–1495, 1495–1993, 1993–2516, 2516–3110, 3110–4100
Slope	0–15.37, 15.37–25.53, 25.53–35.14, 35.14–45.57, 45.57–70.01
Soil	S001, S002, S003, S004, S005, S006
Drainage density	0.092–2.17, 2.17–3.62, 3.62–4.92, 4.92–6.25, 6.25–9.557
Geology	Basic intrusive, Chungthang formation, Gorubathan formation, Lingtse gnesis, Kanchenjunga gnesis/Darjeeling gnesis (undifferential)
Road density	0–0.88, 0.88–2.55, 2.55–4.48, 4.48–6.86, 6.86–11.175
Normalized Difference Vegetation Index	−0.11–0.14, 0.14–0.24, 0.24–0.34, 0.34–0.43, 0.43–0.642
Aspect	Flat, north, north-east, east, south-east, south, south-west, west, north-west
Topographic Position Index	−63.51–14.57, −14.57–4.48, −4.48–4.59, 4.59–15.18, 15.18–65.135
Stream Power Index	0–2.85, 2.85–9.12, 9.12–20.52, 20.52–47.88, 47.88–145.37
Topographic Wetness Index	5.83–8.31, 8.31–9.19, 9.19–10.15, 10.15–11.26, 11.26–15.25
Land use/land cover	Step cultivation, open forest, settlement, bare soil, landslide area, river, dense forest
Rainfall	1874.47–2386.86, 2386.86–2791.41, 2791.41–3096.59, 3096.59–3323.70, 3323.70–3657.28

TABLE 17.5
Monthly Rainfall Distribution in the East Sikkim Area (2009–2015)

Year	January	February	March	April	May	June	July	August	September	October	November	December
2009	5.7	4.2	87.3	251.7	335.4	355.4	408.6	454.1	180.1	201.6	1.7	5.4
2010	5.7	18	187	359.4	272.7	504.6	601	493.8	375.8	95.6	23.6	0.1
2011	21.6	40.5	68.5	14.7	278.8	515.9	587.3	459.1	376.7	44.9	60.8	2.3
2012	17.8	21.5	28.4	312.2	201.6	614.4	481.3	442.2	410.9	72.4	0.1	1
2013	4.3	32.1	128	256.1	409	382.6	412.1	325.1	195.5	191.8	40.7	7.9
2014	0	5.4	68.2	96.1	441.4	472.7	478.7	522.3	273	16.7	2.4	4.2
2015	7.4	17.4	73.3	270.3	387.8	603.1	561	284.7	316.1	99.6	55.8	1

Source: Indian Meteorological Department, Gangtok, Sikkim.

17.3.3   MODELLING LANDSLIDE SUSCEPTIBILITY AND RISK

17.3.3.1   Application of Bivariate Statistical Index Model

The BSI approach is used for the landslide susceptibility and landslide risk assessment modelling in the Rorachu watershed (Table 17.5). The statistical index method is a bivariate statistical approach which has been used by Van Westen et al. (1997), for landslide susceptibility modelling. In recent years, the BSI model has been widely used by different researchers for landslide susceptibility and landslide risk modelling. In a BSI model, a weighted value for each categorical unit is defined as the natural logarithm of the landslide density in the categorical unit divided by the landslide density in the entire study area (Van Westen et al., 1997; Rautela and Lakhera, 2000; Cevik and Topal, 2003). This BSI approach is based on the following equations (Van Westen et al. 1997):

$$W_{BSI} = \ln\left(\frac{Eij}{E}\right) \tag{17.5}$$

$$W_{BSI} = ln\left(\frac{N(Si)}{N(Si)} / \frac{\sum N(Si)}{\sum N(Si)}\right) \tag{17.6}$$

where W_{BSI} is the weight given to a certain class i of parameter j, Eij is landslide density within class i of parameter j, and E is total landslide density within the entire study area. Here, $N(Si)$ is the number of landslide pixels in parameter class i, and $N(Ni)$ is the number of pixels in the same parameter class. In the current research, every landslide's causative factors were cross-checked with the landslide inventory map for determining the density of landslides for each class. The ultimate landslide susceptibility and risk map were produced using the ArcGIS raster calculator tool. Positive W_{BSI} indicates the significant relationship between landslide causative factors and the distribution of landslides. The negative W_{BSI} indicates the relationship between landslide causative factors, and the distribution of landslides is not relevant. In this study, the final LSI map (Figure 17.8) was prepared by the BSI model (Equation 17.7):

$$\left(LSI_{BSI} = \begin{pmatrix} (W_{BSI}*Elevation) + (W_{BSI}*Slope) + (W_{BSI}*Aspect) \\ + (W_{BSI}*Geology) + (W_{BSI}*Soil) + (W_{BSI}*Drainage\,density) \\ + (W_{BSI}*Road\,density) + (W_{BSI}*Rainfall) + (W_{BSI}*TWI) \\ + (W_{BSI}*SPI) + (W_{BSI}*TPI) + (W_{BSI}*NDVI) + (W_{BSI}*LULC) \end{pmatrix}\right) \tag{17.7}$$

17.3.3.2 Weight of Evidence Model

The WOE model is a bivariate statistical method based on the Bayesian approach and was first implemented for non-spatial and quantitative assessment in the medical sciences (Lusted, 1968). After its initial success, it established a wide acceptance in spatial analyses in geosciences for mineral potential mapping (Bonham-Carter et al. 1988) and was finally applied in landslide susceptibility and hazard mapping by different scholars (Lee et al. 2002a; Van Westen et al. 2003; Mathew et al. 2007; Dahal et al. 2008a, 2008b; Regmi et al. 2010). WOE is a Bayesian approach with a log-linear form which uses prior probability and posterior probability (Regmi et al. 2010a). Previously, Van Western (2002) applied the method for landslide susceptibility assessment. The model is based on a log-linear form of Bayesian rule. The Bayesian theorem can be written as

$$P(A|B) = \frac{P(B|A) \times P(A)}{P(B)} \tag{17.8}$$

Thus, the probability of several events A occurring, given that event B has already occurred, $P(A|B)$, is equal to the probability of event B occurring given that event A has occurred, $P(B|A)$, multiplied by the probability of event A occurring, $P(A)$, and divided by the probability of event B occurring, $P(B)$. This method calculates the weight for each landslide predictive factor (B) based on the degree of association in the presence or absence of the landslide (L) within the area (Bonham-Carter et al., 1989) as follows:

$$Wi^+ = \ln\frac{P\{B|A\}}{P\{B|\bar{A}\}} \tag{17.9}$$

$$Wi^+ = \ln\left\{\frac{(Npix1/(Npix1+Npix2))}{(Npix1/(Npix1+Npix2))}\right\} \tag{17.10}$$

$$Wi^- = \ln \frac{P\{\bar{B} \mid A\}}{P\{\bar{B} \mid \bar{A}\}} \tag{17.11}$$

$$Wi^- = \ln \left\{ \frac{\left(Npix2 / \left(Npix1 + Npix2\right)\right)}{\left(Npix4 / \left(Npix3 + Npix4\right)\right)} \right\} \tag{17.12}$$

where P is the probability and ln is the natural log. Similarly, B is the presence of potential landslide predictive factor, $\bar{B}$ s is the absence of a potential landslide predictive factor, A is the presence of landslide, and $\bar{A}$ is the absence of a landslide. The positive weight (Wi^+) indicates that the predictable variable is present at the landslide locations in the sub-category factors, and augmentation of this weight is an indication of positive correlation between predictable variables and landslides. The negative weight (Wi^-) indicates the absence of the predictable variable and shows the level of negative correlation (Dahal et al. 2008a). The difference between two weights is called weighted contrast (C, Equation 17.13). The final LSI is produced by the combination of each landslide causative factor using overlay methods (Equation 17.14):

$$C = \left(W_i^+ - W_i^- \right) \tag{17.13}$$

$$LSI = \sum_{i=1}^{N} C \tag{17.14}$$

Applying the bivariate WOE statistical model, the LSI map was made (Equation 17.15):

$$\left(LSI_{WOE} = \begin{pmatrix} \left(W_{WOE} * Elevation\right) + \left(W_{WOE} * Slope\right) + \left(W_{WOE} * Aspect\right) \\ + \left(W_{WOE} * Geology\right) + \left(W_{WOE} * Soil\right) + \left(W_{WOE} * Drainage\,density\right) \\ + \left(W_{WOE} * Road\,density\right) + \left(W_{WOE} * Rainfall\right) + \left(W_{WOE} * TWI\right) \\ + \left(W_{WOE} * SPI\right) + \left(W_{WOE} * TPI\right) + \left(W_{WOE} * NDVI\right) + \left(W_{WOE} * LULC\right) \end{pmatrix} \right) \tag{17.15}$$

17.3.3.3 Application of Index of Entropy Model

The IOE model is based on bivariate analysis, which computes the weight of all landslide causative factors and is used to determine the landslide susceptibility and landslide risk. The entropy based on Shannon (1948) is an uncertainty measure associated with a random variable, narrating the system's information content. Imbalance, disorder, uncertainty, and instability of a system are determined based on entropy (Yufeng and Fengxiang, 2009). In this IOE model, the weighting process is based on the methodology proposed by Vlcko et al. (1980). The IOE method has been widely used to determine the weight index of natural hazards and has been used for integrated environmental assessments of natural processes, such as sand storms, droughts, debris flows, and landslides (Li et al. 2002; Mon et al. 1994; Ren, 2000; Yi and Shi, 1994; Yang and Qiao, 2009; Devkota et al. 2013; Jaafari et al. 2014; Youssef et al. 2014a, 2015).

In the present study, the weighted parameter was obtained from the defined level of entropy (Table 17.6), representing the boundary where various factors influence the development of a landslide susceptibility and landslide risk. The information coefficient Wj represents the weight value for

TABLE 17.6

Spatial Relationship between Each Landslide Conditioning Factor and Observed Landslides Using Index of Entropy Models for All Landslide Causative Factor Classes

Factors	Class	Class Pixel	Landslide Pixel	P_{IJ}	(P_{IJ})	H_J	H_J max	I_J	P_J	W_{IJ}
Elevation (m)	4100–3110	6,841	218	3.42	0.52	1.502	2.32	0.353	1.308	0.4617
	3110–2516	16,104	338	2.25	0.34					
	2516–1993	20,031	139	0.74	0.11					
	1993–1495	19,285	16	0.09	0.01					
	1495–816	14,545	5	0.04	0.01					
Geology	Gorubathan formation	9,312	6	0.07	0.02	1.45	2.32	0.375	0.773	0.2898
	Lingtse genesis	4,112	0	0	0.00					
	Basic intrusive	16,210	54	0.36	0.10					
	Chungthang formation	14,123	255	1.94	0.53					
	Kanchanjangha formation	33,049	401	1.30	0.36					
Slope (°)	70.09–45.57	7,799	114	1.57	0.30	2.18	2.32	0.064	1.046	0.068
	45.57–35.14	15,677	210	1.44	0.27					
	35.14–25.53	20,253	218	1.15	0.22					
	25.53–15.37	20,229	126	0.67	0.13					
	15.37–0	12,848	48	0.40	0.08					
Soil	Fine skeletal	5,224	1	0.02	0.00	1.73	2.58	0.331	1.238	0.4097
	Coarse loamy distrudeptic	32,878	127	0.41	0.06					
	Coarse loamy holithic	11,997	156	1.39	0.19					
	Fine loamy	6,534	0	0	0.00					
	Loamy skeletal	3,754	126	3.60	0.48					
	Coarse loamy	16,419	306	2.00	0.27					
Drainage density	9.55–6.25	12,033	62	0.55	0.11	2.11	2.32	0.09	0.996	0.08964
	6.25–4.92	18,190	73	0.43	0.09					
	4.92–3.62	17,869	140	0.84	0.17					
	3.62–2.17	16,626	312	2.01	0.40					
	2.17–0.09	12,088	129	1.14	0.23					
Road density	11.17–6.86	2,051	13	0.67	0.09	2.0	2.32	0.142	1.553	0.2204
	6.86–4.48	4,530	100	2.33	0.30					
	4.48–2.55	8,439	252	3.16	0.41					
	2.55–0.88	14,317	157	1.16	0.15					

Factor	Class									
Rainfall (mm)	0.88–0	47,469	194	0.43	0.06	1.8	2.32	0.224	1.653	0.37054
	1847–2386	6,624	263	4.26	0.52					
	2386–2791	5,199	95	1.96	0.24					
	2791–3096	12,076	137	1.22	0.15					
	3096–3323	31,797	195	0.66	0.08					
	3323–3657	21,110	26	0.13	0.02					
Topographic Position Index	15.25–11.26	6,669	67	1.08	0.21	2.31	2.32	0.004	1.047	0.00418
	11.26–10.15	18,765	173	0.99	0.19					
	10.15–9.19	25,823	228	0.95	0.18					
	9.19–8.31	18,956	171	0.97	0.18					
	8.31–5.83	6,593	77	1.25	0.24					
Stream Power Index	145.37–47.88	133	0	0	0	1.96	2.32	0.159	0.819	0.1303
	47.88–20.52	1,060	7	0.71	0.17					
	20.52–9.12	4,966	57	1.23	0.30					
	9.12–2.85	21,030	255	1.30	0.32					
	2.85–0	49,617	397	0.86	0.21					
Topographic Wetness Index	65.13–15.18	5,334	20	0.40	0.09	2.20	2.32	0.056	0.884	0.0495
	15.18–4.59	13,142	65	0.53	0.12					
	4.59 – –4.48	21,264	206	1.04	0.23					
	–4.48–14.57	22,952	263	1.23	0.28					
	–14.57–63.51	14,114	162	1.23	0.28					
Land use/land cover	Step cultivation	1,648	0	0	0	1.40	2.81	0.504	1.056	0.53206
	Dense forest	45,962	309	0.72	0.11					
	Settlement	3,865	7	0.19	0.03					
	Bare soil	3,921	149	4.08	0.64					
	River	1,439	0	0	0.00					
	Open forest	19,971	251	1.35	0.21					
Aspect	Flat	4	0	0	0	2.54	3.32	0.235	0.733	0.17221
	North	2,614	0	0	0					
	North-east	2,042	5	0.26	0.04					
	East	5,587	52	1.00	0.14					
	South-east	10,496	137	1.40	0.19					
	South	14,015	174	1.33	0.18					

(Continued)

TABLE 17.6 (Continued)

Spatial Relationship between Each Landslide Conditioning Factor and Observed Landslides Using Index of Entropy Models for All Landslide Causative Factor Classes

Factors	Class	Class Pixel	Landslide Pixel	P_{ij}	(P_{ij})	H_j	H_j max	I_j	P_j	W_{ij}
	South-west	10,206	207	2.14	0.29					
	West	12,646	116	0.98	0.13					
	North-west	14,206	28	0.21	0.03					
	North	4,990	0	0	0					
Normalized Difference Vegetation Index	0.64–0.43	9,717	205	2.26	0.39	2.11	2.32	0.091	1.158	0.1048
	0.43–0.33	13,283	160	1.29	0.22					
	0.33–0.24	17,463	90	0.55	0.10					
	0.24–0.14	21,584	92	0.46	0.08					
	0.14–0.11	14,759	169	1.23	0.21					

the causative factors, and it was calculated after Bednarik et al. (2010) and Constantin et al. (2011), as follows:

$$Pij = \frac{b}{a}$$ (17.16)

$$(Pij) = \frac{Pij}{\sum_{j=1}^{Sj} Pij}$$ (17.17)

Here, *Hj* and *Hj max* are the entropy values (Equations 17.16 and 17.17), and they are written as

$$Hj = -\sum_{i=1}^{Sj}(Pij)\log 2(Pij), \qquad j = 1,2,...,n$$ (17.18)

$$Hj\,max = \log 2\,Sj \qquad Sj \text{ is the number of classes}$$ (17.19)

Ij is the information coefficient (Equation 17.20), and *Wj* is the resultant weight value for the landslide causative parameter (Equation 17.21):

$$Ij = \frac{Hj\,max - Hj}{Hj\,max} \qquad I = (0, 1)\, j = 1, 2 \ldots n,$$ (17.20)

$$Wj = Ij \times Pj$$ (17.21)

where *a* and *b* are the domain and landslide percentages, respectively, and *(Pij)* is the probability density. The result varies from 0 to 1. The closer the value of 1 is to 1, the greater the slope instability, and vice versa. The complete calculation of weight determination for the individual landslide causative parameters is presented in Table 17.7. The final landslide susceptibility map was prepared by the summation of all individual landslide causative parameter classes. The final landslide susceptibility map was prepared using the following equation:

$$YIOE = \sum_{i}^{n} \frac{z}{mi} \times C \times Wj$$ (17.22)

Y_{IOE} is the sum of all the classes; *i* is the number of particular parametric maps (1, 2, . . ., *n*); *z* is the number of classes within a landslide parametric map with the greatest number of classes; *mi* is the number of classes within a particular landslide parametric map; *C* is the value of the class after secondary classification; and *Wj* is the weight of a parameter (Bednarik et al. 2010; Devkota et al. 2013; Jaafari et al. 2014). Applying the IOE statistical model, the LSI map was made for the Rorachu watershed (Equation 17.23):

$$Y_{IOE} = \begin{pmatrix} (\text{Elevation} * 0.4617) + (\text{Slope} * 0.068) + (\text{Aspect} * 0.1722) + \\ (\text{Geology} * 0.2898) + (\text{Soil} * 0.4097) + (\text{Drainage density} * 0.0896) + \\ (\text{Road density} * 0.2204) + (\text{Rainfall} * 0.3705) + (\text{TWI} * 0.0495) + \\ (\text{SPI} * 0.1303) + (\text{TPI} * 0.0041) + (\text{NDVI} * 0.1048) + (\text{LULC} * 0.532) \end{pmatrix}$$ (17.23)

TABLE 17.7

Spatial Relationship between Each Landslide Conditioning Factor and Observed Landslides Using Bivariate Statistical Index and Weight of Evidence Models

Factors	Class	Class Pixel	Landslide Pixel	W+	W−	C	Bivariate Statistical Index
Elevation (m)	4100–3110	6,841	218	1.23	−0.27	1.50	1.23
	3110–2516	16,104	338	0.81	−0.40	1.22	0.81
	2516–1993	20,031	139	−0.30	0.09	−0.38	−0.30
	1993–1495	19,285	16	−2.42	0.27	−2.69	−2.42
	1495–816	14,545	5	−3.30	0.20	−3.50	−3.30
Geology	Gorubathan formation	9,312	6	−2.67	0.12	−2.79	−2.67
	Lingtse genesis	4,112	0	0	0.06	−0.06	0
	Basic intrusive	16,210	54	−1.03	0.16	−1.19	−1.03
	Chungthang formation	14,123	255	0.66	−0.24	0.90	0.66
	Kanchenjunga formation	33,049	401	0.26	−0.26	0.52	0.26
Slope (°)	70.09–45.57	7,799	114	0.45	−0.07	0.52	0.45
	45.57–35.14	15,677	210	0.36	−0.12	0.48	0.36
	35.14–25.53	20,253	218	0.14	−0.06	0.20	0.14
	25.53–15.37	20,229	126	−0.40	0.11	−0.52	−0.40
	15.37–0	12,848	48	−0.91	0.11	−1.03	−0.91
Soil	Fine skeletal	5,224	1	−3.89	0.07	−3.95	−3.89
	Coarse loamy distrudeptic	32,878	127	−0.88	0.36	−1.24	−0.88
	Coarse loamy holithic	11,997	156	0.33	−0.08	0.41	0.33
	Fine loamy	6,534	0	0.00	0.09	−0.09	0
	Loamy skeletal	3,754	126	1.28	−0.14	1.42	1.28
	Coarce loamy	16,419	306	0.69	−0.32	1.01	0.69
Drainage density	9.55–6.25	12,033	62	−0.59	0.08	−0.67	−0.59
	6.25–4.92	18,190	73	−0.84	0.16	−1.01	−0.84
	4.92–3.62	17,869	140	−0.17	0.05	−0.22	−0.17
	3.62–2.17	16,626	312	0.70	−0.33	1.03	0.70
	2.17–0.09	12,088	129	0.14	−0.03	0.16	0.14
Road density	11.17–6.86	2,051	13	−0.39	0.01	−0.39	−0.39
	6.86–4.48	4,530	100	0.86	−0.09	0.95	0.86
	4.48–2.55	8,439	252	1.16	−0.32	1.48	1.16
	2.55–0.88	14,317	157	0.16	−0.04	0.20	0.16
	0.88–0	47,469	194	−0.82	0.65	−1.47	−0.82
Rainfall (mm)	1847–2386	6,624	263	1.45	−0.37	1.82	1.45
	2386–2791	5,199	95	0.67	−0.07	0.75	0.67
	2791–3096	12,076	137	0.20	−0.04	0.24	0.20
	3096–3323	31,797	195	−0.42	0.22	−0.64	−0.42
	3323–3657	21,110	26	−2.02	0.28	−2.31	−2.02
Topographic Position Index	15.25–11.26	6,669	67	0.07	−0.01	0.08	0.07
	11.26–10.15	18,765	173	−0.01	0.00	−0.01	−0.01
	10.15–9.19	25,823	228	−0.05	0.03	−0.08	−0.05
	9.19–8.31	18,956	171	−0.03	0.01	−0.04	−0.03
	8.31–5.83	6,593	77	0.23	−0.02	0.25	0.23
Stream Power Index	145.37–47.88	133	0	0	0.00	0.00	0
	47.88–20.52	1,060	7	−0.34	0.00	−0.35	−0.34
	20.52–9.12	4,966	57	0.21	−0.02	0.22	0.21

TABLE 17.7 *(Continued)*

Spatial Relationship between Each Landslide Conditioning Factor and Observed Landslides Using Bivariate Statistical Index and Weight of Evidence Models

Factors	Class	Class Pixel	Landslide Pixel	W+	W–	C	Bivariate Statistical Index
	9.12–2.85	21,030	255	0.26	–0.12	0.38	0.26
	2.85–0	49,617	397	–0.15	0.23	–0.38	–0.15
Topographic Wetness Index	65.13–15.18	5,334	20	–0.91	0.04	–0.95	–0.91
	15.18–4.59	13,142	65	–0.63	0.09	–0.73	–0.63
	4.59 — – 4.48	21,264	206	0.04	–0.02	0.05	0.04
	–4.48–14.57	22,952	263	0.21	–0.10	0.31	0.21
	–14.57–63.51	14,114	162	0.21	–0.05	0.26	0.21
Land use/land cover	Step cultivation	1,648	0	0	0.02	–0.02	0
	Dense forest	45,962	309	–0.33	0.35	–0.67	–0.33
	Settlement	3,865	7	–1.64	0.04	–1.68	–1.64
	Bare soil	3,921	149	1.41	–0.18	1.59	1.41
	River	1,439	0	0	0.02	–0.02	0
	Open forest	19,971	251	0.30	–0.13	0.43	0.30
Aspect	Flat	4	0	0	5E-05	–5E-05	0
	North	2,614	0	0	3E-02	–3E-02	0
	North-east	2,042	5	–1.34	0.02	–1.36	–1.34
	East	5,587	52	0.00	0.00	0.00	0.00
	South-east	10,496	137	0.34	–0.07	0.40	0.34
	South	14,015	174	0.29	–0.08	0.36	0.29
	South-west	10,206	207	0.76	–0.19	0.96	0.76
	West	12,646	116	–0.02	0.00	–0.02	–0.02
	North-west	14,206	28	–1.55	0.16	–1.72	–1.55
	North	4,990	0	0	0.07	–0.07	0
Normalized Difference Vegetation Index	0.64–0.43	9,717	205	0.21	–0.06	0.26	0.82
	0.43–0.33	13,283	160	–0.78	0.19	–0.97	0.26
	0.33–0.24	17,463	90	–0.59	0.12	–0.72	–0.59
	0.24–0.14	21,584	92	0.26	–0.06	0.32	–0.78
	0.14–0.11	14,759	169	0.82	–0.20	1.02	0.21

17.3.4 MULTICOLLINEARITY TEST

Due to the statistical testing phenomenon known as multicollinearity, one predictor variable in a multiple regression model can be linearly predicted from the others with a high degree of accuracy. Multicollinearity has no effect on the model's overall predictive power or reliability; it only affects calculations related to certain predictors. Before using any of the landslide causative factors for the LSI or landslide risk modelling, it is essential to evaluate the multicollinearity of those factors (Zhou et al. 2018; Arabameri et al. 2019; Chen et al. 2018). Both tolerance (TOL) and the variance influencing factors (VIF) are crucial indexes for the diagnosis of multicollinearity. Simply put, VIF is the inverse of TOL. A multicollinearity issue is implied by a TOL of less than 0.20 or 0.10 and/or a VIF of 5 or 10 and above (O'Brien, 2007). The table shows that the smallest TOL for the three models (BSI, IOE, and WOE) for the rainfall parameter was, respectively, 0.321, 0.383, and 0.386. These

TABLE 17.8

Multicollinearity Analysis of the Bivariate Statistical Index, Index of Entropy, and Weight of Evidence Approaches

Factors	BSI		WOE		IOE	
	TOL	VIF	TOL	VIF	TOL	VIF
Elevation	0.634	1.577	0.589	1.697	0.524	1.908
Slope	0.568	1.760	0.598	1.672	0.592	1.688
Aspect	0.821	1.218	0.825	1.213	0.818	1.223
Geology	0.571	1.753	0.579	1.728	0.567	1.763
Soil	0.661	1.513	0.663	1.509	0.656	1.523
Drainage density	0.559	1.789	0.772	1.296	0.552	1.811
Road density	0.847	1.181	0.845	1.183	0.844	1.185
TPI	0.705	1.418	0.626	1.598	0.621	1.611
TWI	0.686	1.459	0.527	1.898	0.524	1.910
SPI	0.863	1.158	0.715	1.399	0.714	1.400
NDVI	0.504	1.986	0.505	1.981	0.499	2.003
Rainfall	0.321	3.115	0.386	2.590	0.383	2.610
LULC	0.979	1.021	0.969	1.031	0.979	1.022

Note: BSI, Bivariate Statistical Index; IOE, Index of Entropy; LULC, land use/land cover; NDVI, Normalized Difference Vegetation Index; SPI, Stream Power Index; TOL, Tolerance; TPI, Topographic Position Index; TWI, Topographic Wetness Index; VIF, Variance Influencing Factors; WOE, Weight of Evidence.

models (BSI, IOE, and WOE) have VIF of 3.115, 2.610, and 2.590, respectively, which represent the rainfall parameter. Therefore, there is no multicollinearity between the current research and independent landslide causative elements. The TOL and the VIF are calculated in the following equations:

$$TOL = 1 - Ri^2 \qquad (17.24)$$

$$VIF = \frac{1}{1 - Ri^2} \qquad (17.25)$$

TOL is the tolerance; *VIF* is the variance influencing factors, and it is the coefficient of determination of landslide conditioning factors. The multicollinearity statistics of all the models (BSI, IOE, and WOE) are shown in Table 17.8.

17.3.5 VALIDATION OF THE MODELS

The mapping of landslide susceptibility in the Rorachu watershed was supported by many bivariate statistical models, including BSI, IOE, and WOE. These statistical model performances cannot be used elsewhere in the world for landslide susceptibility evaluations without adequate validation. It is usual to use the number of correctly identified pixels as proof to bolster the result of the landslide investigation. The model in this study was trained using 70% (500) of the observed landslide pixels, while the remaining 30% (216) were utilised for landslide validation. The validation process employed a total of 716 pixels of observed landslides as a result. There are various methods for validating landslide susceptibility models. Validation techniques include the use of the success rate curve, landslide density, and ROC curve (Van Westen et al. 2003; Chung and Fabbri, 1993; Sarkar and Kanungo, 2004). Here, the Rorachu watershed's landslide susceptibility was accurately mapped using all validation techniques.

17.3.6 LANDSLIDE RISK ZONATION MAPPING

The risk is the maximum amount of loss that could possibly occur as a result of landslide events in a particular area and time frame. In order to determine the possibility that an element-at-risk would be harmed by a certain hazard, landslide hazard risk analysis examines the relationship between the frequency of damaging events and the seriousness of the effects (Guzzetti et al. 2009). The risk analysis has a significant impact on how mountainous communities are developed, how highways are constructed, and how land is utilised. Varnes (1984), Fell (1994), Leroi (1996), Yalcin (2008), and Xu et al. (2012) have all produced successful landslide risk assessments. According to Xu et al. (2012), the following is the likelihood that a specific hazard would cause damage to a particular element:

$$Risk = H \times V \qquad (17.26)$$

Here, H denotes the hazard expressed as a probability of occurrence within a reference period, and V defines the physical vulnerability of a particular type of element-at-risk (from 0 = not vulnerable to 1 = vulnerable) for a specific type of hazard and for a specific element-at-risk.

17.4 RESULTS AND DISCUSSION

17.4.1 BIVARIATE STATISTICAL INDEX MODEL, LANDSLIDE SUSCEPTIBILITY ZONES, AND CAUSATIVE FACTORS

The relationship between landslide causative factors and the location of landslides has been analysed using a BSI model (Table 17.7). It is seen that 450 to 700 slope angles have maximum BSI (0.45), and the 00 to 150 slope class has the lowest BSI with negative values of −0.91. On the other hand, the slope degree and slope instabilities are inversely related to each other in the Rorachu watershed. In the case of slope aspect, the south-east, south, and south-west slope facets have positive BSI values (0.34, 0.28, and 0.76), and the north-east, east, west, and north-west slope facets have negative BSI values (−1.33, −0.001, −0.02, and −1.55). Flat and north-facing slope aspects have no landslides. Among all these, the highest BSI values are present in the south-west facing slope aspect. The BSI values at altitude are positive within the elevation range of 2516 to 3110 m, and the 3110 to 4100 m elevation range exhibits the highest BSI value (BSI = 1.23).

The Chungthang formation has the highest BSI value (BSI = 0.66), and the Gorubathan formation has the lowest value (BSI = −2.67). The loamy skeletal, coarse loamy, and coarse loamy holithic have the most positive BSI values (1.28, 0.69, and 0.33, respectively), and the fine skeletal has the lowest BSI values (−3.88). The NDVI factor ranges between −0.11 and 0.14, which represents the highest BSI values (BSI = 0.82), which indicates the most favorable place for the landslide because of the diverse correlation between absence and presence of vegetation with slope instability. In the case of DD, the highest BSI values (0.69) are found in the lowest DD areas. Regarding rainfall, the rainfall ranges from 1,847 to 2,386, 2,386 to 2,791 mm and 3,096 mm/year, having the highest landslide probability. On the other hand, the three classes of RD showed the strongest favor of landslides. These classes are high (BSI = 0.86), moderate (BSI = 1.16), and low RD (BSI = 0.16). In the case of LULC, bare soil and open forest have the highest BSI values (1.40 and 0.29), representing a good correlation with the landslide. Similarly, the low SPI values represent the highest BSI values (0.26), the very low TPI values represent the highest BSI values (0.22), and the very low TWI values represent the strong correlation with landslide activities.

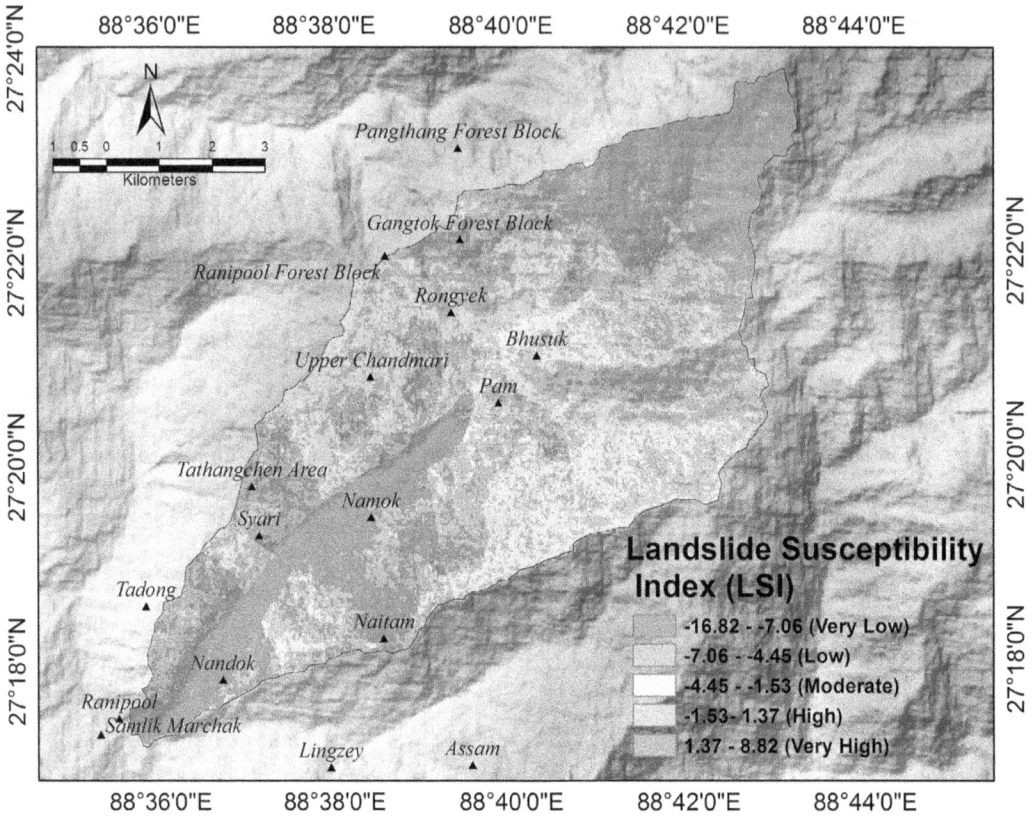

FIGURE 17.8 Landslide susceptibility map generated using Bivariate Statistical Index model.

17.4.2 Weight of Evidence Model, Landslide Susceptibility Zones, and Causative Factors

The landslide susceptibility mapping (LSI) was made using the WOE model (Figure 17.8) using all variable class values (Table 17.7). The WOE model is a bivariate statistical method based on the Bayesian approach and was first implemented for non-spatial and quantitative assessment in the medical sciences (Lusted, 1968). The WOE bivariate statistical model allows us to calculate prior and posterior probability values and enumerate the weight contrast $C = (Wi^+ - Wi^-)$ of each class. The final LSI value has been calculated by combining all weight contrast (C) factors (Figure 17.9). The altitude ranges between 3,110 and 4,100 m, and 2,516 and 3,110 m represent the highest values of contrast weight of 1.50 and 1.22, respectively. The lowest elevation indicates the lowest weight. In the case of slope, the highest contrast (C) weight values (0.52 and 0.48) are found in the classes of 45° to 70° and 35° to 45°, which indicates the highest landslide probability. The slope angle shows the negative weight values, which indicates the low landslide probability. The relationship between slope and weight contrast values is positively correlated. In the case of slope aspect, the south-west (SW), south-east (SE), and south-facing (S) slopes were susceptible and very high risk for landslide occurrence with the highest values (C) of 0.95, 0.40, and 0.36, respectively. All the other slope aspect classes represent negative values, which indicates the lowest landslide probability. The low DD class shows the highest weight values (1.03).

The high DD shows the negative weight values. In terms of geology, it is seen that the Chingthang formation and Kanchenjunga formation represent positive weight values (0.90 and 0.52), and all other geological classes exhibit negative weight values that indicate the lowest probability of landslides. The loamy skeletal and coarse loamy soil classes have a significant probability of landslide susceptibility, which shows the weight values of 1.42 and 1.01, respectively. The relationship between rainfall and landslide occurrence shows that the rainfall classes of 1,847 to 2,386 mm/year and 2,386 to 2,791 mm/year are characterized by the highest (C) probability of landslide occurrence. The moderate and high RD areas reveal the highest weight values of 1.48 and 0.95, respectively. In the Rorachu watershed, the moderate RD class reveals the highest weight values, which indicates the maximum landslide probability along the highway (NH 31A). For LULC, the highest landslide occurrence probability was observed in bare soil (1.59) and open forest (0.43). Other land use classes depict negative correlations between landslide occurrence probabilities and weight contrast values. The relationship between NDVI and landslide contrast weights is negatively correlated, and vice versa. The very low NDVI class has the highest weight values (1.02). In the case of SPI, the minimum values are associated with the highest weight values (0.38), and lower TWI values are accompanied by the highest contrast weight values (0.31). It has also been seen that the very low TPI class represents the highest landslide probability (0.25).

17.4.3 INDEX OF ENTROPY MODEL, LANDSLIDE SUSCEPTIBILITY, AND CAUSATIVE FACTORS

The LSI was calculated by the bivariate IOE model (Figure 17.10). In this model, every landslide causative factor class was represented by a specific landslide probability occurrence density (*Pij*) and, accordingly, the final weight (*Wij*) was calculated. The landslide causative factors' weights (*Wij*) and the probability densities (*Pij*) for each class of landslide causative factors were derived to assess landslide susceptibility (Table 17.6). This bivariate IOE method allows us to calculate the

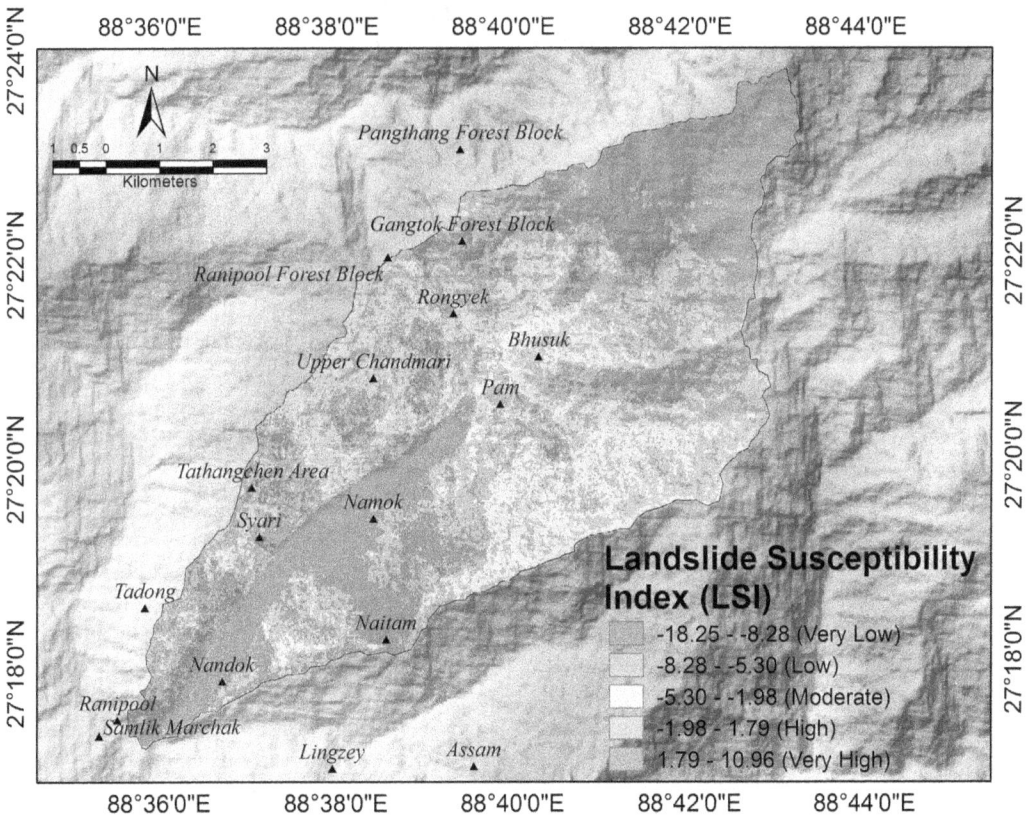

FIGURE 17.9 Landslide susceptibility map emanated by weight of evidence model.

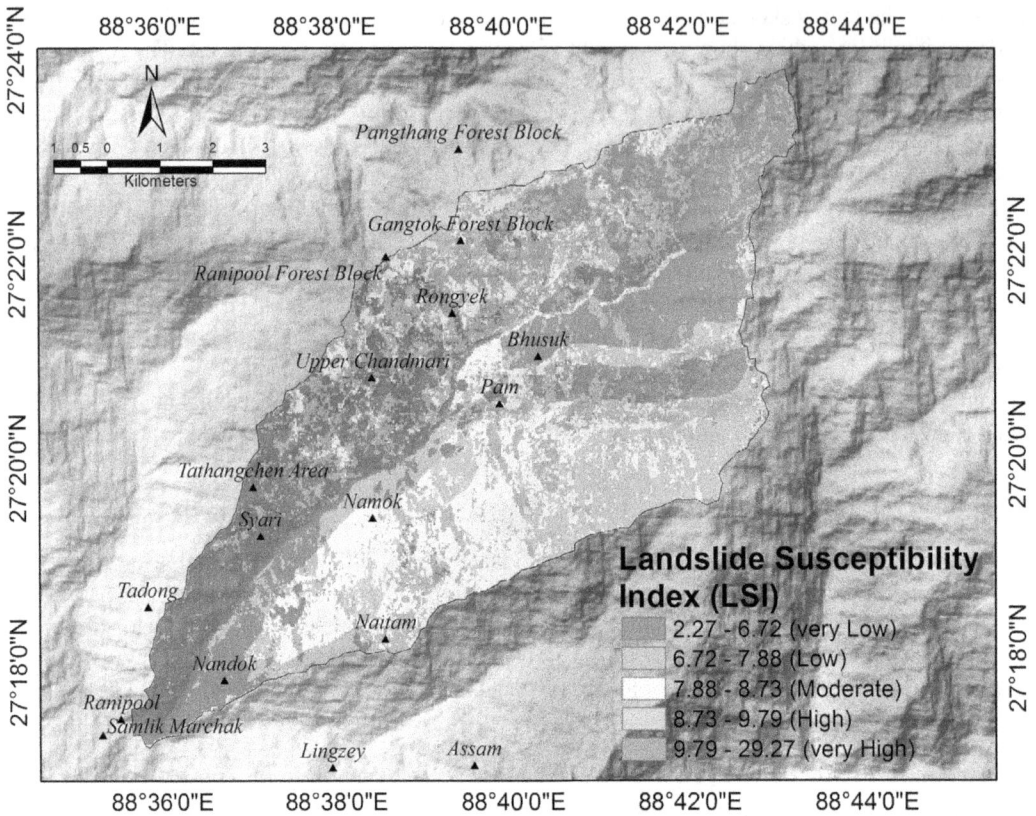

FIGURE 17.10   Landslide susceptibility map emanated by Index of Entropy model.

weight for each landslide causative factor, and the weighting process is based on the methodology proposed by Vlcko et al. (1980). The final LSI values have been derived using the IOE model (Figure 17.10). The calculated weights (Wij) of landslide causative factors indicate (Table 17.6) that the most significant landslide causative factors are LULC (0.53), elevation (0.46), and soil (0.41).

In terms of landslide density (Pij) of every landslide causative factor, the altitude ranges between 3,110 and 4,100 m have the highest landslide density (Pij) values (0.52), and the 1,500-m elevation areas have low landslide density (Pij) values (0.01). It is seen that the slope class of 450 to 700 has the highest probability of landslides where the landslide density (Pij) value is 0.30. In terms of slope aspect, the south-west, south-east, and south-facing slopes were susceptible and very high risk for landslide occurrence, with the highest landslide density values (Pij) of 0.29, 0.19, and 0.18, respectively. The RD shows that moderate to high-density areas have the highest landslide probability (Pij) values of 0.41 and 0.30, respectively. This situation is found in places where one or two roads are very much affected by landslides. It has been observed that the Chungthang formation and Kanchenjunga formation exhibit the highest probability of landslide density (Pij) values of 0.53 and 0.36, respectively. Lingtse gneiss has no landslide density (Pij). In the case of soil, the highest landslide probability (Pij) is found in the classes of loamy skeletal and coarse loamy, where landslide density values (Pij) are 0.48 and 0.27, respectively (Figure 17.11). The rainfall class of 1,847 to 2,386 and 2,386 to 2,791 mm/year represents the highest (Pij) probability of landslide occurrence. In terms of DD, it is seen that low drainage density exhibits the highest (Pij) values of 0.40, and high DD exhibits low landslide probability (Pij) values of 0.09. The highest landslide probability (Pij) is observed in bare soil (0.64) and open forest (0.21). The moderate (0.30) and low (0.32) SPI show the highest probability of landslide density (Pij). The TWI and TPI also influence the

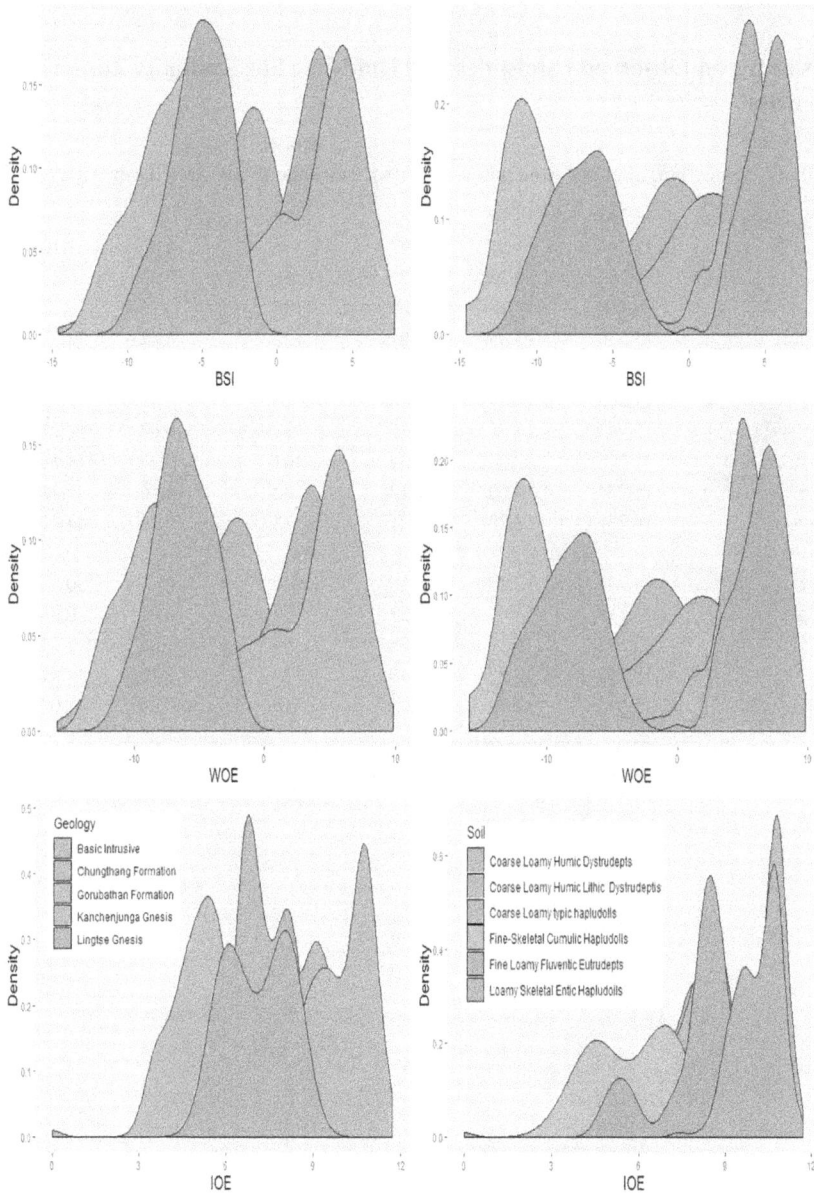

FIGURE 17.11 The relationship between different models (BSI, IOE and WOE) with the geology and soil factors of Rorachu watershed.

landslide probability of occurrence. The susceptibility of the landslide occurrences decreases with the increase of NDVI values, and vice versa.

17.4.4 Result of Model Validations

17.4.4.1 Landslide Density Method and Model Validations

The ratio of actual landslides to the total area of each class that is susceptible to landslides is known as the landslide density (LD) (Table 17.9). From low to high landslide sensitive groups, the LD gradually

TABLE 17.9

Comparisons between Observed Landslides and Landslide Susceptibility Zone with Landslide Density

Model	Susceptibility Zones	No. of Pixels	Area (km²)	Area (%)	No. of Landslide Pixels	Area (km²)	Area (%)	Landslide Density
BSI$_M$	Very low	14,888	13.40	19.38	0	0	0	0
	Low	15,340	13.81	19.97	8	0.01	1.12	0.0005
	Moderate	15,782	14.20	20.55	49	0.04	6.84	0.0031
	High	15,430	13.89	20.09	138	0.12	19.27	0.0089
	Very high	15,366	13.83	20.01	521	0.47	72.77	0.0339
IOE$_M$	Very low	14,883	13.39	19.38	24	0.02	3.35	0.0016
	Low	15,943	14.35	20.76	91	0.08	12.71	0.0057
	Moderate	15,005	13.50	19.54	101	0.09	14.11	0.0067
	High	16,224	14.60	21.12	182	0.16	25.42	0.0112
	Very high	14,751	13.28	19.21	318	0.29	44.41	0.0215
WOE$_M$	Very low	14,741	13.27	19.19	0	0	0	0
	Low	15,566	14.01	20.27	7	0.01	0.98	0.0004
	Moderate	15,573	14.02	20.28	54	0.05	7.54	0.0035
	High	15,352	13.82	19.19	139	0.13	19.41	0.0091
	Very high	15,574	14.02	20.28	516	0.46	72.07	0.0331

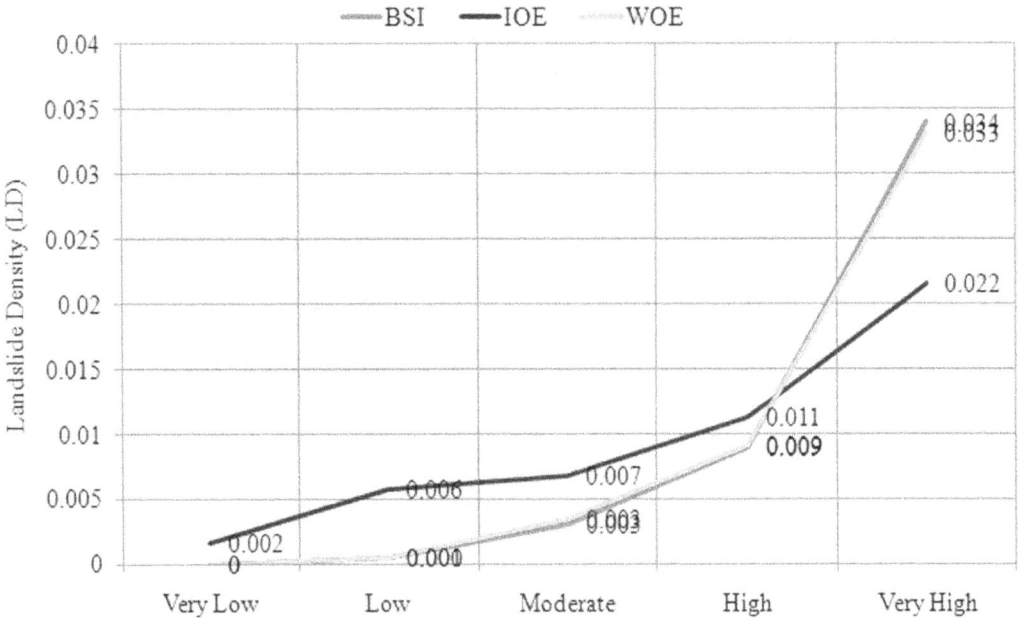

FIGURE 17.12 Landslide density of different susceptible classes in the various Landslide Susceptibility Index models (Bivariate Statistical Index, Index of Entropy, and weight of evidence).

rises. The greatest LD values, 0.0339 (BSI model), 0.0215 (IOE model), and 0.0331 (WOE model), are discovered to be associated with the extremely high landslide sensitive class. Additionally, there was a continual decline in LD estimates from very high to low landslide susceptibility zones (Figure 17.12). Better results than the IOE model are represented by the BSI and WOE models.

17.4.4.2   Result of Success Rate Curve Method and Model Validation

To determine the SRC versus BSI, IOE, and WOE in this study area, the cumulative percentage of recorded landslides was plotted against the cumulative percentage of the LSI susceptibility zonation map (Figure 17.13). The area under the curve (AUC) was calculated for the current study in order to quantitatively assess the accuracy. The AUCs are 0.869, 0.772, and 0.872, respectively, which means that the overall success rates for the BSI, IOE, and WOE models are 86.9%, 77.20%, and 87.20%, respectively. These findings suggest that the BSI provided more accurate landslide susceptibility maps than the IOE and WOE models.

17.4.4.3   Result of Receive Operating Characteristics Curve Method and Model's Validation

The ROC curve can be used to validate the landslide susceptibility maps (Akgun et al. 2012; Regmi et al. 2014; Ozdemir and Altural, 2013). A ROC curve is a graphical diagram that shows how a binary classifier system's diagnostic ability changes with its discrimination threshold. The true-positive (TP) rate and false-positive (FP) rate are plotted against one other at different threshold levels to produce the ROC curve. The quality of probabilistic, deterministic detection, and forecasting systems can be effectively represented using the ROC curve. The quality of probabilistic landslide models for realistically predicting when landslides will occur and when they will not is shown by the AUC. AUC values below 0.5 suggest a random fit and are less accurate for landslide susceptibility models, whereas AUC values between 0.5 and 1 indicate a properly fitted model (Yilmaz, 2009b).

$$\text{Sensitivity} = \text{TP}/(\text{TP} + \text{FN}) \qquad (17.27)$$
$$\text{Specificity} = \text{TN}/(\text{TN} + \text{FP}) \qquad (17.28)$$
$$\text{Specificity} = 1 - \text{Sensitivity} \qquad (17.29)$$

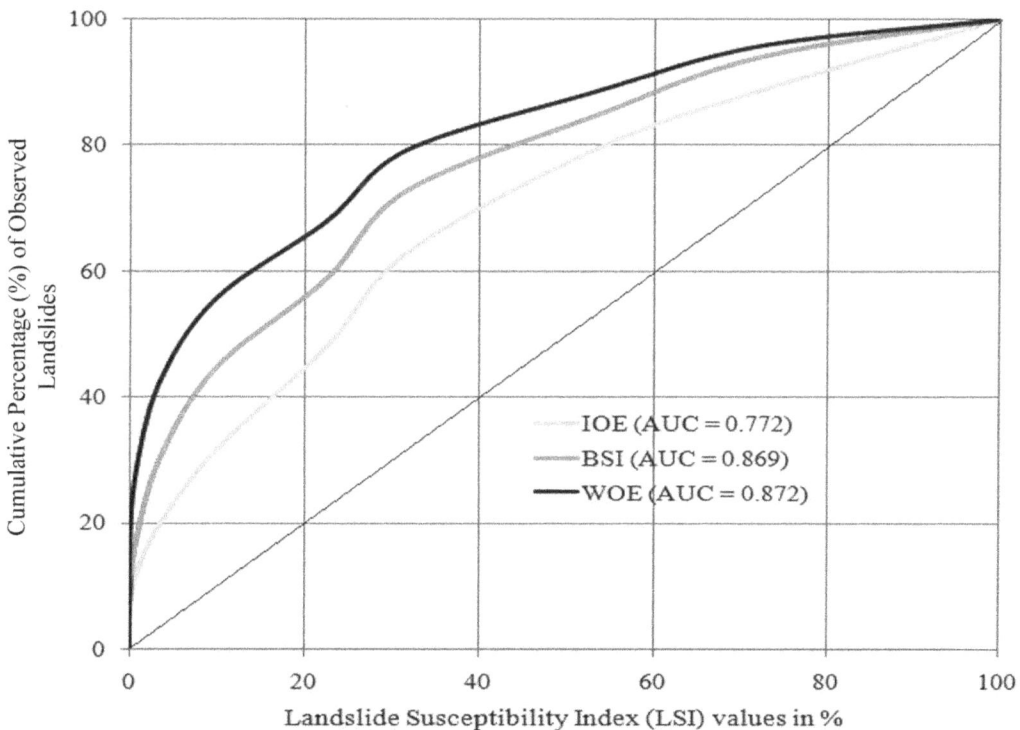

FIGURE 17.13   Success rate curve for the three models (Bivariate Statistical Index, Index of Entropy, and weight of evidence) in the Rorochu watershed.

In this study, the AUC values for the BSI model are 0.858, and we could say that the prediction accuracy for this landslide susceptibility model in the Rorachu watershed is 85.80%. The IOE model presented 68.70% prediction accuracy, and the WOE model revealed 85.80% prediction accuracy of the landslide susceptibility mapping (Figure 17.14) with the standard error of BSI (0.014), IOE (0.019), and WOE (0.014) models (Table 17.10). The BSI model and the IOE model demonstrated the highest levels of accuracy and may be accepted as significant statistical models for landslide susceptibility mapping.

TABLE 17.10

Overall Statistics in the Various Models (Bivariate Statistical Index, Index of Entropy, and Weight of Evidence)

Statistical Model	Area Under Curve	Standard Error	Asymptotic Sig	Asymptotic 95% Confidence Interval	
				Lower Bound	Upper Bound
BSI_M	0.858	0.014	0.000	0.830	0.886
IOE_M	0.687	0.019	0.000	0.649	0.725
WOE_M	0.858	0.014	0.000	0.830	0.886

Note: BSI, Bivariate Statistical Index; IOE, Index of Entropy; WOE, weight of evidence.

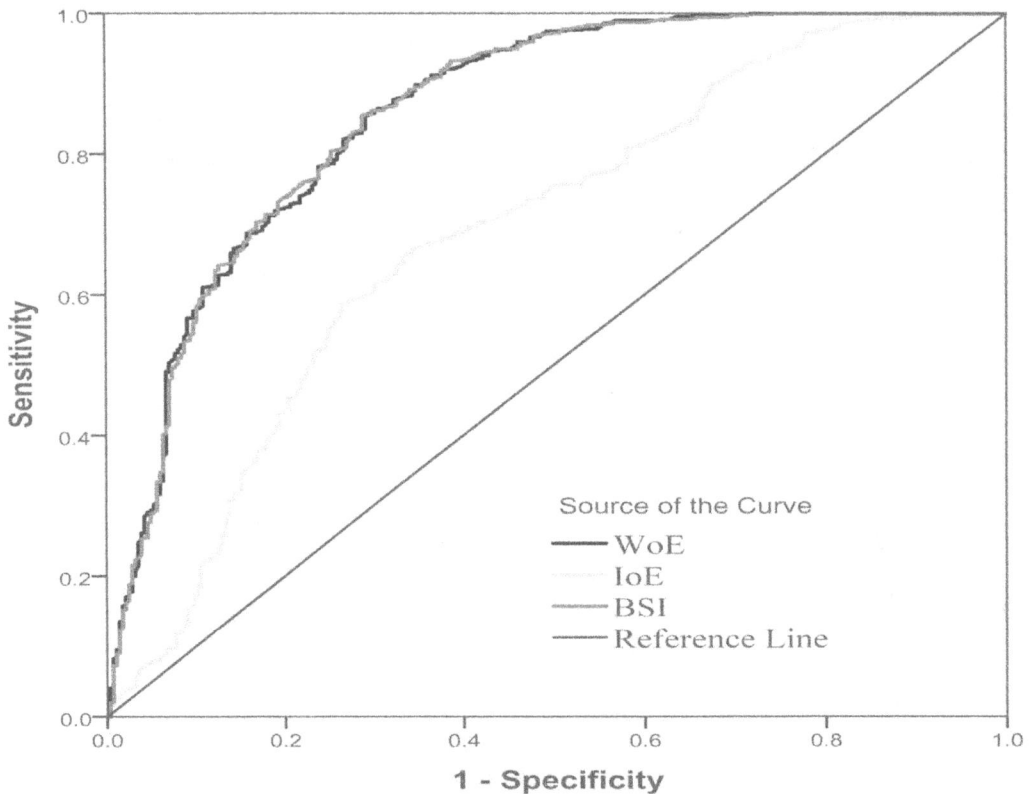

FIGURE 17.14 Receive operating characteristics curve for Bivariate Statistical Index, Index of Entropy, and weight of evidence models.

17.4.5   Analysis of Landslide Risk

Geospatial performances were used to create the landslide risk map for the Rorachu watershed in the Sikkim Himalaya (Figure 17.14). First, a map of the potential for landslides was created using the BSI, IOE, and WOE models. The vulnerability of settlements and highways was used to create the landslide vulnerability map (vulnerability = 1 and non-vulnerability = 0). The Rorachu watershed's final landslide risk map was created by combining the landslide susceptibility and landslide vulnerability of each element-at-risk (Equation 17.26). Five categories—very low, low, moderate, high, and very high—were used to categorize the landslide risk map (Figure 17.15). According to the results of the BSI, IOE, and WOE statistical models, respectively, the generated landslide risk map showed that 20.01% (0.67 km^2), 20.39% (0.68 km^2), and 19.29% (0.64 km^2) of the settlement areas were occupied by very high (VH) risk. According to the BSI, IOE, and WOE models, respectively, 19.01% (0.64 km^2), 19.07% (0.65 km^2), and 17.77% (0.60 km^2) of the road areas are at very high (VH) risk. It has been noted that numerous towns and roads have been constructed in extremely dangerous regions, mostly in the eastern portion of the Rorachu watershed.

17.4.6   Triggering Factors

Several geomorphic, climatic, litho-tectonic, and anthropogenic causes can cause landslides. Despite this, the Rorachu watershed is not exempt from the causes that play a major part in landslides anywhere in the world. The landslide susceptibility mapping is being used by several

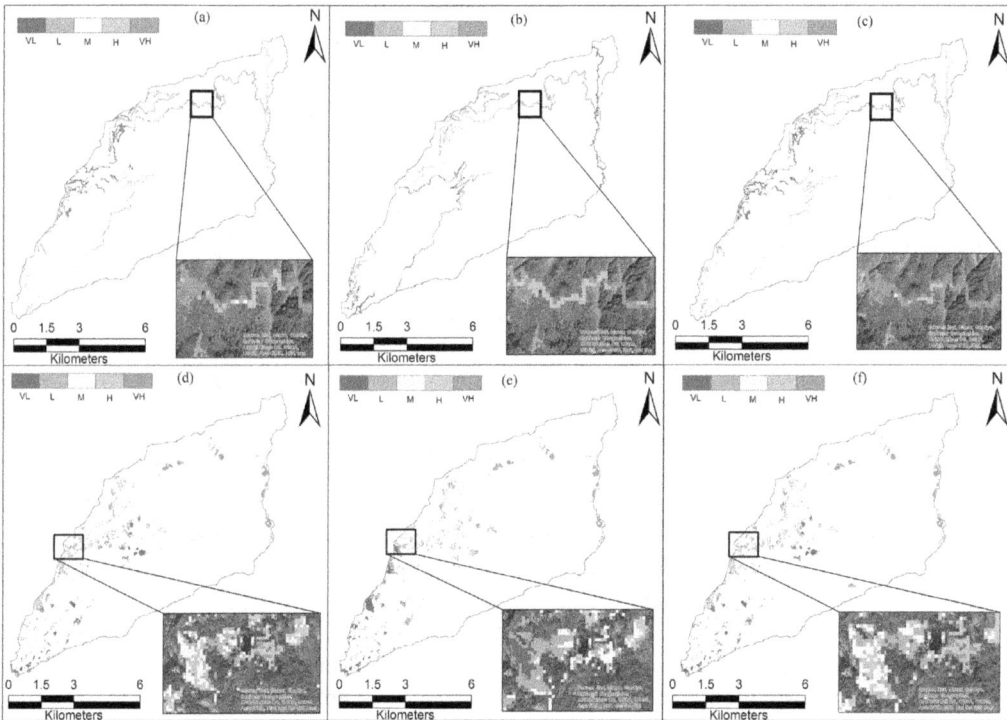

FIGURE 17.15   The landslide risk map of two variables (settlement and road) by the various models: (a) road risk map by weight of evidence model; (b) road risk map by Index of Entropy model; (c) road risk map by Bivariate Statistical Index model; (d) settlement risk map by weight of evidence model; (e) settlement risk map by Index of Entropy model; and (f) settlement risk map by Bivariate Statistical Index model.

TABLE 17.11
Areal Distribution of Bivariate Statistical Index, Index of Entropy, and Weight of Evidence Approaches Based on Landslide Risk Assessment Mapping of Roads

Landslide Risk Zones	BSI		IOE		WOE	
	Area (km²)	Area (%)	Area (km²)	Area (%)	Area (km²)	Area (%)
Very low	0.674	19.86	0.665	19.60	0.671	19.78
Low	0.648	19.09	0.788	23.20	0.639	18.83
Moderate	0.713	21.00	0.644	18.99	0.734	21.61
High	0.714	21.03	0.650	19.15	0.747	22.01
Very high	0.645	19.01	0.647	19.07	0.603	17.77

Note: BSI, Bivariate Statistical Index; IOE, Index of Entropy; WOE, weight of evidence.

TABLE 17.12
Areal Distribution of Bivariate Statistical Index, Index of Entropy, and Weight of Evidence Approaches Based on Landslide Risk Assessment Mapping of Settlement

Landslide Risk Zones	BSI		IOE		WOE	
	Area (km²)	Area (%)	Area (km²)	Area (%)	Area (km²)	Area (%)
Very low	0.6543	19.45	0.6543	19.45	0.6714	19.96
Low	0.6849	20.36	0.6723	19.99	0.6561	19.51
Moderate	0.6669	19.83	0.6804	20.23	0.7065	21.01
High	0.6831	20.31	0.6705	19.94	0.6804	20.23
Very high	0.6741	20.04	0.6858	20.39	0.6489	19.29

Note: BSI, Bivariate Statistical Index; IOE, Index of Entropy; WOE, weight of evidence.

researchers in various parts of the world. According to many researches (Pradhan and Kim, 2014; Lee and Min, 2001; Melchiorre et al. 2008), there are certain parameters that are more useful for landslide susceptibility mapping or slope instability mapping. In slope instability analysis, slope degree is still a crucial characteristic, and it is commonly used to create landslide susceptibility maps (Lee and Min, 2001; Saha et al. 2005; Gorsevski et al. 2012). Because it is controlled by several geologic and geomorphological processes, height continues to be a frequently utilised conditioning factor for landslide susceptibility study (Gorsevski et al. 2012; Pourghasemi et al. 2012b; Pradhan and Kim, 2014). Rainfall and earthquakes are other important variables for determining landslide susceptibility. We used 13 landslide causative elements in this study to map landslide susceptibility and risk.

Expert judgement and experimental data were combined to examine the Rorachu watershed's geology, soil, rainfall, slope, elevation, and aspect as trigger elements for the mapping of the LSI and landslide risk. In this study area, high heights, rainfall, and slope all contribute to the overall chance of landslide vulnerabilities and encumbrance. The altitude (2,500–4,110 m), slope (350–700 m), and rainfall (2,300–3,000 mm) have disproportionately large effects on the landslides in the Rorachu watershed. Landslides in this watershed are notably affected by monsoonal rains (Figure 17.16). We attempt to correlate the Rorachu watershed's landslide susceptibility zones with the research area's triggering elements. We have tried to determine where there is a high- and low-density concentration. According to the association between altitude and the landslide susceptibility model (BSI, IOE, and WOE), the maximum concentration of landslide likelihood is found at an altitude between 2,500 and 4,000 m (Figure 17.16). As a result, the slope between 250 and 450 has the highest concentration of landslide possibility according to the relationships between the slope and the

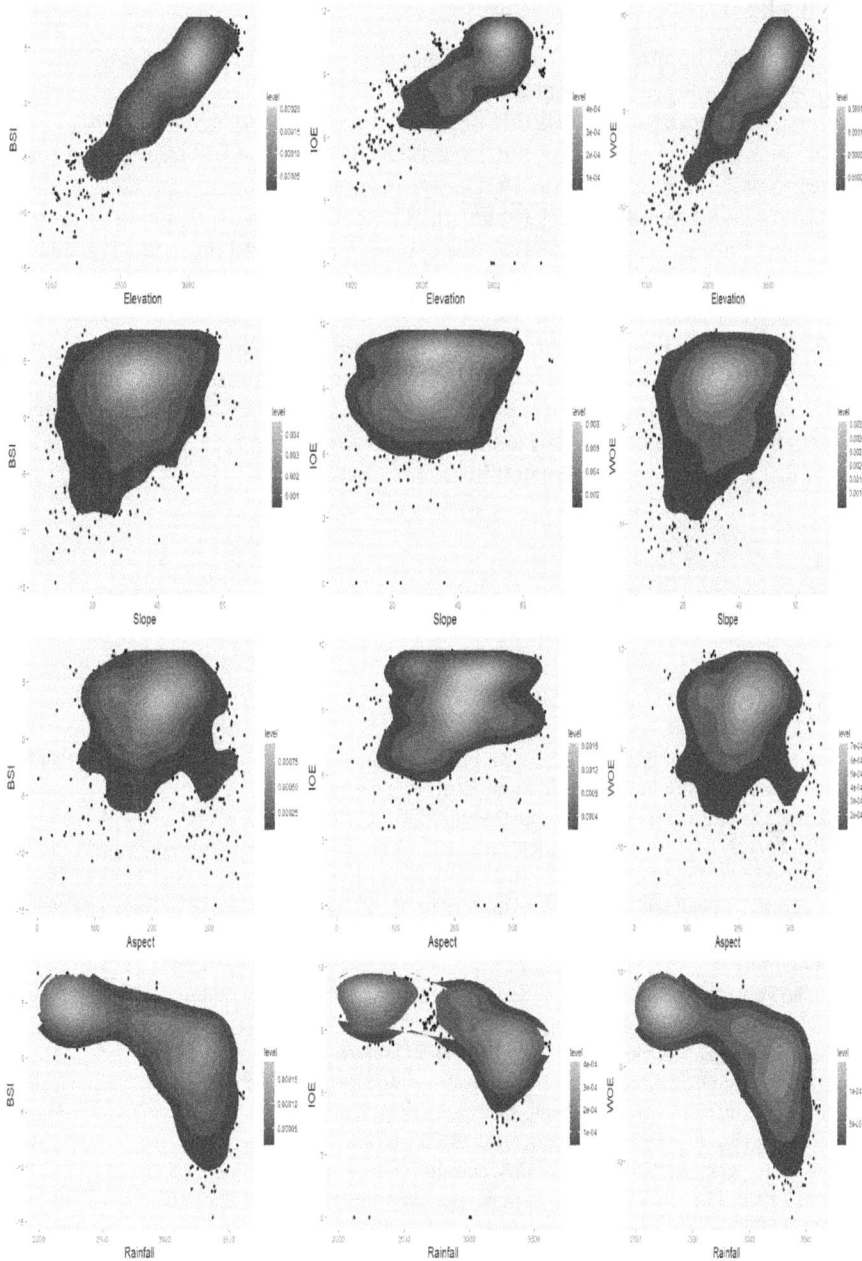

FIGURE 17.16 The correlation and density concentration between the Bivariate Statistical Index, Index of Entropy, and weight of evidence models with the elevations, slope, aspect, and rainfall shown.

landslide susceptibility models (BSI, IOE, and WOE). The maximum concentration of landslide likelihood is seen in the association between rainfall and the bivariate statistical models (BSI, IOE, and WOE) between 2,000 and 2,500 mm and 3,200 and 3,500 mm (Figure 17.16). The BSI, IOE, and WOE statistical models have shown that the south-west, south-east, and south-facing slope aspects significantly increase the vulnerability of landslides. Additionally, the relationship between the landslide-causing variables and the various bivariate statistical models (BSI, IOE, and WOE) was investigated.

17.5   CONCLUSION

Landslides are one of the most frequent natural hazards in the Himalayan Mountain range. Due to their risky nature, landslides present a significant threat to the loss of life and property in the Rorachu watershed. In order to address this problem in the Rorachu River catchment area in East Sikkim, the current study used the BSI, IOE, and WOE models. The LSI map's precision has been examined during validation utilizing the ROC, LD, and SRC methods for the BSI, IOE, and WOE models. The study found that the bivariate statistical model had a decent amount of accuracy when mapping the Rorachu watershed's landslide susceptibility and landslide risk. The triggering variables have a big impact on the landslide activity in this area. The landslide risk zonation map also reveals that NH 31A is the region most vulnerable to landslides. According to the BSI, IOE, and WOE models, a total landslide susceptibility of 27%, 27.80%, and 40.25%, respectively, was shown in the study area. The results of the study indicate that in order to minimize the risk posed by landslides in the Rorachu catchment area, it is possible to use landslide risk maps and LSI. In the future, slope management, land-use plans, city planning, and disaster risk management may be carried out with a map of Rorachu watershed's susceptibility and soil slip risks.

NOTE

The authors declare that they have no known competing financial interests or personal relationships that could have appeared to influence the work reported in this chapter.

REFERENCES

Akgun, A., & Needet, T., 2010. Landslide susceptibility mapping for Ayvalik (Western Turkey) and its vicinity by multi criteria decision analysis. Environ. Earth Sci, 61:595–611.

Akgun, A., Serhat, D., & Fikri, B., 2008. Landslide susceptibility mapping for a landslide-prone area (Findikli, NE of Turkey) by likelihood-frequency ratio and weighted linear combination models. Environ Geol, 54:1127–1143.

Akgun, A., Sezer, E. A., Nefeslioglu, H. A., Gokceoglu, C., & Pradhan, B., 2012. An easy-to-use MATLAB program (mamland) for the assessment of landslide susceptibility using a mamdani fuzzy algorithm. Comput Geosci, 38(1):23–34.

Aleotti, P., & Chowdhury, R., 1999. Landslide hazard assessment: Summary review and new perspectives. Bull Eng Geol Environ, 58:21–44.

Arabameri, A., Pradhan, B., Rezaei, K., Sohrabi, M., & Kalantari, Z., 2019. GIS-based landslide susceptibility mapping using numerical risk factor bivariate model and its ensemble with linear multivariate regression and boosted regression tree algorithms. J Mt Sci, 16(3):595–618.

Ayalew, L., & Yamagishi, H., 2005. The application of GIS-based logistic regression for landslide susceptibility mapping in the Kakuda Yahiko mountains, central Japan. Geomorphology, 65:15–31.

Ayalew, L., Yamagishi, H., Maruib, H., & Takami, K., 2005. Landslide in Sado island of Japan: Part II: GIS-based susceptibility mapping with comparisons of results from two methods and verifications. Eng Geol, 81(4):432–445.

Bednarik, M., Magulova B., Matys, M., & Marschalko, M., 2010. Landslide susceptibility assessment of the Kralovany: Liptovsky Mikulas railway case study. Phys Chem Earth, 35:162–171.

Bednarik, M., Yilmaz, I., & Marschalko, M., 2012. Landslide hazard and risk assessment: A case study from the Hlohovec-Sered landslide area in southwest Slovakia. Nat Hazards. http://dx.doi.org/10.1007/s11069-012-0257-7.

Beven, K. J., & Kirkby, M. J., 1979. A physically based, variable contributing area model of basin hydrology. Hydro Sci Bull, 24:43–69.

Bonham-Carter, G. F., Agterberg, F. P., & Wright, D. F., 1989. Integration of geological datasets for gold exploration in Nova Scotia. In: Nicholas Van Driel, J., & Davis, J. C. (eds) Digital Geologic and Geographic Geological Survey of Canada, 601 Booth Street, Ottawa, Ontario KIAOE8, Information Systems, vol. 10, pp. 15–23.

Cevik, E., & Topal, T., 2003. GIS-based landslide susceptibility mapping for a problematic segment of the natural gas pipeline, Hendek (Turkey). Environ Geol, 44:949–962.

Chen, W., Shahabi, H., Shirzadi, A., Hong, H., Akgun, A., Tian, Y., Liu, J., Zhu, A., & Li, S., 2018. Novel hybrid artificial intelligence approach of bivariate statistical-methods-based kernel logistic regression classifier for landslide susceptibility modeling. Bull Eng Geol Environ, 78(2019):4397–4419. https://doi.org/10.1007/s10064-018-1401-8.

Chung, C. J. F., & Fabbri, A. G., 1993. The representation of geoscience information for data integration. Nonrenewable Resour, 2:122–139.

Chung, C. J. F., Fabbri, A. G., & van Westen, C. J., 1995. Multivariate regression analysis for landslide hazard zonation. In: Carrera, A., & Guzzetti, F. (eds) Geographical Information Systems in Assessing Natural Hazards. KAP, Dordrecht, pp. 107–133.

Constantin, M., Bednarik, M., Jurchescu, M. C., & Vlaicu, M., 2011. Landslide susceptibility assessment using the bivariate statistical analysis and the index of entropy in the Sibiciu basin (Romania). Environ Earth Sci, 63(2):397–406.

Dahal, R. K., Hasegawa, S., Nonomura, A., Yamanaka, M., Dhakal, S., & Paudyal, P., 2008a. Predictive modeling of rainfall-induced landslide hazard in the lesser Himalaya of Nepal based on weights-of-evidence. Geomorphology, 102(3):496–510.

Dahal, R. K., Hasegawa, S., Nonomura, A., et al., 2008b. GIS-based weights-of-evidence modelling of rainfall-induced landslides in small catchments for landslide susceptibility mapping. Environ Geol, 54(2):311–324. doi:10.1007/s00254-007-0818-3.

Dai, F., Lee, C., Li, J., & Xu, Z., 2001. Assessment of landslides susceptibility on the natural terrain of Lantau Island, Hong Kong. Environ Geol, 40:381–391.

Devkota, K. C., Regmi A. D., Pourghasemi, H. R., Yishida, K., Pradhan, B., Ryu, I. C., Dhital, M. R., & althuwaynee, O. F., 2012. Landslide susceptibility mapping using certainty factor. Index of entropy and logistic regression models in GIS and their comparision at Mugling: Narayanghat road section in Nepal Himalaya. Nat Hazards, 65:135–165, DOI: 10.1007/s11069-012-0347-6.

Devkota, K. C., Regmi, A. D., Pourghasemi, H. R., Yoshida, K., Pradhan, B., Ryu, I. C., Dhital, M. R., & Althuwaynee, O. F., 2013. Landslide susceptibility mapping using certainty factor, index of entropy and logistic regression models in GIS and their comparison at Mugling: Narayanghat road section in Nepal Himalaya. Nat Hazards, 65:135–165.

Ercanoglu, M., & Gokceoglu, C., 2002. Assessment of landslide susceptibility for a landslide prone area (north of Yenice, NW Turkey) by fuzzy approach. Environ Geol, 41:720–730.

Fell, R., 1994. Landslide risk assessment and acceptable risk. Can Geotech J, 31:261–272.

Gokceolu, C., Sonmez, H., Nefeslioglu, H. A., Duman, T. Y., & Can, T., 2005. The 17 March 2005 Kuzulu landslide (Sivas, Turkey) and landslide-susceptibility map of its near vicinity. Eng Geol, 81:65–83.

Gorsevski, P. V., Donevska, K. R., Mitrovski, C. D., & Frizado, J. P., 2012. Integrating multi criteria evaluation techniques with geographic information systems for landfill site selection: A case study using ordered weighted average. J Waste Manag, 32(2):287–296.

Guisan, A., Weiss, S. B., & Weiss, A. D., 1999. GLM versus CCa spatial modeling of plant species distribution. Plant Ecol, 143:107–122.

Gupta, R. P., Kanunga, D. P., Arora, M. K., & Sarkar S., 2008. Approaches for comparative evolution of raster GIS-based landslide susceptibility zonation maps. Int J Appl Earth Obs Geoinf, 10:330–341.

Guzzetti, F., Carrara, A., Cardinali, M., & Reichenbach, P., 1999. Landslide hazard evaluation: A review of current techniques and their application in a multi-scale study, Central Italy. Geomorphology, 31:181–216.

Guzzetti, F., Reichenbach, P., Ardizzone, F., Cardinali, M., & Galli, M., 2009. Landslide hazard assessment, vulnerability estimation and risk evaluation: An example from the Collazzone area (central Umbria, Italy). Geogr Fis Dinam Qual, 32:183–192.

Hengl, T., Gruber, S., & Shrestha, D. P., 2003. Digital terrain analysis in ILWIS. Published by International Institute for Geo-Information Science and Earth Observation, Enschede. Netherlands, p. 62. http://www.itc.nl/library/papers_2003.

Horton, R. E., 1932. Drainage basin characteristics. Trans Am Geophys Union, 13:350–361.

Horton, R. E., 1945. Erosional development of streams and their drainage basins: A hydro physical approach to quantitative morphology. Geol Soc Am Bullet, 56:275–370.

Jaafari, A., Najafi, A., Pourghasemi, H. R., Rezaeian, J., & Sattarian, A., 2014. GIS-based frequency ratio and index of entropy models for landslide susceptibility assessment in the Caspian Forest, northern Iran. Int J Environ Sci Tech, 11(4):909–926.

Lee, E. M., & Zones, D. K. C., 2004. Landslide Risk Assessment. Thomas Telford, London.

Lee, S., & Choi, J., 2004. Landslide susceptibility mapping using GIS and the weight-of evidence model. Int J Geogr Inf Sci, 18(8):789–814. doi:10.1080/13658810410001702003.

Lee, S., Choi, J., Chwae, U., & Chang, B., 2002a. Landslide susceptibility assessment using weight of evidence. Geosci Remote Sens Symp, IEEE, 5:2865–2867.

Lee, S., & Min, K., 2001. Statistical analysis of landslide susceptibility at Yongin, Korea. Environ Geol, 40:1095–1113.

Leroi, E., 1996. Landslide hazard: Risk maps at different scales: Objectives, tools, and development. In: Senneset, K. (ed) Landslides: Glissements de Terrain, 7th International Symposium on Landslides. Balkema, Trondheim, Norway, pp. 35–51.

Li, X. J., Chen, Y. N., & Ouyang, H., 2002. Analysis on sand disaster with disaster entropy method. Arid Land Geogr, 25(4):350–353.

Lusted, L. B., 1968. Introduction to Medical Decision Making. Charles C. Thomas, Springfield III, p. 271.

Ma, S., Qiu, H., Hu, S., Pei, Y., Yang, W., Yang, D., & Cao, M., 2020. Quantitative assessment of landslide susceptibility on the Loess Plateau in China. Phys Geogr, 41(6):489–16.

Mandal, S., & Mandal, K., 2017. Bivariate statistical index for landslide susceptibility mapping in the Rorachu river basin of eastern Sikkim Himalaya, India. Spat Inf Res, 26(1):59–75.

Marcini, F., Ceppi, C., & Ritrovato, G., 2010. GIS and statistical analysis for landslide susceptibility mapping in the Daunia area, Italy. Nat Haz Earth Sys Sci, 10:1851–1864. https://doi.org/10.5194/nhess-10-1851-2010.

Mathew, J., Jha, V. K., & Rawat, G., 2007. Weights of evidence modeling for landslide hazard zonation mapping in part of Bhagirathi valley, Uttarakhand. Curr Sci, 92(5):628.

Melchiorre, C., Matteucci, M., Azzoni, A., & Zanchi, A., 2008. Artificial neural networks and cluster analysis in landslide susceptibility zonation. Geomorphology, 94:379–400.

Moore, I. D., Grayson, R. B., & Ladson, A. R., 1991. Digital terrain modelling—A review of hydro hydrological, geomorphological, and biological application. Hydrol Process, 5:3–30.

Mon, D. L., Cheng, C. H., & Lin, J. C., 1994. Evaluating weapon system using fuzzy analytic hierarchy process based on entropy weight. Fuzzy Sets and Systems, 62(2):127–134.

Naithani, A. K., 1999. The Himalayan landslides. Employment News, 23(47):20–26.

O'Brien, R. M., 2007. A caution regarding rules of thumb for variance inflation factors. Qual Quant, 41(5):673–690.

Oh, H. J., & Lee, S., 2011. Landslide susceptibility mapping on Panaon island, Philippines using a geographic information system. Environ Earth Sci, 62(5):935–951. doi:10.1007/s12665-010-0579-2.

Ozdemir, A., & Altural, T., 2013. A comparative study of frequency ratio, weights of evidence and logistic regression methods for landslide susceptibility mapping: Sultan Mountains, SW Turkey. J Asian Earth Sci, 64:180–197.

Poudyal, C. P., Chang, C., Oh, H. J., & Lee, S., 2010. Landslide susceptibility maps comparing frequency ratio and artificial neural networks: A case study from the Nepal Himalaya. Environ Earth Sci, 61:1049–1064.

Pourghasemi, H. R., 2008. Landslide hazard assessment using fuzzy logic (case study: A part of Haraz Watershed). M.Sc Thesis, Tarbiat Modarres University International Campus, Iran, p. 92.

Pourghasemi, H. R., Pradhan, B., Gokceoglu, C., & Deylami Moezzi, K., 2012. Landslide susceptibility mapping using a spatial multi criteria evaluation model at Haraz Watershed, Iran. In: Biswajeet P., Manfred B. (eds) Terrigenous Mass Movements: Detection, Modelling, Early Warning and Mitigation Using Geoinformation Technology, Springer, Berlin: Heidelberg, pp. 23–49.

Pourghasemi, H. R., Pradhan, B., Gokceoglu, C., Mohammadi, M., & Moradi, H. R., 2013b. Application of weights-of-evidence and certainty factor models and their comparison in landslide susceptibility mapping at Haraz watershed. Iran Arab J Geosci, 6:2351–2365.

Pradhan, A. M. S., & Kim, Y. T., 2014. Relative effect method of landslide susceptibility zonation in weathered granite soil: A case study in Deokjeok-ri Creek, South Korea. Nat Hazards, 72(2):1189–1217.

Pradhan, B., Chaudhari, A., Adinarayana, J., & Buchroithner, M. F., 2012. Soil erosion assessment and its correlation with landslide events using remote sensing data and GIS: A case study at Penang Island, Malaysia. Environ Monit Asses, 184(2):715–727.

Pradhan, B., & Lee, S., 2010a. Delineation of landslide hazard areas on Penang Island, Malaysia, by using frequency ratio, logistic regression, and artificial neural network models. Environ Earth Sci, 60(5):1037–1054.

Pradhan, B., Mansor, S., Pirasteh, S., & Buchroithner, M. F., 2011. Landslide hazard and risk analyses at a landslide prone catchment area using statistical based geospatial model. Int J Remote Sens, 32(14):4075–4087.

Pradhan, B., & Youssef, A. M., 2010. Manifestation of remote sensing data and GIS on landslide hazard analysis using spatial-based statistical models. Arab J Geosci, 3:319–326.

Rautela, P., & Lakhera, R. C., 2000. Landslide risk analysis between Giri and Tons rivers in Himachal Himalaya (India). Int J Appl Earth Obs Geoinf, 2:153–160.

Rawat, M. S., & Joshi, V., 2016. Landslide hazard zonation in Rorachu sub watershed of east district of Sikkim. In India, Conference Paper. https://www.researchgate.net/publication/322887068

Regmi, A. D., Devkota, K. C., Yoshida, K., Pradhan, B., Pourghasemi, H. R., Kumamoto, T., & Akgun, A., 2014. Application of frequency ratio, statistical index, and weights-of evidence models and their comparison in landslide susceptibility mapping in Central Nepal Himalaya. Arab J Geosci, 7:725–742.

Regmi, N. R., Giardino, J. R., & Vitek, J. D., 2010. Modeling susceptibility to landslides using the weight of evidence approach: Western Colorado, USA. Geomorphology, 115(1):172–187.

Regmi, N. R., Giardino, J. R., & Vitek, J. D., 2010a. Assessing susceptibility to landslide: Using models to understand observed changes in slopes. Geomorphology, 122:25–38.

Ren, L. C., 2000. Disaster entropy: Conception and application. J Nat Disaster, 9(2):26–31.

Saha, A. K., et al., 2005. An approach for GIS based statistical landslide susceptibility zonation with a case study in the Himalayas. Landslides, 2:61–69.

Sarkar, S., & Kanungo, D. P., 2004. An integrated approach for landslides susceptibility mapping using remote sensing and GIS. PE & RS, 70(5):617–625.

Shannon, C. E., 1948. A mathematical theory of communication. Bull Syst Technol J, 27(3):379–423.

Singh, L. P., van Westen, C. J., Champati Ray, P. K., et al., 2005. Accuracy assessment of inSAR derived input maps for landslide susceptibility analysis: A case study from the Swiss alps. Landslides, 2(3):221–228. doi:10.1007/s10346-005-0059-z.

Soeters, R., & van Westen, C. J., 1996. Slope instability recognition analysis and zonation. In: Turner, K. T., & Schuster, R. L. (eds) Landslides: Investigation and Mitigation. Special Report No. 247. Transportation Research Board National Research Council, Washington, DC, pp. 129–177.

Strahler, A. N., 1952. Hypsometric (area-altitude) analysis of erosional topography. Geol Soc Am Bullet, 63:117–142.

Van Westen, C. J., 2002. Use of Weights of Evidence Modeling for Landslide Susceptibility Mapping. International Institute for Geoinformation Science and Earth Observation, Enschede, The Netherlands, p. 21.

Van Westen, C. J., Rengers, N., & Soeters, R., 2003. Use of geomorphological information in indirect landslides susceptibility assessment. Nat Hazards, 30:399–419.

Van Westen, C. J., Rengers, N., Terlien, M. T. J., & Soeters, R., 1997. Prediction of the occurrence of slope instability phenomenal through GIS-based hazard zonation. Geol Rundsch, 86(2):404–414.

Varnes, D. J., 1984. Landslide Hazard Zonation: A Review of Principles and Practice. United Nations International, Paris.

Vlcko, J., Wagner, P., & Rychlikova, Z., 1980. Evaluation of regional slope stability. Miner Slov, 12(3):275–283.

Wan, S., 2009. A spatial decision support system for extracting the core factors and thresholds for landslide susceptibility map. Eng Geol, 108(3–4):237–251.

Wan, S., & Lei, T. C., 2009. A knowledge-based decision support system to analyze the debris-flow problems at Chen-Yu-Lan river, Taiwan. Knowl Based Syst, 2(8):580–588.

Wu, Z., Wu, Y., Yang, Y., Chen, F., Zhang, N., Ke, Y., et al., 2017. A comparative study on the landslide susceptibility mapping using logistic regression and statistical index models. Arab J Geosci, 10:187. https://doi.org/10.1007/s12517-017-2961-9.

Xu, C., Dai, F., Xu, X. and Lee, Y. H., 2012. GIS-based support vector machine modeling of earthquake-triggered landslide susceptibility in the Jianjiang river watershed, China. Geomorphology, 145:70–80. doi:10.1016/j.geomorphology.2011.12.040.

Yalcin, A., 2008. GIS-based landslide susceptibility mapping using analytical hierarchy process and bivariate statistics in Ardesen (Turkey): Comparisons of results and confirmations. Catena, 72:1–12.

Yang, Z., & Qiao, J., 2009. Entropy-based hazard degree assessment for typical landslides in the three gorges area, China. In: Wang, F., & Li T. (eds) Landslide Disaster Mitigation in Three Gorges Reservoir, China. Environ Sci Engin. Springer, Berlin, pp. 519–529.

Yi, C. X., & Shi, P. J., 1994. Entropy production and natural hazard. J Bej Normal Univ (Natural Science Edition), 30(2):276–280.

Yilmaz, I., 2009a. A case study from Koyulhisar (Sivas-Turkey) for landslide susceptibility mapping by artificial neural networks. B Eng Geol Environ, 68:297–306.

Yilmaz, I., 2009b. Landslide susceptibility mapping using frequency ratio, logistic regression, artificial neural networks and their comparison: A case study from Kat landslides (Tokat—Turkey). Comput Geosci, 35(6):1125–1138.

Youssef, A. M., Al-Kathery, M., & Pradhan, B., 2014a. Landslide susceptibility mapping at al hasher area, Jizan (Saudi Arabia) using GIS-based frequency ratio and index of entropy models. Geosci J, 19(1):113–134.

Youssef, A. M., Pradhan, B., Jebur, M. N., & El-Harbi, H. M., 2015. Landslide susceptibility mapping using ensemble bivariate and multivariate statistical models in Fayfa area, Saudi Arabia. Environ Earth Sci, 73:3745–3761.

Yufeng, S., & Fengxiang, J., 2009. Landslide stability analysis based on generalized information entropy. In: 2009 International Conference on Environmental Science and Information Application Technology, pp. 83–85.

Zhou, C., Yin, K., Cao, Y., Ahmed, B., Li, Y., Catani, F., & Pourghasemi, H. R., 2018. Landslide susceptibility modeling applying machine learning methods: A case study from longju in the three gorges reservoir area, China. Computers & Geosciences, 112:23–37.

18 Using Geospatial Techniques for Detection of Land-Use/ Land-Cover Changes Due to Major Hydropower Projects in Upper Beas River Basin, District Kullu, Himachal Pradesh, India

Nishant Vaidya, Kesar Chand, Jagdish Chandra Kuniyal, Suraj Kumar Singh, and Shruti Kanga

18.1 INTRODUCTION

The land-use/land-cover (LULC) information of an area is essential for proper planning, management and monitoring of natural resources. It is an important input for many geological, hydrological, ecological and agricultural models.

Remote sensing images help in gathering quality LULC information at local, regional and global scales because of its synoptic view, map-like format and repetitive coverage (Csaplovics, 1998; Foody, 2002). Further, in mountainous regions like the Himalayas, particularly in the inaccessible areas due to high altitudes and ruggedness in the terrain, remote sensing images are quite useful for mapping. Due to changes in topographical and environmental conditions, spectral characteristics also change from region to region (Arora and Mathur, 2001). Therefore, the approach for LULC classification that incorporates ancillary data from other sources may be more effective than that is based solely upon multi-spectral data from one sensor. The topographic maps are useful in generating the digital elevation model (DEM), which along with its attributes, such as slope and aspect, provide the basis for multi-source classification. Furthermore, the derivatives of multispectral images like principal components analysis (PCA) and Normalized Difference Vegetation Index (NDVI) may also be useful to improve the LULC classification from remote sensing data in mountainous regions (Eiumnoh and Shrestha, 2000; Saha et al., 2005). In mountainous terrain, such as Himalayas, shadow is the major problem in achieving the accurate LULC classification from remote sensing data. The use of NDVI image as an additional layer for classification has been recommended to overcome this problem, since the band ratio derivatives may help in nullifying the topographic effect to some extent (Holben and Justice, 1981; Apan, 1997). However, NDVI alone may not be able to eliminate the shadow effect completely. Later, Eiumnoh and Shrestha (2000) and Saha et al. (2005) incorporated both NDVI and DEM images as additional layers in the classification process and found a significant improvement in the classification accuracy.

DOI: 10.1201/9781003377825-21

18.2   STUDY AREA

The area is an example of a natural region as it is bounded by north-west–south-east running Pir-Panjal range in the north separating Beas river valley of Kullu from Chandra-Bhaga (Chenab River) valley of Lahaul and Spiti district. Bara Bhangal off-shoot of Pir-Panjal range marks the boundary in the north-west and west while north-west–south-east running Dhauladhar range lies in the south-west. Great Himalayas run along the eastern boundary, while river Satluj marks the southern boundary of the district. The altitude varies from about 1,000–6,200 m above mean sea level (amsl) that increases from south-west to north-east. Drained by glaciers/snow-fed perennial rives and rivulets, the geomorphologic character of Kullu is influenced by both glacial and fluvial processes (Sah and Mazri, 2007). The area can be broadly categorised into glaciers and permanent snow fields, rocky/barren slopes, alpine pastures, forested valley slopes and ridges and main valley floor characterised by dominance of human activities. The climates ranging from subtropical to alpine that have resulted in a diverse vegetation cover consisting of subtropical, temperate and alpine vegetation. The Beas basin is the major sub-basin within the Indus drainage system. The ancient name of the River Beas as mentioned in Rigveda is Vipasa. The river Beas rises from the Beas Kund Glacier in the Pir Panjal range near Atal Tunnel to the north of Manali (4,000 m). The Beas River is a part of an antecedent drainage pattern which has been flowing even after the rise of the Dhaludhar range during the upper Pleistocene (Srikantia and Bhargava, 1998). River Beas, after traversing about 370 km, enters into the plains of Punjab and meets with river Satluj at Harike. The entire Beas basin lies within the Indian Territory having a large catchment area in the state of Himachal Pradesh and a small portion lying in the plains of Punjab.

The Upper Beas basin lies in between 75°2′ E longitudes and 75°50′ E latitudes and 31°7′ N to 32°25′ N which drains a total catchment area of about 4,973 km2. The entire river Beas after its origin near Hanuman Tibba flows through high mountain ranges in the Himalaya. The river Beas has many tributaries among which Solang stands first among others. This glacier-fed stream drains into the mountains in the west of Rohtang Pass and joins river Beas at Palchan. Further, down from the right bank, Manalsu Khad is on the river Beas near Manali village (Figure 18.1). The Phojal and Sarwari are two other large streams of the river Beas on the right bank. The river Beas is joined by important rainfed tributaries rising from the Hamta Pass. Parbati is a left bank tributary of the river Beas. It originates from Mantalai Lake from snow peaks at an elevation of 6,300 m. It transverses in the northerly direction and flows down as a small stream in a narrow valley before it joins the main river Beas at an elevation of 1,097 m near Bhunter.

Kullu is the fifth largest district of Himachal Pradesh in terms of geographical area. Kullu is the most rapidly growing district of the state. It had the second highest decadal population growth (26.17%) in 1991–2001 while it was ranked fourth (14.65%) in 2001–2011. The economy of this area depends on mainly three activities: horticulture, tourism and hydro-power generation. These are all heavily dependent on land resources utilisation.

18.3   METHODOLOGY

LULC change has become a central and important component in current strategies for managing natural resources and monitoring environmental changes. LULC change is a dynamic process taking place on bio-physical surfaces that have developed over a period of time, and space is of enormous importance in natural resource studies. LULC change dynamics are important elements for monitoring, evaluating, protecting and planning for earth resources. LULC changes are the major issues and challenges for the eco-friendly and sustainable development for the economic growth of any area. With the population explosion, human activities such as deforestation, soil erosion, global warming and pollution are very harmful for the environment. This causes LULC changes with the demand and supply of land in different activities. Change detection in

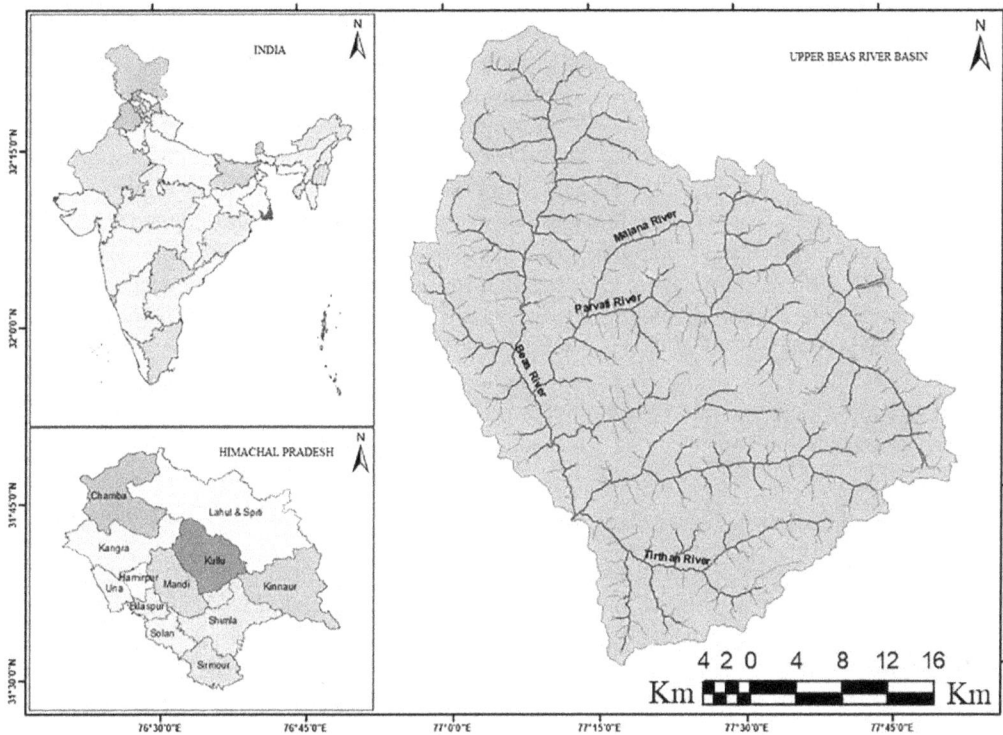

FIGURE 18.1 Study area in the upper Beas River basin.

LULC can be performed on a temporal scale such as a decade to assess landscape change caused due to anthropogenic activities on the land (Gibson and Power, 2000). LULC change is influenced by various natural and human activity processes. In order to improve the economic condition of the area without further deteriorating the bio-environment, every bit of the available land has to be used in the most rational way. This requires the present and the past LULC data of the area (Chaurasia et al., 1996). LULC dynamics are widespread, accelerating and significant processes being driven by human actions but also produce changes that impact humans (Agarwal et al., 2002). Prakasam (2010) studied the LULC change in the river Beas basin in Himachal Pradesh state of India to observe changes during a span of 20 years from 1995 to 2015. With the invention of remote sensing and GIS techniques, LULC mapping is a useful and detailed way to improve the selection of areas designed for agricultural, urban and/or industrial areas of a region (Selcuk et al., 2003). Application of remotely sensed data made it possible to study the changes in land cover in less time, at low cost and with better accuracy (Kachhwala, 1985) in association with GIS that provides a suitable platform for data analysis, update and retrieval (Star et al. 1997; Chilar, 2000). Digital change detection techniques based on multi-temporal and multi-spectral remotely sensed data have demonstrated a great potential as a means to understanding landscape dynamics—detect, identify, map and monitor differences in LULC patterns over time, irrespective of the causal factors. The present study demonstrates the application of multi-temporal satellite imagery in defining LULC dynamics of an area surrounded by two different large hydropower projects in upper Beas river basin located in Kullu district of Himachal Pradesh state in the north-western Himalayan region.

The results quantify the land-cover change patterns in the two large hydropower projects. They demonstrate the potential of multi-temporal Landsat data to provide an accurate, economical means to map and analyse changes in land cover in a spatio-temporal framework that can be used as input for management and policy decisions. With regard to varied themes that link with space such as urbanisation, water management, deforestation, land degradation and so on (Figure 18.2).

The study area, the upper Beas basin, is endowed with enormously rich natural resources that enriched its natural landscape through the ages. But since the last two decades, the upper Beas basin around the Kullu valley, has witnessed remarkable expansion, growth and developmental activities such as increased construction works like large hydropower projects, buildings, houses, hotels and roads along with other anthropogenic activities that resulted in large-scale deforestation, soil erosion and erratic climatic patterns leading to imbalance. This has therefore resulted in increased land consumption as well as modification and alterations in the status of the LULC over time. It is thus necessary to evaluate and detect these changes in land-use pattern with the help of remote sensing data and GIS techniques and make an attempt to predict the possible changes that may occur in the land-fuse status so that planners can have a basic tool for planning.

18.3.1 LAND-USE/LAND-COVER DETECTION AND ANALYSIS

To work out the LULC classification, a supervised classification method with maximum likelihood algorithm was applied in the ERDAS Imagine 2013 software. For better classification results, some indices such as NDVI and Normalized Difference Built-up Index (NDBI) were also applied to classify the satellite images freely available in Google Earth at a resolution of 25 to 0.5 m of 15 October 2004 and 2018. With the help of GPS, ground verification was done for doubtful areas. Based on the

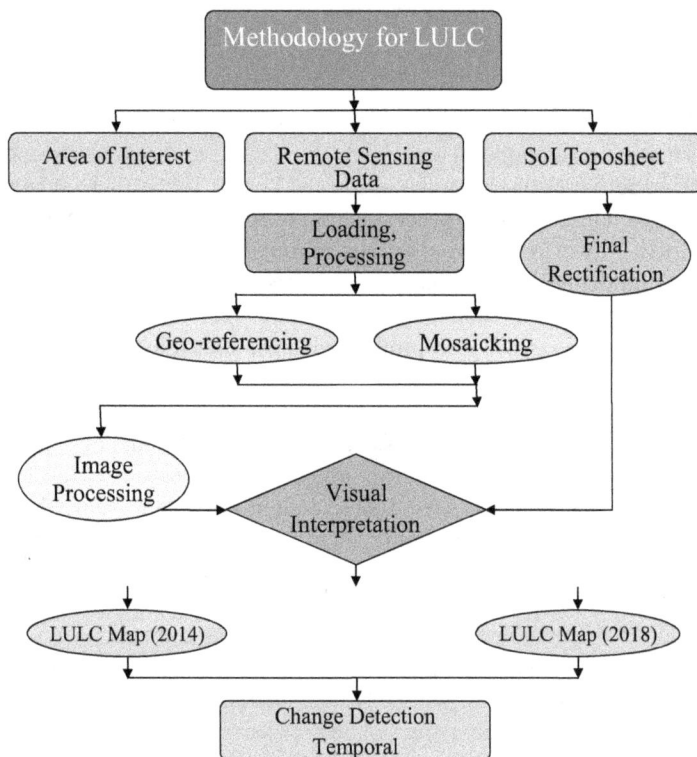

FIGURE 18.2 Methodology for land-use/land-cover change detection.

ground truthing, the misclassified areas were corrected using the recode option in ERDAS Imagine 2013. Seven LULC types are identified and used in this study, namely, (1) built-up land, (2) forest, (3) agricultural land, (4) streams, (5) shrubs, (6) wasteland and (7) plantation.

18.3.2 LAND-USE/LAND-COVER CHANGE DETECTION AND ANALYSIS

For performing LULC change detection, a post-classification detection method was employed. A change matrix (Weng, 2001) was produced with the help of ERDAS Imagine 2013 software. Quantitative areal data of the overall LULC changes as well as gains and losses in each category between 2004 and 2018 were then compiled.

For identification of geographical locations of hydropower projects, we used the latest technology GPS GARMIN. With the help of GPS we were able to identify latitude, longitude and elevation of existing major hydroelectric power (HEP) project sites in the upper Beas basin area **(Photo 18.1)**.

18.4 RESULTS AND DISCUSSION

18.4.1 PARBATI HYDROELECTRIC PROJECTS

The Parbati Hydro Electric Project proposes to harness the Parbati and Sainj Rivers in Kullu valley in three stages. The construction work for Stage II and III is going on. But the work on Stage I has not started because it could not get the environmental clearance from the Ministry of Environment, Forest and Climate Change and CC so far. Stage II is a run of river (ROR) scheme that envisages the utilisation of river Parbati's water and five nallahs (small streams), namely the Jigari, Manihar, Pancha, Hurla and Jiva, to generate 800 MW. The project activities are spread over three valleys, namely the Manikaran, Gharsa and Sainj valleys. The project will have a 91-m concrete gravity dam on the Parbati.

PHOTO 18.1 Identification of latitude, longitude and elevation of HEPs.

PHOTO 18.2   Parbati Stage II hydropower project, Kullu.

The dam will be located just downstream from the confluence of the Tosh Nallah. The project includes a spillway section that is 39 m long with four bays controlled by four radial gates, two 4.5-m intake tunnels, and a 6-m head race tunnel. The Stage III project is proposed on the Sainj. It is again a run-of-the river scheme with a proposed power generation capacity of 520 (130 × 4) MW at Suind.

This project would generate 1,977.23 million units. It envisages utilisation of water released from tail-race of the Parbati Stage II (**Photo 18.2**). After using this water in powerhouse at Suind, it ultimately will be discharged into the Sainj. The project comprises a concrete gravity dam 75 m high, a spillway section 59 m long having four bays controlled by radial gates, two 5.8-m intake tunnels, and a 7.5-m head-race tunnel. An underground powerhouse will be constructed near village Bihali, Sainj.

18.4.2   ALLAIN-DUHANGAN HYDROELECTRIC PROJECT

Rajasthan Spinning and Weaving Mills Limited (RSWML), a private limited company incorporated in India, is setting up Allain-Duhangan Hydroelectric Project (ADHEP) of 2 × 96 MW (192 MW hydropower generation facility on the Allain and the Duhangan tributaries of Beas River in Tehsil Manali, district Kullu, Himachal Pradesh in India). The project is located near village Prini, approximately 3 km south-east of Manali town (**Photo 18.3**). It is a ROR scheme to utilise the combined discharge of the Allain and Duhangan streams.

18.4.3   CHANGE DETECTION IN LAND USE/LAND COVER

In the present study region, we selected two large hydroelectric projects based on altitude and size to show how hydropower affects the LULC of the surrounding area.

PHOTO 18.3   Allain Duhangan hydropower project, Manali, Kullu.

With the help of geospatial technology, we used two time period satellite imageries, i.e. 2004 and 2018 (Figure 18.3). According to the study we able to find out large LULC change in the surroundings of Parbati Stage-II HEP in Sainj, District Kullu (**Photo 18.2**).

This hydroelectric project is located at an elevation of 1,300 m near Sainj River left bank tributary of river Beas. We observed in our study that built-up area increased 0.1% from 2004 to 2018; agricultural area decreased to 2.31% due to constructional activities of HEPs and forest area increased by 5% (Figure 18.4) because of the Compensatory Afforestation Program by the state government of Himachal Pradesh.

Allian Duhangan hydro project is located at an elevation of 2,800 m amsl along the small stream Allain at Prini, Manali (**Photo 18.3**).

As we observed in satellite images, it was also found that the difference in agriculture area was decreased by 2% (Figure 18.5), as we know due to the construction of residential colonies, powerhouse site and other construction activities of hydroelectric project in their surroundings. Open land also decreased by 5% in between the time period of 2004 to 2018; this is due to the construction of the dam site and reservoir which were constructed on the barren land (Figure 18.6). There is a positive sign of increasing forest cover; this is also due to the compensatory program initiated by the government of Himachal Pradesh.

18.5   CONCLUSION

The study demonstrated that the application of GIS helps in studying the changes in land-use pattern in an area. The existing LULC has been dynamic in nature from 2004 to 2018 in large hydropower projects in the upper Beas river basin. The study area has been divided into seven major categories including seven LULC types that are identified and used in this study: (1) built-up land, (2) forest,

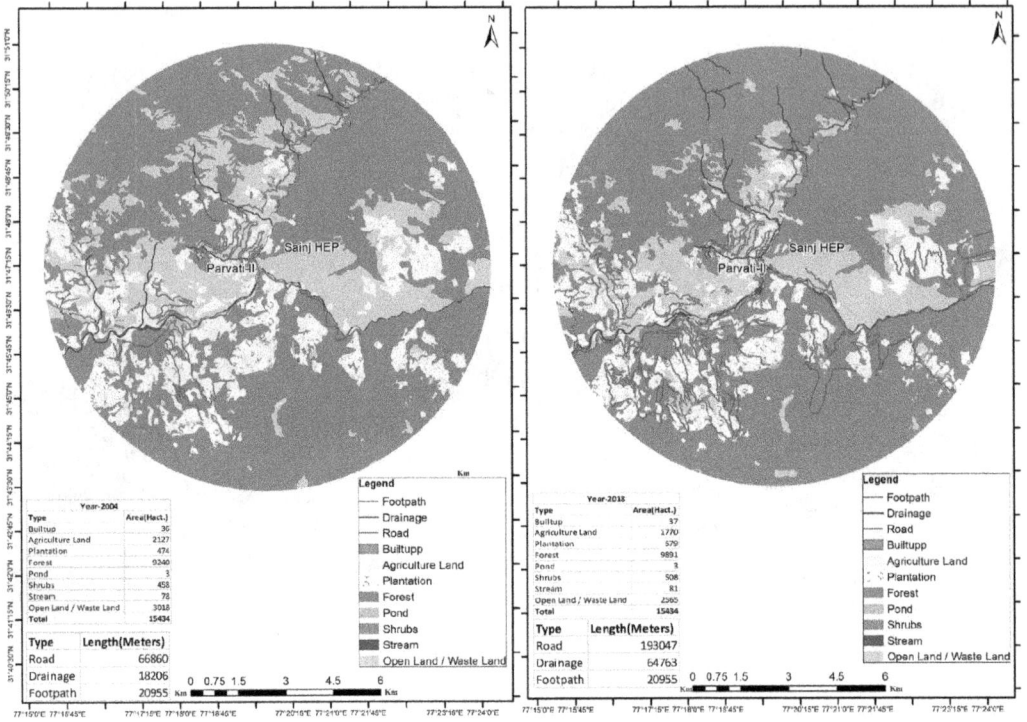

FIGURE 18.3   Land-use/land-cover change detection in the Parbati Stage-II HEP in Sainj, Kullu.

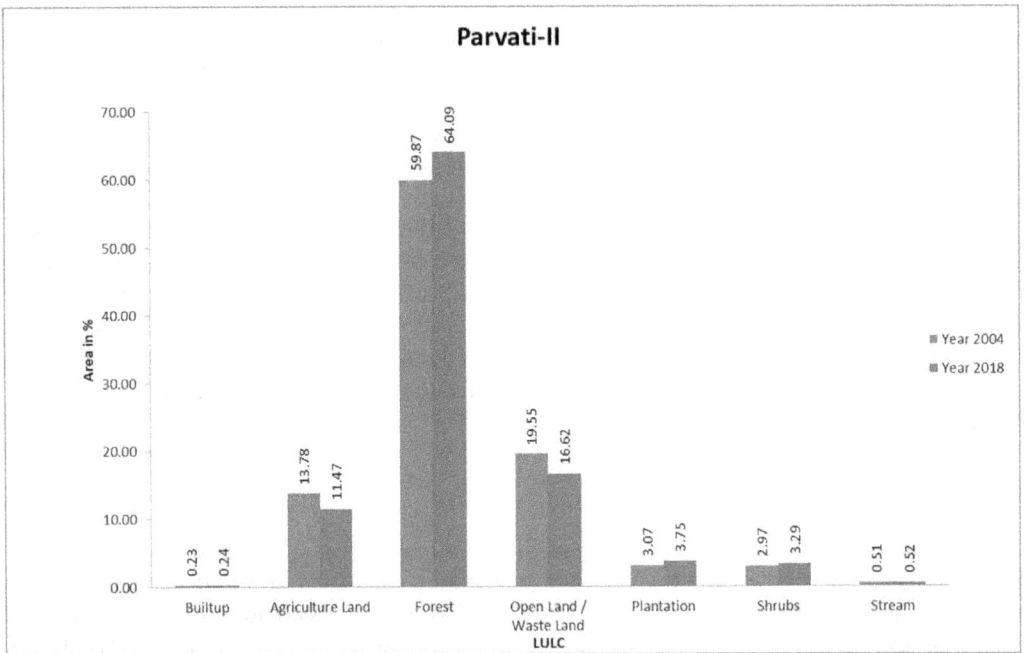

FIGURE 18.4   Analysis of land-use/land-cover change in large HEP Parbati Stage-II.

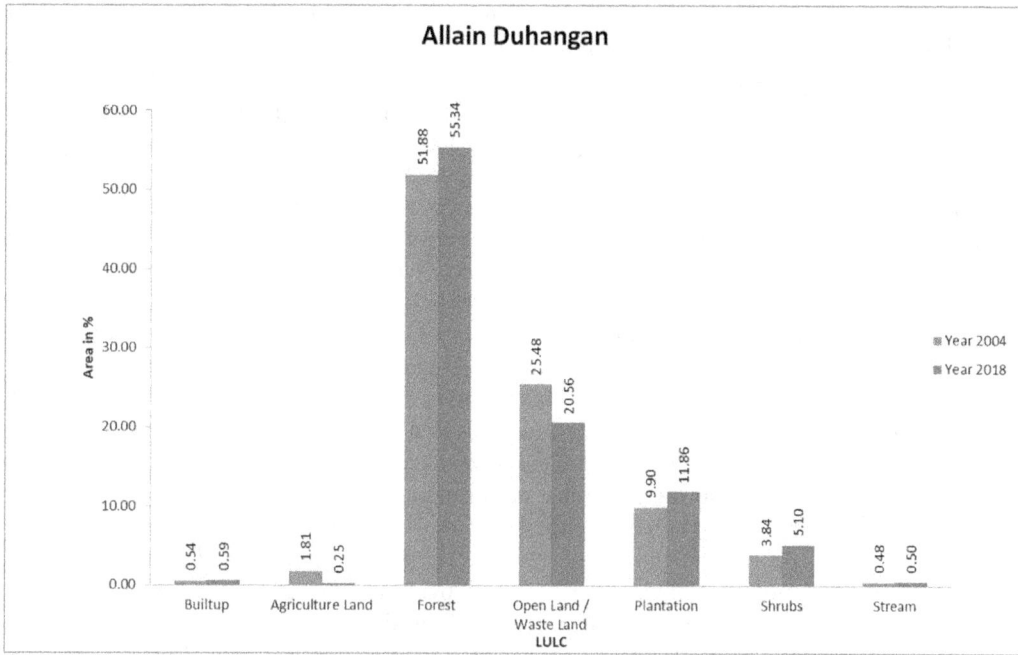

FIGURE 18.5 Analysis of land-use/land-cover change in large HEP Allain-Duhangan in Prini, Manali.

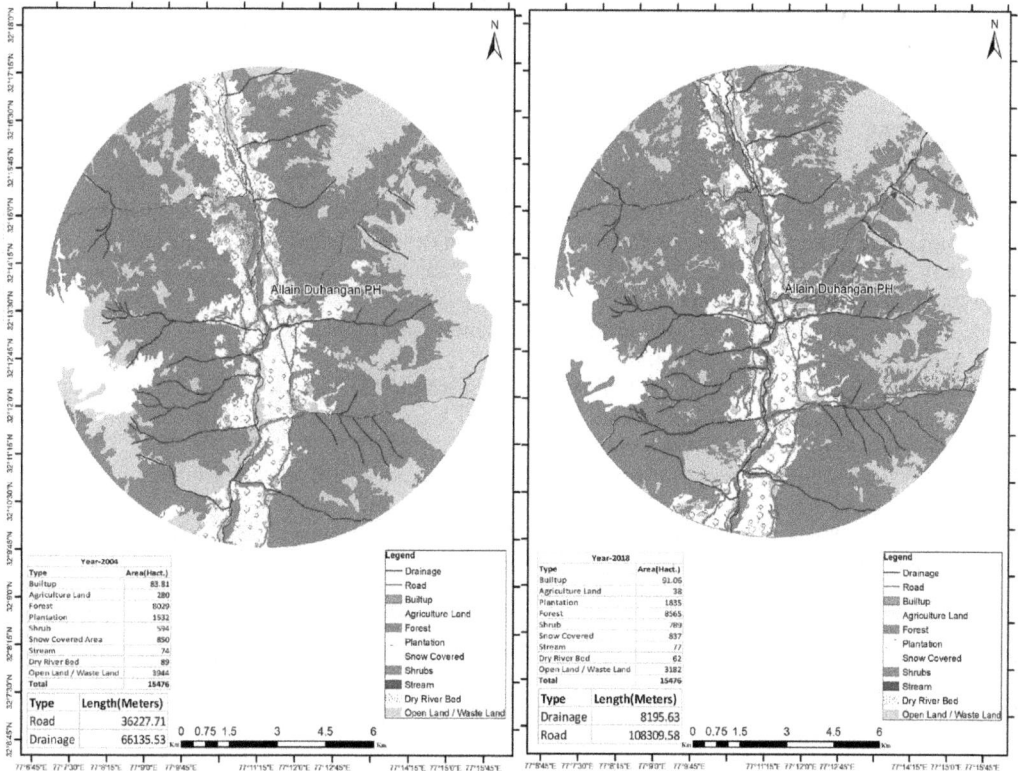

FIGURE 18.6 Land-use/land-cover change detection in Allain-Duhangan, Prini, Mana.

(3) agricultural land, (4) streams, (5) shrubs, (6) wasteland and (7) plantation. Human activities have resulted in vast changes in the LULC of the study area. The surface water bodies have been depleting at a faster rate which is a matter of great concern. These changes are likely to alter the structure, function and complexity of the local ecology with critical implications for the maintenance of the biodiversity, genetic species and landscape. The present study revealed that good agricultural land surrounding the large hydropower projects has been reduced from 2,127 ha to 1,770 ha in Parbati Stage-II project from 2004 to 2018 and from 280 ha to 38 ha in Allain-Duhangan project. Most of the area has shifted into the built-up category due to construction of residential colony and office buildings. It is suggested that the large hydropower projects should be restricted to wasteland or unproductive lands. The study area shows major concerns with increased spatial distribution of the built-up area which was increased from 36 ha in 2004 to 37 ha in 2018, i.e. from 0.23% in 2004 to 0.24% in 2018 to the spatial geographical area of Parbati Stage-II hydropower project, and 83.81 ha in 2004 to 91.06 ha in 2018, i.e. from 0.54% in 2004 to 0.59% in 2018 in Allain-Duhangan project. The study also shows some positive land-use analysis that wastelands are being reduced but the forest area is being increased due to the good initiative taken by the state government regarding an afforestation program. It is also suggested that afforestation programs be intensified, alerting that change from agricultural land to non-agricultural land shall be restricted and the water bodies shall be conserved.

In the present study, an attempt has been made to provide a complete assessment of the prevailing environmental problems along with suggestions and certain recommendations in order to minimise the adverse impacts of HEPs on LULC change in the upper Beas basin, Himachal Pradesh, India. During the study, it was observed that the impact of such projects is not only within the project operational area but also in the adjoining area, both at higher and lower reaches, due to rock blasting, tunnelling and construction activities and the debris dumping.

Geoprocessing techniques employed in this study will assist planners and decision-makers in basin development and management studies and help them identify suitable sites for hydropower projects in the upper Beas basin.

REFERENCES

Agarwal, C., Green, G.M., Grove, J.M., Evans, T.P., and Schweik. C.M., 2002. A Review and Assessment of Land-Use Change Models: Dynamics of Space, Time, and Human Choice, Technical Report. US Department of Agriculture, Forest Service, Northeastern Research Station, Pennsylvania, USA.

Apan, A.A., 1997. Land cover mapping for tropical forest rehabilitation planning using remotely-sensed data. Int. Arch. Photogramm. Remote Sens., 18(5), 1029–1049.

Arora, M.K., and Mathur, S., 2001. Multi-source classification using artificial neural network in a rugged terrain. GeoCarto. Int., 16(3), 37–44.

Chaurasia, R., Closhali, D.C., Dhaliwal, S.S., Minakshi, Sharma, P.K., Kudrat, M., and Tiwari, A.K., 1996. Landuse change analysis for agricultural management: A case study of Tehsil Talwandi Sabo, Punjab. J. Indian Soc. Remote Sens., 24(2), 115–123.

Chilar, J., 2000. Land cover mapping of large areas from satellites: Status and research priorities. Int. J. Remote Sens., 21(67), 1093–1114.

Csaplovics, E., 1998. High resolution space imagery for regional environmental monitoring: Status quo and future trends. Int. Arch. Photogramm. Remote Sens., 32(7), 211–216.

Eiumnoh, A., and Shrestha, P., 2000. Application of DEM data to Landsat image classification: Evaluation in a tropical wet-dry landscape of Thailand. Photogramm. Eng. Remote Sens., 66(3), 297–304.

Foody, G.M., 2002. Status of land cover classification accuracy assessment. Remote Sens. Environ., 80, 185–201.

Gibson, P., and Power, C., 2000. Introductory Remote Sensing: Digital Image Processing and Applications. Routledge, London.

Holben, B., and Justice, C., 1981. An examination of spectral band ratioing to reduce the topographic effect on remotely sensed data, International Journal of Remote Sensing, 2(2), 115–133.

Jaiswal, R.K., Saxena, R., and Mukherjee, S., 1999. Application of remote sensing technology for land use/land cover change analysis. J. Indian Soc. Remote Sens., 27(2), 123–128.

Kachhwala, T.S., 1985. Temporal monitoring of forest land for change detection and forest cover mapping through satellite remote sensing. In: Proceedings of the 6th Asian Conference on Remote Sensing. National Remote Sensing Agency, Hyderabad, pp. 77–83.

Prakasam, C., 2010. Land use and land cover change detection through remote sensing approach: A case study of Kodaikanal taluk, Tamil Nadu. Int. J. Geo. Geosci., 1, 150–158.

Sah, M.P., and Mazari, R.K., 2007. An overview of the geoenvironmental status of the Kullu valley, Himachal Pradesh, India. J. Mountain Sci., 4(1), 3–23.

Saha, A.K., Arora, M.K., Csaplovics, E., and Gupta, R.P., 2005. Land cover classification using IRS LISS III image and DEM in a rugged terrain: A case study in Himalayas. Geocarto. Int., 20(2), 33–40.

Samant, H.P., and Subramanyan, V., 1998. Land use/land cover change in Mumbai-Navi Mumbai cities and its effects on the drainage basins and channels-a study using GIS. J. Indian Soc. Remote Sens., 26(1–2), 1–6.

Selcuk, R., Nisanci, R., Uzun, B., Yalcin, A., Inan, H., and Yomralioglu, T., 2003. Monitoring land-use changes by GIS and remote sensing techniques: Case study of Trabzon, www.fig.net/pub/morocco/proceedings/TS18/TS18_6_reis_el_al.pdf

Srikantia, S.V., and Bhargava, O.N., 1998. Geology of Himachal Pradesh. Geological Society of India, Bangalore, pp. ix + 408.

Star, J.L., Estes, J.E., and McGwire, K.C., 1997. Integration of Geographic Information Systems and Remote Sensing. University Press, Cambridge, New York.

Strahler, A.H., Logan, T.L., and Bryant, N.A., 1978. Improving forest cover classification accuracy from Landsat by incorporating topographic information. In: Proceedings of 12th Symposium Remote Sensing Environment, Ann Arbor, Michigan, US, p. 2, pp. 927–942.

Weng, Q., 2001. A remote sensing-GIS evaluation of urban expansion and its impact on surface temperature in the Zhujiang delta, Southern China. Int. J. Remote Sens., 22 (10), 1999–2014.

19 Forecasting the Effects of Urban Expansion on the Agricultural and Forest Landscape Using MLPNN-Markov Chain Model in Greater Guwahati Metropolitan Area, India

Sanu Dolui and Sayani Chakraborty

19.1 INTRODUCTION

Industrialization and persistent increase in urban population are unquestionably the most important factors in the transformation of biotic and ecological landscapes into urban landscapes (Norton, Briony et al., 2016; Borthakur et al., 2012). Although eco-sensitive landscapes (forest, wetlands, and agricultural land) have many advantages for city environments, including boosting livelihood, preserving biodiversity, reducing pollution levels, and moderating city temperature, unfortunately incessant urban growth and land-cover change make those ecological areas susceptible to conversion (Hassan et al., 2016; Mahmoudzadeh et al., 2022). Numerous cities in India are experiencing fast urbanization, mostly as a result of natural population expansion and rural-to-urban migration. Relocation of city dwellers to urban peripheral areas is a result of a change in lifestyles that prioritize open spaces and environmentally pleasant surroundings (Bhagat, 2011; Sridhar, 2010; Sarkar, 2019). The unpredictability and indeterminate direction in which urban growth took place generated an unpleasant sprawled settlement pattern in which urban landscape extended farther from the city to periphery areas and into rural areas (Shahraki et al., 2011). This destructive pattern of urban growth expansion to excessive levels led to the destruction of vegetation and farmland, fragmented forest areas, produced heat island consequences, and put strain on ecosystems in peri-urban areas (Qiao et al., 2010; Seifollahi-Aghmiuni et al., 2022; Pawe & Saikia, 2018; Perveen et al., 2017; Tewolde & Cabral, 2011). Urban sprawl leads to an enormous quantity of land to be covered in impervious material, which reduces rainwater's ability to percolate to groundwater aquifers, increasing the risk of flooding, including waterlogging and erosion (Bhatta, 2010). The phenomenon of urban sprawl is not only confined to metropolitan cities in India; rather, it is presenting challenges for medium-sized cities also. Although large metropolitan cities have plans to regulate urban expansion, the majority of medium-sized Indian cities are plagued by challenges associated with unplanned land-use/land-cover (LULC) change as a result of insufficient or non-existent planning efforts and financial restraints. As a consequence of this, there has been a general trend, prevalent in the majority of cities of medium size, towards the expansion of built-up areas along the city's periphery. Therefore, the implementation of efficient urban planning strategies by the government and other stakeholders is urgently needed.

 DOI: 10.1201/9781003377825-22

Urban sprawl is the wide-scale growth of urban infrastructure into rural areas, consuming a vast amount of land that was formerly under agriculture or forest cover. Different human activities like logging, farmland expansion, installation of socio-economic and industries infrastructure, degrade the forest areas and fragment the remaining forests into smaller, isolated patches (Referowska-Chodak, 2019). A numerous recent study on Indian cities revealed that urbanization sprawling has prolonged impact on forest and vegetation cover specifically on the edge of the city (Bhat et al., 2017; Shamsudeen et al., 2022; Gumma et al., 2017; Kanga et al., 2022). One of the foremost threats of urban expansion is a continuous loss of agricultural land specially in suburban and rural areas (Zhou et al., 2017; Decocq et al., 2016). To accommodate an ever-increasing population, infrastructure is being developed and expanded on the most fertile lands to continue this urban growth which leads to urban sprawl (Brueckner, 2000; Rafferty, 2023). Farmers that live close to urban centres have thus always been vulnerable to stress from urban growth. Fertile land boosts livelihood opportunities and demand for land for housing and other activities, and that can be easily correlated: lands that are highly suitable for agriculture also have high degree of urbanization, whereas lands that are less suitable for agriculture have low degrees of urbanization. With expectations of higher returns from urban establishments than from value of agricultural products, farmers are frequently persuaded by the possibility of selling their land for urban development instead of continuing the risky practice of cultivating on the outskirts of dwelling units (Tilahun et al., 2022; Coulibaly et al., 2020). Thus, it can be concluded that the agricultural lands and forest cover started decreasing day by day at the edge of cities due to non-existence or loosely regulated urban legislation in those areas.

The present study area of Guwahati city with its plain and hill landscape and pleasant climate with fertile soil make it an ideal location for businesses and development endeavours. In the past two decades, in response to increasing industrialization and urbanization, this city progressively drew immigrants from neighbouring states and emerged into a significant centre for higher learning, government, politics, business, and manufacturing. Eventually that progression extended into peri-urban areas, contributing to urban sprawl (Hemani & Das, 2016). Those ideal conditions led to this town booming in a short span of time but also created headaches for development planners and local admiration to control this unattended development and redirect it in a proper direction. The high-speed transit routes attracted a variety of companies, enterprises, institutions, and commercial entities that built up their constellation along the highway; the majority of these areas had previously been farmland or natural forest habitat. This rapid expansion propelled the city to further expand into peri-urban areas, leading to the phenomenon of urban sprawl and destroying ecologically sensitive areas of forest, wetland, and agricultural areas.

Cities are dynamically complicated adaptive structures that are spatially heterogeneous and coevolve with the settings in which they are located. Because of this, it is hard to completely anticipate or exert control over the dynamic trajectory that cities would undergo. Yet, planning and design efforts that are based on sustainability and urban ecological knowledge should be used to influence or guide cities in more desired directions (Wu, 2014). Hence, LULC study has already attracted a lot of attention from policymakers in order to address the unprecedented challenges that urban growth presents in terms of governance, urban planning, and land use. To properly comprehend the trajectory of LULC, one must be conscious of present land-use patterns and how they have evolved through time (Rahimi, 2016). Remote sensing and geographic information system (GIS) techniques applied in this study offer a valuable perspective for monitoring LULC and are widely acknowledged as being productive tools for mapping land-use change (Shalaby & Tateishi, 2007; Sharma & Joshi, 2013; Rwanga & Ndambuki, 2017). In recent decades several new earth-observing satellites, such as Sentinel-2 and Landsat 8 and 9, were recently launched and provided high-resolution multi-spectral data free of cost which encouraged researchers to classify images with greater accuracy (Harris & Baumann, 2015). Tracking land-use patterns using data collected via satellite images requires sophisticated classification algorithms that enable accurate mapping of heterogeneous land-use categories. The popularity of machine learning algorithms in monitoring of LCLU changes is due to their flexibility, trustworthiness, and better spectral separation ability among the different classes as compared to traditional algorithms (Talukdar et al., 2020). Random Forest (RF) is a powerful machine learning classifier, Due to its flexibility, RF has been used frequently in different studies and achieved higher accuracy than other classifier methods; that is why it has been used in the present study also (Abdi, 2019; Purwanto et al., 2022; Zhang et al., 2021; Melichar et al., 2023).

19.1.1 LITERATURE REVIEW

Modelling is among the strategies used by policymakers to evaluate and regulate land-use changes and their consequences, such as urbanization. Many researches have been conducted over the past few decades in an effort to identify the most effective method for forecasting changes in land cover. Despite the fact that the literature offers a broad spectrum of LULC change models, no model proved its inherent superiority than others, and that is why it is extremely difficult to choose the model that will yield the most promising results (Verburg et al., 2006). In prediction of future dynamics, Markov chain (MC) is indeed a prominent tool which can aggressively predict urban or rural landscape transformation with satellite data (Arsanjani et al., 2013; Dadhich & Hanaoka, 2011; Muller & Middleton, 1994). Though the stochastic Markov model has been frequently implemented to track urban expansion, it is not appropriate for spatial modelling as it fails to acknowledge the spatial dimension. A hybrid CA-MC (cellular automata-Markov chain) model integrates the spatial interaction effect to depict the spatial-temporal pattern of urban transformation. However, the limited capability of the CA-MC model to comprehend urban growth dynamics in relation to prospective driving forces, means its application for land-use modelling is severely curtailed. Following an analysis of the components that contribute to the growth, LULC change models offer a more powerful and indigenous forecast of the future. For this reason, the CA-MC model may be combined with other models such as analytical hierarchy process (AHP) and logistic regression (LR) to enhance the model's ability to predict the future and gain a deeper knowledge of evolving growth patterns (Hassan & Elhassan, 2020; Hamdy et al., 2016; Aburas et al., 2017; Arsanjani et al., 2013; Mustafa et al., 2017). AHP is basically a deterministic MCDM model without any randomness involved in decision-making process and mostly reliant on the expertise knowledge in the relevant field of study. As AHP method is heavily dependent on expert judgements, it was assumed that the result may be influenced by human biasedness. The LR model, on the other hand, is fundamentally a mathematical algorithm developed to determine the linear association among urban growth and its associated factors. However, if the relationship between urban expansion and the influencing factors is non-linear and complicated, it is possible that it may not perform successfully in such instances. Many recent studies, instead of using AHP and the LR model, have employed artificial intelligence (AI) to develop and optimize the models that could predict trends in urban expansion (Xu et al., 2022; Asadi et al., 2022; Sajan et al., 2022). The artificial neural network (ANN) model is influenced by the neurological system of the human brain which can compute non-linear complicated relationships in a short time. Since urban areas are extremely complex systems, modelling urban expansion using ANN is quite pertinent. Application of the ANN model has some advantages over other models: it is capable of quickly spotting the complex non-linear associations inside the database, it free forms human bias, and it requires a minimal amount of data for training the model, which ultimately cuts down the calibration time of the model. After comparing the benefits and drawbacks of the various models under consideration, the authors of this study decided to make use of the multilayer perceptron neural network (MLPNN)-based Markov chain model in order to make an accurate forecast regarding the future pattern of LULC. In the Indian scenario, this hybrid MLPNN-based Markov chain (MLPNN-MC) model has already been tested by numerous researchers in various cities of India providing a high level of accuracy for LULC change prediction in several studies (Thapa & Murayama, 2009; Mozumder & Tripathi, 2014; Mishra & Rai, 2016; Vinayak et al., 2021). The integrated MLPNN-MC method is a robust approach by taking the advantages of both models, where MC (Markov chain) control of the temporal dynamics of land-cover changes by identifying their pattern of growth and ANN models the complicated intertwined relationship among factors of urban growth. The multi-layer perceptron (MLP) neural network is trained by a supervised backpropagation (BP) algorithm and classified remotely sensed imagery through information from training sites and is capable of detangling complex non-linear interactions among factors (Maithani, 2009; Shahi et al., 2020; Singh et al., 2022).

Several studies had already been conducted on urban expansion and its consequences in Guwahati city and its surrounding peri-urban areas (Mahadevia et al., 2014; Deka & Devi, 2017; Pawe & Saikia, 2018; Chetia et al., 2020; Gogoi et al., 2023). Though research has already been conducted in LULC changes and various urban phenomenon (urban sprawl, urban heat island), simulations of urban expansion in this city are still lacking. Therefore, it is indeed necessary to comprehend the urban expansion that the city of Guwahati has undergone to visualize a future scenario of urban sprawl. In light of the aforementioned background, the purpose of this investigation is to analyze the trajectories of LULC change in Greater Guwahati Metropolitan areas from 2000 to 2022 and subsequently, develop scenarios of prospective LULC changes for the years 2035 and 2050. Moreover, this spatial information obtained from the study will be helpful to various stakeholders to formulate strategies required for sustainable development.

19.2   STUDY AREAS

Guwahati has emerged as the most important urban centre in the state of Assam and the most populous city in the entire region of north-eastern India, because of its strategic positioning (gateway to NER India), excellent transportation facilities, rapid population growth, and expanding economy (Pawe & Saikia, 2018; Hemani & Das, 2016). This city is surrounded by a number of Precambrian residual hills with forest cover and wetlands. The study area which has parts of Greater Guwahati Metropolitan Region has a longitudinal extension of 91°27′ E to 91°56′ E and latitudinal extension of 26°0′ N to 26°20′ N which extends over an area of 842 km2 including 262 km2 of Guwahati Metropolitan Area (Figure 19.1). The Guwahati city has a population of 957,352 as of 2011 but it was projected that in the year 2025 the population will grow up to 2.1 million (GMDA Master Plan, 2025). The climate of Guwahati city and associated regions experiences a typical warm and humid summer with temperatures ranging from 22°C to 37°C, and winters remain basically dry and cold, where minimum temperatures go down to 10°C to 14°C (Borthakur & Kr, 2012). The city receives a sufficient amount of rainfall 1,082 mm annually (Tiwari et al., 2017). River Brahmaputra segregates Guwahati city into two segments, where the southern part encompasses an urbanized core and forms a starfish urban layout that extends outward from the core along the transportation corridor, while the northern part represents the fringe and rural hinterland areas. Recent unplanned expansion in the western periphery and southern part along the wetland and forest areas is raising concern for the local administrator (Pawe & Saikia, 2020). Rampant urban growth is typically accomplished through an expansion of the urban fabric, with residential and establishment of new industry occurring sporadically on agricultural land and green fields adding to the problem (Figure 19.1).

19.3   MATERIALS AND METHODOLOGY

19.3.1   Spatial Database

Open data policies of US Geological Survey (USGS) Landsat satellite images along with better spatial and spectral as well as temporal resolution enhance the capacity of land-cover monitoring at intense time intervals. For the purpose of obtaining high-resolution datasets, this research makes use of Landsat 7 (Enhanced Thematic Mapper Plus **ETM+**), Landsat 8 (Operational Land Imager [OLI]-1), and Landsat 9 (OLI-2) sensors, all of which are endowed with a 15-m panchromatic band. Different meteorological and physiographical settings may cause inconsistencies in data when images are collected from different sources at multiple time frames. To avoid cloud distraction and

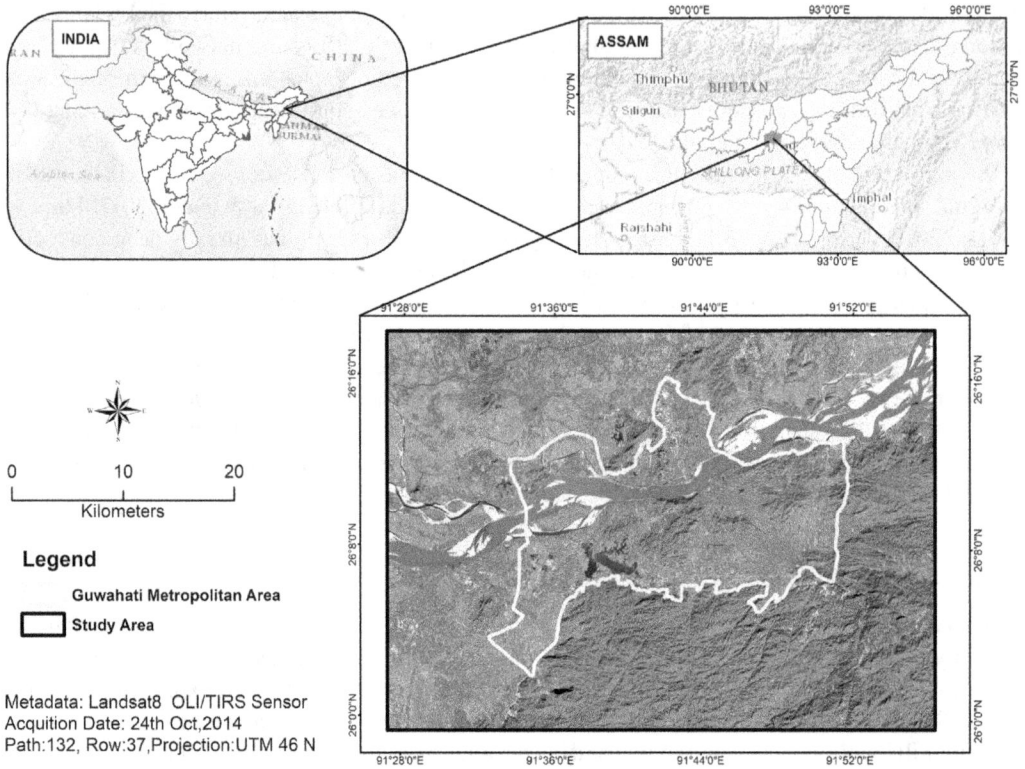

FIGURE 19.1 The study area of Greater Guwahati Metropolitan area, Assam.

TABLE 19.1

Information about the Used Landsat Images in the Study

City Name	Satellite	Acquisition Date	Sensor	Row/Path	Image Resolution (m)	Cloud Cover (%)
Guwahati, Assam, India	Landsat 7	25/10/2000	Enhanced Thematic Mapper Plus (ETM +)	Path:137 Row:42	15	3.00
	Landsat 8	24/10/2014	Operational Land Imager (OLI)	Path:137 Row:42	15	1.83
	Landsat 9	23/11/2022	Operational Land Imager 2 (OLI2)	Path:137 Row:42	15	0.70

ensure seasonal consistency, images were mainly captured during the winter period based on the absence of cloud cover. These images were obtained from the USGS for 2000, 2014, and 2022 at intervals of 14 and 8 years (http://earthexplorer.usgs.gov) (Table 19.1).

Drivers are important in any LULC change prediction models, and here, 12 driving variables were selected for the simulation of urban growth: elevation (C1), slope (C2), distance from forest areas (C3), distance from agricultural land (C4), proximity to river distance (C5), proximity to road distance (C6), proximity to railway track (C7), distance from airport (C8), distance from settlement areas (C9), distance from central business district (C.B.D) area (C10), distance from industrial areas

(C11), and city development plan 2025 (C12). Standard GIS operations like Euclidian distance, proximity analysis was carried out to prepare the training geo-database for the ANN model. All map layouts have been prepared using ArcGIS 10.6.1 which was resampled at 15 × 15 pixels in order to match the database with image resolution (www.esri.com). All the sources of spatial database are outlined in Table 19.2 (see also Figure 19.2).

19.3.2 IMAGE PRE-PROCESSING

To minimize impurities from raw Landsat satellite images, it should be geometrically, radiometrically and atmospherically corrected. The image processing software ENVI 5.4 (www.harris.com) was used in image preprocessing. To eliminate visual and locational imperfections, satellite images were geometrically rectified with the topo-sheet published by the Survey of India and further re-projected to UTM 46N, WGS84 datum. Radiometric calibration was performed in ENVI to correct radiance, reflectance, and temperature brightness effects (Chander et al., 2009). A FLAASH module of the ENVI software was implemented in order to rectify atmospheric imperfections present in images (Bagheri & Tousi, 2017; Nguyen et al., 2015; López-Serrano et al., 2016; Moravec et al., 2021). To remove salt-and-pepper effects from images, median filter tools of ENVI software were used. In this study, Gram-Schmidt sharpening algorithms have been applied to enhance the spectral properties of multispectral images with a high-resolution panchromatic band over the same area. To classified images, a modified National Remote Sensing Centre classification scheme for India was adopted (level I) for different categories of LULC. After monitoring, the physical and cultural fabric of the study area was segmented into five land-use classes, viz. agriculture, forest cover, vegetation, water bodies, and urban areas (Table 19.3). The size of land-cover classes was kept to a minimum to ensure spectral separability among the classes and to define a clear boundary between two classes (Saikia et al., 2013). Finally, calibrated images (2000, 2014, and 2022) were imported to ESA SNAP software for Random Forest classification.

TABLE 19.2
Data Used and Its Source

Serial number	Criteria name	Data source	Techniques used	Spatial Resolution (m)
C1	Elevation	National Remote Sensing Centre	Elevation Map	30
C2	Slope	Cartosat DEM Data (https:// bhuvan.nrsc.gov.in/)	Slope Map	30
C3	Forest	European Space Agency (ESA)	Calculate	10
C4	Agricultural land	Sentinel-2A Imagery, https://	distance	
C5	River	scihub.copernicus.eu/		
C6	Road	Open Street Map (OSM)	Proximity	30
C7	Railway track	https://www.openstreetmap.org	analysis	
C8	Airport			
C9	Settlement areas	European Space Agency (ESA)	Euclidean	10
C10	C.B.D area	Sentinel-2A Imagery, https:// scihub.copernicus.eu/	distance	
C11	Industrial areas	Open Street Map (OSM) https://www.openstreetmap.org	Proximity analysis	30
C12	City development plan 2025	Guwahati Metropolitan Development Authority (Assam) https://gmda.assam.gov.in/	Identified planned township areas	15

Note: All those raster images were resampled into 15 × 15 m to match the LULC image resolution.

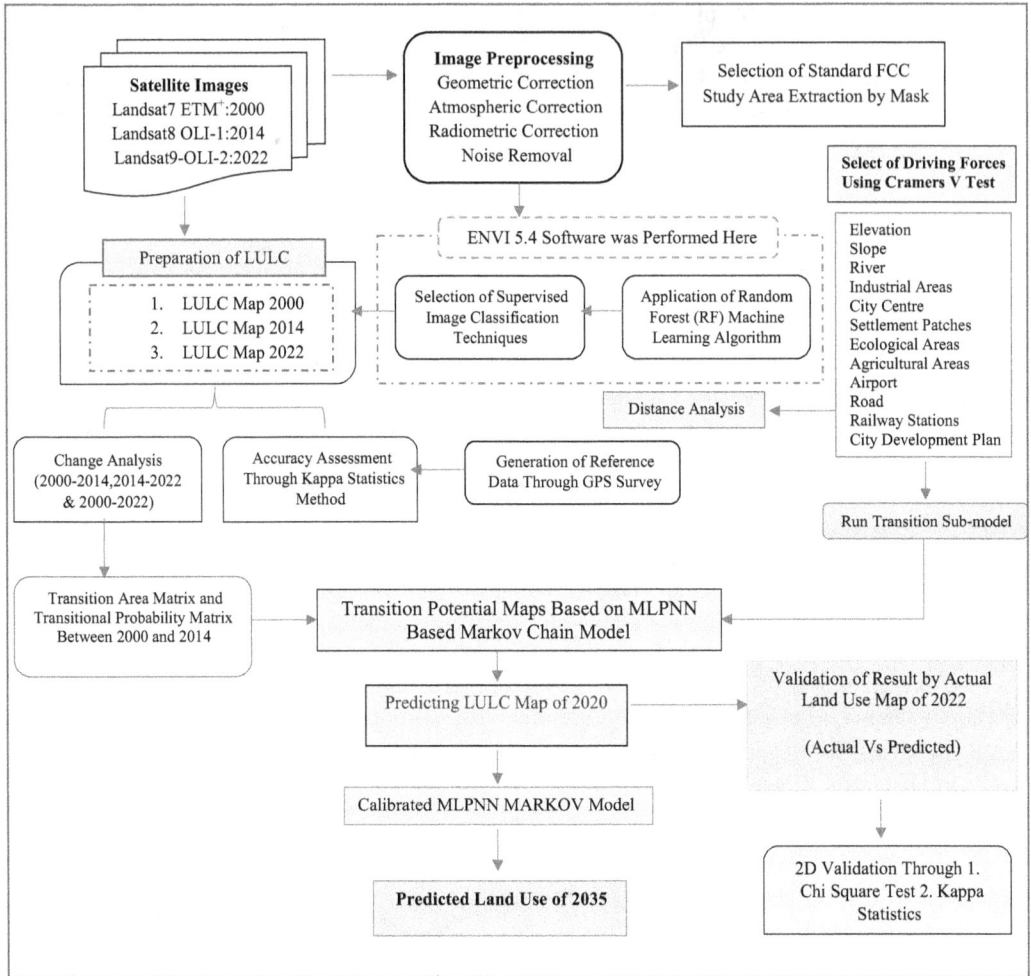

FIGURE 19.2 Workflow of the methodology.

TABLE 19.3
LULC Classes Identified in the Study Area

Class name	Class types	Class description
C1	Agriculture	Cultivation of crops like rice, potato, and others vegetables
C2	forest	Natural Forested areas including evergreen, deciduous, and mixed forests
C4	vegetation	Areas covered with shrubs, bushes and small trees, Grassland as well as a bare land that has very little or no grass cover, Aquatic vegetation other than forest
C4	water bodies	River, permanent open water, lakes, ponds, canals, permanent/seasonal wetlands, low-lying areas, marshy land, and swamps
C5	Urban areas	All the man-made continuous structure, residential, commercial and industrial unit, compact settlement patches, villages, roads network, bridges, mixed urban-natural landuse, airport, and townships

19.3.3 LAND-USE/LAND-COVER CLASSIFICATION USING RANDOM FOREST (RF) ALGORITHM

Random Forest is a new non-parametric ensemble machine-learning algorithm developed by Breiman based on a decision tree (Breiman, 2001). Numerous advantages of Random Forest include efficiency in dealing with the non-parametric nature of data; outperformance of other machine learning algorithms in terms of accuracy; ability to compute missing values in the absence of data; flexibility to classify with continuous or categorical data values and capability to be used for multiple applications including regression, classification, and prediction analysis. All these advantages make Random Forest a powerful algorithm (Boulesteix et al., 2012; Amini et al., 2022). RF classifier is based on the aggregation of a large number of decision trees in order to overcome the weaknesses of a single decision tree (Maxwell et al., 2018, 2020). RF is based on the bagging algorithm and uses an ensemble learning technique that produced higher accuracy than other machine learning algorithms because it generates multiple classifications of the same data and performs more accurately than any single classifier (Rodriguez-Galiano et al., 2012; Basheer et al., 2022). RF is a stable algorithm which can handle non-linearity in a dataset, is methodologically sound to outlier values, and can automatically compute missing values (Junaid et al., 2023; Woznicki et al., 2019). RF's adaptability has led to its adoption in a wide assortment of earth science applications, such as modelling forest resources (Waśniewski et al., 2020), land use (Amini et al., 2022), and environmental and natural hazard management (Ramo & Chuvieco, 2017; Akinci et al., 2020; Farhadi & Najafzadeh, 2021). After reviewing the previously mentioned advantages of RF, this algorithm was used in this study for extracting LULC information from satellite images of 2000, 2014, and 2022.

19.3.4 ACCURACY ASSESSMENTS

The accuracy of the thematic maps must be evaluated, as inaccurate geo-referencing and poor selection of training samples, cloud covers, inconsistent pixels, and algorithms selection errors may impact classification results. For this assessment, the outcomes from the classified image were matched with additional sources of dataset, which would be assumed to be truthful or directly verified through ground survey. After development of the error-matrix table to obtain accuracy assessment, producer's accuracy, user's accuracy, overall accuracy, and kappa coefficient (T) are common algorithms to calculate (Manandhar et al., 2009; Sari et al., 2021). Cohen's kappa coefficient values can vary from 0 to 1, with 0 indicating that there is no agreement between two classes and 1 indicating that there is perfect agreement among classes (Maxwell et al., 2020; Baig et al., 2022). In the present study, accuracy of all the thematic maps (2000, 2014, and 2022) was assessed through the development of an error matrix. If overall accuracies as recommended by Anderson et al. (1976) should exceed the minimum standard of 85%, then the results obtained can be considered as reliable. For accuracy assessments, ground truth data were collected through three different sources: (1) GPS survey of city surrounding through personal visit by using handheld GPS-type Garmin eTrex 30s, (2) Google Earth images, and (3) interview with local inhabitant about unchanged landscape for historical LULC cover information. Finally, 300 distinct reference points were generated with a minimum sample number of at least 50 for each class to have sufficient representation of each LULC class (Table 19.4).

19.3.5 HYBRID MODEL OF MLPNN-MARKOV CHAIN

MC and ANN models were combined to simulate future land use of the study area, which offers certain advantages when compared with traditional techniques. Historical changes in land use that have occurred in Guwahati Metropolitan areas were taken as basic input to replicate and anticipate the future spatial pattern in LULC. In This MLPNN-MCM model, the land-change model (LCM) of TerrSet 18.31, which was developed by Clark Lab, was utilized for the purpose of predicting future LULC for the years 2035 and 2050 . The LCM model has been used effectively in a number

TABLE 19.4

Interpretation of Kappa Statistics Values

Cohen's Kappa Statistic (k)	Strength of agreements
<0.00	No agreements
Below 0.40	Poor agreement
0.41–0.60	Moderate agreement
0.61–0.75	good agreement
0.75–0.90	Excellent agreement
0.91 and above	Almost perfect agreement

of studies for a variety of applications (Hasan et al., 2020; Leta, 2021; Mas et al., 2014). To predict the future land transformation using LCM, the following steps are necessary: (1) LULC change analysis, (2) testing potential of the selected variables, (3) preparation of transition potential maps using MLPNN, (4) calibration and validation of simulated LULC map, and (5) change prediction for 2035 and 2050.

19.3.5.1 Land-Use/Land-Cover Change Analysis: Analyzing Past Land-Cover Change

In the change analysis step, historical changes between two respective time periods (T2-T1) have taken into consideration for analysis the trend of LULC changes. For extracting LULC information, Landsat ETM⁺/OLI-1/OLI-2 images of years 2000, 2014, and 2022 were classified, respectively. The area changes from one LULC class to another during a particular time period can be determined spatially and quantitatively by this analysis (gain, loss, net changes, rates of changes, etc.). Optionally, a basis roads layer and elevation may be used in the prediction process. In this study, cross-tabulation analysis was carried out to quantify LULC changes during 2000–2014 (period 1), 2014–2022 (period 2), and 2000–2022 (period 3), respectively. LULC transitions between 2000 and 2014 were used to simulate the LULC of 2022 for validation and calibration of model, while transitions between 2014–2022 and 2000–2022 were used to predict the LULC simulation of 2035 and 2050.

19.3.5.2 Testing Potential of Selected Variables

There is no single, generally accepted elaboration of what causes changes in LULC, and numerous experts have argued that the applicability of various factors may differ from scenario to scenario (Ansari & Golabi, 2019; Dang & Kawasaki, 2017). The suitability of a particular place in terms of its appropriateness for urban growth is judged based on the factors presented in this context (Ahmed & Ahmed, 2012; Nouri et al., 2014). Explanatory variables associated with urban sprawl have been explored in this research in order to simulate the trend using an MLP neural network. In this study, 12 variables were added to the model either as static (cannot be changed easily, i.e. elevation, slope) or dynamic components (change through time, i.e. settlements, agricultural land). Cramer's V test was employed to validate the factors and evaluate their potential explanatory power. Association among the variables and land-use type were determined by the Cramer V values, which can vary from 0 to 1, with values close to 1 indicating a significant correlation and values close to 0 indicating a poor correlation. A Cramer's coefficient value greater than 0.15 suggests the variable is sufficient in explaining a relationship, whereas a value greater than 0.40 indicates that the independent variable has an excellent capacity of describing the dependent variable.

19.3.5.3   Modelling Transitional Potential Map through Multi-layer Perceptron

After testing the correlation and assessing the skill of the driving variables, components were added to the LCM model to generate the transition potential map. Within the MLP model, transition potential maps have been developed with an appropriate level of accuracy and weighting for each factor responsible. This algorithm repeatedly adjusts the neural network's weights to ensure the minimal amount of inaccuracy between the node's output and the desired outputs. As the major objective of the study is to predict future urban growth, the probability of other land use converting to built-up land was the key concern. Areas with higher transition potential values reveals places with higher suitability to be converted, whereas lower transition values indicate lowest chance of conversion. In this model, four major transition potential maps (changes from one LULC class to other classes) were considered: (1) agricultural land to urban, (2) forest cover to urban, (3) vegetation to urban, and (4) water bodies to urban (Figure 19.5). To improve the functionality of the MLP neural network, the significant transitions were integrated into the transition sub-model. Due to usefulness of the MLPNN has been selected to model these transitions (Figure 19.5).

The subsequent modifications that have transpired between 2000 and 2014 were used to predict the transition potentials for 2022 in order to validate the MLPNN-MC model. The effectiveness of MLP in transition potential mapping was evaluated using the metrics root mean square (RMS) error, accuracy rate, and skill measure. The MLPNN has produced an accuracy rate after 10,000 iterations, which is a significant amount. The goal of this method is to produce an association between transition probability and driving factors. These sample cells were split into two groups: 50% were employed for training and 50% for validation. The determined weight assists in reducing error and boosting precision during this procedure. It would be acceptable to obtain a transitional map after the accuracy is above 80% (Vinayak et al., 2021).

19.3.5.4   Modeling with the Combination of Multilayer Perceptron and Markov Chain Methods

In the change prediction tab, after the development of transition potential maps, a Markov transitional matrix was produced to compute the likelihood of each transition for the years 2035 and 2050. This tab allowed for an accurate prediction of the future land cover by combining ANN and the MC model. The MC model determined the expected changes that would be anticipated in the future by contrasting the two LULC maps for generating a transitional probability matrix (Eastman, 2009; Rimal et al., 2018) and using the multi-layer perceptron-artificial neural network (MLP-ANN) model to link up the association between the driving factors and the different land-use types for producing a prospective transitional map (Sang et al., 2011). By analyzing how land use has changed in the past, the MLPNN-MC model was able to accurately forecast the trend, direction, and magnitude of land-use changes in the future.

19.3.5.5   Validation of Model

In order to assess a model's dependability, calibration and validation are essential steps to be performed, where simulated land-use maps needed to be compared with actual land-use maps. In this procedure, the simulated LULC map for 2022 was validated against the actual LULC obtained for the same year 2022, which was retrieved from the satellite image (Kamusoko et al., 2009; Mungai et al., 2022). Standard Cohen's kappa indices such as kappa index for no information (K_{no}), kappa index for grid cell level location ($K_{location}$), and kappa for stratum-level location ($K_{locationStrata}$) and kappa standard ($K_{standard}$) which is the same as k_{no} (Pontius, 2000; Cohen, 1960) were used to compare the agreements of the two maps with the validate module. Furthermore, the actual and simulated images with 75% or higher similarity will enable the execution of subsequent simulations with satisfactory accuracy. The procedure of calibrating is carried out if there is less than a 75% overlap between the projected and actual images.

19.4 RESULTS AND DISCUSSION

19.4.1 RESULTS

19.4.1.1 Trend of Land-Use Change with Emphasis on Forest and Agricultural Land

Spatial trend of change analysis, which visualizes land-cover maps at specific intervals, is a useful technique for illuminating and depicting general trajectories of land-use change. The dynamics of the distribution of major LULC was observed during the study period. The LULC classed maps for the years 2000, 2014, and 2022 are evaluated for their degree of correctness and presented in Table 19.5. The accuracy and convenience of the aforementioned classified images were evaluated by computing four kinds of accuracies: user, producer, overall accuracy, and kappa coefficient value. Table 19.5 depicts an accuracy assessment for the LULC classified maps of the years 2000, 2014, and 2022. The overall accuracy for 2000, 2014, and 2022 LULC maps was achieved at 92.33%, 94.67%, and 96.67%, correspondingly, which meet the minimum acceptable accuracy requirement of more than 80% as advised by several researchers. The kappa coefficients that were determined for every classed map are as follows: 90.40% (2000), 93.32% (2014), and 95.82% (2022). Because it demonstrates that the maps are sufficiently accurate and obtained an accuracy of more than 90%, it was ensured that the classed maps were competent for further investigation.

Guwahati is one of the fastest expanding cities in India, and its population has registered a growth from 823,000 in 2001 to 968,000 in 2011 and is projected to reach 1,135,000 in 2021. It is anticipated that the greater Guwahati metropolitan area will eventually contain a population of 2.1 million people (GMA 2025, projection). These incessant population growth rates show strong evidence of the expansion of metropolitan areas over the past two decades. Changes in different kinds of LULC categories are analyzed later, with special emphasis placed on agricultural and forest alterations.

The Guwahati city is surrounded by dense forest and hillocks in the Nilachal hills, Jalukbari hills, Hengrabari, Jalukbari and Amchang reserve forest, Narakasur hills, Kharghuli hills and Sonaighuli hills, North Kalapahar reserve forest, Sarania reserve forest, and Khanapara reserve forest. The study areas experienced a steady decline in forested areas from the year 2000 where 37.96% (319.75 km2) of the total geographical areas was covered by dense forest and open forest, respectively, which decline to 33.41% (281.45 km2) in 2022 (Figure 19.3). Between 2000 and 2022 the loss of forest covers 1.74 km2 consecutively. According to a study using satellite imagery, some of the previous study outlined that significant loss of forest cover in this area was due to socio-economic and infrastructural developments (Saikia, 2014). The population expansion in this region has led to an increase in human activity in the protected and non-protected forest areas, resulting in forest loss and fragmentation. The border areas of Assam-Meghalaya states in NH 27, adjacent to Guwahati city experience massive socio-economic development along this highway causing forest losses (Table 19.6).

Additionally, a diminishing propensity can also be observed over the agricultural class, where sizable amounts were converted into dwelling units among the respective time frame (2000–2022). In 2000, a 172.40 km2 (20.46%) area was covered by agriculture, which declined to 140.37 km2 (16.54%) in 2020, which means there was an overall reduction of 32.03 km2 in agricultural land between 2000 and 2022. At the present time, farming activities are largely concentrated in the city's outskirts, mainly surrounding areas of *Deepar Beel* Lake (Ramsar site) as well as areas adjacent to Guwahati airport. Apart from that in the northern part of the city along the bank of the Brahmaputra River where extensive agricultural lands were found, inhabitants are mainly engaged in cultivation and fishing. Cultivable lands in this study area are decreasing due to city's infrastructural developments, river bank erosion, and illegal conversion of agricultural to commercial and industrial uses; these are the root causes of conversion. The agricultural lands that encircle urban areas in this city are crucially relevant for the livelihood of local farmers and long-term viability of locally

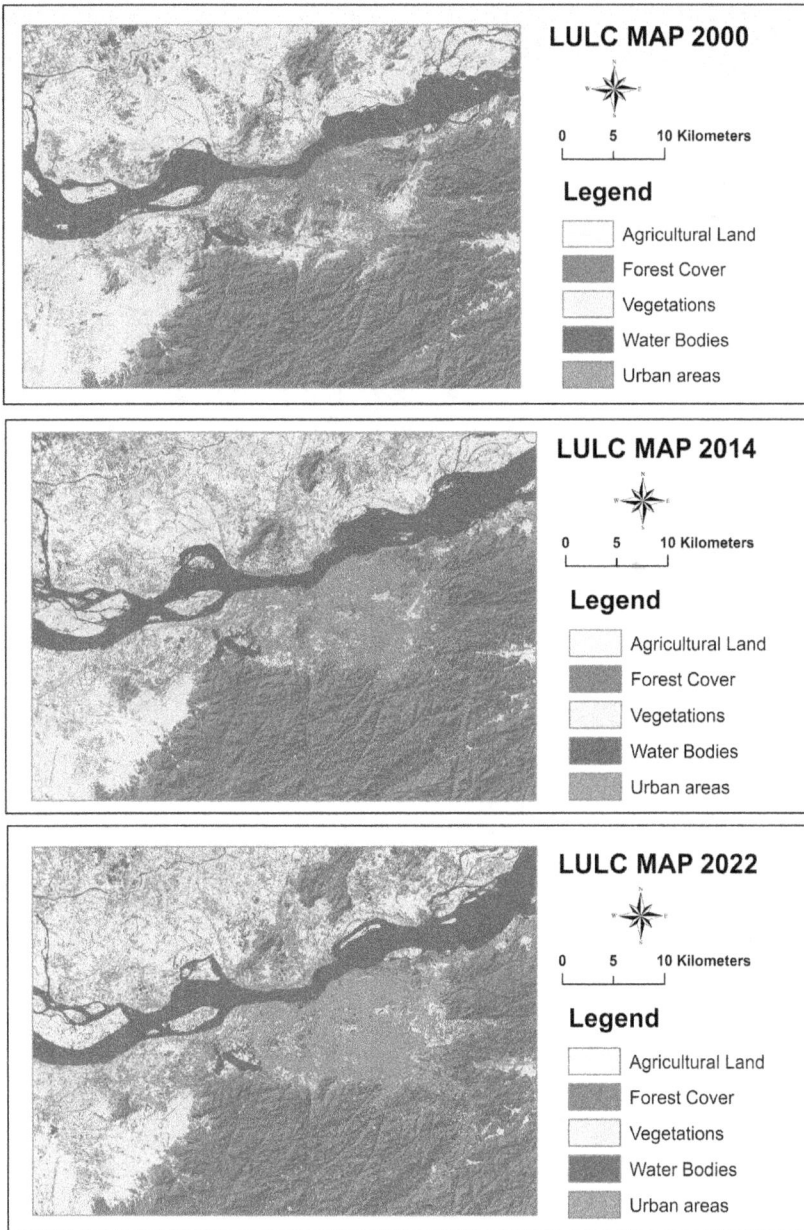

FIGURE 19.3 Land use/land cover in the study area in 2000, 2014, and 2022.

noteworthy natural and indigenous landscapes. The negative impact of urban sprawl is initially apparent on farmland because of its physical proximity to urban areas.

Guwahati is a main administrative, industrial, trading, educational, and socio-cultural center of the north-eastern states of India, and massive urban expansion has been observed in this city in the last two decades, as noted by various researchers (Chetia, Saikia, Basumatary et al., 2020; Pawe & Saikia, 2020). It can be easily understandable from the fact that total area under built-up land was 94.74 km2 (11.25%) in 2000, whereas it almost doubled 190.92 km2 (23.26%) in 2022. The built-up area grew annually by 4.25 km2 during period 1 (2000–2014), while during period 2 (2014–2022)

TABLE 19.5
Accuracy Amounts of the Random Forest (RF) Supervised Classification for 2000, 2014, and 2022

Land Use Categories	2000				2014				2022			
	User Accuracy (%)	Producer Accuracy (%)	Overall Accuracy	Kappa Coefficient (T) (%)	User Accuracy (%)	Producer Accuracy (%)	Overall Accuracy (%)	Kappa Coefficient (T) (%)	User Accuracy (%)	Producer Accuracy (%)	Overall Accuracy	Kappa Coefficient (T) (%)
Agriculture	91.30	91.30	92.33	90.40	94.29	94.29	94.67	93.32	97.10	95.71	96.67	95.82
Forest	89.09	96.08			92.45	96.08			96.15	98.04		
Vegetation	91.38	89.83			94.92	93.33			96.67	96.67		
Water Bodies	91.07	92.73			92.86	94.55			94.64	96.36		
Urban Areas	98.39	92.42			98.39	95.31			98.41	96.88		

TABLE 19.6
Area under Different Land Use Classes and Magnitude, Percentage Change and Annual Rate of Change for 2000, 2014, and 2022

Class Name	Area in km²						Magnitude of LULC changes (2000–2014, 2014–2022, 2000–2022)					
	LULC 2000 (km²)	%	LULC 2014 (km²)	%	LULC 2022 (km²)	%	2000–2014 (km²)	Rates of change per years	2014–2022 (km²)	Rates of change per years	2000–2022 (km²)	Rates of change per years
Agriculture	172.40	20.46	158.67	18.83	140.37	16.54	–13.73	–0.98	–18.30	–2.29	–32.03	–1.46
Forest	319.75	37.96	292.45	34.71	281.45	33.41	–27.30	–1.95	–11.00	–1.38	–38.30	–1.74
Vegetation	161.63	19.19	148.23	17.60	142.67	16.46	–13.40	–0.96	–5.56	–0.70	–18.96	–0.86
Water Bodies	93.93	11.15	88.90	10.55	87.04	10.33	–5.03	–0.36	–1.86	–0.23	–6.89	–0.31
Urban Areas	94.74	11.25	154.20	18.30	190.92	23.26	59.46	4.25	36.72	4.59	96.18	4.37
Total Areas	842.45	100.00	842.45	100.00	842.45	100.00						

Sources: Landsat images (2000, 2014, 2022).

it grew by 4.59 km2, and during period 3 (2000–2022) it grew by 4.37 km2 per annum. The city's unique geophysical structure restricts its spatial expansion in southern and eastern parts (hills and dense forest); hence, the urban growth of the city has been highly unplanned and sprawling in the western and northern parts. After 2000, significant built-up growth in suburban areas as compared to the inner part of city was observed from satellite images.

In this study area, very minimal changes were seen in the area of the Brahmaputra River during these study periods due to variation in sand depositions. But Deepor Beel, a perennial freshwater lake and former channel of the Brahmaputra located on the western side of Guwahati city, has shrunk due to its conversion to residential, institutional, industrial, tourism development, and agricultural uses reported in various studies (Ahmed et al., 2021; Bhattacharyya & Kapil, 2009). Unauthorized dredging of wetland and low-laying areas without considering ecological values and legitimate regulations led to the sudden and anomalous horizontal expansion of the city, where city development plans were completely circumvented.

Other LULC categories apart from dense forest cover, scattered forest, hardwood tree cover, shrubs, plantations, wetlands vegetation, grasses, and mosses also decrease in this region from 161.63 km2 (19.19%) in 2000 to 142.67 km2 (16.46%) in 2022. Over the years there was an increasing trend of built-up land in the peripheral areas of the city while semi-natural vegetation land diminished. Clearing of the vegetation cover for agricultural and settlement established in the outskirts of the city occurred in a disorderly manner.

Finally, land-use change by intense urbanization is a human-induced phenomenon, characterized by intertwined patterns and magnitude that amplify multiple detrimental effects like deforestation and decline in biodiversity (habitat destruction, fragmentation). Guwahati city is situated in an area with a high seismic activity level (V), making it more vulnerable to several natural hazards like earthquake, flooding, and increased risk of landslide in hill areas.

19.4.1.2 Drivers of Urban Growth

The spatial arrangement of present land use of a particular city is an illustration of how geographical factors, historical events, and economical and social factors interact. In order to improve trustworthiness of urban simulation, different driving variables were included in the model; as there is no universally accepted recommendation for variables, different researchers suggested different factors based on geophysical characteristics of different study areas (Thapa & Murayama, 2012). Although in most of the studies similar variables were employed in the majority of the research, the outcomes varied. Previous research highlighted the key variables that influence potential urban expansion and land-use changes (Morshed et al., 2023; Hamdy et al., 2016; Salem et al., 2021; Al Qadhi et al., 2021; Shahi et al., 2020; Gharaibeh et al., 2020; Mansour et al., 2023; Deep & Saklani, 2014). This research determines the driving variables from three sources: (1) review of related literature, (2) analysis of the official comprehensive plan, and (3) prior knowledge of the subject region. The importance of variables is as follows.

Elevation and slope (C1 and C2): Elevation and slope are always detrimental in positing settlement, both are considered as restrictive drivers where dwelling units are rapidly expanding and highly concentrated in flat places, while high elevation and steepness in topography increase construction cost (Zhang et al., 2019). *Distance from forest areas (C3)*: Considering the ecological values of forest, it should be preserved. But at present increased population growth and urbanization in the city forced more people moved into forested areas, which will inevitably place additional pressure on the resources in those areas and fragment the forest cover (Han et al., 2021; Geacu & Grigorescu, 2022). *Agricultural land (C4)*: Though preservation of agricultural land is crucial for food production, the development of cities in previously agricultural areas has been attributed to a reduction in the total area of farmland (Radwan et al., 2019). *River distance (C5)*: Most of the largest cities of the world are located either on riverside or near the ocean as cities need large amounts of water for industrial and other needs (Gidey et al., 2023). *Road distance (C6)*: The development of high-speed roadways is a significant factor in the spatial determination of urban expansion. Road networks support urban integration, ensure speedy supply of industrial logistics, and connect various functional areas that help in industrial investment and economic development.

In the absence of sound development policy, urban areas expands in a linear fashion along main roads, primarily on agricultural land. This results in the appearance of a ribbon-like urban sprawl (Zhao et al., 2017; Shi et al., 2019). *Railway network (C7)*: Railway lines have played a key role in the urbanization process; in this city Guwahati railway expansion helped in improving economic layout and accelerated the pace of urbanization in an unprecedented way. The railway provides better accessibility, connecting the city with surrounding villages and small towns and enhancing livelihood opportunities of inhabitants with better mobility (Varquez et al., 2020). *Airport (C8)*: Recent expansion of air transportation services in India now offers better connectivity for even distant and inaccessible and strategically important locations. This linkage has an important influence on the growth of the economy. An airport can be installed on barren and unused land outside the city to improve connectivity with other cities and make the city globally connected (Sheard, 2018). *Settlement patches (C9)*: The growth of urban areas and urban land-use changes are closely related to present settlement patches as settlement patches already have existing infrastructure, and that is why new settlements are constructed to avail those facilities (Gidey et al., 2017; Yang et al., 2019). *Distance from CBD (C10)*: As one moves farther away from the CBD, accessibility decreases and transport costs increase and that is why most Indian cities have high clustering of settlement near CBD areas. High clustering of banking, educational institutions, commercial areas, business opportunities, civic amenities, efficient transportation, and high standard of living attract more people into CBD areas (Mahmoudzadeh et al., 2022; Kim & Kim, 2022). *Industrial areas (C11)*: Historically, it was proven that the development of industrial property spurs economic expansion and employment opportunities, luring residents of rural areas towards urban areas. A large number of cities in India experienced rapid development in recent decades due to industrial installation. *City development plans (C12)*: City development plans are generally planned for the next 20–30 years based on projected population and extending area of urban agglomeration taken into consideration. As all those initiatives and developments were based on projected population growth, they had a significant impact on how cities will develop in the future (Figure 19.4).

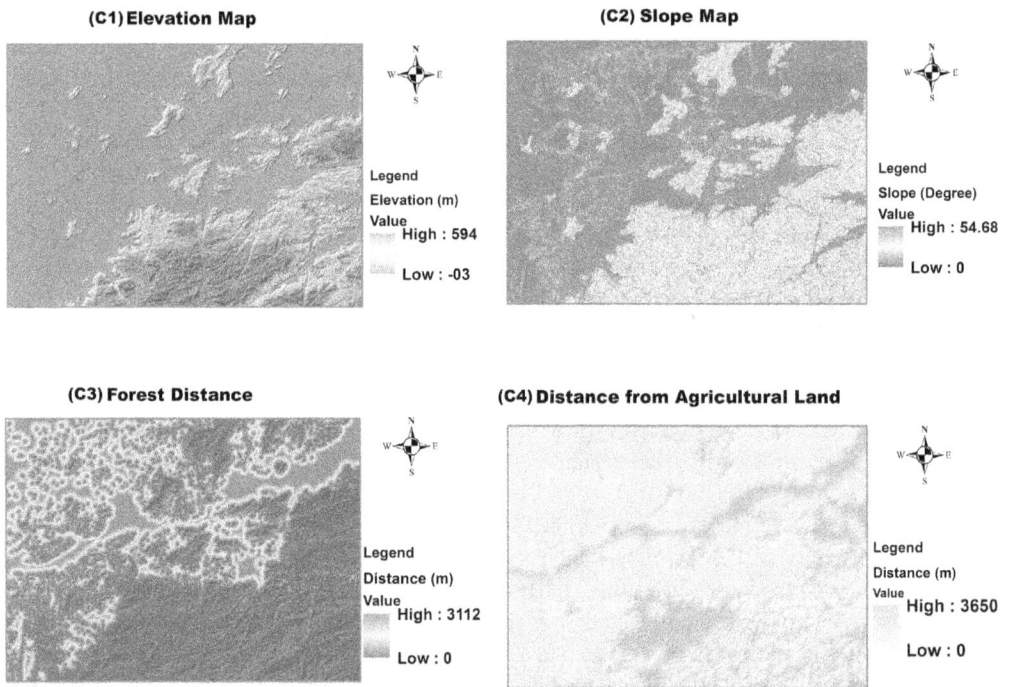

FIGURE 19.4   (C1) Elevation map, (C2) slope map, (C3) distance from forest areas, (C4) distance from agricultural land, (C5) proximity to river distance, (C6) proximity to road distance, (C7) proximity to railway track, (C8) distance from airport, (C9) distance from settlement areas, (C10) distance from central business district area, (C11) distance from industrial areas, and (C12) city development plan 2025.

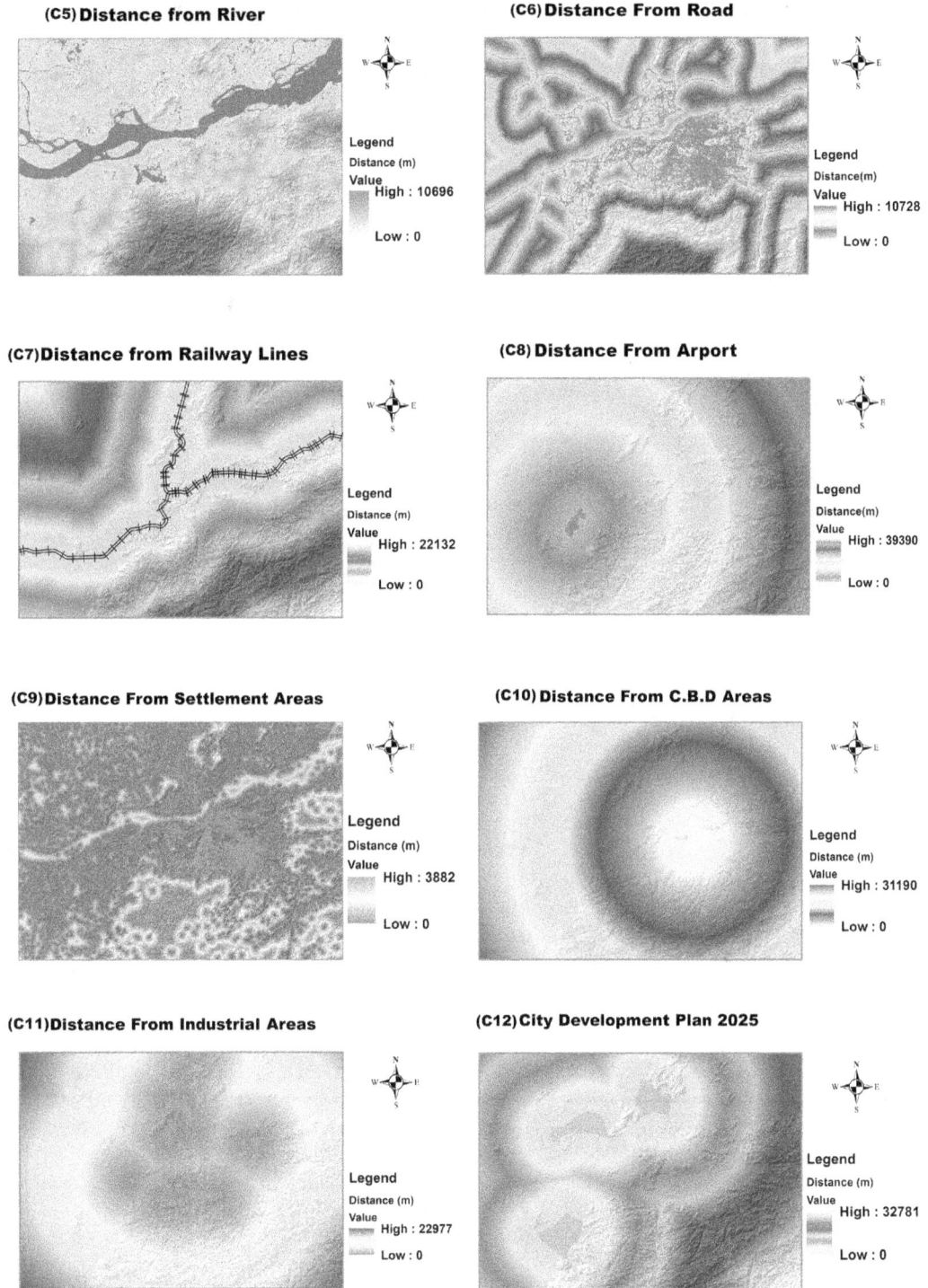

(C5) Distance from River

Legend
Distance (m)
Value
High : 10696

Low : 0

(C6) Distance From Road

Legend
Distance(m)
Value
High : 10728

Low : 0

(C7) Distance from Railway Lines

Legend
Distance (m)
Value
High : 22132

Low : 0

(C8) Distance From Airport

Legend
Distance(m)
Value
High : 39390

Low : 0

(C9) Distance From Settlement Areas

Legend
Distance (m)
Value
High : 3882

Low : 0

(C10) Distance From C.B.D Areas

Legend
Distance (m)
Value
High : 31190

Low : 0

(C11) Distance From Industrial Areas

Legend
Distance (m)
Value
High : 22977

Low : 0

(C12) City Development Plan 2025

Legend
Distance (m)
Value
High : 32781

Low : 0

FIGURE 19.4   (Continued)

Industrialization in a region results in the opening of a variety of business opportunities, including opportunities for retailers, manufacturers, and service providers. This results in the creation of even more jobs and increases the demand for housing, which in turn contributes to the process of urbanization (Qiu et al., 2015).

19.4.1.3 Markov Chain Analysis

The transition probability matrix was calculated using MC analysis for the time period of 2000–2014 (T1), 2014–2022 (T2), 2000–2022 (T3) for the prediction of LULC for 2022, 2035, and 2050 as shown in Table 19.7. Further, it is noted from the table that the probability of change of agricultural land to built-up area in the future is 15.96% (2022), 20.55% (2035), and 24.44% (2050), while the probability of agricultural land to remain as agricultural land is only 0.7642 (T1), 0.7049 (T2), and 0.6298 (T3). Therefore, this huge difference between probabilities reveals the intense decrease of agricultural land in the Patna district. Vegetation cover also has a high probability of conversion to other classes, The chances of remaining unchanged as vegetation classes is 67.27% (2022), 61.02% (2035), and 53.16% (2050). Urban class is the most stable class as it is a well-known fact that there is a lesser chance of conversion to other LULC types: the chances of urban class remaining as urban class is 97.36% (T1), 95.51% (T2), 93.10% (T3). Forest should be considered as a stable class, but a decent decrease in forest cover was observed in this study area. The probability of forest cover to remains as forest class is 94.30% (2022), 91.35% (2030), and 87.40% (2050). Table 19.8 includes Markov transition probability matrices for the LULC for the years 2035 and 2050. One is able to detect that there is a quick urban expansion in Guwahati City, Assam, through the use of visual and

TABLE 19.7
Matrix of Primary Transition Probability of LU/LC Types in MLP-ANN Model

	MLP-Markov	Agriculture	Forest	Vegetation	Water Bodies	Urban Areas	Total	Loss
Probability value of	Agriculture	**0.7642**	0.0000	0.0241	0.0521	0.1596	1.0000	0.2358
2022 based on	Forest	0.0035	**0.9430**	0.0311	0.0001	0.0223	1.0000	0.0570
transition matrix	Vegetation	0.1030	0.0125	**0.6727**	0.0047	0.2071	1.0000	0.3273
of 2000–2014	Water Bodies	0.0057	0.0042	0.0046	**0.8685**	0.1170	1.0000	0.1315
	Urban Areas	0.0000	0.0111	0.0000	0.0153	**0.9736**	1.0000	0.0264
	Total	0.8764	0.9708	0.7325	0.9407	1.4796	5.0000	
	Gain	0.1122	0.0278	0.0598	0.0722	0.5060		
Probability value of	Agriculture	**0.7049**	0.0056	0.0166	0.0674	0.2055	1.0000	0.2951
2035 based on	Forest	0.0117	**0.9135**	0.0433	0.0008	0.0307	1.0000	0.0865
transition matrix	Vegetation	0.0387	0.0806	**0.6102**	0.0141	0.2564	1.0000	0.3898
of 2014–2022	Water Bodies	0.0334	0.0077	0.0193	**0.8133**	0.1263	1.0000	0.1867
	Urban Areas	0.0186	0.0000	0.0004	0.0259	**0.9551**	1.0000	0.0449
	Total	0.8073	1.0074	0.6898	0.9215	1.5740	5.0000	
	Gain	0.1024	0.0939	0.0796	0.1082	0.6189		
Probability value of	Agriculture	**0.6298**	0.0173	0.0224	0.0861	0.2444	1.0000	0.3702
2050 based on	Forest	0.0229	**0.8740**	0.0387	0.0030	0.0614	1.0000	0.1260
transition matrix	Vegetation	0.0708	0.0039	**0.5316**	0.0266	0.3671	1.0000	0.4684
of 2000–2022	Water Bodies	0.0484	0.0130	0.0387	**0.7413**	0.1586	1.0000	0.2587
	Urban Areas	0.0069	0.0362	0.0001	0.0258	**0.9310**	1.0000	0.0690
	Total	0.7788	0.9444	0.6315	0.8828	1.7625	5.0000	
	Gain	0.1490	0.0704	0.0999	0.1415	0.8315		

Note: Bold figures on diagonal are probability of remaining unchanged.

TABLE 19.8

Test Explanatory Power of Selected Variables by Cramer's V Test

Factors	Driving Variable Class	Cramer's V
1	Elevation Map	0.1902
2	Slope Map	0.3172
3	Distance from Forest areas	**0.4807**
4	Distance from Agricultural land	**0.4322**
5	Proximity to River distance	**0.4001**
6	Proximity to Road distance	**0.5792**
7	Proximity to Railway track	0.3015
8	Distance from Airport	0.3344
9	Distance from Settlement areas	0.3864
10	Distance from C.B.D area	0.3472
11	Distance from Industrial areas	0.3197
12	City Development plan 2025	0.2792

Note: Bold numbers show the drivers with strong associations used for the LULC modelling.

TABLE 19.9

Description of Parameters and Performance for MLP test

Input layer neurons	12
Hidden layer neurons	10
Output layer neurons	8
Requested samples per class	10000
Final learning rate	0.0001
Momentum Factors	0.5
Sigmoid constant	1
Acceptable RMS	0.01
Iteration	10000
Training RMS	0.1920
Testing RMS	0.1960
Accuracy rate	96.95%
Skill Measure	0.9390

quantitative analyses of categorized maps of different years. This rapid urban growth, as opposed to typical urban growth, has to be analyzed and simulated further (Table 19.6).

19.4.1.4 Simulation of Potential Transition Changes

In this work, based on the trial-and-error analysis, 12, 10, and eight neurons were defined on input, hidden layer, and output layer neurons, respectively. The momentum factor and sigmoid constant were set at 0.5 and 1, respectively. Four transition probability maps, including agricultural, vegetation, forest, and water bodies, were mainly developed to quantify likelihood of land transformation into urban areas. Due to its capability to make numerous transitions (up to nine) simultaneously, MLP—a strong neural network–based technique—was employed in constructing the transition layers. The minimum number of cells that transitioned from 2000 to 2014 was 349,92 for MLP to run with 10,000 iterations.

Following the application of MLPNN, MLP performance displayed an accuracy rate of 96.95%, which is significantly higher than the cutoff percentage of 80% proposed by Sangermano et al. (2010), exhibiting exceptional model accuracy. Based on this function, the maps of transitional potential were produced. Therefore, these generated transition potentials maps were used to forecast future LULC in the following section according to an evaluation of the skill measure statistic, where a value of 1 denotes shows that flawless forecasting has been achieved, and a value of −1 denotes no skill. The high skill measure value of 0.9390 (93.90%) indicates that the model has significantly performed well in simulation of urban dynamics. The results affirm an appropriate selection of driving forces of urban expansion in Guwahati. The accuracy rate and skill measure for each of the different transition sub-models that were generated using MLP simulation are presented in Tables 19.7 and 19.8.

19.4.1.5 Relative Importance of Variables

After training the model with all of the variables, we compared their relative levels of importance to figure out which one had the most influence and which one had the least. This is done so that variables 1 through 12 are (1) distance from settlements, (2) proximity to road distance, (3) distance to CBD area, (4) river distance, (5) railway track, (6) industrial area, (7) slope, (8) elevation, (9) agricultural area, (10) airport distance, (11) city development plan, and (12) forest areas. The outcome shows that the most influential driving force on potential transition is variable 10 (settlement) while variable 8 (distance from forest) was the least influential (Table 19.9). The model proceeded to test every possible combination of variables in order to determine which combination had the least impact on the skill when held constant. According to Table 19.8, distance from settlement patches (most influential), road (second most influence), and distance from CBD (third most influential) are most influential for urban development which indicates closeness to those areas ensures high probability of urbanization and distance means the intensity of urbanization would decrease. On the other hand, agricultural land (ninth), airport distance (tenth), and city development plan (11th) are found to be less associated with urbanization. Agricultural land and forest should be preserved for their ecological significance. City development still has less influence on city urbanization pattern, as the airport (24-km away) is located far from city central areas and surrounded by scattered patches of settlements.

TABLE 19.10
Sensitivity of Model to Forcing Independent Variables To Be Constant

Model	Variables Name	Accuracy (%)	Skill measure	Influence order
With all variables		96.95	0.9390	N/A
Var. 1 constant	Elevation	96.96	0.9392	8
Var. 2 constant	Slope	96.96	0.9392	7
Var. 3 constant	River Distance	96.81	0.9362	4
Var. 4 constant	C.B.D area	96.72	0.9344	3
Var. 5 constant	Railway Tracks	96.89	0.9378	5
Var. 6 constant	Agricultural areas	97.00	0.9400	9
Var. 7 constant	City Development Plan	97.02	0.9404	11
Var. 8 constant	Forest Areas	97.21	0.9442	12 (least influential)
Var. 9 constant	Industrial area	96.95	0.9390	6
Var. 10 constant	Settlement Distance	58.68	0.1735	1 (most influential)
Var. 11 constant	Road Distance	93.48	0.8696	2
Var. 12 constant	Airport Distance	97.02	0.9404	10

19.4.1.6 Explanation of Potential Transitional Maps

The transition maps of potential areas and the likelihoods of urban transformation are illustrated in Figure 19.6. Projecting the land-use dynamics for the years 2035 and 2050 of Greater Guwahati Metropolitan area were main objective of this chapter. The potential transition map indicated a high probability of conversion of agricultural, forest, and vegetation land to settlement areas. In the northern part of the study area where industrial development was recently installed along the roads, there is a higher probability of conversion from agricultural land to urban areas. On the other hand, areas near deeper Bheel and the airport surrounding areas have a higher probability for conversion of vegetation cover to settlement patches. But in the southern and eastern part of the study areas there is a lesser probability of conversion due to the existence of dense forest cover, although along the NH 27 on the Assam-Meghalaya border some forest lands were cleared for different development purposes including industrial and educational institutions, resorts, and other activities.

19.4.1.7 Land-Use/Land-Cover Prediction and Validation

The MLP-MC technique was first used to forecast LULC patterns in 2022 utilizing the LULC maps from the years 2000 and 2014. This was done in order to make a reliable prediction for a future LULC scenario. Using kappa index statistics, the predicted LULC map for 2022 was compared with the actual LULC map for 2022 for validation purposes. A kappa statistic for amount and location is produced from the comparison of the projected LULC of 2022 with the observed LULC map of 2022. The statistics shows that K no value is 0.9548, K location value is 0.9555, K location Strata value is 0.9555, and K standard value is 0.9525 (overall kappa), respectively (Figure 19.6). All kappa index values were found to be higher than 0.80, demonstrating good agreement between the predicted and observed LULC maps. Overall, the validation process's results make it quite evident how much the two maps resemble

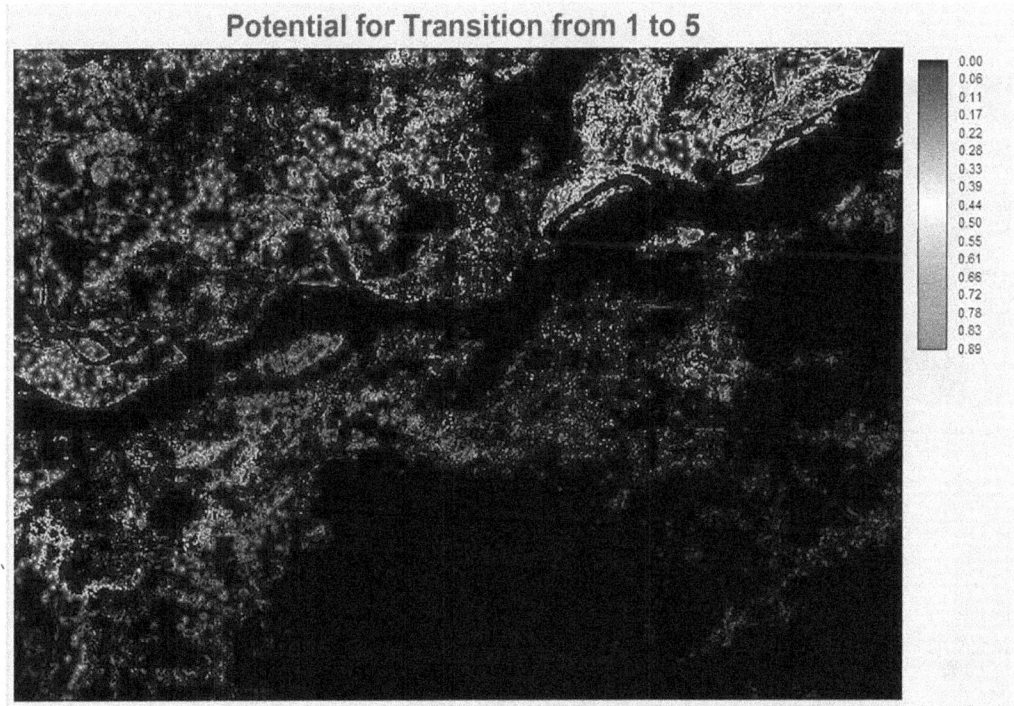

(a)

FIGURE 19.5 Transitional probability map: (a) agricultural land to urban, (b) forest cover to urban, (c) vegetation to urban, and (d) water bodies to urban.

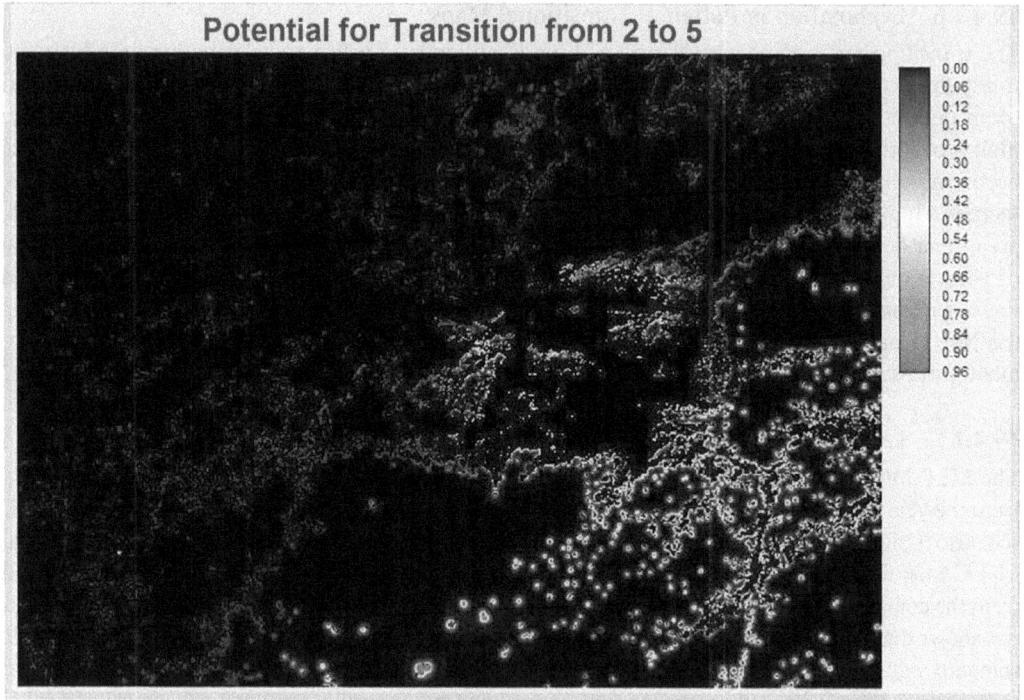

(b)

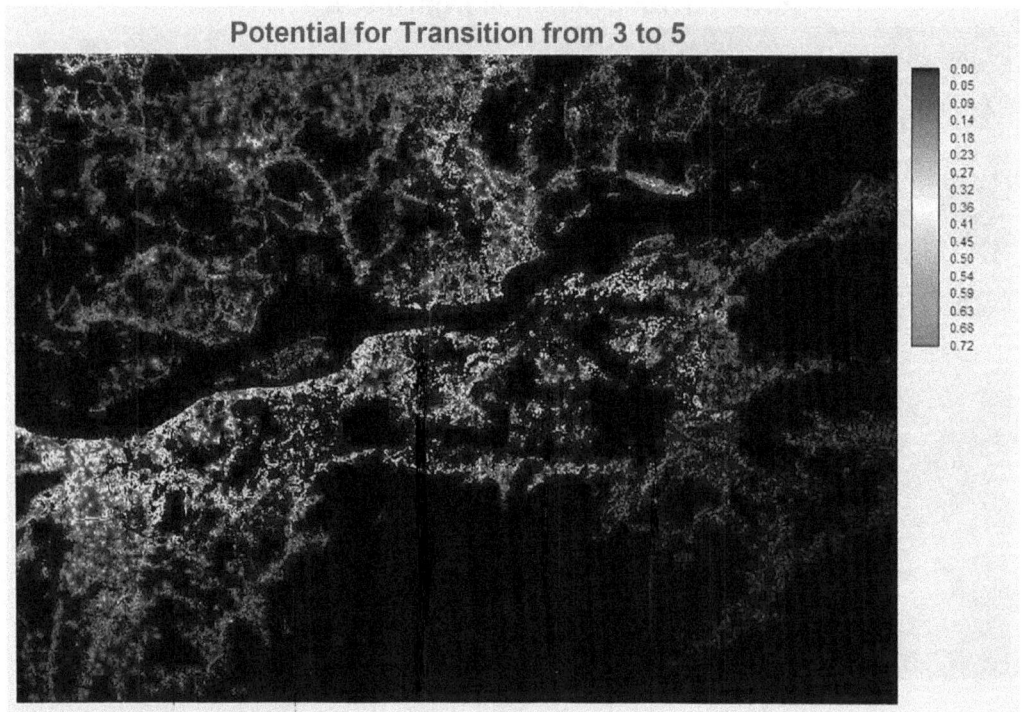

(c)

FIGURE 19.5   (Continued)

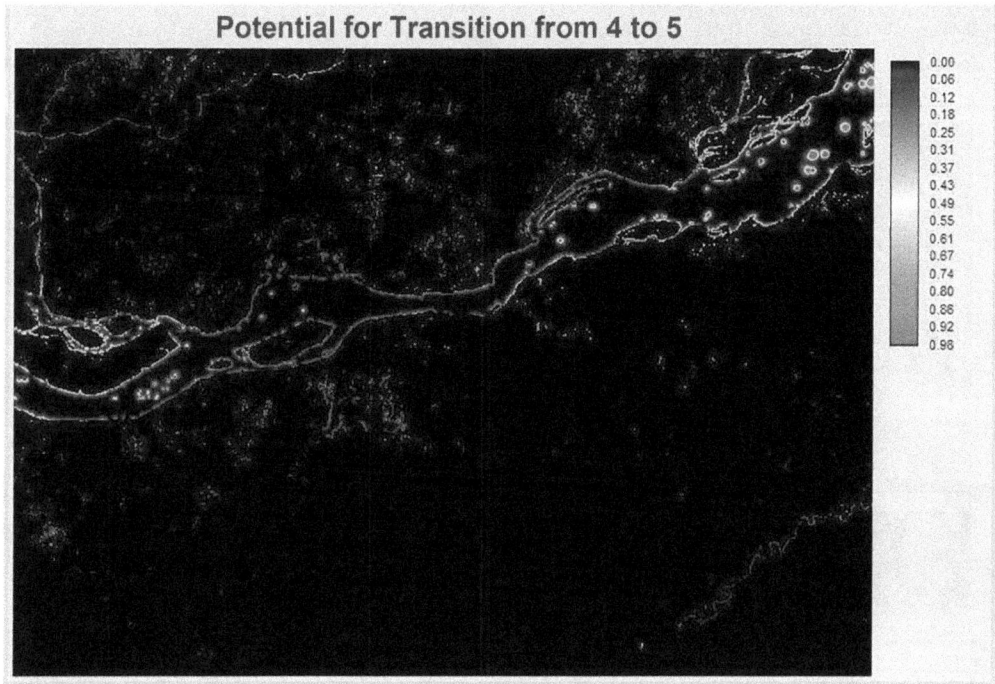

(d)

FIGURE 19.5 (Continued)

TABLE 19.11
Validation through Kappa Statistics

Classification agreement/disagreement	Kappa Values
Agreement Chance = 0.2000	K_{no} = 0.9548
Agreement Quantity = 0.0397	K location = 0.9555
Agreement Strata = 0.0000	K location Strata = 0.9555
Agreement Grid cell = 0.7242	K standard = 0.9525
Disagree Grid cell = 0.0337	
Disagree Strata = 0.0000	
Disagree Quantity = 0.0024	

---

one another. As a result, the model was robust, performed significantly better, and was highly accurate in predicting potential LULC change situations in Guwahati (Table 19.10, Figures 19.7 and 19.8).

The MLP-MCA–based prediction showed that built-up land would be increased significantly from 2022 (190.92 km2) to 2035 (244.68 km2), and for 2050 it would be anticipated to increase 289.97 km2. Within this 28 years, built-up land would be double what it is at present. An added area of 99.05 km2 urban land to the present city would be of increasing concern for the development authority. In the period 2022–2050, agricultural land is projected to undergo significant loss from 140.37 km2 (16.54%) to 106.47 km2 (12.64%). Inevitably, the reduction in the amount of agricultural land and vegetation will come at the expense of an unchecked increase in the amount of built-up area. The vegetation-covered area would also be decreased from 142.67 to 108.52 km2 between 2022 and 2050. Following the same trend, forest cover would decrease from its present status 281.45 km2 (33.41%) in 2022 to 268.69 km2 (31.89%) and in 2050 to 252.82 km2 (30.01%). Water bodies are the only class anticipated to remain stable in 2050 (Table 19.11; Figures 19.7 and 19.8).

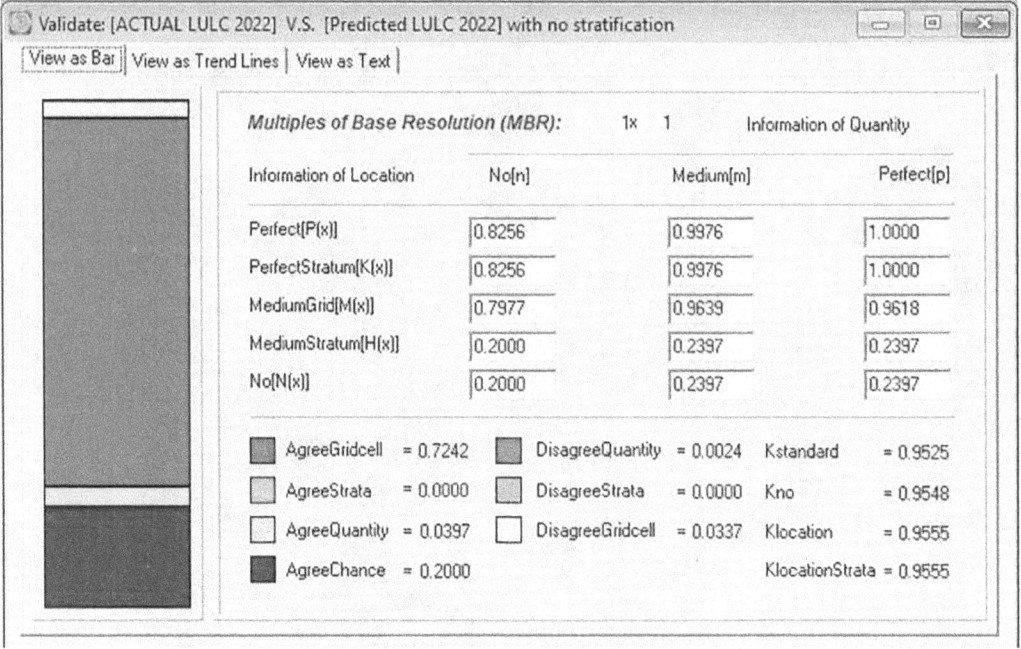

FIGURE 19.6   Validation through kappa statistics.

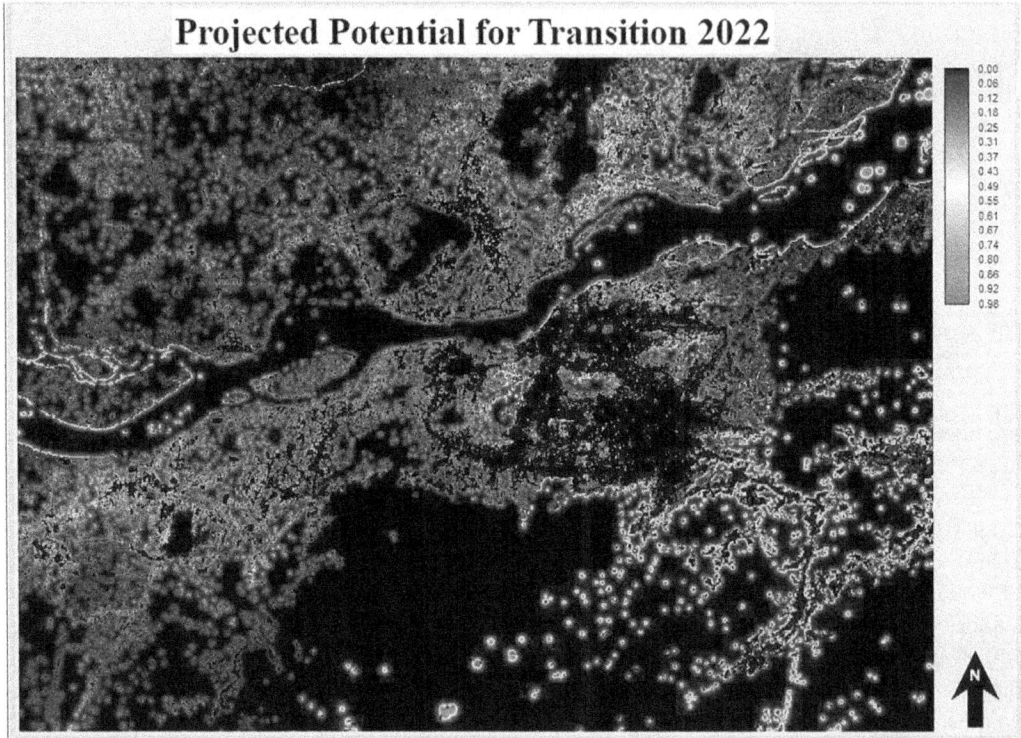

FIGURE 19.7   Conditional probabilty image for the years (a) 2022, (b) 2035, and (c) 2050.

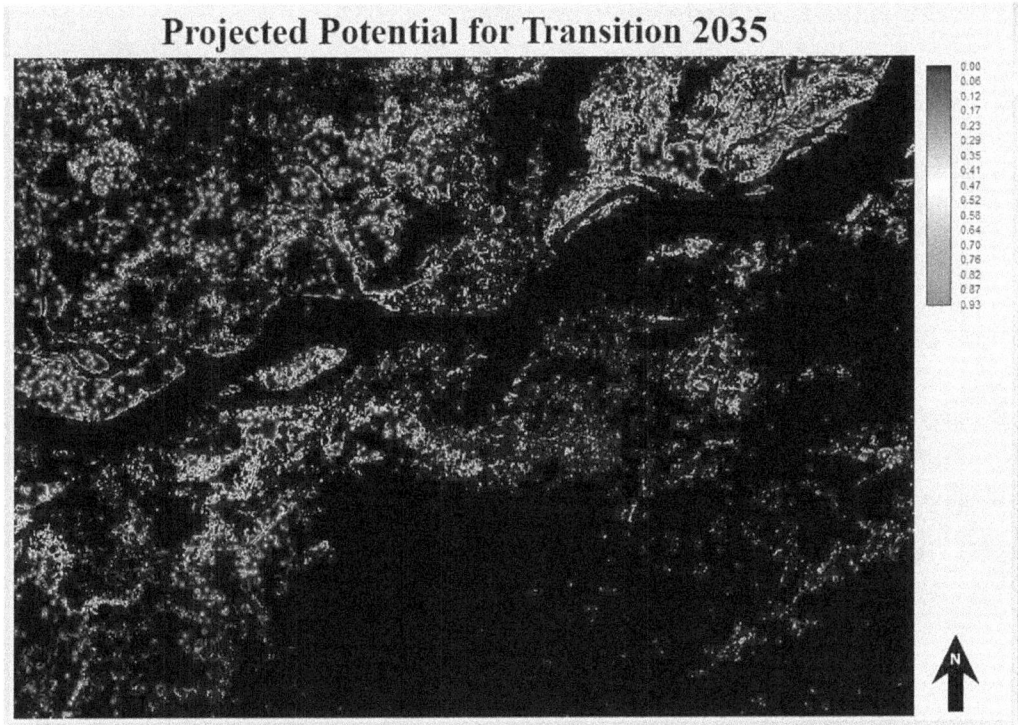

Projected Potential for Transition 2035

(b)

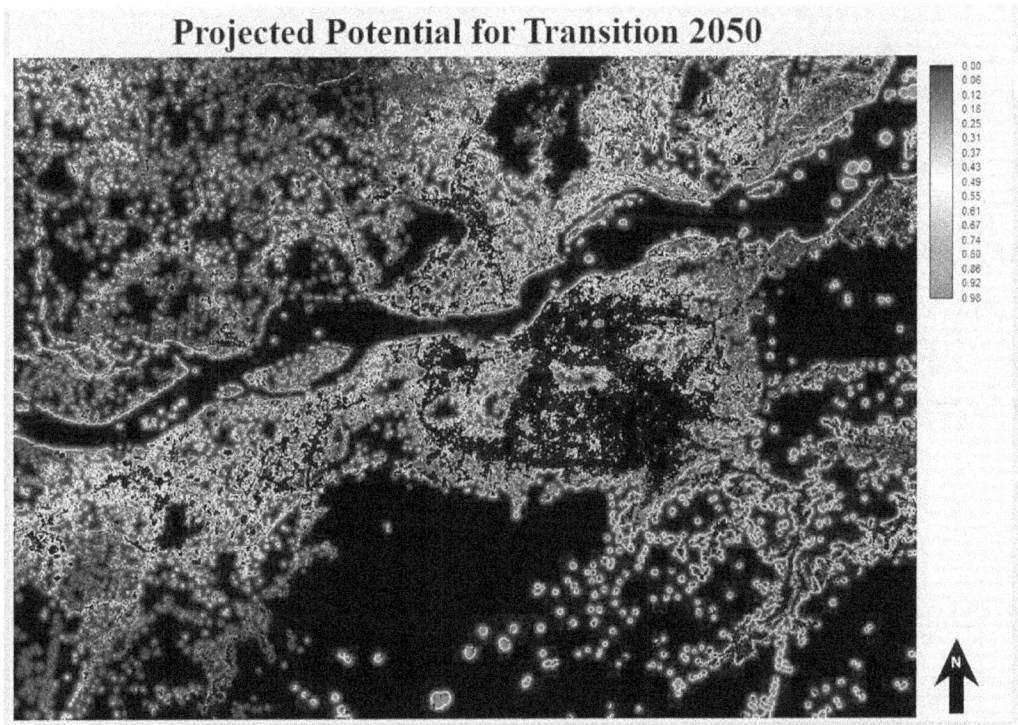

Projected Potential for Transition 2050

(c)

FIGURE 19.7   (Continued)

Projected Land Use Land Cover Map For 2035

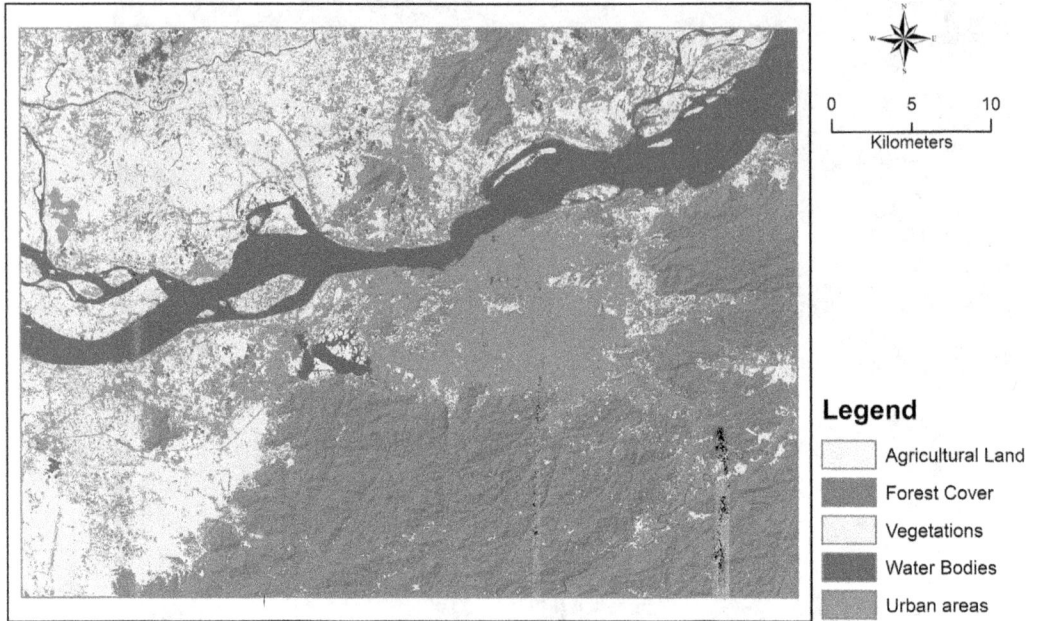

Projected Land Use Land Cover Map For 2050

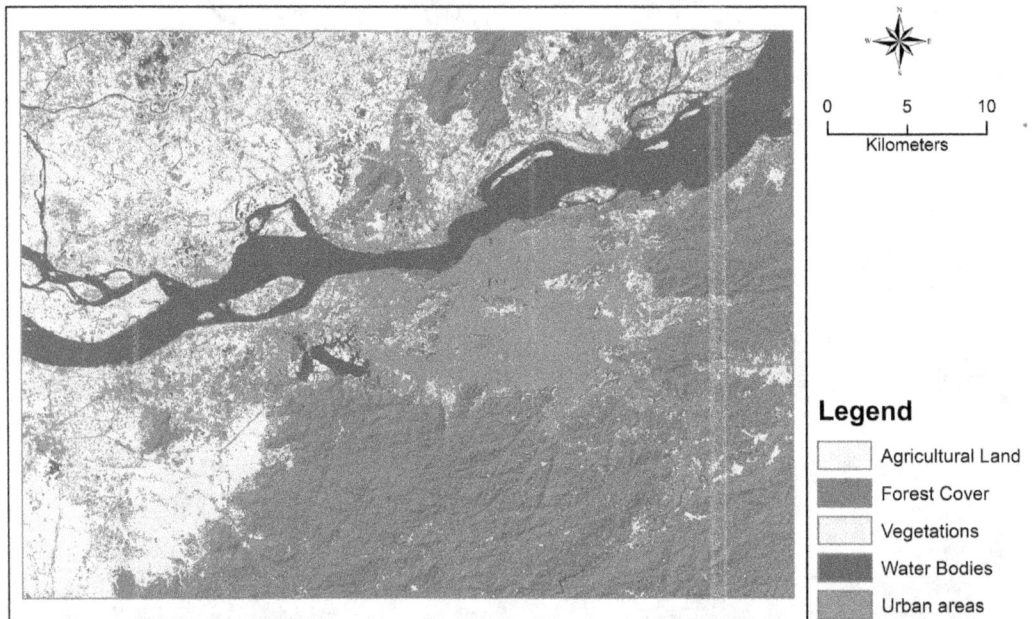

FIGURE 19.8   Simulated land-use/land-cover map for 2035 and 2050.

TABLE 19.12
Simulated LULC Changes for the Years 2035 and 2050

Class Name	2022		2035		2050		Quantity of change			Quantity of change		
	Km²	%	Km²	%	Km²	%	2022–2035 (km²)	%	Change per Annum	2022–2050 (km²)	%	Change per Annum
Agriculture	140.37	16.54	124.37	14.76%	106.47	12.64%	−16.00	−11.40%	−1.23	−33.90	−24.15%	−1.21
Forest	281.45	33.41	268.69	31.89	252.82	30.01	−12.76	−4.53	−0.98	−28.63	−10.17	−1.02
Vegetation	142.67	16.46	119.67	14.21	108.52	12.88	−23.00	−16.12	−1.77	−34.15	−23.94	−1.22
Water Bodies	87.04	10.33	85.04	10.09	84.67	10.05	−2.00	−2.30	−0.15	−2.37	−2.72	−0.08
Urban Areas	190.92	23.26	244.68	29.04	289.97	34.42	53.76	28.16	4.14	99.05	51.88	3.54
Total Areas	842.45	100.00	842.45	100.00	842.45	100.00						

19.4.2 DISCUSSION

In the past two decades, the city of Guwahati has experienced a surge of construction activities, which has resulted in the conversion of the majority of the city's ecological area into artificial land that has exacerbated urban sprawling. The ecosystem and ecology of the city have been negatively impacted by rapid population growth, high migration rates, and changes in land-use patterns. Inadvertent city growth led to the current chaotic civic situation in which the natural environment is steadily being destroyed for the planned construction of buildings and roads, the release of industrial wastes into drainage channels, and other factors that have seriously affected the depletion of natural resources and environmental pollution. Moreover, these unregulated or illegal settlements did not follow existing city development plans and haphazardly sprouted on hillsides, riverbanks, moist swampy plains, and the edges of the forest areas, The present study monitored and projected the changing pattern of land-use practices in Guwahati between 2000 to 2050. It can be anticipated that there will be a continuous decrease in the forest area, agriculture, and vegetation cover along with a tremendous increase in built-up area that should degrade the city's life and environment. Therefore, it is crucial that urban and environmental planners cooperate in order to improve urban living in Guwahati. In places that have already lost their forests, new afforestation initiatives should be implemented, and social tree-planting initiatives should be supported. In order to promote the sustainable growth of the overpopulated city of Guwahati, the government should also support satellite towns or alternative economic hubs. Intense rainfall with deforestation causing frequent landslides in surrounding cities have been reported in recent times.

Application of MLPNN increased substantially over the last several years because the advances in computing performance could efficiently handle complicated patterns like urban expansion and alterations in land use. Additionally, it offers non-linearities and can deal with ambiguous or incomplete data (Aburas et al., 2019). In this study, MLPNN-Markov model achieved 95% accuracy in predicting future LULC. Similar methodology is followed by Mishra and Rai (2016) as they applied the MLPNN model in **the** Patna district (Bihar), India, and achieved a successes rate of 83.35%. In their study, Rimal, Bhagawat et al. (2020) based the historical trend of land-use change during 1989–2016 and predicted urban expansion by 2026 and 2036 using ANN and MC for Nepal's Tarai region. The evaluation implied that the value of 88% verified the accuracy of the model. Morshed et al. (2023), in their study simulating future intra-urban land-use patterns of Jashore, Bangladesh, predicted a LULC pattern of 2030, 2040, 2050 and achieved an 87% in the MLPNN MC model. Mansour, Shawky et al. (2023) used spatiotemporal modelling of urban expansion and its effects, particularly on Egyptian coastal areas for the years 2030 and 2050. While validating, the MLPNN model achieved an accuracy of 87.30% for the year 2020.

1. The city of Guwahati benefits from strategic position, consisting of a multitude of natural features, including rivers, marsh, forest-covered wetlands, and hillocks. Yet Guwahati's spontaneous, urban growth has altered the city's arrangement of land use. Built-up areas have increased at a rate of 4.37 km2 per annum between 2000 and 2022; following the same trend it would be anticipated that in 2022–2050, built-up land will increase 3.54 km2 annually. In addition, the analysis finds that from 2022 to 2050, built-up land will grow by 51.88% in the study areas. Deforestation and the subsequent changing LULC increase the land surface temperature, since human-made structures mainly composed of asphalt and concrete reduce the chances of heat transmission and increase the chances of heat island–like phenomena. People are settling in the hills and outskirts of the cities surrounding due to the intense overcrowding and lack of space, and they are progressively expanding to the adjacent regions. Generally, outside of the main city is less controlled by municipal laws, due to the incompetence of local authorities to accommodate the growing population. Numerous developers and entrepreneurs find that these areas are advantageous for developments due to cheap cost compared to the inner city (Chetia et al., 2020).

2. Forest areas decreased at a rate of 1.74 km2 per annum between 2000 and 2022, and it would be anticipated that between 2022 and 2050, the forest will shrink at a rate of 1.21 km2. With a significant population growth, people are residing on hills and foothills of city forest. This burgeoning population first settled in the easily accessed plain areas around the city's core before extending into the surrounding hills. Several anthropogenic activities and unauthorized encroachment of forest patches increases the chances of landslides and floods in Guwahati city.

3. The agricultural areas that are located on the outskirts of cities act as transition zones between the natural and rural ecosystems and the urban landscapes. Agricultural land decreased at a rate of 1.46 km2 between 2000 and 2022, and further, it would be projected that from 2022 to 2050 the rate will be 1.21 km2 per annum. Vegetation areas are decreased at a rate of 0.86 km2 per annum, between 2000 and 2022, and the same trend would be estimated to continue at 1.22 km2 between 2022 and 2050. The removal of vegetation cover as a result of growing construction in the city does not follow the master plan of 2025.

4. Encroachment of water bodies and shrinking of Ramsar site-*Dipar Beel* wetland due to earth filling and garbage dumping is described as a noteworthy problem in Guwahati that needs to be addressed in the next master plan. Flooding incidents have increased in frequency in Guwahati over the past few years as a result of an overflow of water from the river Brahmaputra and obstructions in the drainage system caused by natural drainage systems, which most likely produce floods regularly every monsoon.

5. After monitoring present scenario of urban expansion in Guwahati city, it was appread that eastern and southern portions of the city surrounded by dense forest cover of Amchang Wildlife sanctuary and other forest areas forbid further urban expansion in those areas. Whereas the simulated LULC map of Greater Guwahati metropolitan area showed that after saturation of city administrative limits, urban areas futher anticipated to be expand further in south-west part, areas surroundings of Deepar Beel lake and adjacent areas to the airport. Northern part of river Brahmaputra, recent development in industrial and educational institutions along the **NH-27** and **SH-2** areas is triggering further expansion in that region.

6. By 2050, the urban class will cover the area of 289.97 km2, which is 34.42% of the entire study area, while forest cover, agricultural land, vegetation cover, and water categories will occupy 30.01%, 12.64%, 12.88%, and 10.05%, respectively. The majority of previous and likely future urban sprawl is strongly correlated with the major roadways, indicating that the road network is one of the most important driving forces behind urban growth.

19.5 CONCLUSION

After monitoring urban growth in Guwahati town, it appears that the intensity of urban sprawl has accelerated in recent years. The proportion of built-up areas shows noticeable variations, which may be the result of unregulated urban growth and exorbitant infrastructure development. If urban sprawl continues to spread in its current trajectory, Guwahati's vulnerable eco-sensitive and peri-urban areas could be in serious jeopardy. As far as the authors are aware, the performance of the MLPNN-MCM has not been assessed previously over this study area; therefore, this research will also help to assess the performance of the MLPNN-MCM approach over Guwahati and its surrounding region. Although the studied neural network model is straightforward and practical in use, it can handle extremely complex aspects of urban systems. The ANN demonstrates robustness by automatically calculating the parameter values throughout the training process, which can then be simply imported back into GIS to simulate urban change and cut down on calibration time. The GIS database makes it easy to retrieve spatial data for the mode's calibration and testing. Secondary cities have unfortunately received scant attention in urban policy research, and Guwahati epitomizes urban woes in a developing country. The result of this study also depicts the usefulness of MLPNN-based MCM to quantify and visualize the future LULC patterns with higher accuracy during valida-tion periods. Important driving factors including climate unpredictability, vulnerability to hazards, and potential government initiatives were not included in this model. This is one of the study's shortcomings and a focus for future research by considering the more varied socio-economic and political causes for urban expansion. Therefore, there will be room to improve these models in the future for greater accuracy. In this study, rather than producing a short-term projection, the LULC prediction for 2030 and 2050 was carried out to allow local authorities enough of a timeline for policy-making, revision, and implementation with a foresight of its utility in sustainable planning.

REFERENCES

Abdi, A. M. (2019). Land cover and land use classification performance of machine learning algorithms in a boreal landscape using Sentinel-2 data. *GIScience & Remote Sensing*, *57*(1), 1–20. https://doi.org/10.10 80/15481603.2019.1650447

Aburas, M. M., Ahamad, M. S. S., & Omar, N. Q. (2019). Spatio-temporal simulation and prediction of land-use change using conventional and machine learning models: A review. *Environmental Monitoring and Assessment*, *191*, 1–28. https://doi.org/10.1007/s10661-019-7330-6

Aburas, M. M., Ho, Y. M., Ramli, M. F., & Ash'aari, Z. H. (2017). Improving the capability of an integrated CA-Markov model to simulate spatio-temporal urban growth trends using an analytical hierarchy process and frequency ratio. *International Journal of Applied Earth Observation and Geoinformation*, *59*, 65–78. https://doi.org/10.1016/j.jag.2017.03.006

Ahmed, B., & Ahmed, R. (2012). Modeling urban land cover growth dynamics using multi-temporal satellite images: A case study of Dhaka, Bangladesh. *ISPRS International Journal of Geo-Information*, *1*(1), 3–31. https://doi.org/10.3390/ijgi1010003

Ahmed, I. A., Shahfahad, S., Baig, M. R. I., Talukdar, S., Asgher, M. S., Usmani, T. M., Ahmed, S., & Rahman, A. (2021). Lake water volume calculation using time series LANDSAT satellite data: A geospatial analy-sis of Deepor Beel Lake, Guwahati. *Frontiers in Engineering and Built Environment*, *1*(1), 107–130. http://dx.doi.org/10.1108/FEBE-02-2021-0009

Akinci, H., Kilicoglu, C., & Dogan, S. (2020). Random forest-based landslide susceptibility mapping in coastal regions of Artvin, Turkey. *ISPRS International Journal of Geo-Information*, *9*(9), 553. https://doi.org/10.3390/ijgi9090553

AlQadhi, S., Mallick, J., Balha, A., Bindajam, A. A., Singh, C. K., & Hoa, P. V. (2021). Spatial and decadal prediction of land use/land cover using Multi-Layer Perceptron-Neural Network (MLP-NN) algorithm for a semi-arid region of Asir, Saudi Arabia. *Earth Science Informatics*, *14*, 1547–1562. https://doi.org/10.1007/s12145-021-00633-2

Amini, S., Saber, M., Rabiei-Dastjerdi, H., & Homayouni, S. (2022). Urban land use and land cover change analysis using random forest classification of Landsat time series. *Remote Sensing*, *14*(11), 2654. https://doi.org/10.3390/rs14112654

Anderson, J. R., et al. (1976). *A land use and land cover classification system for use with remote sensor data* (p. 28). Geological Survey Professional Paper No. 964, U.S. Government Printing Office, Washington, DC.

Ansari, A., & Golabi, M. H. (2019). Prediction of spatial land use changes based on LCM in a GIS environment for Desert Wetlands: A case study: Meighan Wetland, Iran. *International Soil and Water Conservation Research*, *7*, 64–70. https://doi.org/10.1016/j.iswcr.2018.10.001

Arsanjani, J. J., Helbich, M., Kainz, W., & Boloorani, A. D. (2013). Integration of logistic regression, Markov chain and cellular automata models to simulate urban expansion. *International Journal of Applied Earth Observation and Geoinformation*, *21*, 265–275. https://doi.org/10.1016/j.jag.2011.12.014

Asadi, M., Oshnooei-Nooshabadi, A., Saleh, S.-S., Habibnezhad, F., Sarafraz-Asbagh, S., & Van Genderen, J. L. (2022). Urban sprawl simulation mapping of Urmia (Iran) by comparison of cellular automata: Markov chain and Artificial Neural Network (ANN) modeling approach. *Sustainability*, *14*(23), 15625. MDPI AG. http://dx.doi.org/10.3390/su142315625

Bagheri, B., & Tousi, S. N. (2017). An explanation of urban sprawl phenomenon in Shiraz Metropolitan Area (SMA). *Cities*, *73*, 71–90. https://doi.org/10.1016/j.cities.2017.10.011

Baig, M. F., Mustafa, M. R. U., Baig, I., Takaijudin, H. B., & Zeshan, M. T. (2022). Assessment of land use land cover changes and future predictions using CA-ANN simulation for Selangor, Malaysia. *Water*, *14*(3), 402. https://doi.org/10.3390/w14030402

Basheer, S., Wang, X., Farooque, A. A., Nawaz, R. A., Liu, K., Adekanmbi, T., & Liu, S. (2022). Comparison of land use land cover classifiers using different satellite imagery and machine learning techniques. *Remote Sensing, 14*(19), 4978. https://doi.org/10.3390/rs14194978

Bhagat, R. B. (2011). Emerging pattern of urbanization in India. *Economic and Political Weekly, 46*. www.jstor.org/stable/23017782

Bhat, P. A., Shafiq, M. U., Mir, A. A., & Ahmed, P. (2017). Urban sprawl and its impact on landuse/land cover dynamics of Dehradun City, India. *International Journal of Sustainable Built Environment, 6*, 513–521. https://doi.org/10.1016/J.IJSBE.2017.10.003

Bhatta, B. (2010). Causes and consequences of urban growth and sprawl. In B. Bhatta (Ed.), *Analysis of urban growth and sprawl from remote sensing data* (pp. 17–36). Springer-Verlag, Berlin, Heidelberg. https://doi.org/10.1007/978-3-642-05299-6_2

Bhattacharyya, K. G., & Kapil, N. (2009). Impact of urbanization on the quality of water in a natural reservoir: A case study with the Deepor Beel in Guwahati city, India. *Water and Environment Journal, 24*. https://doi.org/10.1111/j.1747-6593.2008.00157.x

Borthakur, M., & Kr, B. (2012). A study of changing urban landscape and heat island phenomenon in Guwahati Metropolitan Area. *International Journal of Scientific and Research Publications, 2*(11), ISSN 2250-3153.

Boulesteix, A. L., Janitza, S., Kruppa, J., & König, I. R. (2012). Overview of random forest methodology and practical guidance with emphasis on computational biology and bioinformatics. *Wiley Interdisciplinary Reviews: Data Mining and Knowledge Discovery, 2*, 493–507. http://dx.doi.org/10.1002/widm.1072

Breiman, L. (2001). Random forests. *Machine Learning, 45*, 5–32. http://dx.doi.org/10.1023/A:1010933404324

Brueckner, J. K. (2000). Urban sprawl: Diagnosis and remedies. *International Regional Science Review, 23*, 160–171. https://doi.org/10.1177/016001700761012710

Chander, G., Markham, B. L., & Helder, D. L. (2009). Summary of current radiometric calibration coefficients for Landsat MSS, TM, ETM+, and EO-1 ALI sensors. *Remote Sensing of Environment, 113*(5), 893–903. https://doi.org/10.1016/j.rse.2009.01.007

Chetia, S., Saikia, A., Basumatary, M., et al. (2020). When the heat is on: Urbanization and land surface temperature in Guwahati, India. *Acta Geophys., 68*, 891–901. https://doi.org/10.1007/s11600-020-00422-3

Cohen, J. (1960). A coefficient of agreement for nominal scales. *Educational and Psychological Measurement, 20*, 37–46. http://dx.doi.org/10.1177/001316446002000104

Coulibaly, B., & Li, S. (2020). Impact of agricultural land loss on rural livelihoods in peri-urban areas: Empirical evidence from Sebougou, Mali. *Land, 9*(12), 470. https://doi.org/10.3390/land9120470

Dadhich, P. N., & Hanaoka, S. (2011). Spatio-temporal urban growth modeling of Jaipur, India. *Journal of Urban Technology, 18*, 45–65. https://doi.org/10.1080/10630732.2011.615567

Dang, A. N., & Kawasaki, A. (2017). Integrating biophysical and socio-economic factors for land-use and land-cover change projection in agricultural economic regions. *Ecological Modelling, 344*, 29–37. https://doi.org/10.1016/J.ECOLMODEL.2016.11.004

Decocq, G., Andrieu, E., Brunet, J., et al. (2016). Ecosystem services from small forest patches in agricultural landscapes. *Curr. Forestry Rep., 2*, 30–44. https://doi.org/10.1007/s40725-016-0028-x

Deep, S., & Saklani, A. (2014). Urban sprawl modeling using cellular automata. *Egyptian Journal of Remote Sensing and Space Science, 17*, 179–187. https://doi.org/10.1016/j.ejrs.2014.07.001

Deka, P. P., & Devi, M. K. (2017). Problems and prospects of development in Guwahati, Assam. In P. Sharma & S. Rajput (Eds.), *Sustainable smart cities in India*. The Urban Book Series. Springer, Cham. https://doi.org/10.1007/978-3-319-47145-7_7

Eastman, J. R. (2009). *IDRISI taiga guide to GIS and image processing: Clark labs for cartographic technology and geographic analysis*. Clark University, Worcester.

Farhadi, H., & Najafzadeh, M. (2021). Flood risk mapping by remote sensing data and random forest technique. *Water, 13*(21), 3115. https://doi.org/10.3390/w13213115

Geacu, S., & Grigorescu, I. (2022). Historical changes in urban and peri-urban forests: Evidence from the Galați Area, Romania. *Land, 11*(11), 2043. https://doi.org/10.3390/land11112043

Gharaibeh, A., Shaamala, A., Obeidat, R., & Al-Kofahi, S. (2020). Improving land-use change modeling by integrating ANN with cellular automata-Markov Chain model. *Heliyon, 6*(9), e05092. https://doi.org/10.1016/j.heliyon.2020.e05092

Gidey, E., Dikinya, O., Sebego, R., Segosebe, E., & Zenebe, A. (2017). Modeling the spatio-temporal dynamics and evolution of land use and land cover (1984–2015) using remote sensing and GIS in Raya, Northern Ethiopia. *Modeling Earth Systems and Environment, 3*, 1285–1301. https://doi.org/10.1007/s40808-017-0375-z

Gidey, E., Dikinya, O., Sebego, R., Segosebe, E., Zenebe, A., Mussa, S., Mhangara, P., & Birhane, E. (2023). Land use and land cover change determinants in Raya Valley, Tigray, Northern Ethiopian Highlands. *Agriculture*, *13*(2), 507. https://doi.org/10.3390/agriculture13020507

Gogoi, D., Bhaskaran, G., & Gogoi, A. (2023). An analysis of land dynamics in relation to urban sprawl in the Guwahati city of Assam, India. *Ecocycles*, *9*(1), 49–60. ISSN 2416–2140

Gumma, M., Mohammad, I., Nedumaran, S., Whitbread, A., & Lagerkvist, C. (2017). Urban sprawl and adverse impacts on agricultural land: A case study on Hyderabad, India. *Remote Sensing*, *9*(11), 1136. https://doi.org/10.3390/rs9111136

Hamdy, O., Zhao, S., Osman, T., Salheen, M., & Eid, Y. (2016). Applying a hybrid model of Markov chain and logistic regression to identify future urban sprawl in Abouelreesh, Aswan: A case study. *Geosciences*, *6*(4), 43. MDPI AG. http://dx.doi.org/10.3390/geosciences6040043

Han, J., Dong, Y., Ren, Z., Du, Y., Wang, C., Jia, G., Zhang, P., & Guo, Y. (2021). Remarkable effects of urbanization on forest landscape multifunctionality in urban peripheries: Evidence from Liaoyuan city in Northeast China. *Forests*, *12*(12), 1779. https://doi.org/10.3390/f12121779

Harris, R., & Baumann, I. (2015). Open data policies and satellite Earth observation. *Space Policy*, *32*, 44–53. https://doi.org/10.1016/j.spacepol.2015.01.001

Hasan, S., Shi, W., Zhu, X., Abbas, S., & Khan, H. U. A. (2020). Future simulation of land use changes in rapidly urbanizing South China based on land change modeler and remote sensing data. *Sustainability*, *12*(11), 4350. https://doi.org/10.3390/su12114350

Hassan, M., & Elhassan, S. (2020). Modelling of urban growth and planning: A critical review. *Journal of Building Construction and Planning Research*, *8*, 245–262. doi: 10.4236/jbcpr.2020.84016.

Hassan, Z., Shabbir, R., Ahmad, S. S., Malik, A. H., Aziz, N., Butt, A., & Erum, S. (2016). Dynamics of Land Use and Land Cover Change (LULCC) using geospatial techniques: A case study of Islamabad Pakistan. *SpringerPlus*, *5*(1), 812. https://doi.org/10.1186/s40064-016-2414-z

Hemani, S., & Das, A. K. (2016). City profile: Guwahati. *Cities*, *50*, 137–157. https://doi.org/10.1016/j.cities.2015.08.003

Junaid, M., Sun, J., Iqbal, A., Sohail, M., Zafar, S., & Khan, A. (2023). Mapping LULC dynamics and its potential implication on forest cover in Malam Jabba region with Landsat time series imagery and random forest classification. *Sustainability*, *15*(3), 1858. https://doi.org/10.3390/su15031858

Kamusoko, C., Aniya, M., Adi, B., & Manjoro, M. (2009). Rural sustainability under threat in Zimbabwe-simulation of future land use/cover changes in the Bindura district based on the Markov-cellular automata model. *Applied Geography*, *29*, 435–447. https://doi.org/10.1016/j.apgeog.2008.10.002

Kanga, S., Meraj, G., Johnson, B. A., Singh, S. K., PV, M. N., Farooq, M., Kumar, P., Marazi, A., & Sahu, N. (2022). Understanding the linkage between urban growth and land surface temperature: A case study of Bangalore city, India. *Remote Sensing*, *14*(17), 4241. https://doi.org/10.3390/rs14174241

Kim, H., & Kim, D. (2022). Changes in urban growth patterns in Busan Metropolitan city, Korea: Population and urbanized areas. *Land*, *11*(8), 1319. https://doi.org/10.3390/land11081319

Leta, M. K., Demissie, T. A., & Tränckner, J. (2021). Modeling and prediction of land use land cover change dynamics based on Land Change Modeler (LCM) in Nashe Watershed, Upper Blue Nile Basin, Ethiopia. *Sustainability*, *13*(7), 3740. https://doi.org/10.3390/su13073740

López-Serrano, P., Corral-Rivas, J., Díaz-Varela, R., Álvarez-González, J., & López-Sánchez, C. (2016). Evaluation of radiometric and atmospheric correction algorithms for aboveground forest biomass estimation using Landsat 5 TM data. *Remote Sensing*, *8*(5), 369. https://doi.org/10.3390/rs8050369

Mahadevia, D., Desai, R., & Mishra, A. (2014). City profile: Guwahati. Working Paper 24. Centre for Urban Equity, CEPT University, Ahmedabad.

Mahmoudzadeh, H., Abedini, A., & Aram, F. (2022). Urban growth modeling and land-use/land-cover change analysis in a metropolitan area (Case study: Tabriz). *Land*, *11*(12), 2162. https://doi.org/10.3390/land11122162

Maithani, S. (2009). A neural network based urban growth model of an Indian city. *J. Indian Soc. Remote Sens.*, *37*, 363–376. https://doi.org/10.1007/s12524-009-0041-7

Manandhar, R., Odeh, I., & Ancev, T. (2009). Improving the accuracy of land use and land cover classification of Landsat data using post-classification enhancement. *Remote Sensing*, *1*(3), 330–344. https://doi.org/10.3390/rs1030330

Mansour, S., Ghoneim, E., El-Kersh, A., Said, S., & Abdelnaby, S. (2023). Spatiotemporal monitoring of urban sprawl in a coastal city using GIS-based Markov chain and artificial neural network (ANN). *Remote Sensing*, *15*(3), 601. MDPI AG. http://dx.doi.org/10.3390/rs15030601

Mas, J. F., Kolb, M., Paegelow, M., Olmedo, M. T. C., & Houet, T. (2014). Inductive pattern-based land use/cover change models: A comparison of four software packages. *Environmental Modelling & Software*, *51*, 94–111. https://doi.org/10.1016/j.envsoft.2013.09.010

Maxwell, A. E., Sharma, M., Kite, J. S., Donaldson, K. A., Thompson, J. A., Bell, M. L., & Maynard, S. M. (2020). Slope failure prediction using random forest machine learning and LiDAR in an eroded folded mountain belt. *Remote Sensing*, *12*(3), 486. https://doi.org/10.3390/rs12030486

Maxwell, A. E., Warner, T. A., & Fang, F. (2018). Implementation of machine-learning classification in remote sensing: An applied review. *International Journal of Remote Sensing*, *39*(9), 2784–2817. https://doi.org/10.1080/01431161.2018.1433343

Melichar, M., Didan, K., Barreto-Muñoz, A., Duberstein, J. N., Jiménez Hernández, E., Crimmins, T., Li, H., Traphagen, M., Thomas, K. A., & Nagler, P. L. (2023). Random forest classification of multitemporal Landsat 8 spectral data and phenology metrics for land cover mapping in the Sonoran and Mojave deserts. *Remote Sensing*, *15*(5), 1266. https://doi.org/10.3390/rs15051266

Mishra, V. N., & Rai, P. K. (2016). A remote sensing aided multi-layer perceptron-Markov chain analysis for land use and land cover change prediction in Patna district (Bihar), India. *Arabian Journal of Geosciences*, *9*, 1–18. https://doi.org/10.1007/s12517-015-2138-3

Moravec, D., Komárek, J., López-Cuervo Medina, S., & Molina, I. (2021). Effect of atmospheric corrections on NDVI: Intercomparability of Landsat 8, Sentinel-2, and UAV sensors. *Remote Sensing*, *13*(18), 3550. https://doi.org/10.3390/rs13183550

Morshed, S. R., Fattah, M. A., Hoque, M. M., et al. (2023). Simulating future intra-urban land use patterns of a developing city: A case study of Jashore, Bangladesh. *GeoJournal*, *88*, 425–448. https://doi.org/10.1007/s10708-022-10609-4

Mozumder, C., & Tripathi, N. K. (2014). Geospatial scenario-based modelling of urban and agricultural intrusions in Ramsar wetland Deepor Beel in Northeast India using a multi-layer perceptron neural network. *Int. J. Appl. Earth Obs. Geoinformation*, *32*, 92–104. https://doi.org/10.1016/j.jag.2014.03.002

Muller, M. R., & Middleton, J. (1994). A Markov model of land-use change dynamics in the Niagara Region, Ontario, Canada. *Landscape Ecology*, *9*, 151–157. https://doi.org/10.1007/BF00124382

Mungai, L. M., Messina, J. P., Zulu, L. C., Qi, J., & Snapp, S. (2022). Modeling spatiotemporal patterns of land use/land cover change in Central Malawi using a neural network model. *Remote Sensing*, *14*(14), 3477. https://doi.org/10.3390/rs14143477

Mustafa, A., Heppenstall, A., Omrani, H., Saadi, I., Cools, M., & Teller, J. (2018). Modelling built-up expansion and densification with multinomial logistic regression, cellular automata and genetic algorithm. *Computers, Environment and Urban Systems*, *67*(Complete), 147–156. https://doi.org/10.1016/j.compenvurbsys.2017.09.009

Nguyen, H., Jung, J., Lee, J., Choi, S.-U., Hong, S.-Y., & Heo, J. (2015). Optimal atmospheric correction for above-ground forest biomass estimation with the ETM+ remote sensor. *Sensors*, *15*(8), 18865–18886. https://doi.org/10.3390/s150818865

Norton, B. A., Evans, K. L., & Warren, P. H. (2016). Urban biodiversity and landscape ecology: Patterns, processes and planning. *Current Landscape Ecology Reports*, *1*, 178–192. https://doi.org/10.1007/s40823-016-0018-5

Nouri, J., Gharagozlou, A., Arjmandi, R. et al. (2014). Predicting urban land use changes using a CA–Markov Model. *Arabian Journal for Science and Engineering*, *39*, 5565–5573. https://doi.org/10.1007/s13369-014-1119-2

Pawe, C. K., & Saikia, A. (2018). Unplanned urban growth: Land use/land cover change in the Guwahati metropolitan area, India. *Geografisk Tidsskrift-Danish Journal of Geography*, *118*, 88–100. https://doi.org/10.1080/00167223.2017.1405357

Pawe, C. K., & Saikia, A. (2020). Decumbent development: Urban sprawl in the Guwahati metropolitan area, India. *Singapore Journal of Tropical Geography*. https://doi.org/10.1111/sjtg.12317

Perveen, S., Kamruzzaman, M., & Yigitcanlar, T. (2017). Developing policy scenarios for sustainable urban growth management: A delphi approach. *Sustainability*, *9*(10), 1787. https://doi.org/10.3390/su9101787

Pontius, R. G. (2000). Quantification error versus location error in comparison of categorical maps. *Photogrammetric Engineering and Remote Sensing*, *66*, 1011–1016.

Purwanto, A. D., Wikantika, K., Deliar, A., & Darmawan, S. (2022). Decision tree and random forest classification algorithms for mangrove forest mapping in Sembilang National Park, Indonesia. *Remote Sensing*, *15*(1), 16. https://doi.org/10.3390/rs15010016

Qiao, Z., Tian, G., Zhang, L., & Xu, X. (2014). Influences of urban expansion on urban heat island in Beijing during 1989–2010. *Advances in Meteorology, 2014,* 1–11. https://doi.org/10.1155/2014%2F187169

Qiu, R., Xu, W., & Zhang, J. (2015). The transformation of urban industrial land use: A quantitative method. *Journal of Urban Management, 4*(1), 40–52. ISSN 2226–5856, Elsevier, Amsterdam. https://doi.org/10.1016/j.jum.2015.07.001

Radwan, T. M., Blackburn, G. A., Whyatt, J. D., & Atkinson, P. M. (2019). Dramatic loss of agricultural land due to urban expansion threatens food security in the Nile Delta, Egypt. *Remote Sensing, 11*(3), 332. https://doi.org/10.3390/rs11030332

Rafferty, J. P. (2023, March 31). *Urban sprawl: Encyclopedia britannica.* www.britannica.com/topic/urban-sprawl

Rahimi, A. (2016). A methodological approach to urban land-use change modeling using infill development pattern: A case study in Tabriz, Iran. *Ecological Processes, 5,* 1–15. https://doi.org/10.1186/s13717-016-0044-6

Ramo, R., & Chuvieco, E. (2017). Developing a random forest algorithm for MODIS global burned area classification. *Remote Sensing, 9*(11), 1193. https://doi.org/10.3390/rs9111193

Referowska-Chodak, E. (2019). Pressures and threats to nature related to human activities in European urban and suburban forests. *Forests, 10*(9), 765. https://doi.org/10.3390/f10090765

Rimal, B., Sloan, S., Keshtkar, H., Sharma, R., Rijal, S., & Shrestha, U. B. (2020). Patterns of historical and future urban expansion in Nepal. *Remote Sensing, 12*(4), 628. https://doi.org/10.3390/rs12040628

Rimal, B., Zhang, L., Keshtkar, H., Haack, B., Rijal, S., & Zhang, P. (2018). Land use/land cover dynamics and modeling of urban land expansion by the integration of cellular automata and Markov chain. *ISPRS International Journal of Geo-Information, 7*(4), 154. https://doi.org/10.3390/ijgi7040154

Rodriguez-Galiano, V. F., Ghimire, B., Rogan, J., Chica-Olmo, M., & Rigol-Sánchez, J. P. (2012). An assessment of the effectiveness of a random forest classifier for land-cover classification. *ISPRS Journal of Photogrammetry and Remote Sensing, 67,* 93–104. https://doi.org/10.1016/j.isprsjprs.2011.11.002

Rwanga, S., & Ndambuki, J. (2017). Accuracy assessment of land use/land cover classification using remote sensing and GIS. *International Journal of Geosciences, 8,* 611–622. doi: 10.4236/ijg.2017.84033.

Saikia, A. (2014). Drivers of forest loss. In A. Saikia (Ed.), *Over-exploitation of forests: Springer briefs in geography.* Springer, Cham, Switzerland. https://doi.org/10.1007/978-3-319-01408-1_7

Saikia, A., Hazarika, R., & Sahariah, D. (2013). Land-use/land-cover change and fragmentation in the Nameri Tiger Reserve, India. *Geografisk Tidsskrift-Danish Journal of Geography, 113,* 1–10. https://doi.org/10.1080/00167223.2013.782991

Sajan, B., Mishra, V. N., Kanga, S., Meraj, G., Singh, S. K., & Kumar, P. (2022). Cellular automata-based artificial neural network model for assessing past, present, and future land use/land cover dynamics. *Agronomy, 12*(11), 2772. https://doi.org/10.3390/agronomy12112772

Salem, M., Bose, A., Bashir, B., Basak, D., Roy, S., Chowdhury, I. R., Alsalman, A., & Tsurusaki, N. (2021). Urban expansion simulation based on various driving factors using a logistic regression model: Delhi as a case study. *Sustainability, 13*(19), 10805. https://doi.org/10.3390/su131910805

Sang, L., Zhang, C., Yang, J., Zhu, D., & Yun, W. (2011). Simulation of land use spatial pattern of towns and villages based on CA: Markov model. *Mathematical and Computer Modelling, 54*(3–4), 938–943. https://doi.org/10.1016/j.mcm.2010.11.019

Sangermano, F., Eastman, J. R., & Zhu, H. (2010). Similarity weighted instance-based learning for the generation of transition potentials in land use change modeling. *Transactions in GIS, 14*(5), 569–580.

Sari, I. L., Weston, C. J., Newnham, G. J., & Volkova, L. (2021). Assessing accuracy of land cover change maps derived from automated digital processing and visual interpretation in tropical forests in Indonesia. *Remote Sensing, 13*(8), 1446. https://doi.org/10.3390/rs13081446

Sarkar, R. (2019). Urbanization in India before and after the economic reforms: What does the census data reveal? *Journal of Asian and African Studies, 54*(8), 1213–1226. https://doi.org/10.1177/0021909619865581

Seifollahi-Aghmiuni, S., Kalantari, Z., Egidi, G., Gaburova, L., & Salvati, L. (2022). Urbanisation-driven land degradation and socioeconomic challenges in peri-urban areas: Insights from Southern Europe. *Ambio, 51*(6), 1446–1458. https://doi.org/10.1007/s13280-022-01701-7

Shahi, E., et al. (2020). Monitoring and modeling land use/cover changes in Arasbaran protected area using and integrated Markov chain and artificial neural network. *Modeling Earth Systems and Environment, 6,* 1901–1911. https://doi.org/10.1007/s40808-020-00801-1

Shahraki, S. Z., Saurí, D., Serra, P., Modugno, S., Seifolddini, F., & Pourahmad, A. (2011). Urban sprawl pattern and land-use change detection in Yazd, Iran. *Habitat International, 35,* 521–528. https://doi.org/10.1016/j.habitatint.2011.02.004

Shalaby, A., & Tateishi, R. (2007). Remote sensing and GIS for mapping and monitoring land cover and land-use changes in the Northwestern coastal zone of Egypt. *Applied Geography*, *27*, 28–41. https://doi.org/10.1016/J.APGEOG.2006.09.004

Shamsudeen, M., Padmanaban, R., Cabral, P., & Morgado, P. (2022). Spatio-temporal analysis of the impact of landscape changes on vegetation and land surface temperature over Tamil Nadu. *Earth*, *3*(2), 614–638. https://doi.org/10.3390/earth3020036

Sharma, R., & Joshi, P. K. (2013). Monitoring urban landscape dynamics over Delhi (India) using remote sensing (1998–2011) inputs. *Journal of Indian Society of Remote Sensing*, *41*, 641–650. http://dx.doi.org/10.1007/s12524-012-0248-x

Sheard, N. (2018). Airport size and urban growth. *London School of Economics and Political Science*, *86*(342), 300–335, April. https://doi.org/10.1111/ecca.12262

Shi, G., Shan, J., Ding, L., Ye, P., Li, Y., & Jiang, N. (2019). Urban road network expansion and its driving variables: A case study of Nanjing city. *International Journal of Environmental Research and Public Health*, *16*(13), 2318. https://doi.org/10.3390/ijerph16132318

Singh, A., Kushwaha, S., Alarfaj, M., & Singh, M. (2022). Comprehensive overview of backpropagation algorithm for digital image denoising. *Electronics*, *11*(10), 1590. https://doi.org/10.3390/electronics11101590

Sridhar, K. S. (2010). Determinants of city growth and output in India. *Review of Urban & Regional Development Studies*, *22*, 22–38. https://doi.org/10.1111/j.1467-940X.2010.00167.x

Talukdar, S., Singha, P., Mahato, S., Shahfahad, Pal, S., Liou, Y.-A., & Rahman, A. (2020). Land-use land-cover classification by machine learning classifiers for satellite observations: A review. *Remote Sensing*, *12*(7), 1135. MDPI AG. http://dx.doi.org/10.3390/rs12071135

Tewolde, M. G., & Cabral, P. (2011). Urban sprawl analysis and modeling in Asmara, Eritrea. *Remote Sensing*, *3*(10), 2148–2165. https://doi.org/10.3390/rs3102148

Thapa, R. B., & Murayama, Y. (2009). Urban mapping, accuracy, & image classification: A comparison of multiple approaches in Tsukuba city, Japan. *Applied Geography*, *29*, 135–144. http://dx.doi.org/10.1016/j.apgeog.2008.08.001

Tilahun, D., Gashu, K., & Shiferaw, G. T. (2022). Effects of agricultural land and urban expansion on peri-urban forest degradation and implications on sustainable environmental management in Southern Ethiopia. *Sustainability*, *14*(24), 16527. https://doi.org/10.3390/su142416527

Tiwari, S., Dumka, U. C., Gautam, A. S., Kaskaoutis, D. G., Srivastava, A. K., Bisht, D. S., Chakrabarty, R. K., Sumlin, B. J., & Solmon, F. (2017). Assess-ment of PM2.5 and PM10 over Guwahati in Brahmaputra River Valley: Temporal evolution, source apportionment and meteorological dependence. *Atmos. Pollut. Res.*, *8*(1), 13–28. https://doi.org/10.1016/j.apr.2016.07.008

Varquez, A. C. G., Dong, S., Hanaoka, S., & Kanda, M. (2020). Improvement of an urban growth model for railway-induced urban expansion. *Sustainability*, *12*(17), 6801. https://doi.org/10.3390/su12176801

Verburg, P. H., Kok, K., Pontius, R. G., & Veldkamp, A. (2006). Modeling land-use and land-cover change. In E. F. Lambin & H. Geist (Eds.), *Land-use and land-cover change*. Global Change—The IGBP Series. Springer, Berlin, Heidelberg. https://doi.org/10.1007/3-540-32202-7_5

Vinayak, B., Lee, H. S., & Gedem, S. (2021). Prediction of land use and land cover changes in Mumbai city, India, using remote sensing data and a multilayer perceptron neural network-based Markov chain model. *Sustainability*, *13*(2), 471. MDPI AG. http://dx.doi.org/10.3390/su13020471

Waśniewski, A., Hościło, A., Zagajewski, B., & Moukétou-Tarazewicz, D. (2020). Assessment of Sentinel-2 satellite images and random forest classifier for rainforest mapping in Gabon. *Forests*, *11*(9), 941. https://doi.org/10.3390/f11090941

Woznicki, S. A., Baynes, J., Panlasigui, S., Mehaffey, M., & Neale, A. (2019). Development of a spatially complete floodplain map of the conterminous United States using random forest. *The Science of the Total Environment*, *647*, 942–953. https://doi.org/10.1016/j.scitotenv.2018.07.353

Wu, J. (2014). Urban ecology and sustainability: The state-of-the-science and future directions. *Landscape and Urban Planning*, *125*, 209–221. https://doi.org/10.1016/j.landurbplan.2014.01.018

Xu, T., Zhou, D., & Li, Y. (2022). Integrating ANNs and cellular automata—Markov chain to simulate urban expansion with annual land use data. *Land*, *11*(7), 1074. https://doi.org/10.3390/land11071074

Zhang, T., Su, J., Xu, Z., Luo, Y., & Li, J. (2021). Sentinel-2 satellite imagery for urban land cover classification by optimized random forest classifier. *Applied Sciences*, *11*(2), 543. https://doi.org/10.3390/app11020543

Zhang, Y., Long, H., Tu, S., Ge, D., Ma, L., & Wang, L. (2019). Spatial identification of land use functions and their tradeoffs/synergies in China: Implications for sustainable land management. *Ecological Indicators*, *107*, 105550. https://doi.org/10.1016/j.ecolind.2019.105550

Zhao, G., Zheng, X., Yuan, Z., & Zhang, L. (2017). Spatial and temporal characteristics of road networks and urban expansion. *Land*, *6*(2), 30. https://doi.org/10.3390/land6020030

Zhou, W., Zhang, S., Yu, W., Wang, J., & Wang, W. (2017). Effects of urban expansion on forest loss and fragmentation in six megaregions, China. *Remote Sensing*, *9*(10), 991. https://doi.org/10.3390/rs9100991

20 Soil Erosion and Digital Soil Mapping, a Dossier of Garhbeta and Its Adjoining Areas of West Medinipur

Swaagat Ray, Divyadyuti Banerjee,
and Lakshminarayan Satpati

20.1 INTRODUCTION

The badland topography of Garhbeta is carved out by numerous gullies which provide a distinct geomorphic characteristic to the region and at the same time poses a serious threat to the livelihood of the people residing in the region. This chapter mainly focuses upon the badland topography caused by the gully erosion in the region and how it can be mitigated using digital soil mapping (DSM) technique. Maps have been prepared showing the gully erosion in the Garhbeta and its adjoining areas using satellite images of different years, viz. 1998, 2009 and 2021, respectively. The maps show that the incidence of gully erosion increased over the years. However, steps have been taken to ameliorate the problem to some extent.

Sarkar in his paper 'Badland denudation process and management: A case study of Garhbeta badland, West Medinipur, West Bengal, India' (2020) focused on the processes responsible for soil erosion and its management in Garhbeta region. The use of remote sensing techniques and various other data helped him to provide a sketch of the badland topography in the Garhbeta area. In 'Spatial modelling of gully erosion: A new ensemble of classification and regression tree (CART) and general linear model (GLM) data-mining algorithms', Gayen and Pourghasemi (2019) and Gayen et al. (2019) used CART and GLM models, individually and combined in the Pathro river basin area of Jharkhand, India, in order to assess their efficiency and performance to demarcate the gully erosion in the region. Gayen et al. (2019) in 'Gully erosion susceptibility assessment and management of hazard-prone areas in India using different machine learning algorithms' used different machine learning algorithms like support vector machine, flexible discriminant analysis, multivariate additive regression splines, etc. to know the susceptibility of gully erosion to manage hazard-prone areas in the Pathro river basin area of Jharkhand, India. Kumar et al. (2020) in their article 'Spatial extent, formation process, reclaimability classification system and restoration strategies of gully and ravine lands in India' discusses that land debasement is a global issue affecting food security, climate and the environment. Gorges, particularly those in India, are a significant concern due to their unpredictable formation and high seepage thickness. In 1976, India's gorge land was 3.67 million ha, but it has decreased to around 60%. Factors contributing to gorge growth include extreme precipitation, poor agrarian practices, vegetation expulsion and overgrazing. To recover these socio-financially vital lands, strategies include land formation, terracing and riser adjustment. For gorges over 3 m deep, a minimal bund can be built to control overflow, and the recovery process includes adjusting ravine heads, gorge beds and side inclines. Sen et al. (2004) in their publication 'Geomorphological investigation of badlands: A case study at Garhbeta, West Medinipur district, West Bengal, India' is of the opinion that the Garhbeta barren wilderness, also known as 'Ganganir Danga', is a 3.40 km² Pleistocene lateritic upland in West Bengal, India.

DOI: 10.1201/9781003377825-23

Researchers have used maps, flying photos, satellite pictures and past writing to study the physics of provincial landform advancement and landform changes in the area.

Most of the research discusses soil erosion problems in different areas but these studies do not suggest the remedies to such problems. Similar problems have also been observed in studies dealing with the Garhbeta badland topography of the West Medinipur district of West Bengal. The gullied topography of Garthbeta disrupts the livelihood of the people residing in the area.

Agricultural production has lessened, structures have been demolished, electric poles have been damaged, etc. in the region. The objectives of this chapter are to analyze the extent of gully erosion in the region and how this can be mitigated using DSM techniques or what suggestions can be prepared using the different maps produced on gully erosion.

20.2 METHODOLOGY AND DATA

20.2.1 DATA SOURCE

In order to attain the aforementioned objectives in the paper, the data (multi-dated satellite images) have been downloaded from the US Geological Services (USGS) EarthExplorer website as shown in Table 20.1:

TABLE 20.1
Catalogue of Data Sources

Data Type		Sub-type	Nature	Source	Purpose
Earth observatory images (secondary data)	Sensor	Thematic Mapper (TM)	$p = 139, r = 44$ December 1998	www.usgs. gov	Image classification for land-cover change, NDBI, NDSI, BSI, NDMI, NDVI
		Thematic Mapper (TM)	$p = 139, r = 44$ November 2009		
		Operational Land Imager-Thermal Infrared Sensor (OLI-TIRS)	$p = 139, r = 44$ December 2021		

Note: BSI, Bare Soil Index; NDBI, Normalized Difference Built-up Index; NDMI, Normalized Difference Moisture Index; NDSI, Normalized Difference Soil Index; NDVI, Normalized Difference Vegetation Index; OLI, Operational Land Imager; TIRS, Thermal Infrared Sensor; TM, Thematic Mapper.

20.2.2 METHODOLOGY

A detailed description of methodology is provided in Table 20.2.

20.3 OVERVIEW OF THE AREA

Garhbeta badland (22°49′ N and 87°27′ E) is located on the concave right bank of Shilabati river in Garhbeta-I C.D. Block of West Medinipur district in the Indian state on West Bengal. It falls under the Gangani mouza of Garhbeta-I C.D. Block. The Garhbeta region covers a total surface area of 3.40 km². It is a Pleistocene lateritic upland with a mesmerizing beauty of badlands that is difficult to express in words. Most of the year, it experiences a sub-humid monsoon type of climate with a hot dry summer season (March–May), wet monsoon season (June–October) and dry, cool winter (December–January).

20.4 RESULTS AND DISCUSSION

Land-use/land-cover (LULC) maps of Garhbeta and its adjoining areas (Figure 20.1) have been prepared for different years (1998, 2009 and 2021, respectively).

TABLE 20.2
Methodology

Parameters	Purpose	Methods
Land-use land classification (LULC)	To determine the amount of land surface occupied by each physical and cultural entity	Supervised classification
Gully erosion	To determine the total land area subjected to gully erosion	Single-class classification
Normalized Difference Moisture Index (NDMI)	To show the open water features in a more enhanced way	(B2 – B5)/(B2 + B5) (LANDSAT-5) (B3 – B6)/(B3 + B6) (LANDSAT-9)
Normalized Difference Vegetation Index (NDVI)	To estimate the density of vegetation of an area and assess the changes in plant health	(B4 – B3)/(B4 + B3) (LANDSAT-5) (B5 – B4)/(B5 + B4) (LANDSAT-9)
Bare Soil Index (BSI)	To show the soil variations	(B7 + B3) – (B4 + B1)/(B7 + B3) + (B4 + B1) (LANDSAT-5) (B7 + B4) – (B5 + B2)/(B7 + B4) + (B5 + B2) (LANDSAT-9)
Normalized Difference Soil Index (NDSI)	To separate the soil from impervious surfaces and vegetation	(B7 – B2)/(B7 + B2) (LANDSAT-5) (B7 – B3)/(B7 + B3) (LANDSAT-9)
Normalized Difference Built-up Index (NDBI)	To estimate the human-made features of an area	(B5 – B4)/(B5 + B4) (LANDSAT-5) (B6 – B5)/(B6 + B5) (LANDSAT-9)

Note: BSI, Bare Soil Index; NDBI, Normalized Difference Built-up Index; NDMI, Normalized Difference Moisture Index; NDSI, Normalized Difference Soil Index; NDVI, Normalized Difference Vegetation Index.

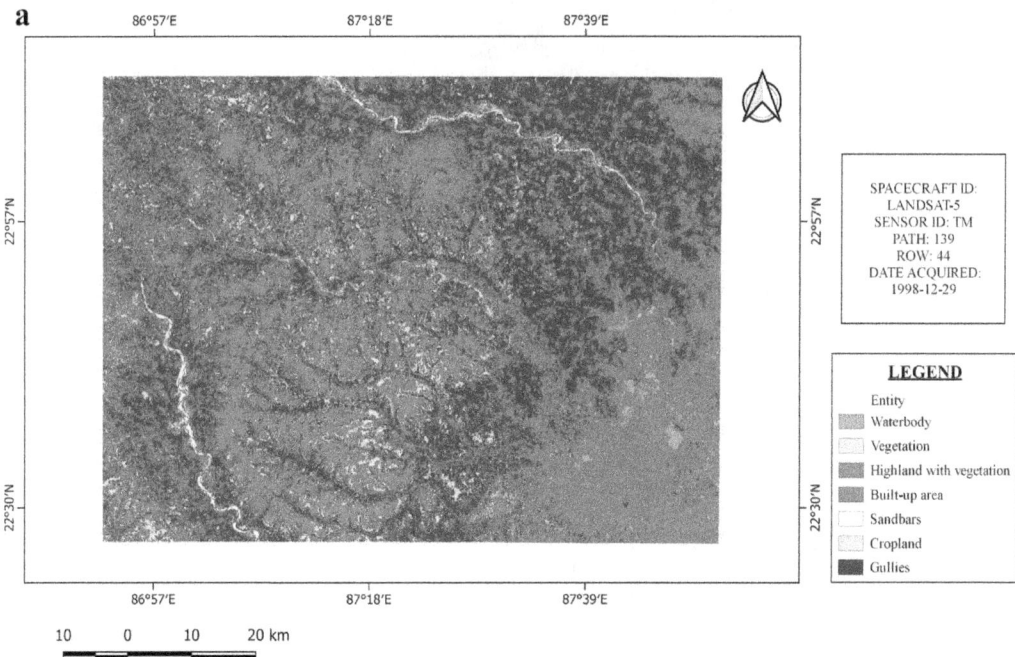

FIGURE 20.1 Land-use/land-cover maps: (a) 1998, (b) 2009, (c) 2021.

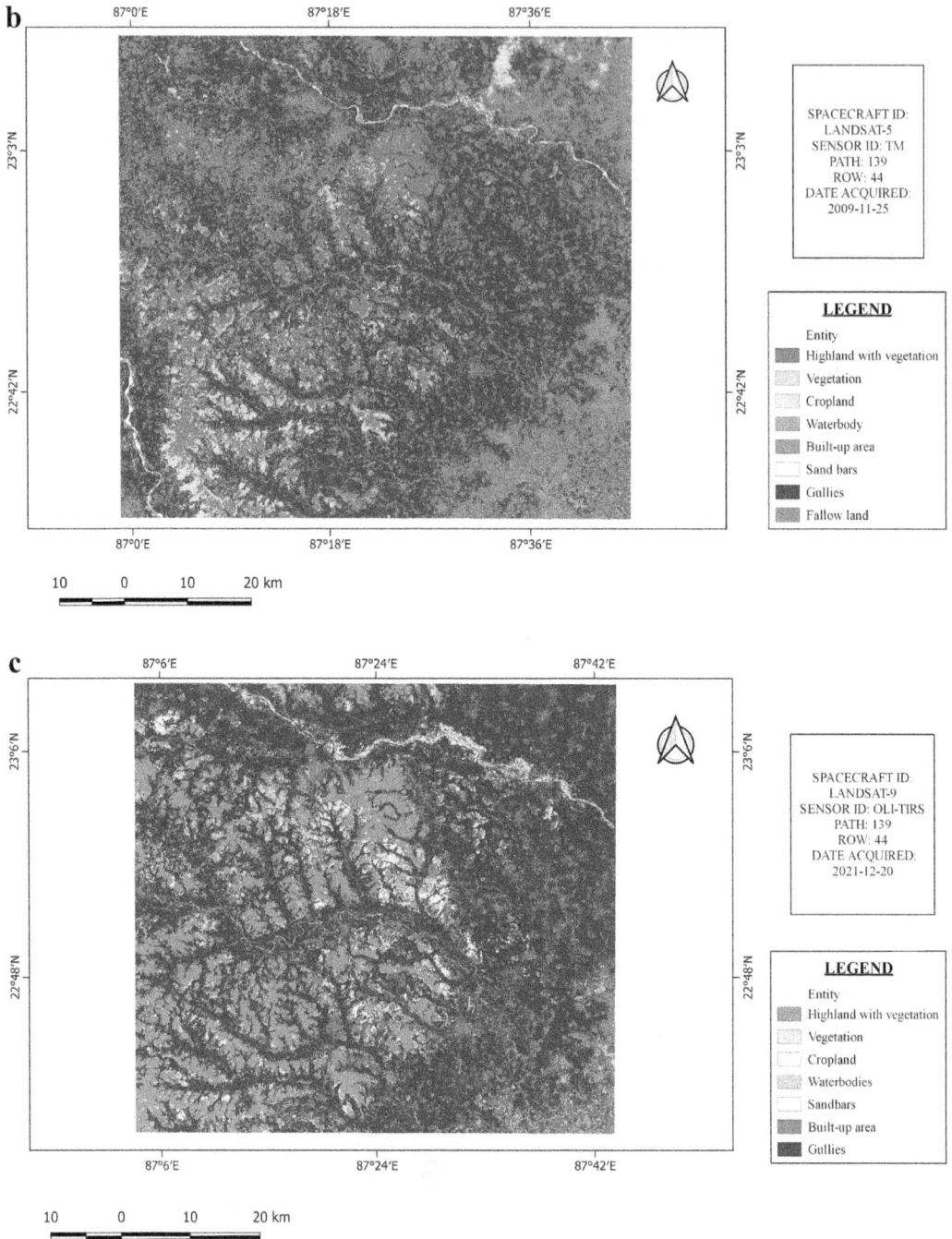

FIGURE 20.1 (Continued)

In order to find out the proportion of gullied topography in the region, gully erosion maps (Figure 20.2) for the different years have also been prepared.

It can be seen from the maps that the proportion of the gullies has increased over the years. A number of factors are responsible for the formation of gully erosion in the area. These include high-intensity rainfall, loose friable soil devoid of organic carbon and vegetation, irrational construction

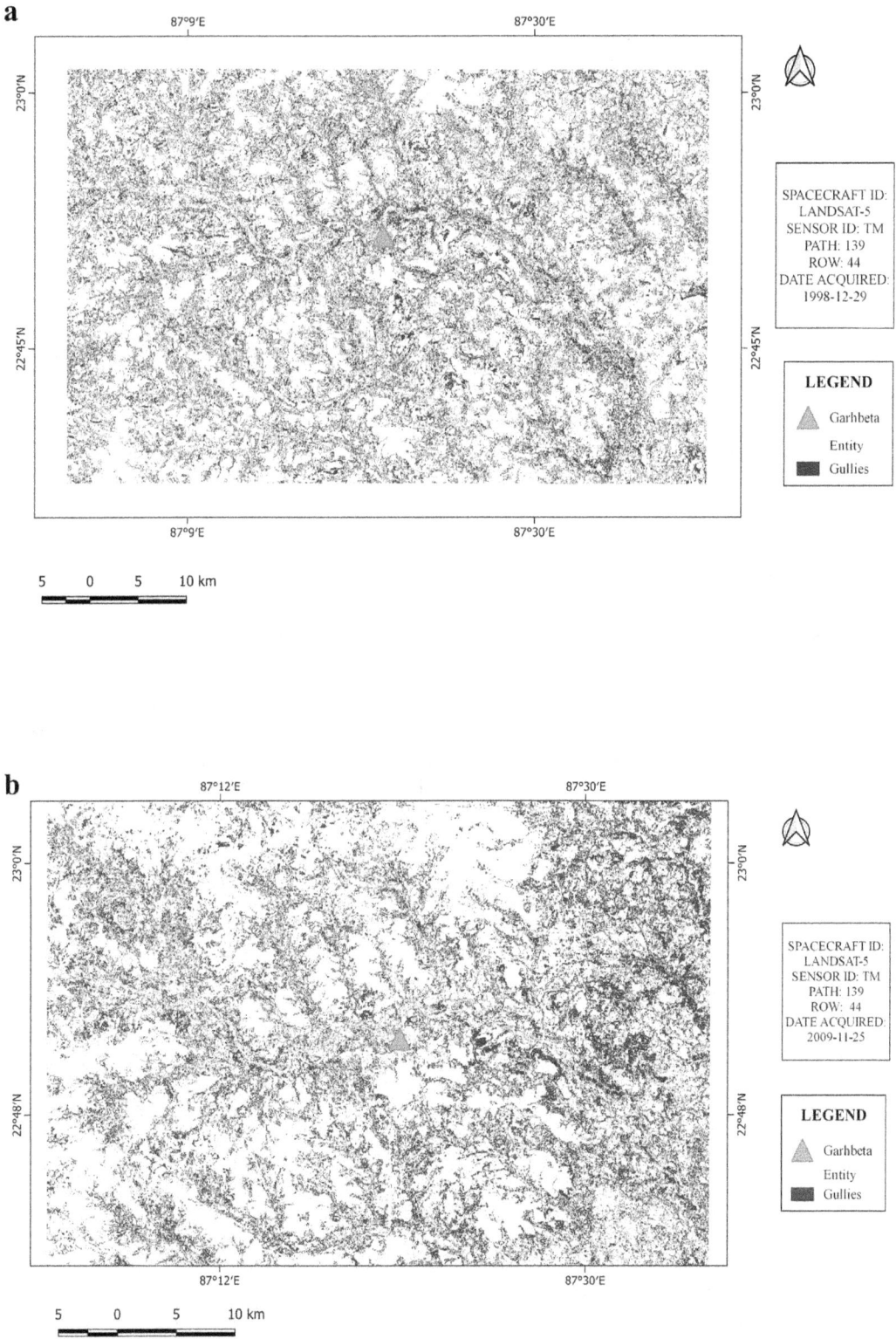

FIGURE 20.2 Gully erosion maps: (a) 1998, (b) 2009, (c) 2021.

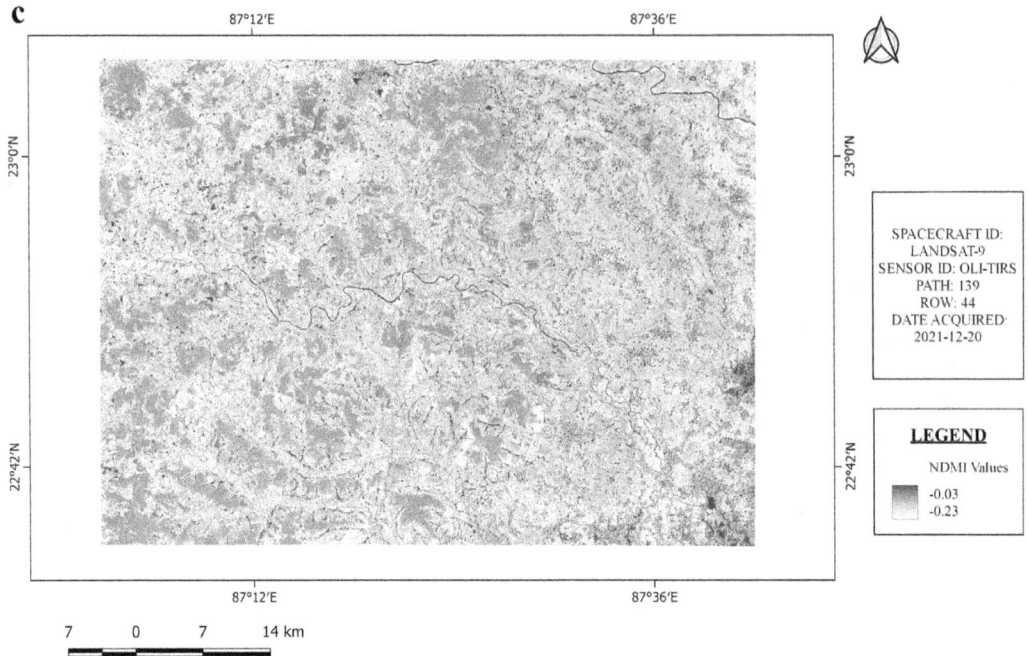

FIGURE 20.2   (Continued)

at sites, faulty agricultural practices, vegetation removal and overgrazing. However, in Garhbeta, the most pertinent factors include high-intensity rainfall, faulty agricultural practices and irrational construction at sites. Garhbeta and its associated areas are characterized by very high drainage density with several rills and gullies. Garhbeta badland is associated with active riverine processes with intensive gully erosion over the lateritic escarpment facing the river Shilabati. The prevalence of the gullies exists over the entire area. However, the gullies are seen to be more congested in the north-eastern and eastern parts of the map area. In the other portions of the map area, the gullied topography in Garhbeta and its adjoining areas covers a total surface area of 643.64 km² in 1998, 547.02 km² in 2009 and 977.62 km² in 2021, respectively. Therefore, the prevalence of gully erosion decreased from 1998 to 2009, and it again increased from 2009 to 2021. This may have occurred due to the increasing pressure of population in the region and occurrence of urbanization at a rapid pace which has led to the removal of vegetation from the region and made the soil particles much more loose and friable. Due to intensive rainfall in the region, the soils have been washed away leading to the formation of several rills. As time went by, the size of the rills increased due to the expansion and contracting process, leading to the formation of gullies and ravines.

The Modified Normalized Difference Water Index or Normalized Difference Moisture Index (NDMI) shows the open water features in a more enhanced way. It also eliminates the built-up features that are associated with open water in other indices.

NDMI maps for different years (Figure 20.3) have been prepared for Garhbeta and its adjoining area using Landsat satellite images from USGS EarthExplorer. It can be seen that in 1998 (Figure 20.3a), the highest index value (greater than −0.08) is observed in the northern, western and southern portions, while the least index value (less than −0.46) is primarily concentrated in the central, western and a little bit in the southern parts of the area. The mid-values are found throughout the area are mainly concentrated in the central, southern and eastern portions. In 2009 (Figure 20.3b), the highest index value (greater than −0.05) is found in the northern, central and some portions of the south while the least value (less than −0.25) is prominent in the southern, central and in a scattered fashion in the western, northern and eastern parts of the map area. The mid-values are found in a scattered manner over the entire region. In 2021 (Figure 20.3c), the highest index value (greater than −0.03) is present in the eastern and south-eastern pockets while the least index value (less than −0.23) is present almost over the entire region, mostly concentrated in the north-west and the western parts of the

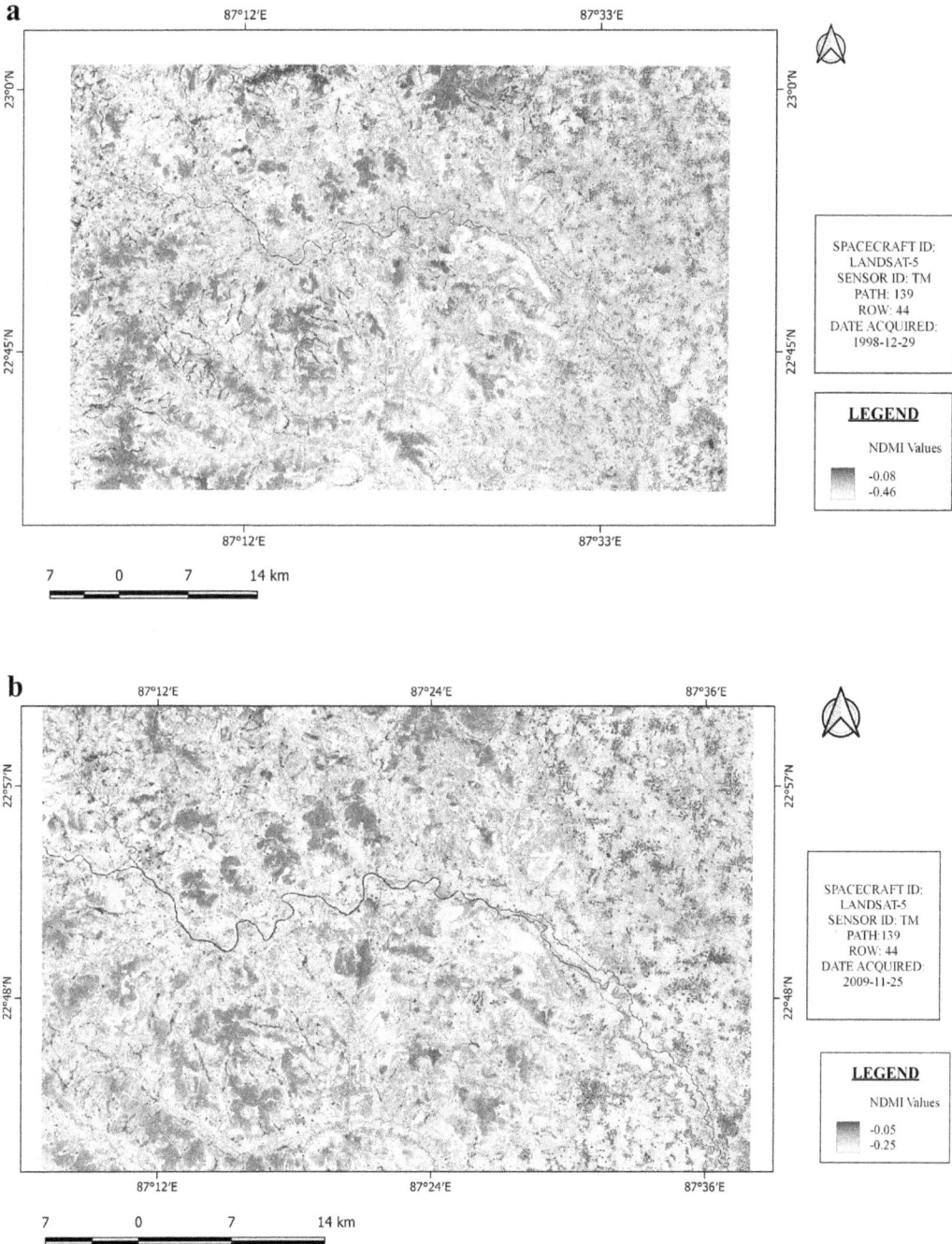

FIGURE 20.3 Normalized Difference Moisture Index: (a) 1998, (b) 2009, (c) 2021.

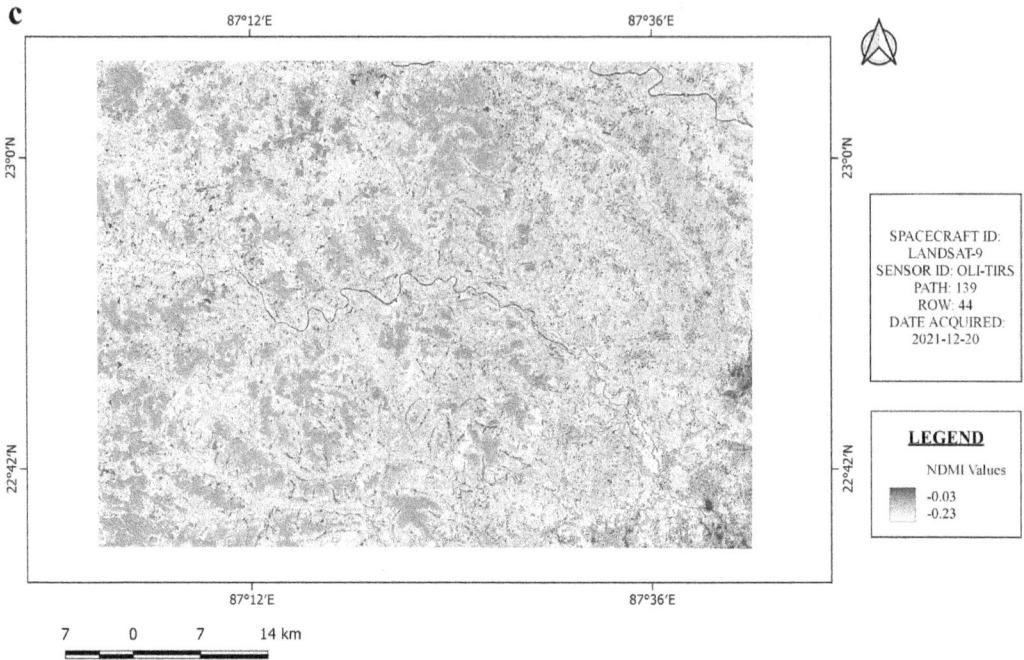

FIGURE 20.3   (Continued)

region. The mid-values are present throughout the map area, concentrated in different parts of the region. Thus, it can be noticed from the different NDMI maps that the index values have increased over the years, i.e. from 1998 to 2021, respectively.

The Normalized Difference Vegetation Index (NDVI) is used to estimate the density of vegetation of an area and assess the changes in plant health. The value of NDVI ranges between −1 to 1. Low values indicate vegetation stress and poor vegetation health, while higher values indicate healthy dense vegetation.

NDVI maps for different years (Figure 20.4) have been prepared for Garhbeta and its adjoining area using Landsat satellite images from USGS EarthExplorer. It can be observed that in 1998 (Figure 20.4a), the highest index value (>0.26) is predominant in the southern and western portions and scattered in the central, northern and north-eastern portions, while the lowest index value (<0.06) is predominant in the north-eastern, eastern and south-eastern portions and stretches across the central portion of the map area. In the 2009 map (Figure 20.4b), it can be seen that the highest index value (>0.24) is mostly present in the southern and central portions stretching to the north, and the lowest index value (<0.12) is mostly dominant in the eastern portions of the map area covering the northern and southern portions as well. From the 2021 map (Figure 20.4c), it can be seen that the highest index value (>0.29) is found in the southern, western and a few pockets in the northern portions of the map area. The lowest index value (<0.14) is mostly concentrated in the eastern, north-eastern and south-eastern parts of the map area. Thus, it can be seen that the lowest index values have increased over the years and the highest index value have decreased from 1998 to 2009 and it again increased from 2009 to 2021.

The Bare Soil Index (BSI) is a numerical indicator that uses blue, red, near-infrared (NIR) and short-wave infrared (SWIR) spectral bands to show the soil variations. The SWIR and red spectral bands quantify the soil mineral composition, while the blue and NIR spectral bands are used to enhance the presence of vegetation.

BSI maps for different years (Figure 20.5) have been prepared for Garhbeta and its adjoining area using Landsat satellite images from USGS EarthExplorer. From the 1998 map (Figure 20.5a), it can be seen that the highest index value (>0.16) is found alongside the gullies and in the northern and south-eastern portions of the map area. The lowest index value (<0.05) is found dispersed in the central, northern and southern pockets of the map area. The mid-index values are present throughout the map area. In the 2009 map (Figure 20.5b), it can be seen that the highest index value (>0.39) is found scattered in the eastern, southern, northern and central portions of the map area, while the lowest index value (<0.32) is found in the central, southern and northern portions of the map area

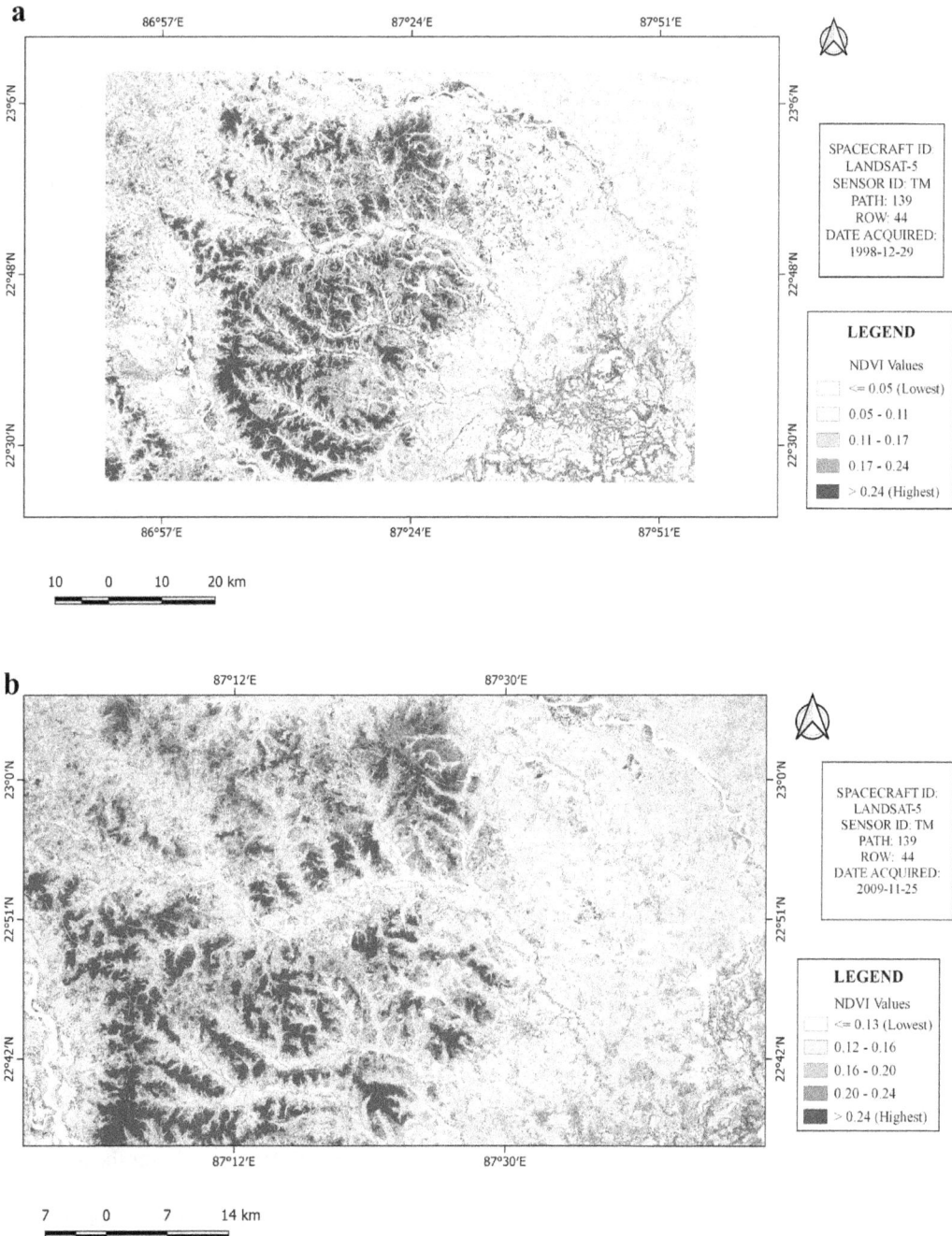

FIGURE 20.4 Normalized Difference Vegetation Index: (a) 1998, (b) 2009, (c) 2021.

c

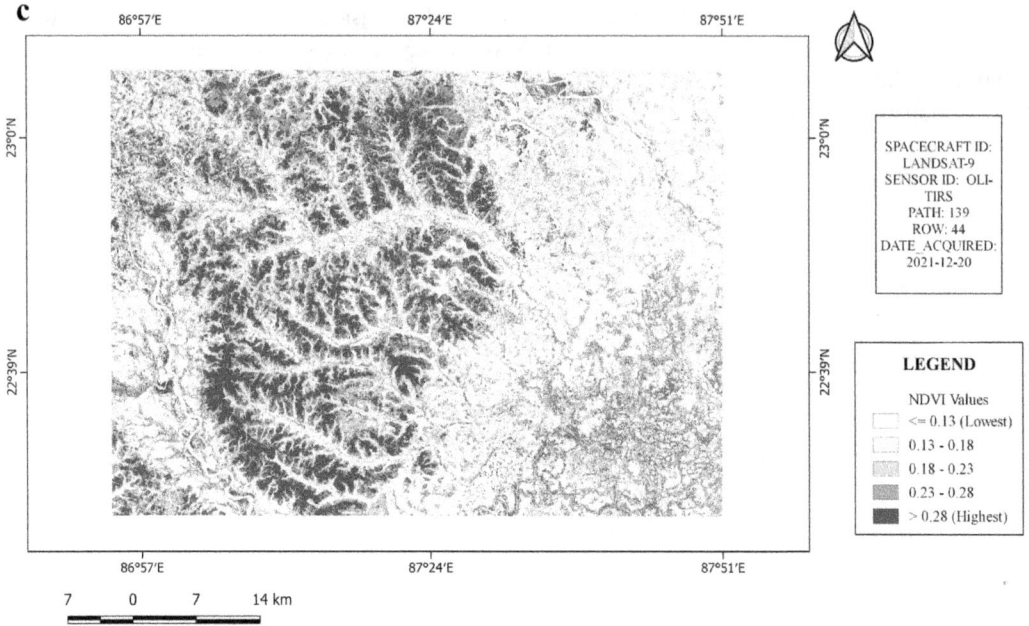

FIGURE 20.4   (Continued)

a

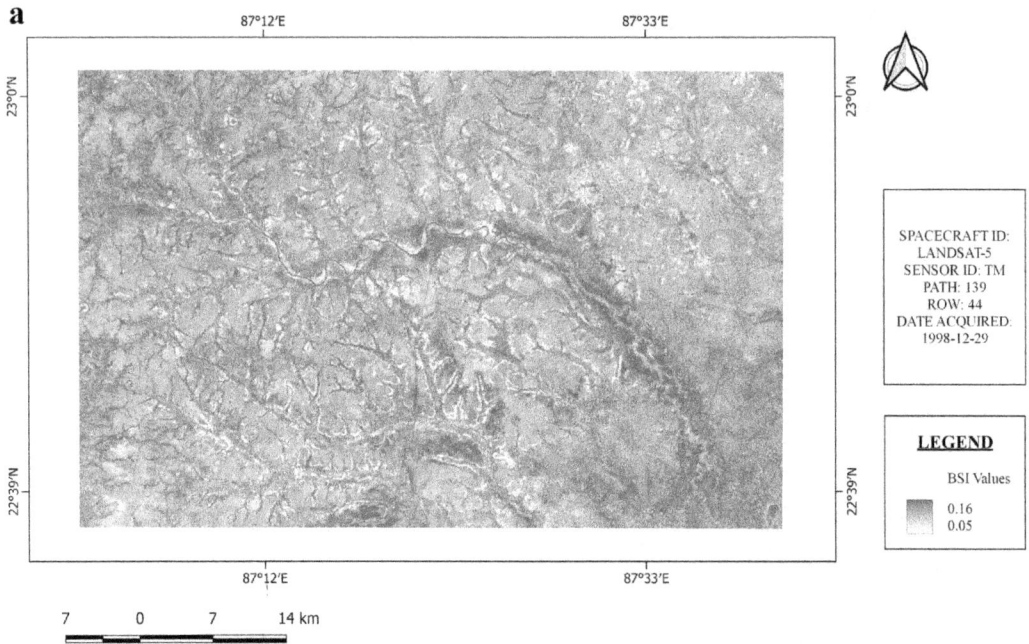

FIGURE 20.5   Bare Soil Index: (a) 1998, (b) 2009, (c) 2021.

in a scattered fashion. The mid-index values are present over the entire map area. From the 2021 map (Figure 20.5c), it can be seen that the highest index value (>0.05) is found all over the map area and primarily concentrated alongside the gullies while the lowest index values (less than −0.17) are observed mainly in the central portion and scattered to the north of the region. Thus, it can be said the BSI of the region has decreased with the highest and the lowest index values increasing from 1998 to 2009 and decreasing in 2021.

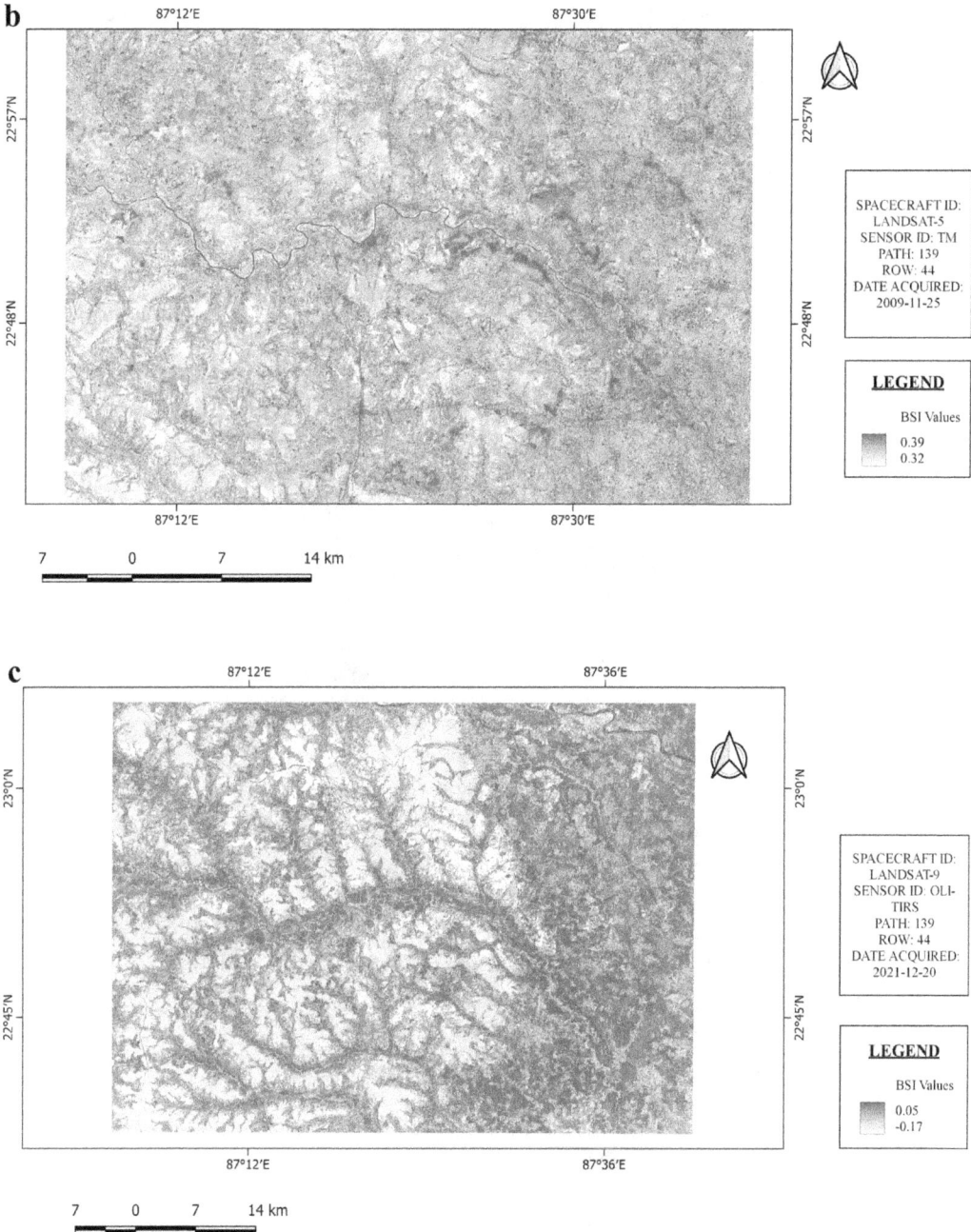

FIGURE 20.5 (Continued)

The Normalized Difference Soil Index (NDSI) is used to separate the soil from impervious surfaces and vegetation. NDSI maps for different years (Figure 20.6) have been prepared for Garhbeta and its adjoining area using Landsat satellite images from USGS EarthExplorer. From the 1998 map (Figure 20.6a), it can be seen that the highest index value (>0.62) is mainly found in the northern, central and southern portions of the map area while the lowest index value (<0.5) is mainly present in the eastern half of the map area. The mid-index values are scattered over the entire map area. In the 2009 map (Figure 20.6b), it can be seen that the highest index value (>0.63) is mainly found in the northern, central and southern portions of the map area while the lowest index value (<0.55) is mainly present in a few pockets of the eastern half of the map area. The mid-index values are scattered over the entire region. From the 2021 map (Figure 20.6c), it can be seen that the highest index value (>0.16) is present all over the map area, especially alongside the gullies, whereas the least index value (less than −0.04) is in a few pockets in the southern portion of the map area. Thus, it

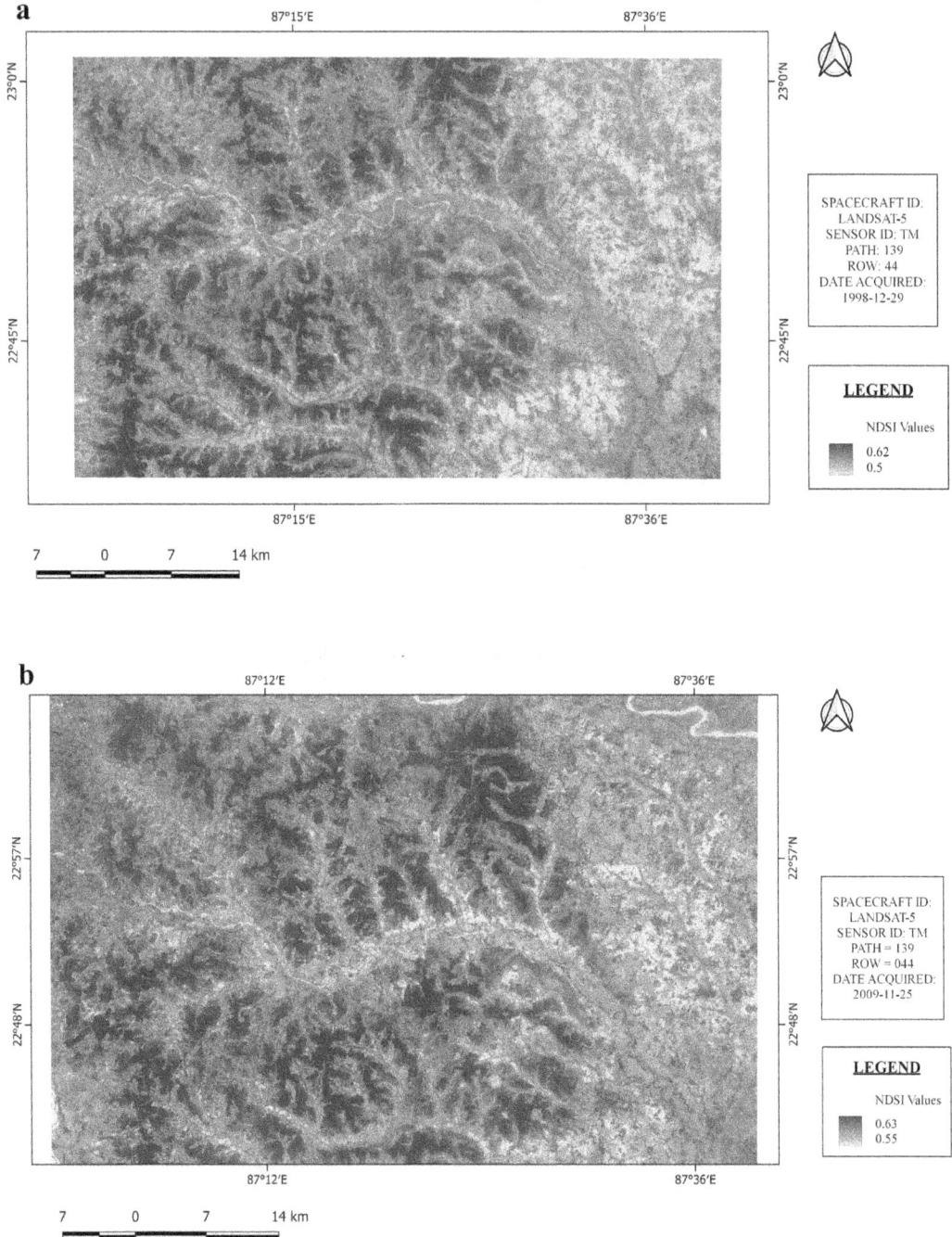

FIGURE 20.6 Normalized Difference Soil Index: (a) 1998, (b) 2009, (c) 2021.

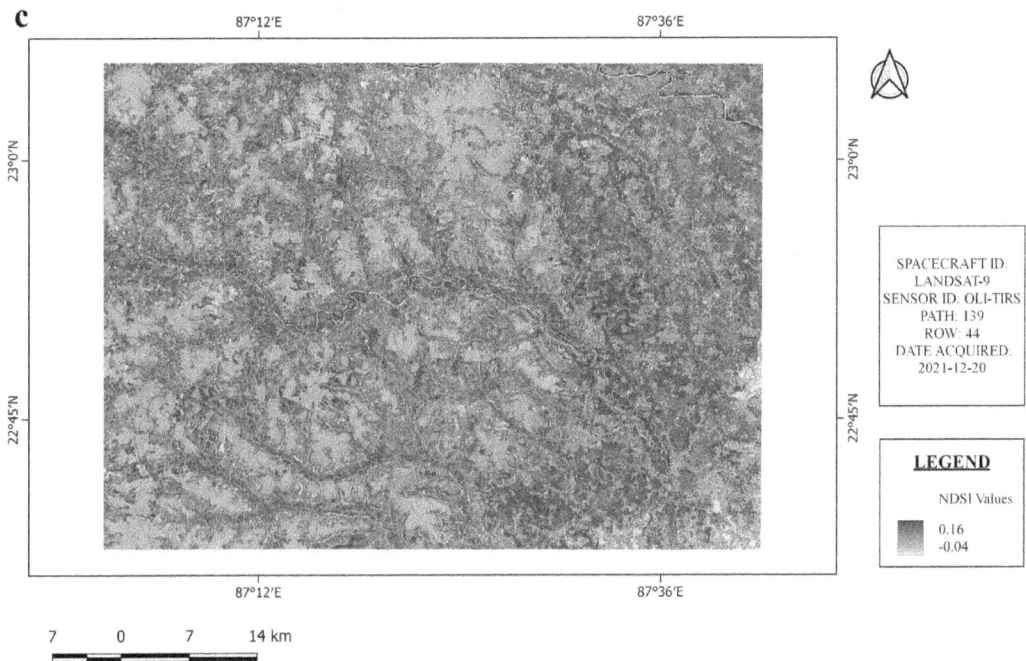

FIGURE 20.6 (Continued)

can be seen that the NDSI values have decreased over the years from 1998 to 2021 with the lowest index value reaching down to negative.

The Normalized Difference Built-up Index (NDBI) is used to estimate the human-made features of an area. The built-up areas and bare soil surface reflect more SWIR wavelength than NIR. However, in case of a greener surface, the reflectance of NIR is more than SWIR. NDBI maps for different years (Figure 20.7) have been prepared for Garhbeta and its adjoining area using Landsat satellite images from USGS EarthExplorer. From the 1998 map (Figure 20.7a), it can be seen that the highest index value (>0.31) is found scattered all over the map area, whereas the lowest index value (less than −0.11) is mainly found in small portions of the northern, central and southern portions of the map area. The mid-index values are present throughout the map area. From the 2009 map (Figure 20.7b), it can be seen that the highest index value (>0.1) is predominant along the gullies and in the northern and southern portions, while the lowest index value (less than −0.15) is found in the southern and western portions of the map area. The mid-index values are present throughout the map area. From the 2021 map (Figure 20.7c), it can be seen that the highest index value (>0.06) is present all over the map area, mainly along the gullied land, while the lowest index value (less than −0.19) is found only in a few pockets of the northern part of the map area. The mid-index values are scattered throughout the map area. Thus, it can be said that the NDBI of the region has decreased during the time period.

Thus, from the maps it can be analyzed that except for the NDMI and NDVI, all other indices have decreased.

- Most of the people of Medinipur live a rural livelihood. They mainly engage themselves in primary economic activities like agriculture, animal husbandry, poultry, etc. Some of them engage themselves in secondary economic activities while a very few of them are involved in tertiary economic activities. Agriculture serves the backbone of the economy of West Medinipur. As a consequence, plots of lands have been deforested to cultivate crops in the region.

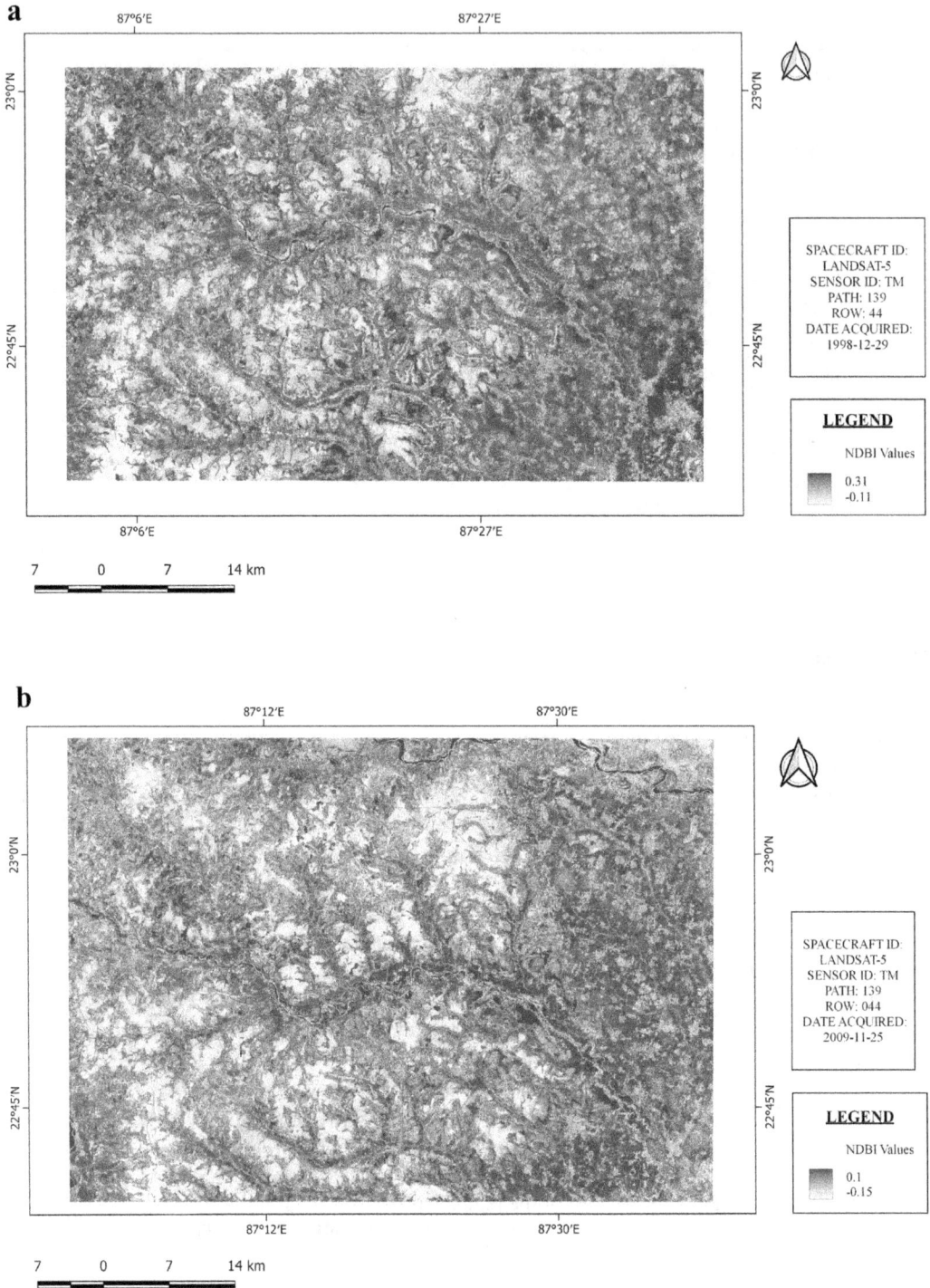

FIGURE 20.7 Normalized Difference Built-up Index: (a) 1998, (b) 2009, (c) 2021.

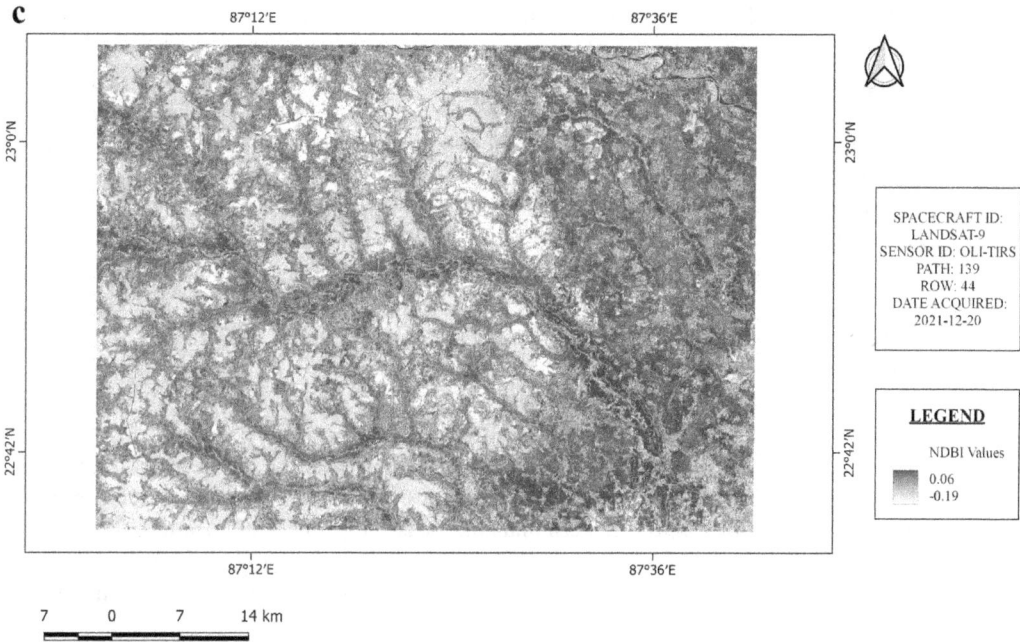

FIGURE 20.7 (Continued)

- Urbanization is a global phenomenon, and the people of Medinipur are no strangers to the process. In order to facilitate the residents with better amenities, markets, educational institutions, offices, hospitals, etc. have been constructed.
- Tourism in West Medinipur is a popular economic activity. A major section of the populace depends upon the tourism sector to earn their daily bread. The tourists are enthralled by the spectacular scenery of the Garhbeta badland. Therefore, several hotels, resorts and restaurants have been opened to facilitate the tourists. The NDBI maps also show that several built-up areas have emerged in close proximity to the gullied topography. Due to all such activities, forests have been removed from the region. The soils have become much more loose and friable.
- Garhbeta region is characterized by high-intensity rainfall and high drainage density. Rainfall in the region has washed away the loose topsoil leading to the formation of rills. As time went by, the rills gradually widened, resulting in the formation of gullies.
- Also, faulty agricultural practices, illegal construction at sites, overgrazing and several other activities act as catalyst to the process of gully development.
- The BSI map of 1998 indicates that soil is extremely fertile along the Shilabati river. The soil becomes unproductive if we move farther away from the river basin. The soil present in the gullies is the least productive. However, the soil present in the region is more or less suitable for cultivation. Therefore, the vegetation of the region is primarily concentrated along the banks of the Shilabati river and in the gap areas between the gullies as is evident from the NDVI map. The vegetation of the region mainly includes mixed deciduous forest.
- The NDSI map of 1998 also shows that the soils present in between the gullies and along the riverbank are covered with vegetation. The highlands present in Garhbeta and its associated areas provide favourable soils for the growth of vegetation as is evident from the LULC map.

- Moisture index, i.e. NDMI, is also high along the riverbank and in the gap areas. So the concentration of built-up areas is also prevalent along the banks of river Shilabati. The concentration of urban landscape is less in the gap areas between the gullies owing to the undulating badland topography, dense concentration of vegetation, inaccessible terrain and unfavourable slope. The concentration is dense in the highland region as it can be witnessed from the LULC map of 2009. But the concentration of dense vegetation has decreased, and that of less vegetal cover has increased from 1998. The NDSI map of 2009 also shows that the soils present in between the gullies and along the riverbank are covered with vegetation. However, the concentration of soil suitable for vegetation growth is more compared to that of 1998.
- The concentration of built-up areas is also prevalent along the banks of river Shilabati. The presence of urban landscape is less in the gap areas between the gullies, but the urban landscape has decreased from 1998 to 2009.
- The concentration of soil with vegetation has reduced significantly over the present years. Moisture index, i.e. the NDMI map of 2021 shows that the presence of water bodies in the highland areas is more than the rest of the area. However, the highest index value is present in the Shilabati River, in the eastern and south-eastern pockets and scattered over the entire map area.
- The NDMI values have increased over the years. The urban concentration is primarily centred along the banks of river Shilabati and where the water availability is found in the map area. But the values have decreased over the time period. Thus, it can be said that the NDSI map shows the areas along the banks of river Shilabati fit for human use. Soil erosion primarily happens in the gaps of the gullies. The topsoils have been removed due to high-intensity rainfall and drainage density. The index values have decreased from 1998 to 2021 indicating that the soil has been continuously leached.
- The areas where soil was fit for human use in 1998 have deteriorated in 2021. The vegetation cover has increased along with the areas where vegetation growth is not possible from 1998 to 2021 which has been clearly depicted in NDVI maps. The NDMI maps for 1998, 2009 and 2021 show an increase in the index values. Thus, it is evident that where vegetation growth is possible, moisture content is also high. However, the areas where moisture availability was high in 1998 and 2009 are now having a moderate moisture capacity. The built-up areas have gradually decreased from 1998 to 2021.

But the concentration remains same along the Shilabati riverbanks. BSI maps show that the index values have decreased from 1998 to 2021 with the least index value reaching down to negative.

Gully erosion is like a daily phenomenon in the Garhbeta region. Due to this, a lot of damages have been incurred by the people in the region. Croplands have been wiped out, electric poles have been damaged, buildings have been demolished, trees have been uprooted, roads have been damaged, etc. All these activities have caused a disruption of their livelihood. In order to ameliorate the problems, the local authorities and the government have taken a number of steps such as afforestation, minimizing the construction activities and reducing the activity of cutting down forests. All these activities are done to maintain the ecological balance of the area.

20.5 CONCLUSION

Soil degradation also means degradation of soil quality. The area was once covered by dense mixed deciduous forest (in the 1940s). Rapid deforestation has made this area less vegetative. So it is essential to undertake afforestation programs to stop deforestation. Several models and maps can be prepared using satellite images in order to understand the soil erosion and its susceptibility and formulate programs and policies. Some management strategies were also undertaken in this area for protecting soil erosion and degradation of land. The local panchayat samitee and soil conservation department launched an afforestation program at a soil erosion–prone area of the badland, but it failed due to unscientific plantation and improper management. The allotted amount per year for the plantation program was near about two lakhs which is not sufficient for the total area of the badland. Construction of small check dam in the gullies for resisting soil erosion is needed in every gully channel. A check dam can arrest eroded and transported sediment. A big check dam with reservoir was installed by the Western Region Development ministry fund of the Government of West Bengal at the end of the main gully channel where it meets with the river. It became very successful to check soil erosion and development of badland. The Department of Forest, Government of West Bengal has constructed a check dam in a selected gully channel for trial. It is necessary to construct a small check dam in every gully channel from the head of the gully to the end. A siltation pond was also installed successfully. Jute textiles have been found useful for controlling surface soil erosion. It can reduce soil erosion, arrest essential nutrient, maintain soil moisture with control of temperature and enhance suitable conditions for plant growth and development. Application of the bioengineering process (jute textile technology) has become a success in this area. Various field trials have been conducted for soil erosion control and establishment of vegetation cover using jute-geotextiles since 1999. It was observed that rapid root development, enhancement of quality of soil was possible through decomposing mulch. This produced humus required by new plants. Besides protecting erosion during rain, it is also very useful to protect soil from wind erosion during storms. Some sediment-catch ponds are located at different parts of the badland for storm runoff and overland flow during monsoon. It also is essential to install a number of catchment ponds in different sections of the badland. Lastly, nothing is possible without the active participation of the people residing in the region.

REFERENCES

Gayen, A., Pourghasemi, H.R. (2019). 30: Spatial modelling of gully erosion: A new ensemble of CART and GLM data-mining algorithms. In H.R. Pourghasemi, C. Gokceoglu (Eds.). *Spatial modelling in GIS and R for earth and environmental sciences* (653–669). www.sciencedirect.com/science/article/pii/B9780128152263000302

Gayen, A., Pourghasemi, H.R., Saha, S., Keesstra, S., Bai, S. (2019). Gully erosion susceptibility assessment and management of hazard-prone areas in India using different machine learning algorithms. *Science of the Total Environment, 668*, 124–138. www.sciencedirect.com/science/article/pii/S0048969719309489

Kumar, G., Adhikary, P.P., Dash, C.J. (2020). Spatial extent, formation process, reclaimability classification system and restoration strategies of gully and ravine lands in India. In Shit, P., Pourghasemi, H., Bhunia, G. (eds) *Gully erosion studies from India and surrounding regions* Advances in Science, Technology & Innovation. Springer, Cham. https://doi.org/10.1007/978-3-030-23243-6_1

Sarkar, A. (2020). *Badland denudation process and management: A case study of Garhbeta badland, West Medinipur, West Bengal, India.* https://www.researchgate.net/publication/338762436

Sen, J., Sen, S., Bandyopadhyay, S. (2004). Geomorphological investigation of badlands: A case study at Garhbeta, West Medinipur district, West Bengal, India. In S. Singh, H.S. Sharma, S.K. Dey (Eds.). *Geomorphology and environment* (204–234). Kolkata, India: ACB Publications.

21 Geographical Analysis of Animal Health Facilities in Nagaur District of Rajasthan

Govind Singh

21.1 INTRODUCTION

Livestock husbandry is the most important economic activity in western Rajasthan where the Nagaur district is contributing significantly. As Mahatma Gandhi quotes, the greatness of a nation can be judged by the way it treats its animals. Animal food and fodder, water and veterinary amenities are the primary factors to measure the animal health of an area, and animal health is a milestone in dairy development. Animal health is essential for quality of animal life, improvement in breed, reproduction and preparation of animal power for rapid growth and development. The feed and fodder of animals generally contain one or two locally available concentrated feed ingredients, grasses and crop residues which makes the feed imbalanced. Hence, imbalanced feeding is caused due to inadequate proteins, energy, minerals; vitamins etc. and adversely affects the health and productivity of animals (20th Livstock Census 2019). Water is required for digestion of food, distribution of absorbed nutrients to particular parts, excretion of harmful and toxic elements through urine and for keeping stable body temperature. There should be free access to fresh and clean drinking water round the clock for an animal.

The Food and Agriculture Organization of the United Nations (FAO 2021a, 2021b) has reported that the rate of annual occurrence of climatic change disasters like extreme weather events, forest fires, floods and droughts alongside biological hazards like COVID-19 during the second decade of the twenty-first century has risen threefold as compared to that during the decades of the 1970s and 1980s. The agriculture and allied livestock sector bears around 63% of the overall impact caused by these disasters. Damages to standing crops and livestock amounted to more than $100 billion during the last decade. Such damage is disproportionately borne by landless agricultural laborers and smallholders primarily subsisting upon their livestock.

The disease outbreak is a big obstacle for animal production. The timely detection of disease reduces the treatment cost, mortality rate and makes the recovery faster. During the field visit, it was explored in the views of various livestock farmers that they have lost their productive animals due to lack of proper reach of animal health facilities during the deadly spread of Lumpy Virus. From ancient times, animal husbandry was a practice of prosperity. There are so many ancient sources, like Brahmanand Purana, Mahabharata, Skand Purana, Padma Purana, Arthashatra of Kautilya, etc. who talk about traditional health-care wisdom of Indians of the time. But nowadays, the time has been changed and traditional knowledge vanished due to lack of documentation. Hence, the present animal health practices are mainly dependent on modern animal health-care facilities, so it is quite useful to analyze the availability and accessibility of modern animal health facilities in the study area.

21.2 OBJECTIVES OF THE STUDY

The major objectives of this study are as follows:

- To analyze the level of animal health-care facilities in each panchayat samiti and find out the backward panchayat samitis of the district

DOI: 10.1201/9781003377825-24

- To evaluate the spatial variation in animal health-care facilities of Nagaur District
- To identify various factors behind emergence and growth of variation/disparities
- To suggest some remedial measures

21.3   STUDY AREA

The study area of the present work is Nagaur District, which is situated between 26°25′ N to 27°40′ N latitudes and 73°10′ E to 75°15′ E longitudes. The district is bounded by Bikaner District to the north-west, Churu District to the north, Sikar District to the north-east, Jaipur District to the east,

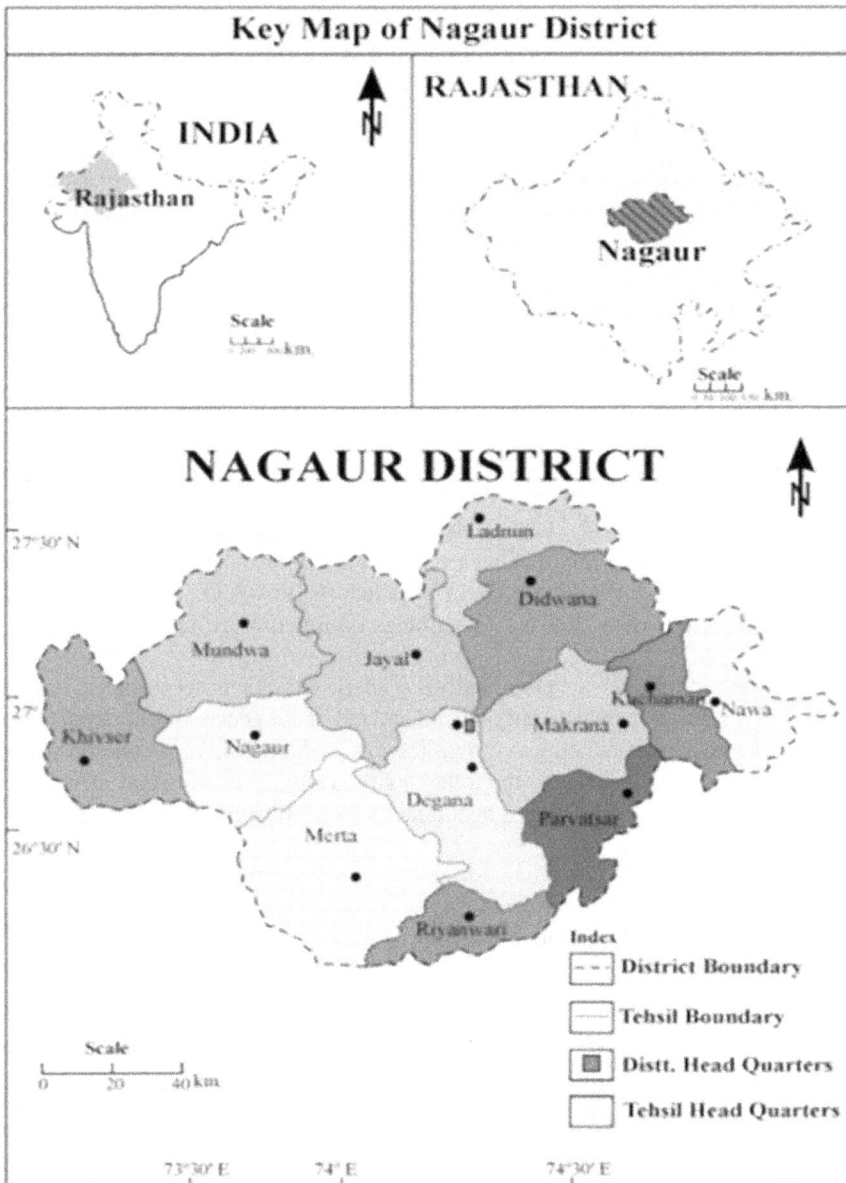

FIGURE 21.1   Site and situation of the study area.

Ajmer District to the south-east, Pali District to the south and Jodhpur District to the south-west and west. According to the census 2011, Nagaur District has a population of 3,307,743 and area equal to 17,718 km2. Density of population is 187 people per square kilometer, sex ratio is 948 and literacy rate is 64.08% (Census of India 2011). According to the latest livestock census 2019, the total animal population of the district is 27,649,718 and density of animal population is 156 animal per square kilometer. Presently in 2022 the district consists of 14 tehsils, namely Nagaur, Kheenvsar, Jayal, Merta, Degana, Didwana, Ladnun, Parbatsar, Makrana, Kuchaman City, Nawa, Riyanbadi, Mundwa and Sanju, and 1,607 villages. According to the data received (2017–2018) from the Animal Husbandry Department, Government of Rajasthan, the District Nagaur is divided into two district offices, namely Nagaur and Kuchaman City. The Nagaur office covers seven tehsils and seven panchayat samitis, namely Nagaur, Jayal, Mundawa, Khinwsar, Merta, Degana and Riyanbadi. The Kuchaman City office also serves six tehsils and seven panchayat samitis, namely Kuchaman, Nawa, Parbatsar, Makrana, Ladnun, Didwana and Molasar, where the first six are tehsils also, but Molasar is the only panchayat samiti which is not an independent tehsil and part of the Didwana tehsil.

21.4 RESEARCH METHODOLOGY

Mainly the secondary data, i.e. Census of India (2011), Statistical Handbook of Nagaur, District Statistical Outline of Nagaur 2009–2019, etc. collected from various government offices and departments of the district have been used in this research, but some primary inputs found from animal keepers and villagers have been inculcated in this study, whose animals got affected from Lumpy Skin disease recently. The following secondary datasets have been used:

- District Census Handbook of Nagaur 2011
- Animal Census of Nagaur District 2007
- Animal Census of Nagaur District 2019
- Nagaur District Statistical Outline 2006 to 2019 (District Statistical Outline-Nagaur 2009 and 2015)
- Administrative Report 2020–2021, Administrative Report, Directorate of Animal Husbandry, Rajasthan. These data has been recieved from variouse sources like Economic Reviwe (2020–21), Jaya Blok-A.O., (2015), Kuchaman Blok-A.O., (2015), Statistical Year Book (2019), and Tripathi (2000) etc. Other block information are available in bibliography.

For measuring disparities or analyzing regionally any single variable is not sufficient to evaluate the complex characteristics which are generally hidden and not in our general observation. A composite picture may be generated by calculating a composite score with the help of wisely chosen indicators. Hence, after going through the datasets, the following indicators have been considered for detailed analysis of variation or disparity in animal health facilities of Nagaur:

1. X_1: Number of veterinary hospitals
2. X_2: Number of veterinary dispensaries
3. X_3: Number of artificial insemination centres
4. X_4: Number of mobile veterinary clinics
5. X_5: Number of veterinary sub-centres
6. X_6: Number of animals sterilized

The biasness of scale and determination of weightage of variables may cause this analysis to be insignificant, so by removing these two obstacles the standardization process has been opted. In this process, each variable is subtracted by its mean and then divided by standard deviation, thus the obtained value is the Z-score, and the process is called standardization. For this study, all the indicators of animal health facilities among the tehsils have been processed with the help of SPSS Statistics and standard score calculated by Z-score technique as follows:

$$Z - \text{score} = (x_i - \bar{x}) \div \text{S.D.}$$

where $\bar{x}$ is the mean of all x of the variables concerned; S.D. is the standard deviation.

$$\bar{x} = \frac{\sum x}{N}$$

$$SD = \sqrt{\frac{\sum d^2}{N}} \quad \text{where } d = x - \bar{x}$$

C.S. is the average of Z-scores of all indicators.

21.5 RESULTS AND DISCUSSION

The data received from various government offices have been tabulated as veterinary facilities in 2007–2008 (Table 21.1) and veterinary facilities in 2017–2018 separately (Table 21.2).

With the help of Table 21.1 the means were calculated for each indicator, i.e. veterinary hospital X_1, veterinary dispensaries X_2, artificial insemination centres X_3, mobile veterinary clinics X_4, veterinary sub-centres X_5 and animals sterilized X_6 for 2007–2008, and then deviation from mean for each indicator and standard deviation have been calculated for the same year. A similar process has been repeated for 2017–2018 with Table 21.2. All these calculation results generated through SPSS software were found within a very short span of time without any manual calculations.

Through SPSS software Z-scores ZX_1, ZX_2, ZX_3, ZX_4, ZX_5 and ZX_6 were calculated for each indicator X_1, X_2, X_3, X_4, X_5 and X_6 for assessment year 2007–2008 (Table 21.3) and year 2017–2018 (Table 21.4). The composite score was obtained by adding all these Z-scores for year 2007–2008 and the same process again repeated for year 2017–2018 (Table 21.2). Further, for composite scores of 2007–2008 the mean is 0.0000006 with standard deviation 0.712366 and for composite scores of 2017–2018 the mean is −0.0000001 with standard deviation 0.548626. Standardized values are as in Table 21.3 for 2007–2008 and Table 21.4 for 2017–2018.

TABLE 21.1
Veterinary Facilities in 2007–2008

S. No.	Name of Panchayat Samiti	Number of Veterinary Hospitals X_1	Number of Veterinary Dispensaries X_2	Number of Artificial Insemination Centres X_3	Number of Mobile Veterinary Clinics X_4	Number of Veterinary Sub-centres X_5	Number of Animals Sterilized X_6
1	Ladnun	7	0	3	0	3	0
2	Degana	8	1	8	0	6	0
3	Merta	6	1	7	0	6	0
4	Riyan	4	1	5	0	2	0
5	Makrana	5	0	3	0	6	0
6	Didwana	7	1	5	0	10	0
7	Jayal	4	1	5	0	4	0
8	Parbatsar	6	0	6	0	8	0
9	Kuchaman	11	3	12	0	10	30,180
10	Nagaur	4	2	6	1	12	32,885
11	Mundwa	9	2	11	0	4	0

Source: District Statistical Outline-Nagaur (2009).

TABLE 21.2

Veterinary Facilities in 2017–2018

S. No.	Name of Panchayat Samiti	Number of Veterinary Hospitals X_1	Number of Veterinary Dispensaries X_2	Number of Artificial Insemination Centres X_3	Number of Mobile Veterinary Clinics X_4	Number of Veterinary Sub-centres X_5	Number of Animals Sterilized X_6
1	Ladnun	14	0	16	1	17	5,419
2	Degana	8	3	20	1	14	3,396
3	Merta	6	0	14	1	13	2,778
4	Riyan	9	1	14	1	6	2,161
5	Makrana	11	1	15	1	14	3,746
6	Didwana	9	1	20	1	21	5,039
7	Jayal	12	0	19	0	15	4,013
8	Parbatsar	10	2	10	1	13	8,965
9	Kuchaman	11	0	19	2	15	5,783
10	Nagaur	8	2	17	1	12	3,396
11	Mundwa	5	1	11	0	15	2,367
12	Khinwsar	8	0	15	1	11	1,544
13	Molasar	9	0	11	0	16	3,359
14	Nawa	10	0	16	1	10	2,594

Source: District Statistical Outline-Nagaur (2019).

TABLE 21.3

Standardized Veterinary Facilities (Z-Score) in 2007–2008

S. No.	2006–2007 Panchayat Samiti (PS)	ZX_1	ZX_2	ZX_3	ZX_4	ZX_5	ZX_6	C.S.	Rank
1	Ladnun	0.24218	−1.15577	−1.1868	−0.30151	−1.07782	−0.44896	−0.65478	11
2	Degana	0.68618	−0.09631	0.53094	−0.30151	−0.14182	−0.44896	0.038087	4
3	Merta	−0.20182	−0.09631	0.18739	−0.30151	−0.14182	−0.44896	−0.16717	6
4	Riyan	−1.08981	−0.09631	−0.49971	−0.30151	−1.38983	−0.44896	−0.63769	9
5	Makrana	−0.64581	−1.15577	−1.1868	−0.30151	−0.14182	−0.44896	−0.64678	10
6	Didwana	0.24218	−0.09631	−0.49971	−0.30151	1.10619	−0.44896	0.000313	5
7	Jayal	−1.08981	−0.09631	−0.49971	−0.30151	−0.76582	−0.44896	−0.53369	8
8	Parbatsar	−0.20182	−1.15577	−0.15616	−0.30151	0.48218	−0.44896	−0.29701	7
9	Kuchaman	2.01816	2.0226	1.90513	−0.30151	1.10619	1.91442	1.444165	1
10	Nagaur	−1.08981	0.96314	−0.15616	3.01511	1.73019	2.12624	1.098118	2
11	Mundwa	1.13017	0.96314	1.56159	−0.30151	−0.76582	−0.44896	0.356435	3

Source: Computed by author based on Table 21.1.
Note: C.S. Composite Score

It is evident that in the year 2007–2008 the composite score varies from 1.44 to −0.65. The highest score is recorded by Kuchaman (1.44) followed by Nagaur (1.1) and Mundwa (0.36). Panchayat Samiti Kuchaman has scored top in each indicator except in mobile veterinary clinic where Nagaur PS is the only winner.

TABLE 21.4

Standardized Veterinary Facilities (Z-Score) in 2017–2018

S. No.	Panchayat Samiti (PS)	ZX 1	ZX 2	ZX 3	ZX 4	ZX 5	ZX 6	C.S	Rank
1	Ladnun	2.01928	−0.80589	0.15049	0.26726	0.93994	0.7959	0.56116	3
2	Degana	−0.55071	2.27115	1.35444	0.26726	0.08173	−0.26209	0.52696	4
3	Merta	−1.40737	−0.80589	−0.45148	0.26726	−0.20433	−0.58529	−0.53118	10
4	Riyan	−0.12238	0.21979	−0.45148	0.26726	−2.20681	−0.90796	−0.5336	11
5	Makrana	0.73428	0.21979	−0.15049	0.26726	0.08173	−0.07904	0.17892	6
6	Didwana	−0.12238	0.21979	1.35444	0.26726	2.08421	0.59716	0.73341	2
7	Jayal	1.16261	−0.80589	1.05345	−1.60357	0.3678	0.06059	0.03916	8
8	Parbatsar	0.30595	1.24547	−1.65542	0.26726	−0.20433	2.65037	0.43488	5
9	Kuchaman	0.73428	−0.80589	1.05345	2.13809	0.3678	0.98626	0.74566	1
10	Nagaur	−0.55071	1.24547	0.45148	0.26726	−0.4904	−0.26209	0.11016	7
11	Mundwa	−1.83571	0.21979	−1.35444	−1.60357	0.3678	−0.80023	−0.83439	14
12	Khinwsar	−0.55071	−0.80589	−0.15049	0.26726	−0.77647	−1.23064	−0.54116	12
13	Molasar	−0.12238	−0.80589	−1.35444	−1.60357	0.65387	−0.28144	−0.58564	13
14	Nawa	0.30595	−0.80589	0.15049	0.26726	−1.06254	−0.68151	−0.30437	9

Source: Computed by author based on Table 21.2.
Note: C.S. Composite Score

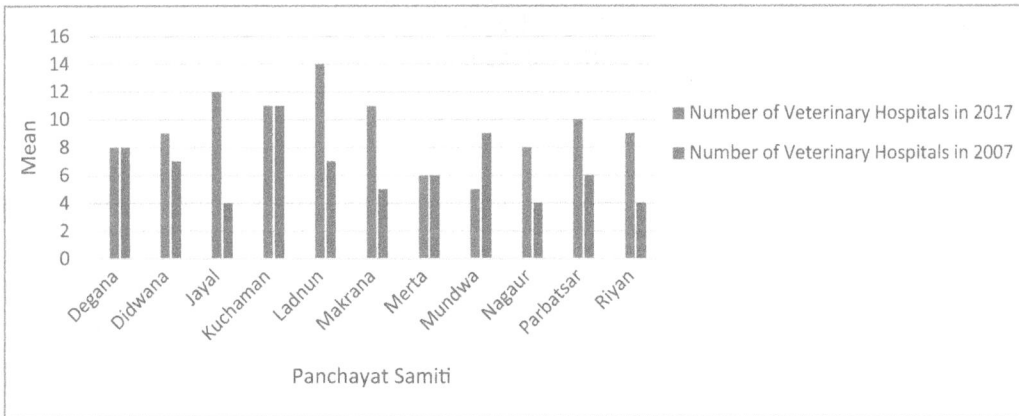

FIGURE 21.2 Panchayat samiti-wise veterinary hospitals in 2007 and 2017.

Similarly for the year 2017–2018 the composite score varies between 0.75 and −0.83. The highest score is again recorded by Kuchaman (0.75) followed by Didwana (0.73) and Ladnu (0.56). This time Panchayat Samiti Kuchaman is not the top in each indicator.

21.5.1 Distribution of Veterinary Hospitals

It has been observed that there are 71 veterinary hospitals distributed in 11 panchayat samitis of the district in 2007–2008 where a maximum of 11 were situated in Kuchaman and a minimum of four in Nagaur, Jayal and Riyan each, with skewness 0.715 and kurtosis 0.024, which means the distribution is moderately uneven. Further, in 2017–2018 the number of hospitals recorded by 130 with an increase of about 83% distributed in 14 panchayat samitis with maximum improvement shown by Ladnun with the highest score 2.0193 and the least score −1.83571 by Mundwa.

21.5.2 DISTRIBUTION OF VETERINARY DISPENSARIES

It is observed that there are only 12 veterinary dispensaries distributed over 11 panchayat samitis of the district in 2007–2008 where maximum three were situated in Kuchaman and followed by Nagaur and Mundawa having two each with skewness 0.663 and kurtosis 0.199. Makrana, Parbatsar and Ladnun have no dispensary. Further in 2017–2018, the number of dispensaries recorded by 11 with a decrease of about 8.33% distributed in 14 panchayat samitis where no significant improvement was shown by the district but an improvement was shown by Degana with highest score 2.27115. In 2017–2018 the skewness is found to be 1.073 which shows highly uneven distribution of the facilities.

21.5.3 DISTRIBUTION OF ARTIFICIAL INSEMINATION CENTRES

It is observed that there are 71 artificial insemination centres distributed over 11 panchayat samitis of the district in 2007–2008 where a maximum of 12 are situated in Kuchaman followed by Mundawa having 11 centres with skewness 0.874 and kurtosis 0.115. Makrana and Ladnun have the least number of centres at three each. Further, in 2017–2018 the number of artificial insemination centres was recorded at 217 with a massive increase of about 206% distributed in 14 panchayat samitis where maximum improvement was shown by Degana and Didwana with highest score 1.35444 and least score −1.6554 by Parbatsar. The data also reveal that the government has a focus on artificial insemination techniques rather than developing natural indigenous breeding techniques. On one hand this facility makes it easy for the animal keeper to expand the animal family as well as the quantity of milk, meat and wool, but on the other hand due to a lack of information mixed breeds are increased unscientifically and the level of satisfaction of animals suffers which will affect the quality of products.

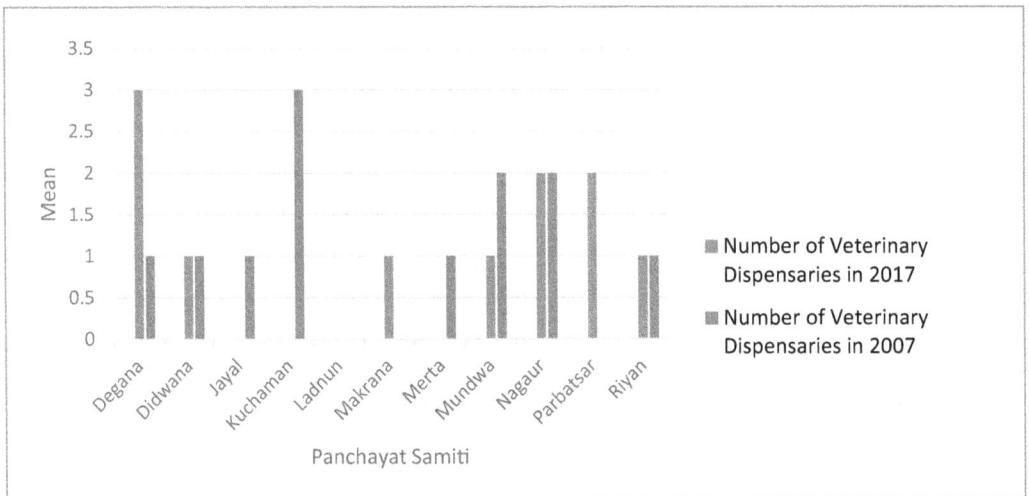

FIGURE 21.3 Panchayat samiti-wise veterinary dispensaries in 2007 and 2017.

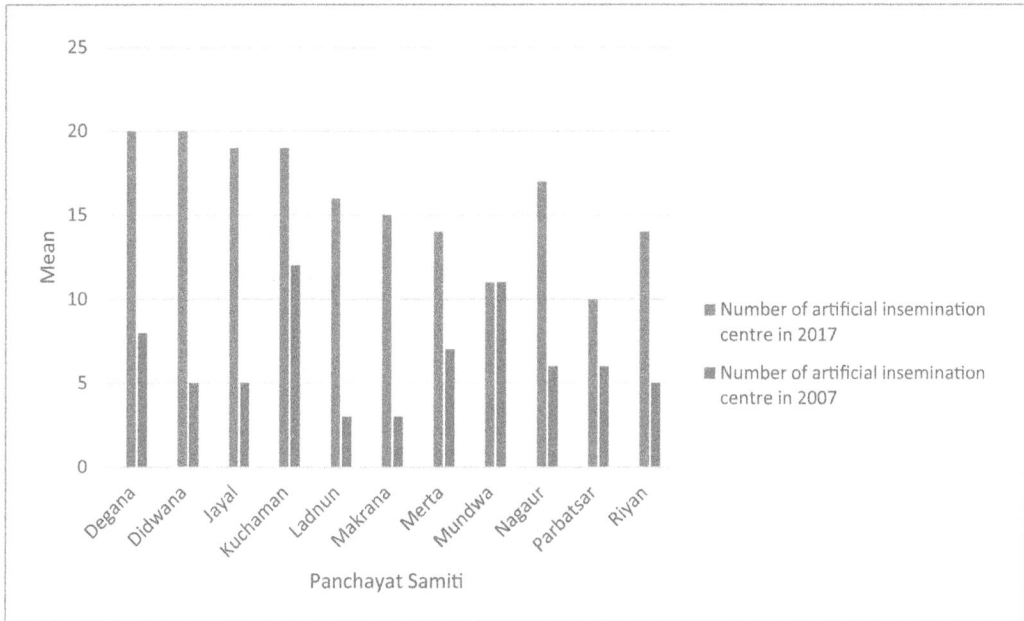

FIGURE 21.4 Panchayat samiti-wise artificial insemination centre in 2007 and 2017.

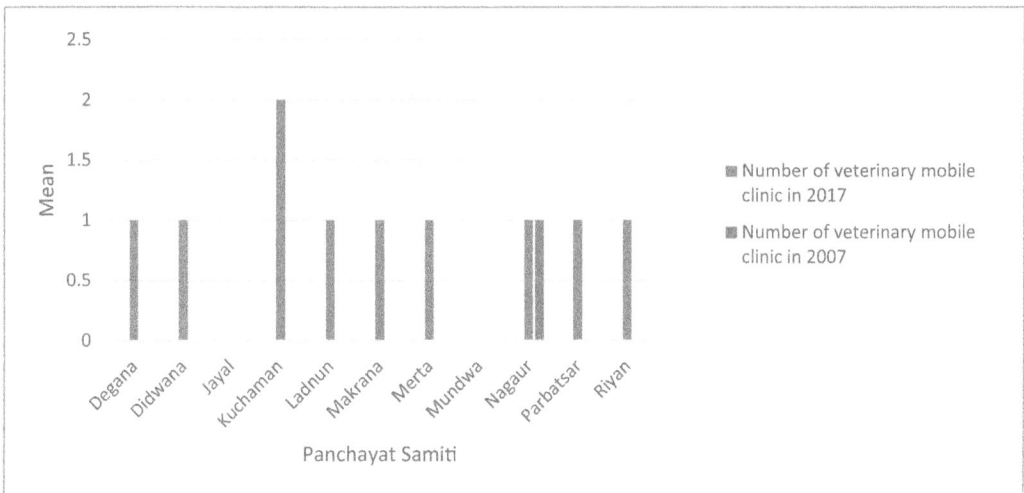

FIGURE 21.5 Panchayat samiti-wise veterinary mobile clinic in 2007 and 2017.

21.5.4 DISTRIBUTION OF MOBILE VETERINARY CLINICS

It is observed that in 2006–2007 there is only one mobile veterinary clinic which is situated in Nagaur and the rest of the areas lack this facility with skewness 3.317 and kurtosis 11 which shows the highest asymmetry. Further, in 2016–2017 the number of mobile veterinary clinics recorded by 12 with an increase of 1,100% substantially skewed distribution over 14 panchayat samitis with the highest score by Kuchaman, i.e. 2.13809. Although the number of mobile veterinary clinics increased from 1 (2007–2008) to 12 (2017–2018), this number is like cumin in a camel's mouth. The huge area and large number of livestock population demands this facility in the hundreds.

21.5.5   DISTRIBUTION OF VETERINARY SUB-CENTRES

It is observed that there are 71 veterinary sub-centres distributed over 11 panchayat samitis of the district in 2007–2008 where maximum 12 are situated in Nagaur followed by Kuchaman and Didwana having 10 centres each with skewness 0.386 and kurtosis −0.907. Riyan and Ladnun have the least centres at two and three, respectively. Further, in 2016–2017 the number of veterinary sub-centres recorded by 192 with a massive increase of about 170% distributed in 14 panchayat samitis with maximum improvement shown by Didwana with highest score 2.08421 and least score −2.20681 by Riyan.

21.5.6   DISTRIBUTION OF STERILIZATION FACILITY

The data show that 63,065 animals were sterilized in the district in 2006–2007 with the maximum 32,885 in Nagaur Panchayat Samiti followed by 30,180 in Kuchaman, and the rest facing a lack of this facility with skewness 1.933 and kurtosis 2.110. Further, in 2017–2018 the number of sterilized animals was recorded as 54,560 with a decrease of about 13.48% distributed in 14 panchayat samitis where maximum improvement was shown by Parbatsar with the highest score 2.65037 and the least score −1.23064 by Khinwsar (Khinwsar Block An Overviwe 2015).

21.5.7   OVERALL DISTRIBUTION OF ANIMAL HEALTH-CARE FACILITIES

For evaluating and assessing the overall distribution of animal health-care facilities, all six indicators have been clubbed together and calculate the composite index. Table 21.1 reveals that in 2007–2008 Kuchaman (1.444165) was on top, whereas the least composite index was scored by Ladnun (−0.65478). In 2017–2018 Kuchaman (0.74566) maintained its top position but Ladnun (0.56116) showed the highest improvement in its position and reached the top third, whereas Mundawa (−0.83439) slipped to the bottom with the highest decline of 11 points. Increasing animal health-care facilities were recorded in Ladnun (seven points), Makrana (four points), Didwana (three

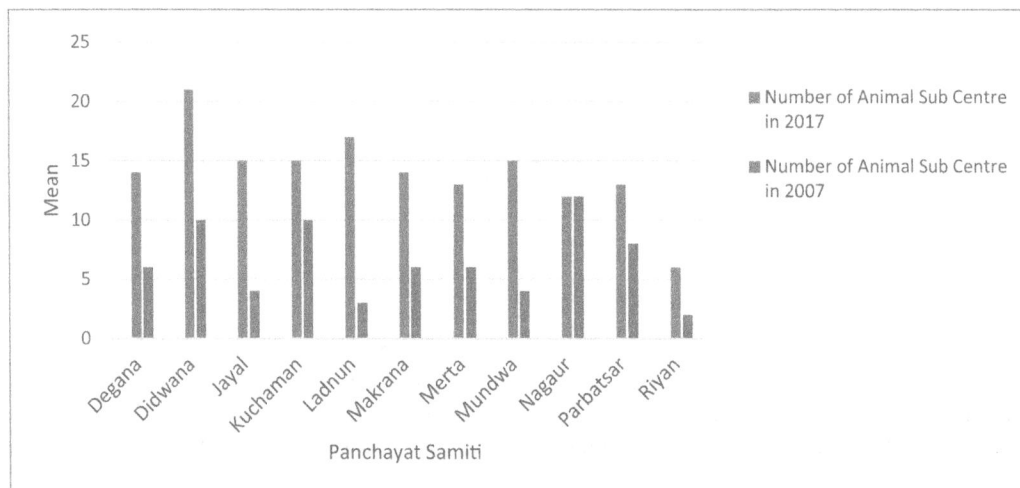

FIGURE 21.6   Panchayat samiti-wise animal sub-centre in 2007 and 2017.

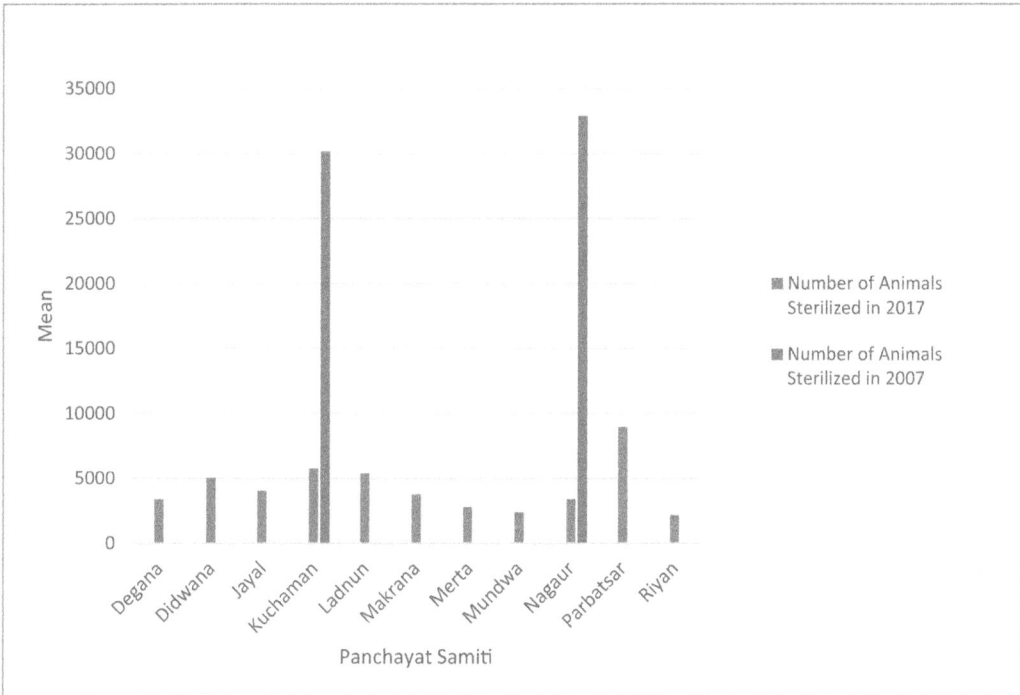

FIGURE 21.7 Panchayat samiti-wise animals sterilized in 2007 and 2017.

TABLE 21.5

Rank Comparison with Respect to Animal Health Care during 2007–2008 and 2017–2018

S. No.	Name of Panchayat Samiti	Rank in 2007–2008	Rank in 2017–2018	Improvement/ Change in Position
1	Ladnun	11	3	7
2	Degana	4	4	0
3	Merta	6	10	−4
4	Riyan	9	11	−2
5	Makrana	10	6	4
6	Didwana	5	2	3
7	Jayal	8	8	0
8	Parbatsar	7	5	2
9	Kuchaman	1	1	0
10	Nagaur	2	7	−5
11	Mundwa	3	14	−11
12	Khinwsar	N.A.	12	N.A.
13	Molasar	N.A.	13	N.A.
14	Nawa	N.A.	9	N.A.

Source: Compiled by author based on Table 21.3 and Table 21.4.

---

points) and Parbatsar (two points), whereas declining animal health-care facilities were recorded in Mundawa (11 points), Nagaur (five points), Merta (four points) and Riyan (two points). At the same time, Kuchaman, Degana and Jayal remained at the same position in 2007 as well as in 2017. The reconstruction of tehsils or panchayat samitis is also a factor of disparity.

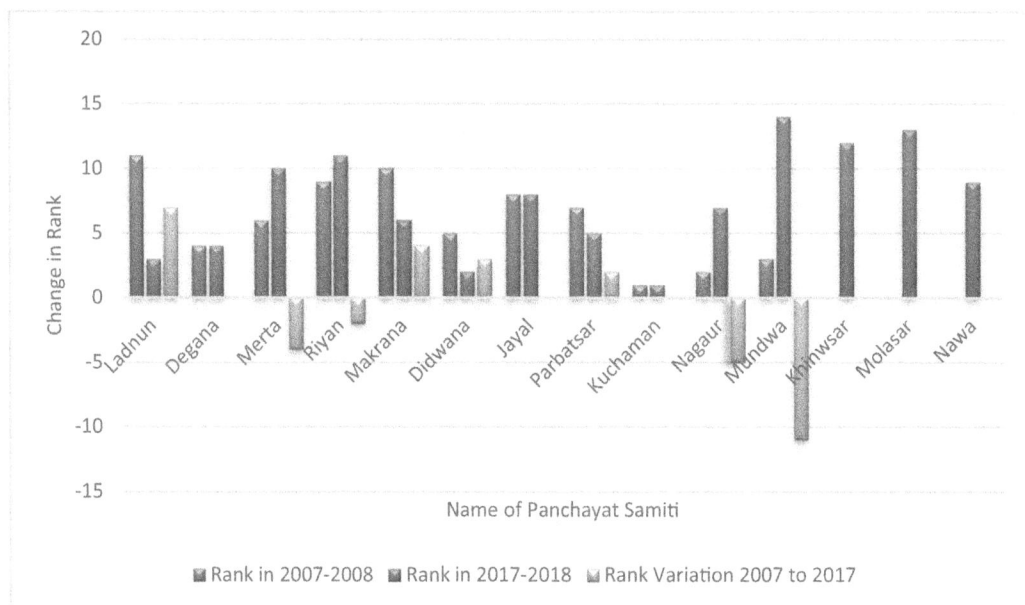

FIGURE 21.8   Panchayat samiti-wise change in ranks with respect to animal health-care facilities during 2007–2008 to 2017–2018.

TABLE 21.6
Categorization for Level of Animal Health-Care Facilities

S. No.	Category	C.I. Value 2007–2008	Panchayat Samiti Included 2007–2008	C.I. Value 2017–2018	Panchayat Samiti Included 2017–2018
1	Satisfactory	Above 0.7124	1. Kuchaman 2. Nagaur	Above 0.5486	1. Kuchaman 2. Didwana 3. Ladnun
2	Above average	Between 0.7124 and 0.3562	1. Mundwa	Between 0.5486 and 0.2743	1. Degana 2. Parbatsar
3	Average	Between 0.3562 and −0.3562	1. Degana 2. Didwana 3. Merta 4. Parbatsar	Between 0.2743 and −0.2743	1. Makrana 2. Nagaur 3. Jayal
4	Below average	Between −0.3562 and −0.7124	1. Jayal 2. Riyan 3. Makrana 4. Ladnun	Between −0.2743 and −0.5486	9. Nawa 10. Merta 11. Riyan 12. Khinwsar
5	Poor	Below −0.7124	—	Below −0.5486	1. Molasar 2. Mundwa

Source: Compiled by author based on results of the research.
Note: C.I. Composite Index

In 2007–2008, eight of 11 panchayat samitis, i.e. 73% fell in to the average or below average category of animal health-care facilities. Similarly, in 2017–2018 a total of nine out of 14 panchayat samitis were in the average/below average or poor category from which two, namely Molasar and Mundawa, were in the poor category. Only two samitis, namely Kuchaman and Nagaur, were equipped with satisfactory facilities in 2007–2008 and the same could not be maintained by Nagaur in 2017–2018, and the category was occupied by Didwana and Ladnun, hence Kuchaman could maintain itself on the top.

Summarily Ladnun, Makrana, Didwana, Parbatsar, Degana and Jayal have improved their stage and entered into an upward category, but Nagaur, Mundawa and Merta lost their position and entered into a downward category. It is evident by analyzing the performance of the district offices of Nagaur and Kuchaman City, in 2017–2018 the top six positions are occupied by Kuchaman City district office and only the Molasar is in poor condition where the new reconstruction is a push factor. Out of the seven panchayat samitis served by Nagaur district office, five are at below average or

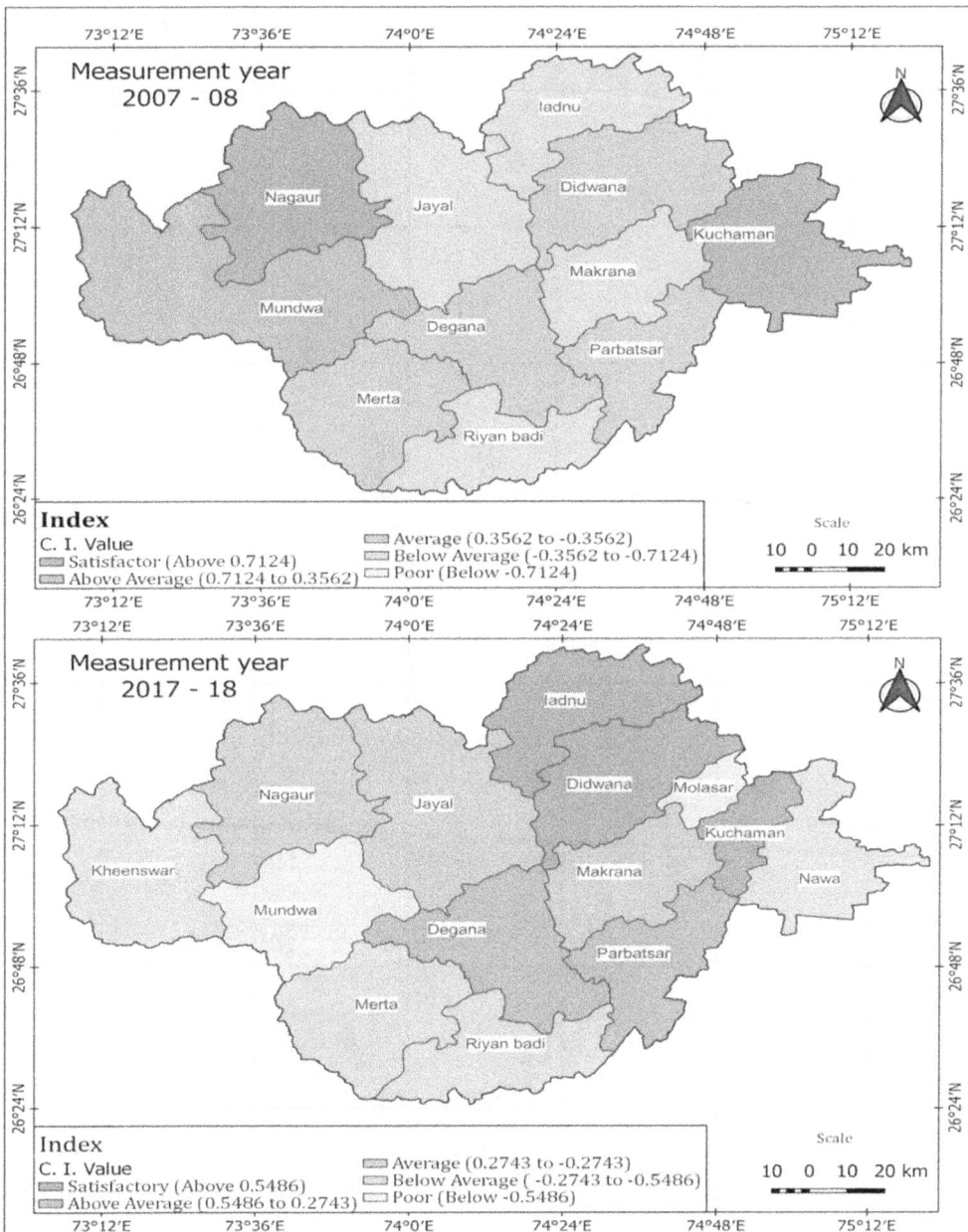

FIGURE 21.9 Spatial variation in animal health-care facilities in Nagaur, 2007–2008 and 2017–2018.

poor level, only two, namely Nagaur and Jayal, are at average level and none are at satisfactory level. The level of animal health facilities of Nagaur is far behind Kuchaman City. The reasons behind these disparities are large serving area, tough terrain, comparably unpleasant climate, undeveloped agriculture, lack of infrastructure, lack of political will power, unawareness, etc. which make several negative factors for Nagaur compared to Kuchaman City.

21.6 CONCLUSION AND SUGGESTIONS

In this chapter, it was presented that approximately 70% of animal keepers are facing a lack of animal health-care facilities due to lack of information, lesser availability, lesser accessibility,

FIGURE 21.10 Insights for animal care from traditional knowers during field visit 26 October 2022.

FIGURE 21.11 Insights for animal care from traditional knowers during field visit 29 October 2022.

FIGURE 21.12 Meeting with camel knower Sualal of Parbatsar dressing the camel in Ramdev Animal Fair on 22 January 2023.

ignorance, large serving area, large livestock population, lesser governmental priorities, etc. At the same time, it recognizes the tension and pressure of workload on the veterinary institutes that serve a huge animal population in the rural areas.

In this analytical study, it was found that the role of veterinary amenities is very important to take care of animal health, and the key findings and suggestions for improving animal health-care facilities are as follows:

1. The study area had a huge livestock population distributed over 14 panchayat samitis, but the number of veterinary amenities is much less with respect to animal population and area, so these facilities are under pressure, especially the Nagaur district office which should be divided into one or more offices like Kuchaman City.
2. Kuchaman is leading the district from a decade in context of animal health-care facilities because Kuchaman City is nearer to the state capital and also has its organizational setup from years ago and has less area to serve as a district which improves the quality of work.
3. During the field visit it was reflected that animal keepers or farmers are generally not aware of animal health-care facilities, hence they try traditional methods for curing their animals.
4. It was the peak time of spreading lumpy skin disease, and we found a huge loss of livestock especially milch cows in the study area. Animals and animal keepers are helpless due to lack of animal health-care facilities, either these are unavailable or out of reach.
5. So, now it is necessary to expand these facilities in the area, and government should try to make them easily available and accessible to the common man.
6. Artificial insemination is not used wisely which results in a huge herd of non-descript unproductive mixtures.

BIBLIOGRAPHY

20th Livestock Census (2019), All India Report, Department of Animal Husbandry and Dairying under Ministry of Fisheries, Animal Husbandry and Dairying, GOI, New Delhi.

Census of India (2011), Indian Census in Its Perspective General, Ministry of Home Affairs India, New Delhi.

Chandna, R.C. (2016), Geography of Population, Kalyani Publishers, New Delhi.

Degana Block-An Overview (2015), Office of Block Statistical Officer, Panchayat Samiti, Degana.

Dhawan, B.D. (2003), Studies in Traditional and Modern Irrigated Agriculture, Commonwealth Publishers, New Delhi.

Didwana Block-An Overview (2015), Office of Block Statistical Officer, Panchayat Samiti, Didwana.

District Statistical Outline-Nagaur (2009), Office of Asst. Director, Economics and Statistics Department, Nagaur.

District Statistical Outline-Nagaur (2019), Office of Asst. Director, Economics and Statistics Department, Nagaur.

Economic Review (2020–21), Directorate of Economics and Statistics, Govt. of Rajasthan, Jaipur.

Epstein, S.T. (1962), Economic Development and Social Change in South Asia, Oxford University Press, London.

FAO (2021a), The Impact of Disasters and Crises on Agriculture and Food Security: 2021, Rome, FAO. https://doi.org/10.4060/cb3673en

FAO (2021b), The State of Food Security and Nutrition in the World 2021. Transforming Food Systems for Food Security, Improved Nutrition and Affordable Healthy Diets for All. Rome, FAO. https://doi.org/10.4060/cb4474en

Gujar, M.L. and Varghese, K.A. (2005), "Structural Charges Overtime in Cost of Cultivation of Major Rubi Crops in Rajasthan," Indian Journal of Agriculture Economics, Vol. 60, No. 3, April–June, p. 249.

Jayal Block-An Overview (2015), Office of Block Statistical Officer, Panchayat Samiti, Jayal.

Khinwsar Block-An Overview (2015), Office of Block Statistical Officer, Panchayat Samiti, Khinwsar.

Kuchaman Block-An Overview (2015), Office of Block Statistical Officer, Panchayat Samiti, Kuchaman City.

Ladnun Block-An Overview (2015), Office of Block Statistical Officer, Panchayat Samiti, Ladnun.

Makrana Block-An Overview (2015), Office of Block Statistical Officer, Panchayat Samiti, Makrana.

Merta Block-An Overview (2015), Office of Block Statistical Officer, Panchayat Samiti, Merta.

Mundawa Block-An Overview (2015), Office of Block Statistical Officer, Panchayat Samiti, Mundawa.

Nagaur Block-An Overview (2015), Office of Block Statistical Officer, Panchayat Samiti, Nagaur.

Nawa Block-An Overview (2015), Office of Block Statistical Officer, Panchayat Samiti, Nawa.

Parbatsar Block-An Overview (2015), Office of Block Statistical Officer, Panchayat Samiti, Parbatsar.

Riyan Block-An Overview (2015), Office of Block Statistical Officer, Panchayat Samiti, Riyan.

Statistical Year Book (2019), Directorate of Economics & Statistics, Govt. of Rajasthan, Jaipur.

Tripathi, R.K. (2000), Population Geography, Commonwealth Publishers, New Delhi. www.Censusindia.gov.in

22 Morphometric Parameters-Based Flood Susceptibility Mapping and 1D Simulation Modelling of Adyar Watershed, Chennai Basin: A Geo-Spatial Approach

Sabirul Sk, Asraful Alam, and Govindarajan Bhaskaran

22.1 INTRODUCTION

Floods are classed as natural calamities and are often caused by heavy rains and overflowing rivers (Douben, 2006). They can cause major property damage and human casualties, as well as infrastructure failure (Duan et al., 2022). In tropical and subtropical regions of the world with a predominance of monsoon rains, flood risks are among the most deadly, prevalent, and common natural hazards (Hirabayashi et al., 2013). Over 80% of the population in Asia, where the monsoon is dominant, is vulnerable to flooding, a recurring natural hazard (Sanyal & Lu, 2004). Because of its rugged terrain and wet, cyclonic environment, the Adyar watershed is susceptible to flooding (Sundaram et al., 2023). Due to the frequency of cyclones in this region, these locations are susceptible to flooding. Adyar watershed is an urban area. Because of the increased risk of catastrophes, the potential for economic loss and human casualties due to natural disasters, and the need for sustainable communities in urban areas, integrating urban planning and hazard mitigation efforts is essential (Sundaram et al., 2023). Consequently, one of the strategies for scenario modeling for urban planning can use the susceptibility map (Tehrany et al., 2014). It is a recurrent natural calamity that affects worldwide, and it ranks among the most dangerous and frequent threats in tropical and subtropical regions that are dominated by monsoons. It refers to water overflows from its channel over the dry area. Generally, alluvial areas are more vulnerable to floods as compared to other types of soil (Saint-Laurent et al., 2019). Sometimes extreme flood occurrence leads to a change in the morphology of the rivers and their floodplain (Ralph & Hesse, 2010). Every major flood has significant challenges to watershed management, mostly in urban areas (Niemczynowicz, 1999). Management of every challenge related to flood control, floodplain modulation, and effects on land-use modifications all need a comprehension of the recurrence and size of flooding occasions and how these are changing throughout time (Asdak et al., 2018). The watershed plays a very important role in every urban place during flood and high rainfall peak times (Rozalis et al., 2010). It drains all water in a catchment basin (Smith et al., 2013). Floods generally happen in low-lying deltaic regions (Tian et al., 2015). Because of the rapid buildup and release of runoff waters from upstream to downstream brought on by extremely heavy rainfall, floods occur (Tian et al., 2015). There are many factors for the flood which differ from area to area, it depends upon the geographical location. It is very hard to control floods completely. So, flood vulnerability mapping is one of the most

DOI: 10.1201/9781003377825-25

helpful tools for detecting flood-susceptible areas. Researchers use the analytic hierarchy process (AHP) for preparing flood vulnerability/susceptibility and flood risk zone using geographic information systems (GIS) (Hoque et al., 2019). AHP is considered one of the most effective techniques for vulnerability (Ghosh & Kar, 2018). Some historical evidence showed that floods occurred in the Adyar watershed of India in 2015 and 2021 because of heavy rainfall (Sundaram et al., 2023). The major objective of this study is to identify and apply some quantitative techniques in combination with GIS for flood vulnerability mapping and to quantify areas under risk in the Adyar watershed, Tamil Nadu, India.

As indicated by verifiable records of the Adyar watershed, more than 60% of the populace has consistently lived under significant danger of flooding. A literature study reveals the paucity of evaluation of the geographical position and features of the Adyar watershed in terms of its high flood risk.

Using remotely sensed data, this study aims to acquire geo-information about the Adyar watershed in Tamil Nadu and to describe the region's vulnerability to flooding. The Sentinel-2 image dataset of the study region and the Advanced Space-Borne Thermal Emission and Reflection Radiometer (ASTER) Digital Elevation Model (DEM) data serve as the primary sources of information.

The main objectives of this work are as follows:

- To identify the spatiotemporal pattern and variation of rainfall in the Adyar watershed
- To show the effect of land use/land cover (LULC) on flood in the study area
- To compute the morphometric parameters of the Adyar watershed to assess flood vulnerability
- To prepare a flood vulnerability map of the Adyar watershed using Saaty's AHP method
- To develop a one-dimensional (1D) steady flow flood simulation model of the Adyar River to adopt flood management practices using the US Army Corps of Engineers' Hydrologic Engineering Center's River Analysis System (HEC-RAS)

In this chapter, we examine future flooding scenarios in the Adyar watershed at various levels of flooding. Both physical and economic elements of the environment will be considered in the analysis. In the current environment of haphazard spatial urban growth, the study will aid in determining the level of harm that flooding can cause. Flooding mostly disrupts people's lives and leads to economic losses. Because the population density of the area is exceptionally high, it is critical to determine the geographical extent of floods because the afflicted population multiplies with modest changes in the extent (Sundaram et al., 2023). In this context, there is a growing need to conduct a flood-related demographic impact assessment. It has also been noted in the recent past that the magnitude of the floods has been getting increasingly higher. These issues provide the researcher with the scope for conducting the study.

22.2 STUDY AREA

The Adyar River basin is situated in the southern region of India (Figure 22.1). It is located between 79°50′00″ E to 80°15′00″ E and 12°40′00″ N to 13°10′00″ N. This river originates from Chembarambakkam Lake and runs through Kancheepuram, Tiruvallur, and Chennai districts in Tamil Nadu, India's Kanchipuram district (Sundaram et al., 2023). At Adyar in Chennai, it joins with the Bay of Bengal. The Adyar basin's total catchment area is 680.0789 km², and the river's length is 42 km. Adyar watershed is a part of the Chennai basin. Chennai City experienced flooding in the years 1943, 1976, 1985, 1996, 2005, 2015, and 2021. Flooding is a frequent issue in the Adyar River basin. The majority of Chennai's sewage water is discharged into this river. This basin has a subtropical climate (Suriya & Mudgal, 2012). The basin has 1,104 mm of yearly rainfall on average. Three sizable neighborhoods in Chennai include Adyar, Thiruvanmiyur, and Besant Nagar.

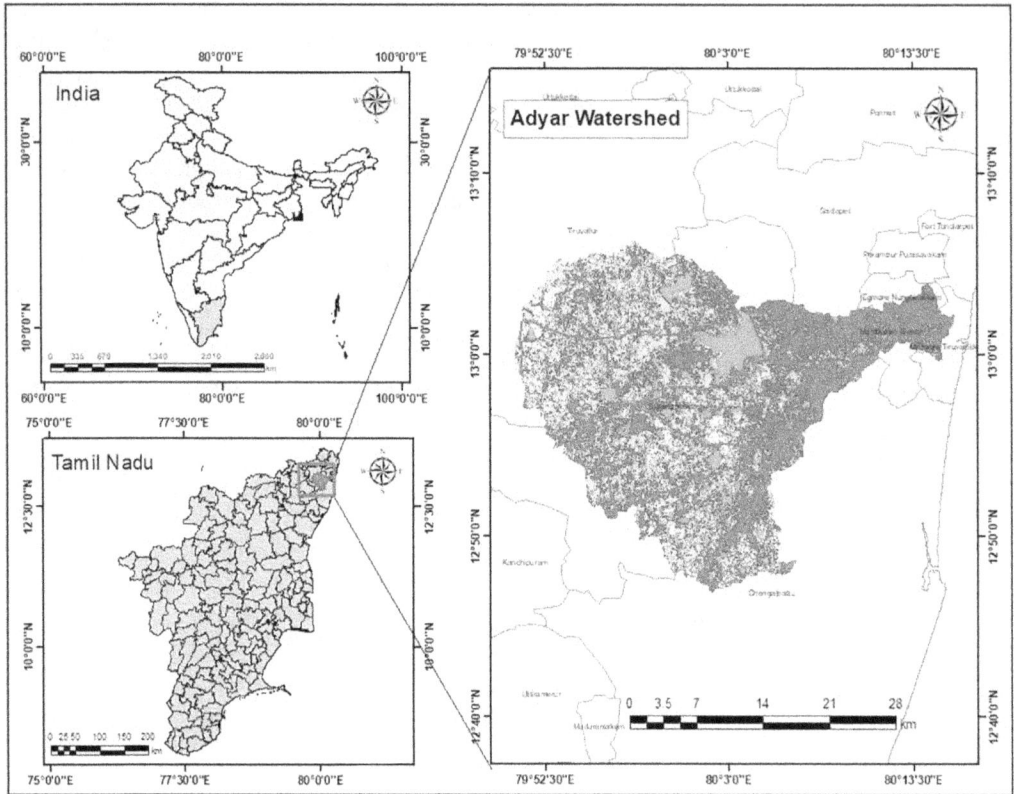

FIGURE 22.1   Location map of the Adyar watershed.

The Adyar watershed, which is situated in an urban area, has been extensively utilized over the last 20 years by all significant businesses, new educational institutions, and housing developments (Sundaram et al., 2023). Because of huge road concretization and unplanned projects that have flattened the natural topography, over 90% to 100% of rainwater in Chennai City falls into storm drains rather than being absorbed into the earth. Infiltration will be further reduced, runoff will increase, and floods will occur in the downstream region due to fast population growth and land-use change. Furthermore, the majority of water bodies within the watershed area have been encroached upon, resulting in reduced water storage. The unplanned or uncoordinated growth of Chennai City further complicates the Adyar watershed's intricate natural drainage system, which impacts urban floods and flooding.

22.3   DATABASE AND METHODOLOGY

A methodological design that starts with determining the research problems and ends with the validation of the derived results through a series of steps has been used in the current work (Figure 22.2).

22.3.1   DATA COLLECTION

With the help of Sentinel-2 remote sensing data, Aster DEM data, and rainfall data, the flood susceptibility map for the Adyar watershed was developed. Maps of drainage frequency, slope,

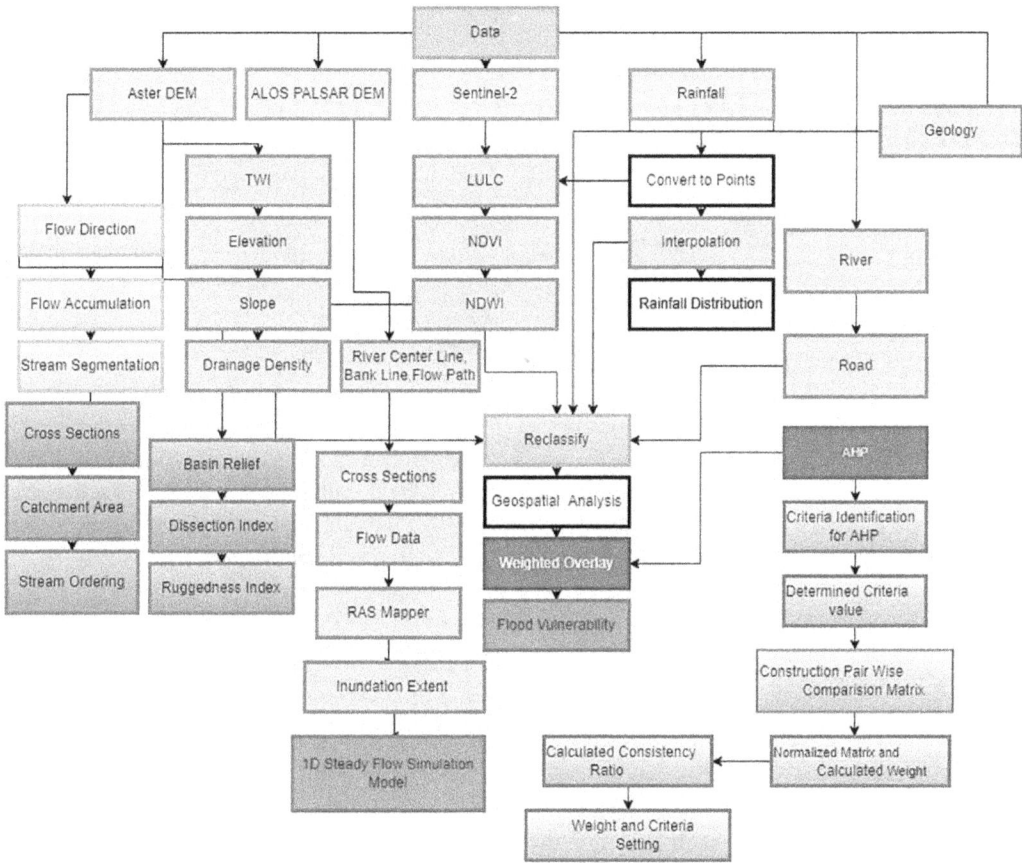

FIGURE 22.2 Methodological flowchart for flood vulnerability mapping and one-dimensional steady flow modelling.

TABLE 22.1
Database and Source

Serial Number	Data Types	Sources
1	ASTER Digital Elevation Model (DEM)	www.earthdata.nasa.gov
	ALOS PALSAR	https://bhuvan-app3.nrsc.gov.in
2	Rainfall	https://dsp.imdpune.gov.in
		https://power.larc.nasa.gov
3	Sentinel-2, Landsat-5	https://earthexplorer.usgs.gov
4	Geology	https://bhukosh.gsi.gov.in
5	Soil map	https://bhukosh.gsi.gov.in
6	Road	https://diva-gis.org

elevation, and topographic moisture index were all produced using the Aster DEM, as well as LULC, Normalized Difference Vegetation Index (NDVI), and Normalized Difference Water Index (NDWI) maps. A 1D steady flow simulation model was created in HEC-RAS using an ALOS PALSAR DEM, and several maps were created using the Aster DEM for morphometric study.

22.3.2 Methods of Data Processing

22.3.2.1 Topographic Wetness Index

The Topographic Wetness Index (TWI) is a useful tool for estimating where water will collect in a region with varying elevations. It is determined by the slope and the area contributing upstream (Beven & Kirkby, 1979):

$$TWI = ln\left(\frac{\alpha}{tan\beta}\right) \tag{22.1}$$

where α is the upslope contributing area (m²) and β is the slope in radians.

22.3.2.2 Normalized Difference Vegetation Index

To quantify vegetation (which absorbs light), the NDVI calculates the difference between near-infrared (NIR) light, which vegetation strongly reflects, and red light (Soltani et al., 2021). This list goes from −1.0 to 1.0, basically portraying greens, with negative qualities beginning from mists, water, and snow, and values near zero starting from rocks and exposed soil (Berhanu & Bisrat, 2018):

$$NDVI = \frac{\left(NIR - Red\right)}{\left(NIR + Red\right)} \tag{22.2}$$

where NIR is the reflection in the near-infrared spectrum and Red is the reflection in the red range of the spectrum.

22.3.2.3 Normalized Difference Water Index

The NDWI is a technique for finding and increasing the clarity of open water features in remotely sensed photographic images (Gao, 1996). The NDWI uses reflected green and NIR light to obliterate soil and terrestrial vegetation while enhancing the presence of such features. The researchers claim that the NDWI could provide estimates of the turbidity of water bodies utilizing remotely sensed digital data (McFeeters, 1996). Using the NDWI to assess fire danger, the moisture content in vegetation cover is measured. A high NDWI score denotes enough moisture, while a low value denotes water stress (McFeeters, 2013):

$$NDWI = \frac{\left(Green - NIR\right)}{\left(Green + NIR\right)} \tag{22.3}$$

where NIR is the reflection in the NIR spectrum and Green is the reflection in the green range of the spectrum.

22.3.2.4 Maximum Likelihood Classification

One of the most used supervised classification techniques in remote sensing is maximum likelihood classification (MLC), which places the class that makes sense for a given pixel based on its likelihood (Sisodia et al., 2014). The likelihood Lk is used to define the posterior probability that a pixel belongs to class k. MLC determines a pixel's likelihood of belonging to a particular class by employing normally distributed statistics for each class in each band (Strahler, 1980). All pixels are classified if there is no probability threshold specified (Settle & Briggs, 1987). The class with the highest probability is given to each pixel, also known as the greatest likelihood. The pixel remains unclassified if the highest likelihood is less than the threshold you set (Otukei & Blaschke, 2010).

22.4   RESULTS AND DISCUSSION

22.4.1   RAINFALL VARIABILITY

Adyar watershed covers Poonamallee, Nungambakkam, Saidapet, Guindy, Meenambakkam, Chembarambakkam, Sriperumbudur, Tambaram, and Kattankulathur rain gauges stations of three districts of Tamil Nadu.

While local rainfall is essential for pluvial flooding, the study discovered that rainfall volumes in upstream catchments also raise the risk of flooding and other risks associated with rivers (Arnaud et al., 2002). Due to the small size of the research areas, both local and upstream rainfalls were taken into account in the analysis. By examining and extrapolating the distribution of rainfall patterns for the 2022 monsoon season, the initial objective is to produce continuous raster rainfall data inside the research region (Figure 22.3), because data on rainfall for 2022 were used to create the vulnerability map. The generated raster layer was divided into four classes using the same spacing.

One of the many intricate and ever-changing elements of the environment that human cultures depend on for life is the climate (Tripathi et al., 2019). The Adyar watershed has a tropical environment with high humidity and temperatures. It was found that the eastern and north-eastern regions of the basin get noticeably higher rainfall than their western and southern equivalents during the north-east monsoon season (October–December), which accounts for the majority of the study area's annual rainfall (Anandharuban & Elango, 2021).

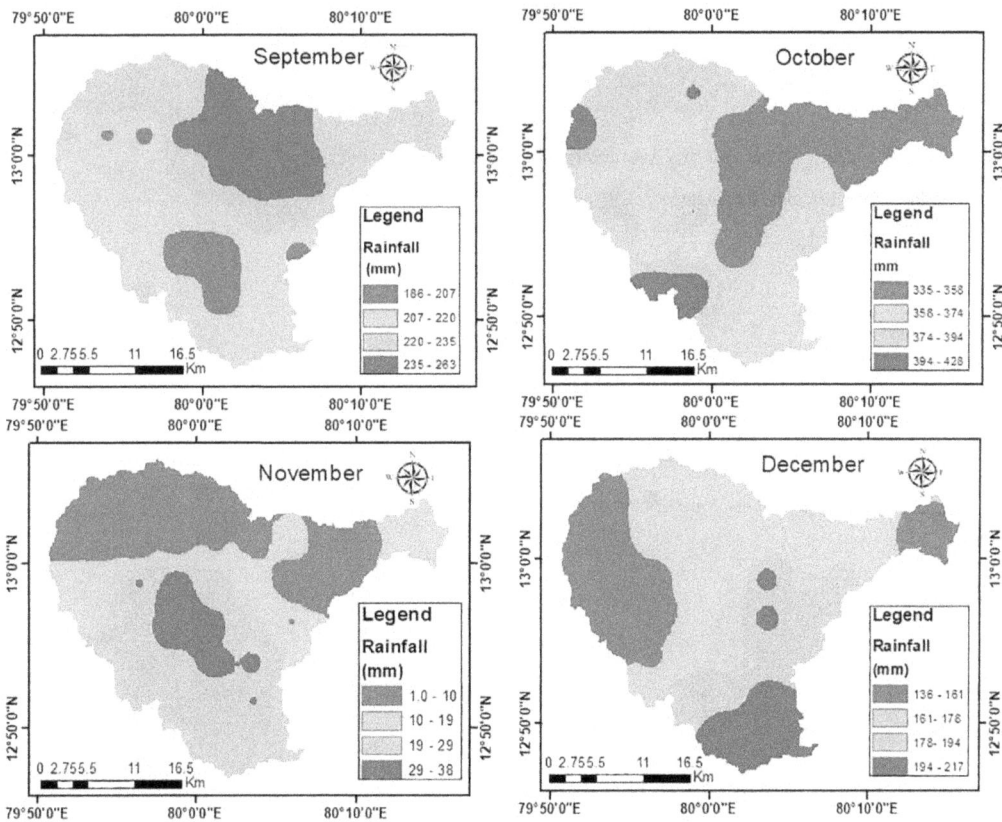

FIGURE 22.3   Monsoon seasonal rainfall map of the Adyar watershed of 2020.

Source: Indian Meteorological Department.

Using data from 2001 to 2020 across the research area, this study evaluates and estimates long-term spatio-temporal changes in rainfall. The rainfall trend was studied using the Mann-Kendall (MK) test and Sen's innovative trend analysis (Ali et al., 2019). Twenty years of monthly and yearly average rainfall have been shown below (Figures 22.4 and 22.5).

Analysis of the rainfall data for 20 years (Figures 22.4 and 22.5) shows Chennai receives the highest amount of rainfall in November. According to previous data analysis, Chennai was flooded between 2005 and 2015. Here also shows the highest amount of rainfall that took place between 2005 and 2015.

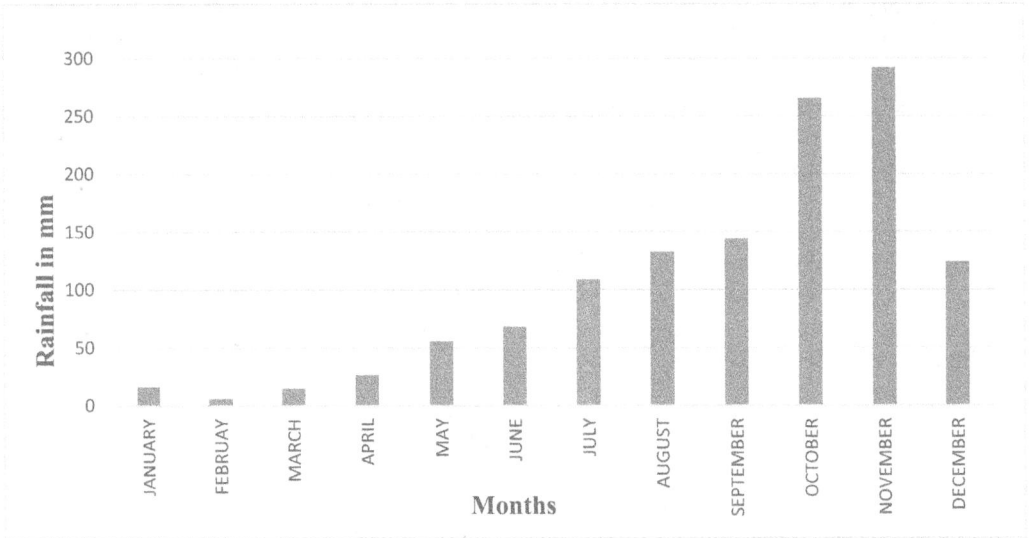

FIGURE 22.4   Years of the monthly average rainfall of Chennai district.

Source: Indian Meteorological Department.

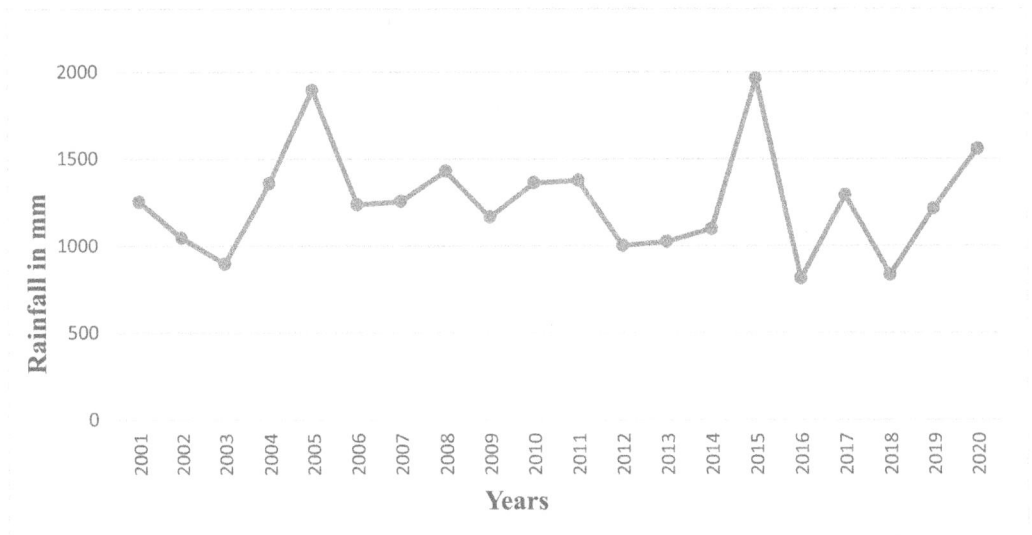

FIGURE 22.5   Years of the annual average rainfall of Chennai district.

Source: Indian Meteorological Department.

The return period has been calculated for 10 and 25 years. A return period is a prediction of how long it will take for rainfall events of a certain magnitude to occur again (Viglione & Blöschl, 2009):

$$Return\,Period\,(T) = \frac{1}{P} \tag{22.4}$$

where P is the probability of exceedance. Here, 10- and 25-year return periods have been calculated.

It is crucial to realize that this is based on statistics; there is no guarantee that this amount of rain will not occur again for another 10 or 25 years (Viglione et al., 2009). Every 10 and 25 years, 1,753.77 mm and 2,087.69 mm of rainfall could happen in the study area. There should be four such downpour events in 100 years of records (an average of once every 25 years). These events could take place over several years or even in a single year. There is also no certainty that such an occurrence will not occur in a nearby location.

22.4.2 Effect of Land Use/Land Cover on Flood

One of the most crucial elements of mapping flood hazards is LULC management since it not only represents the land's current usage but also its relevance for soil stability and infiltration (Zope et al., 2016).

Permanent grassland or other types of crops that cover the soil influence the soil's ability to retain water (Sugianto et al., 2022). On bare fields as opposed to fields with good crop cover, rainwater runoff occurs much more frequently. Runoff is decreased by the slowing down of water from the sky to the ground caused by dense plant cover. Other impermeable surfaces, such as concrete, on the other hand, rarely take in water. Houses, highways, and slum areas all use up land, which makes it harder for water to penetrate the soil and makes it run faster (Brody et al., 2014). To put it another way, certain types of land use serve as coverings that resist water, shortening the amount of time it takes for water to accumulate; additionally, they frequently raise the peak water discharge, resulting in more severe flooding. This indicates that the likelihood of flooding is greatly influenced by LULC (Brody et al., 2015).

For the year 2021, supervised classification using the MLC was performed on Sentinel-2 images and Landsat 5 images have been used for 2010. MLC assigns a classification to the image based on the information found in the samples. The layer "Land use/Land cover" has been allocated to it. Agriculture, woods, barren terrain, scrublands, water features, and built-up areas are among the LULC accessible (Figure 22.6).

22.4.3 Morphometric Analysis

Morphometric analysis is essential for understanding and analyzing basin geometry, rock hardness, and the morphological and geological scenario. It provides a solid mathematical and statistical explanation of the volume, characteristics, altitude of the basin, and channel profiles of the rivers. Aster-DEM with a spatial resolution of 30 × 30 m has been used to study the Adyar River basin, using measures of its linear, areal, and relief features.

22.4.3.1 Stream Ordering and Stream Number

According to the number of tributaries upstream, stream order is a technique for categorizing stream segments (Horton, 1945). It is among the most crucial steps in a river basin's morphometric examination (Figure 22.7). This aids in defining the character of the drainage basin.

Using the ArcGIS platform, the total length of streams in the Adyar River basin was computed and measured (Table 22.3). From the Aster DEM, a total stream length of 612.42 km has been obtained.

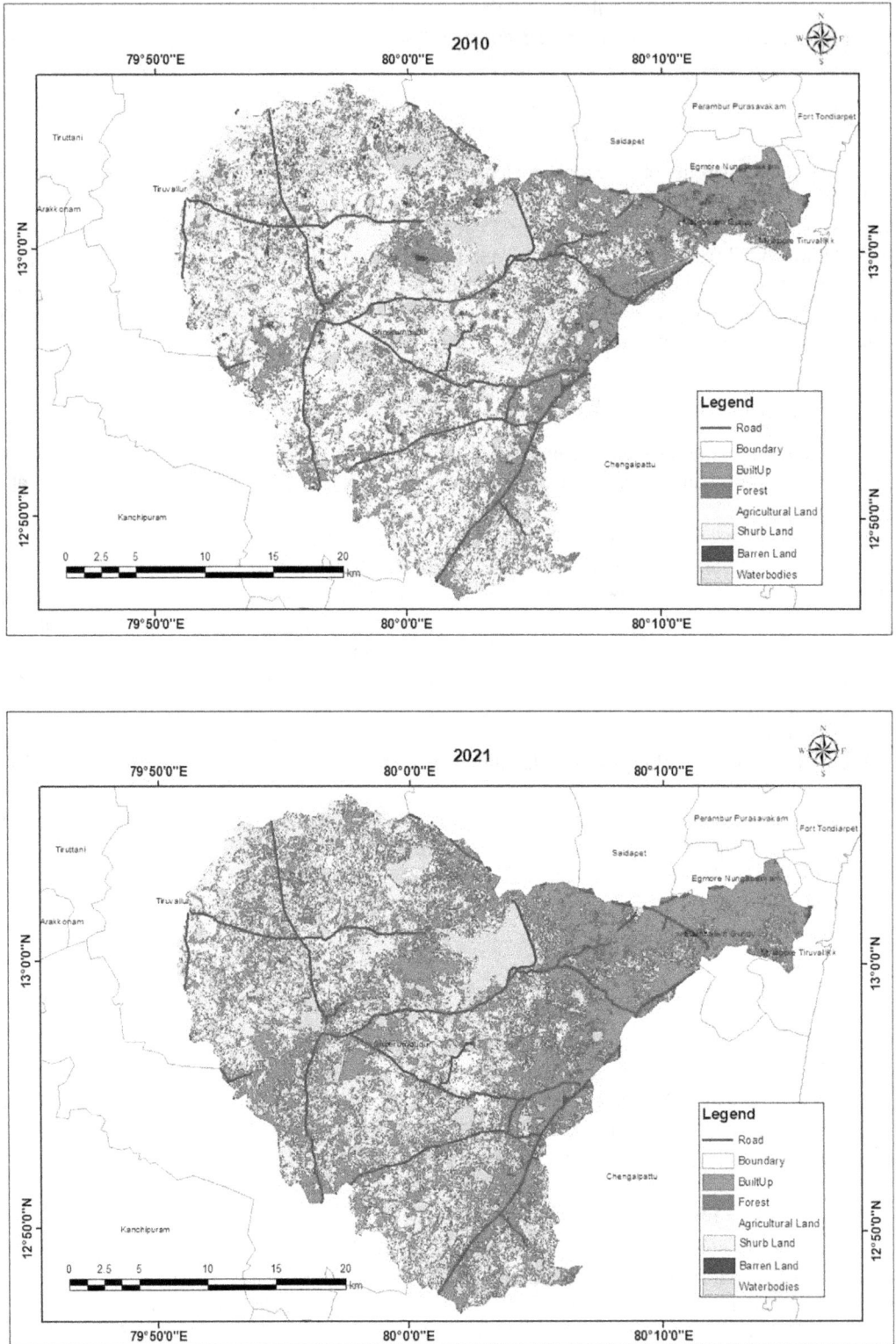

FIGURE 22.6   Land-use/land-cover map of Adyar watershed: 2010 and 2021.

Source: USGS EarthExplorer.

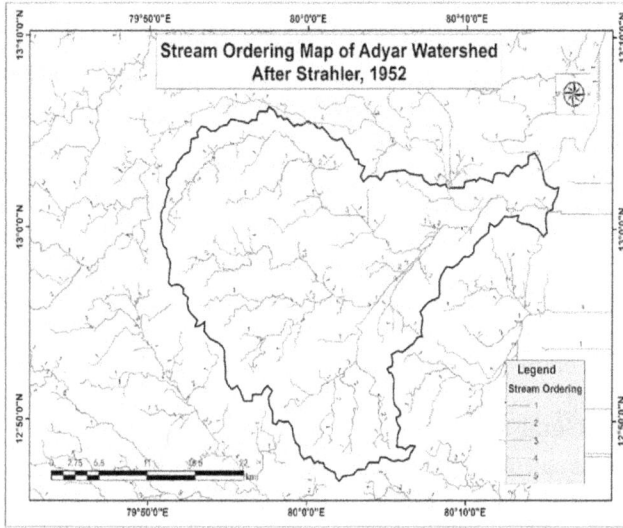

FIGURE 22.7 Stream ordering map of Adyar watershed.

TABLE 22.2
Description of Morphometric Parameters

Aspects	Morphometric Parameters	Formula and Descriptions	References
Linear Aspects	Stream order	Hierarchical order	(Strahler, 1957)
	Stream number	$Nu = N1 + N2 + \ldots nn$	(Horton, 1945)
	Stream length (Lu)	Length of the stream	(Horton, 1945)
	Bifurcation ratio (Rb)	$Rb = Nu/Nu + 1$, where Nu is the total number of stream segment of order 'u'; $Nu + 1$ is the number of segment of next higher order	(Schumm, 1956)
Aerial Aspects	Drainage density (Dd)	$Dd = L/A$, where L is the total stream length, A is an area	(Horton, 1945)
	Stream frequency (Fs)	$Fs = Nu/A$, where N is the total number of streams, and A is an area of the watershed	(Horton, 1945)
	Infiltration number (If)	$If = Fs \times Dd$	(El-Magd et al., 2010)
	Length of overland flow (Lg)	$Lg = 1/2\ Dd$, where Dd is the drainage density	(Horton, 1945)
Relief Aspects	Basin relief (R)	$R = H - h$, where H is maximum elevation, and h is minimum elevation within the basin	(Schumm, 1956)
	Relief ratio (Rr)	$Rr = R/Lb$	(Schumm, 1956)
	Ruggedness number (Rn)	$Rn = R/Dd$	(Strahler, 1957)
	Dissection Index (Di)	$DI = R/Ra$, where Ra is an absolute relief	(Kumar Dubey, 2015)

TABLE 22.3
Calculation Table for Stream Order (Su), Stream Number (Nu)

Su	Nu
1st	152
2nd	32
3rd	15
4th	2
5th	1
Total	202

22.4.3.2 Drainage Density

According to Horton (1945) and Strahler (1957), drainage density is the proportion of a basin's total area to its total length of streams. It is a quantitative assessment of the basin's dissection characteristics and potential for surface runoff. The drainage density of the watershed regulates the surface infiltration limit of the soil and vegetation.

Drainage is a key component for managing risks (Choubin et al., 2019) since the density of the soil determines the composition of the soil and its geotechnical properties. This means that when the population density rises, the catchment area becomes more vulnerable to erosion, resulting in sedimentation on the lower land. A watershed with adequate drainage discharge should have a drainage density of five, whereas moderate and poor watersheds should have drainage density classes of one to five. Dd is 1.08 km/km^2 in the research area, which is considered a low value. Figure 22.15g shows the drainage density map of the Adyar River basin.

22.4.3.3 Infiltration Number

The infiltration number is calculated by multiplying the drainage density (Dd) and stream frequency (Sf) (Agarwal, 1998). The formula for it is $If = Dd \times Sf$, where D stands for drainage density and Sf for stream frequency. A 0.785 infiltration number applies to the Adyar watershed. Higher values of If also imply higher infiltration and little runoff with shallow depth to the water, while lower values of If suggest the opposite. A greater infiltration value suggests lower infiltration capability and higher runoff (Ziegler & Giambelluca, 1997). As a result, there is more runoff and less capacity for infiltration in this area (Elkins et al., 1986).

22.4.3.4 Dissection Index

By rationing between the current river dissections and the river's potential level up to the base level of erosion, the dissection index can be computed (Kumar Dubey, 2015). Dissection levels that are high or low correspond to the river basin's young or old stages, respectively. Dissection value runs from 0 to 1; Di value 1 indicates a mountainous area with severe erosion, whereas Di value 0 indicates a flatter topography with no vertical erosion (Rai et al., 2018) (Figure 22.8).

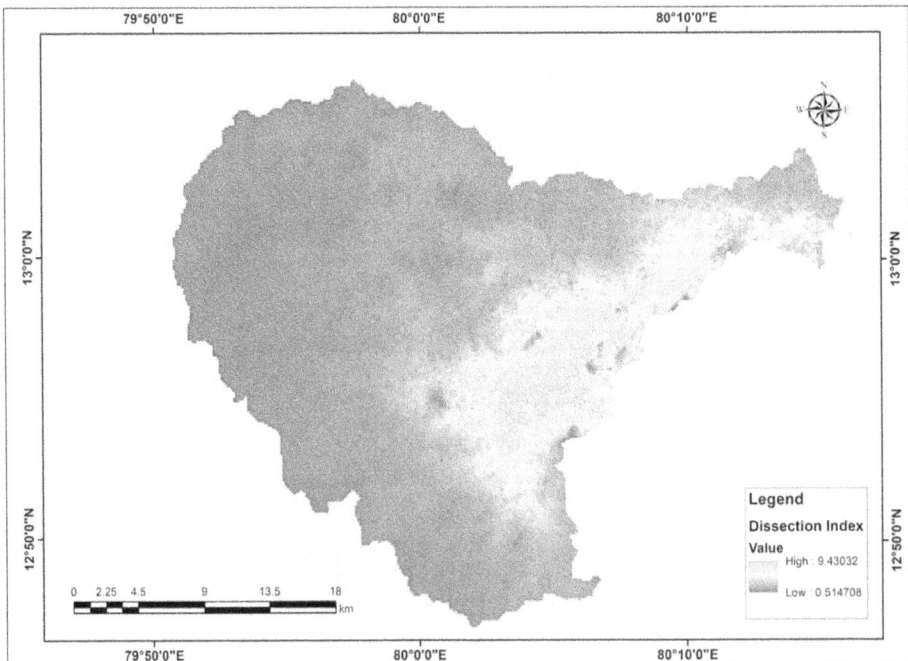

FIGURE 22.8 Dissection index map of Adyar watershed.

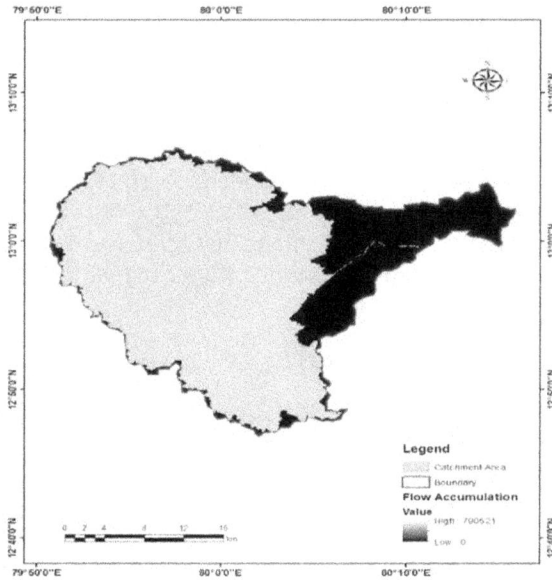

FIGURE 22.9 Catchment area map of Adyar watershed.

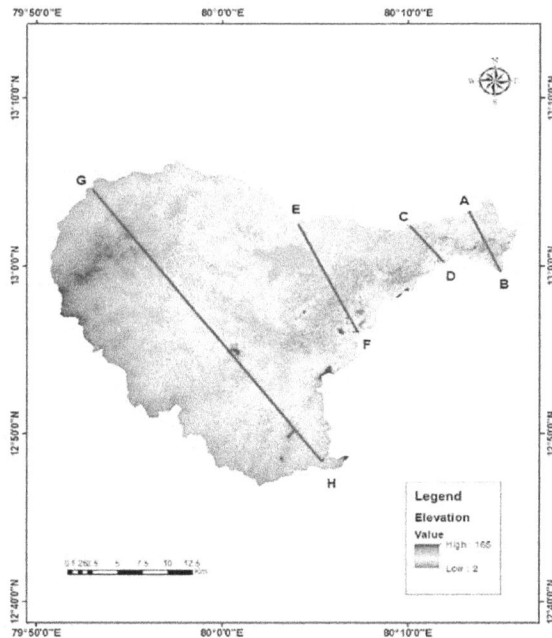

FIGURE 22.10 Locations of the cross sections of the Adyar watershed.

22.4.3.5 Catchment Area Calculation

A watershed catchment is an area of land where all precipitation falls and finally flows to a single exit, with less water wasted to evaporation and deep aquifer recharging (Jones, 2002). A watershed is made up of both surface and subsurface water drainage components that contribute to stream discharge. It is the area of land where surface runoff is removed by a single drainage system and is defined by watersheds that empty into a river, basin, or reservoir. The total catchment area of this basin is 626.0941 km^2 (Figure 22.9).

22.4.3.6 Cross Sections of Adyar Watershed

Cross sections have been extracted from Aster DEM (Sarkar et al., 2020). Figure 22.10 shows the four cross sections of Adyar watershed. Results have shown that the eastern part of the watershed is less elevated than the western part. The A-B cross section showed below 30-m height (Figure 22.11) but the G-H cross section showed above 60-m height (Figure 22.14). From A-B to G-H height is increasing (Figures 22.12–22.14). So, it shows the eastern part of the watershed is more vulnerable to flood.

22.4.4 FLOOD SUSCEPTIBILITY MAPPING

Thematic layers were developed utilizing multiple sources of available datasets based on knowledge of the causes of floods in the study area (Vojtek & Vojteková, 2019). Thematic layers of the selected criteria were combined to generate a map of the research area's flood risk. The flood susceptibility analysis took into consideration nine factors: the research area's elevation, slope, NDVI, NDWI, TWI, rainfall, drainage density, distance from the road, and LULC (Tehrany et al., 2014). Reclassification has been done using the AHP method in ArcGIS.

Using ArcGIS software, elevation and slope maps (Figure 22.15a and 22.15b) of the research region were created from the DEM. The two most crucial variables for determining flood susceptibility are slope and elevation. When creating a map of flood susceptibility, land elevation is a crucial and effective factor (Bui et al., 2020). The variation in the height of the land there controls the climate of that area. Different altitudinal zones experience varying levels of warmth and rainfall, which results in a variety of vegetation cover and soil formations. One of the main causes of flooding is the slope. Surface water flow, vertical infiltration, and the rate of soil erosion are all influenced by a region's land slope (Avand et al., 2021). The relationship

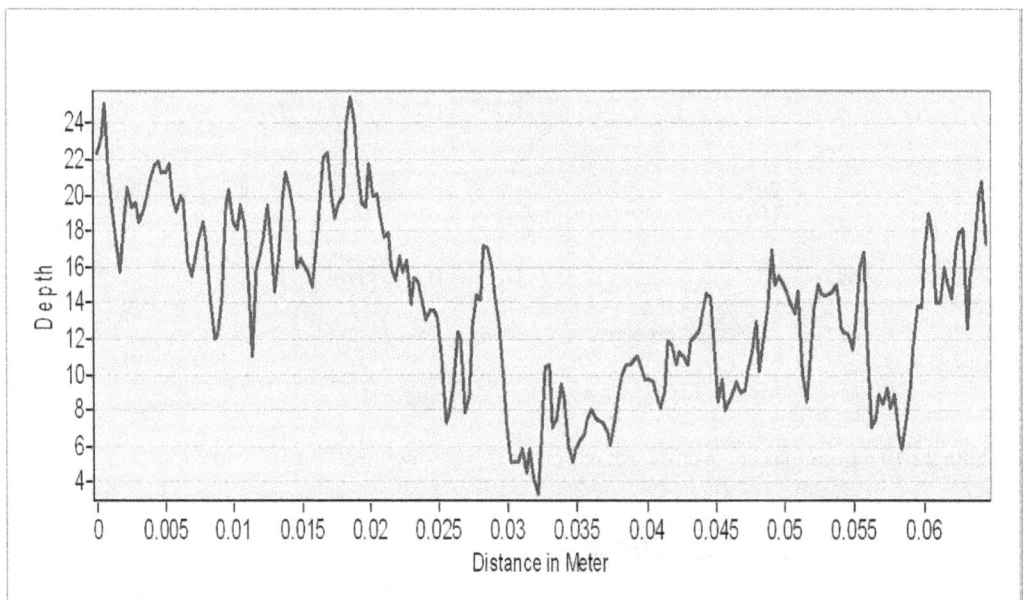

FIGURE 22.11   Cross sections along the line A-B.

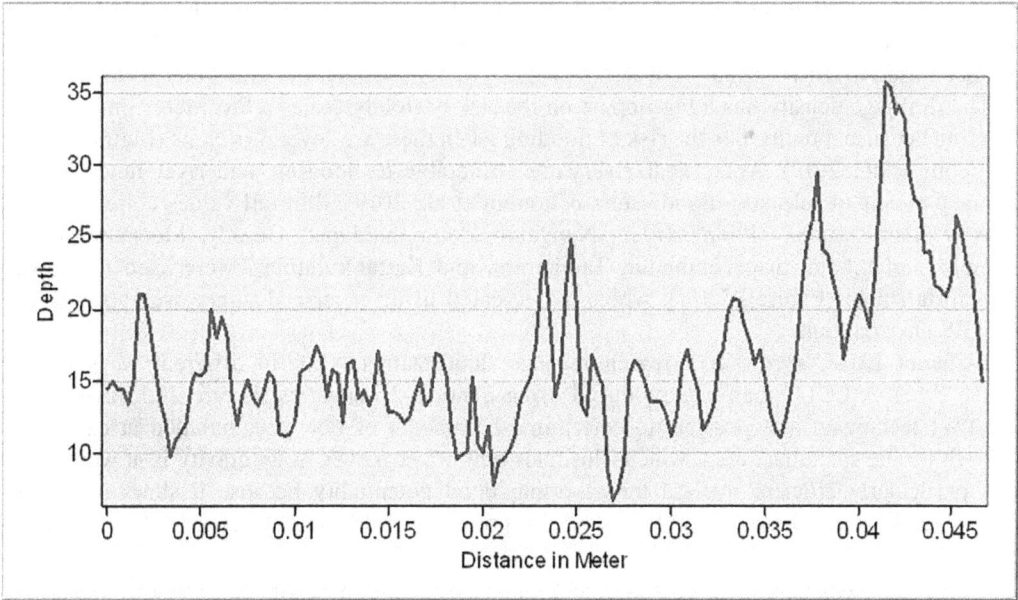

FIGURE 22.12 Cross sections along the line C-D.

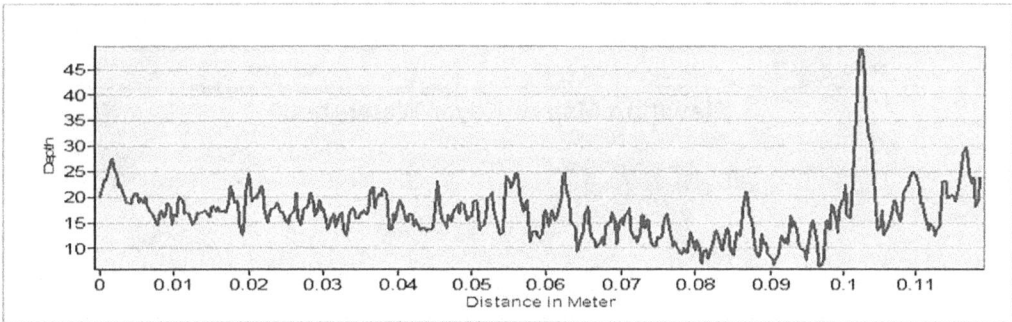

FIGURE 22.13 Cross sections along the line E-F.

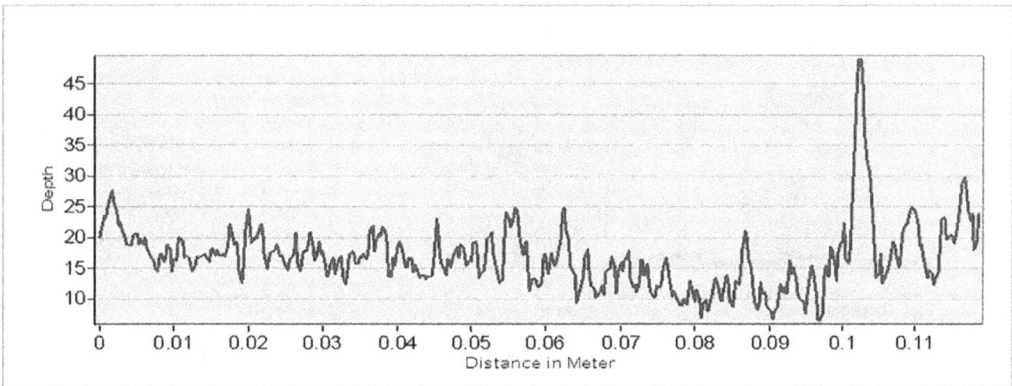

FIGURE 22.14 Cross sections along the line G-H.

between vegetation cover and the rate of overland flow can be understood by looking at how topography affects a catchment area's rate of wetness (Farhadi & Najafzadeh, 2021). The NDWI has been used to differentiate between the classes of soil, vegetation, and water (Gao, 1996).

The drainage density has a big impact on the risk of floods. It has a favorable connection to flooding because it multiplies the risk of flooding when there are several streams (Figure 22.15g) (Choubin et al., 2019). Areas near rivers are vulnerable to flooding, and river flows are the primary means of releasing floodwaters (Choubin et al., 2019). Rainfall values collected from nine weather stations—Poonamallee, Nungambakkam, Saidapet, Guindy, Meenambakkam, Chembarambakkam, Sriperumbudur, Tambaram, and Kattankulathur—were used to prepare the rainfall map (Figure 22.15f), which was created using inverse distance weighting in the ArcGIS environment.

Different LULC types in a region manage flood vulnerability in different ways (Brody et al., 2015). A LULC map was produced using Sentinel-2 imageries (Figure 22.15i). By using the TWI technique, it is possible to determine the amount of flow accumulation at each location within the catchment area, which illustrates how water moves when gravity is at work. TWI is a particularly efficient method for assessing flood potentiality because it slows down surface water flow and recharges groundwater by promoting infiltration (Figure 22.15e) (Rahmati et al., 2016).

All of the major roads in the Adyar watershed were manually extracted from a Google Earth image and loaded into ArcGIS to create a road density map. Because roads act as a made-up barrier to floods, they have a significant impact on flooding. High road density slows down the flood's devastation process to some extent.

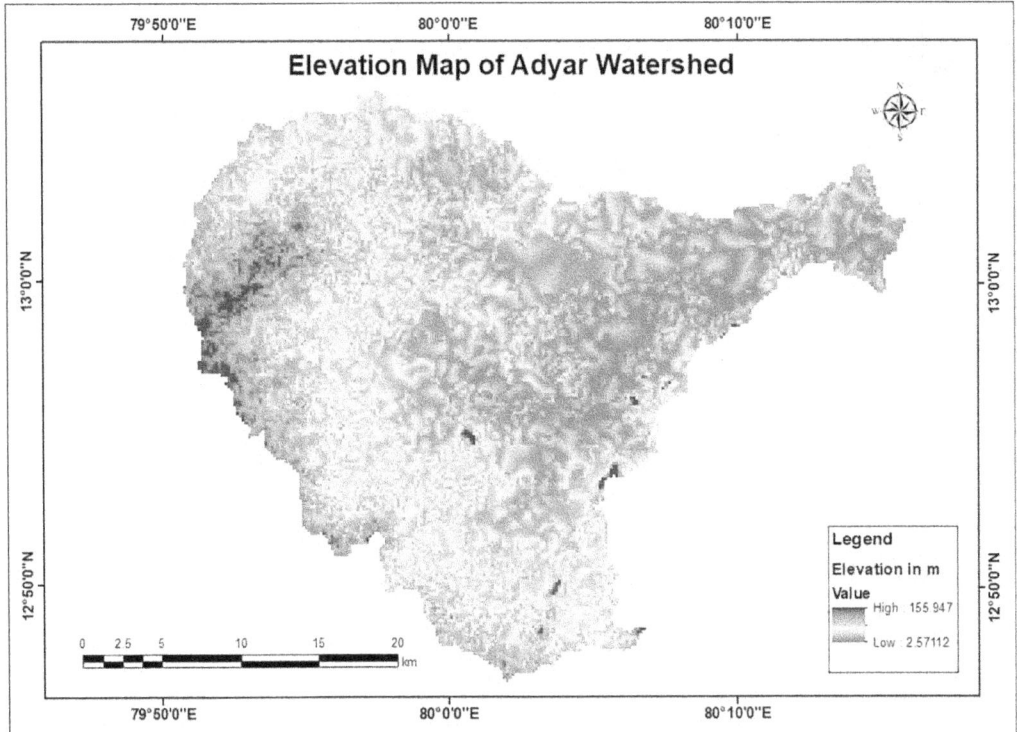

(A)

FIGURE 22.15 Flood susceptibility factors: (a) elevation, (b) slope (c) Normalized Difference Vegetation Index, (d) Normalized Difference Water Index, (e) Topographic Wetness Index, (f) rainfall, (g) drainage density, (h) distance from road, and (i) land use/land cover.

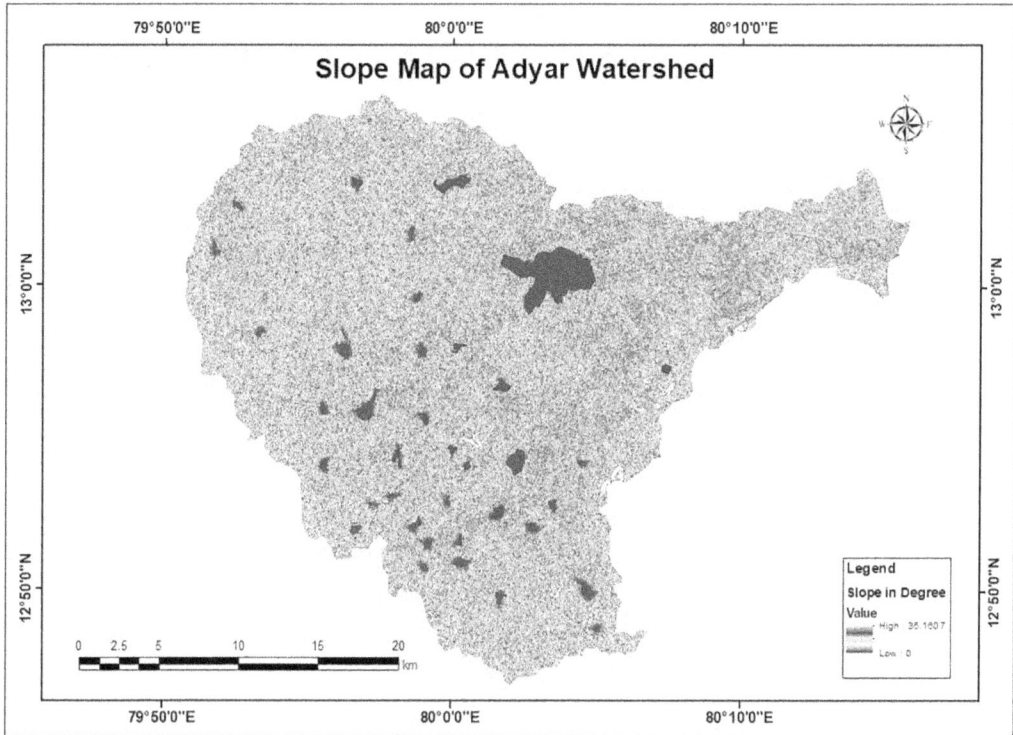

(B)

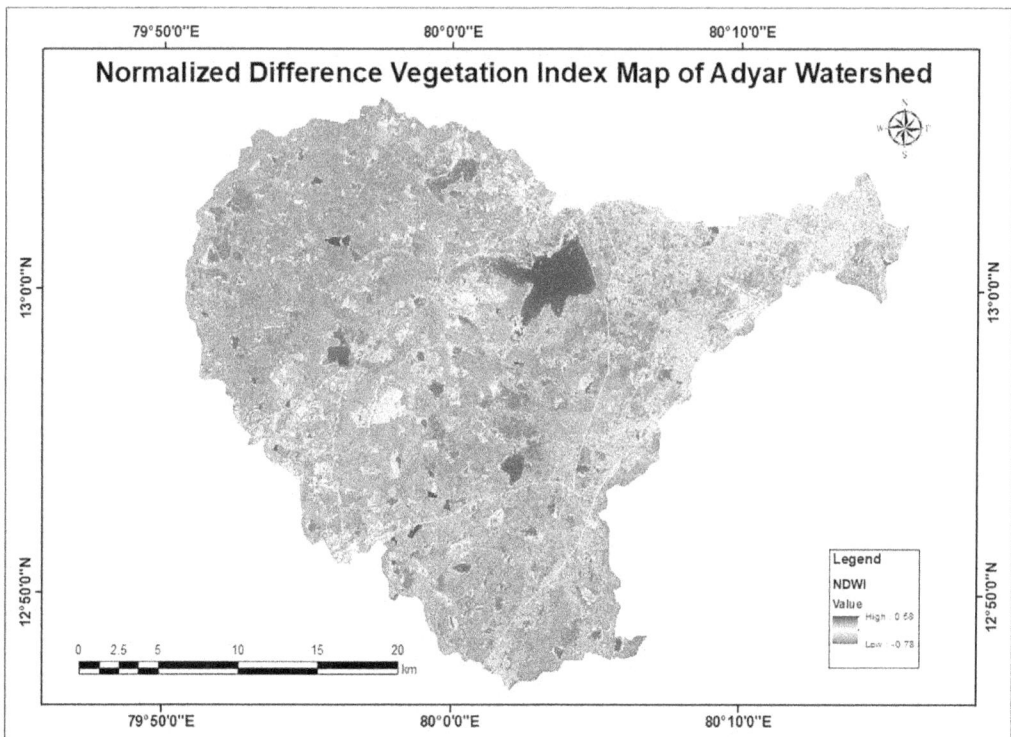

(C)

FIGURE 22.15   (Continued)

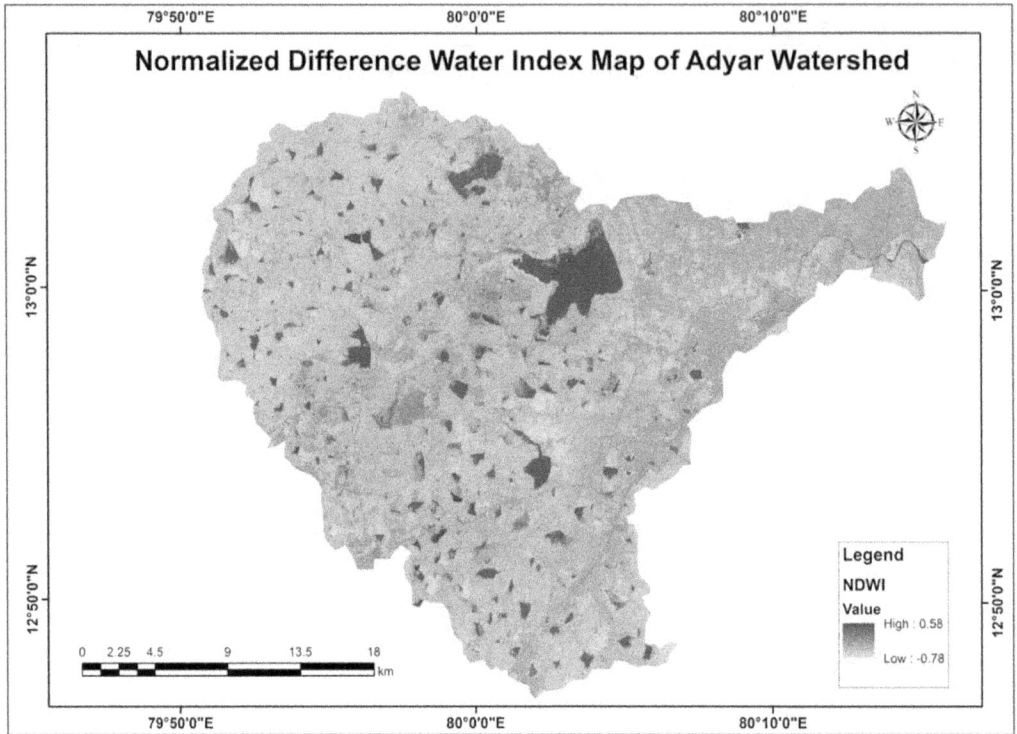

(D)

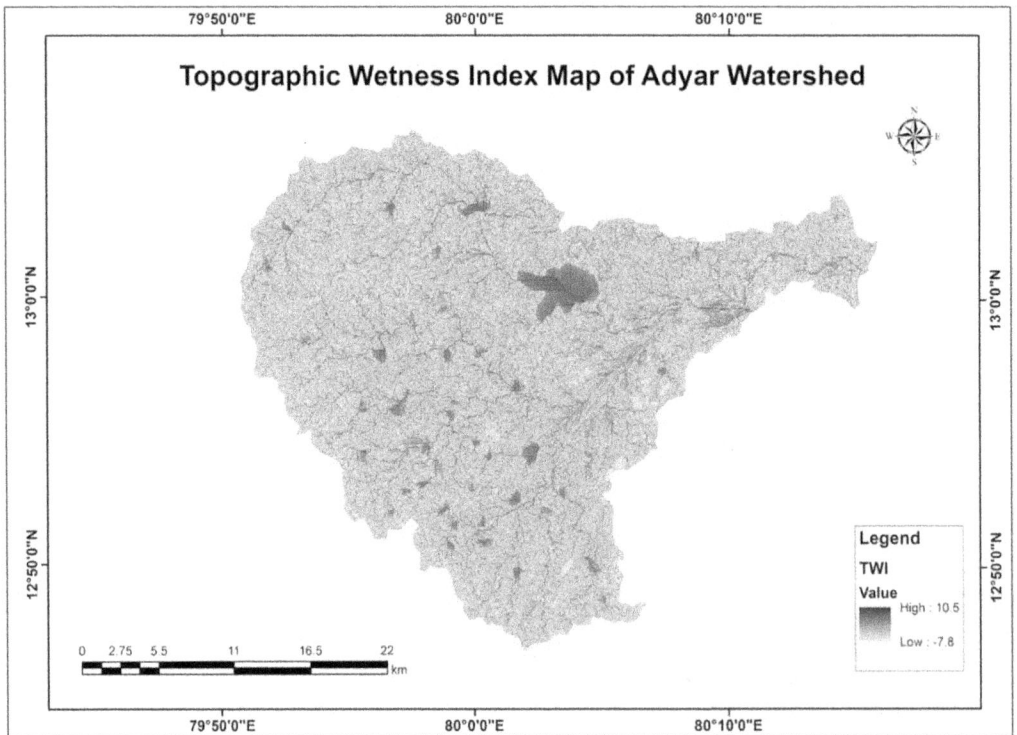

(E)

FIGURE 22.15   (Continued)

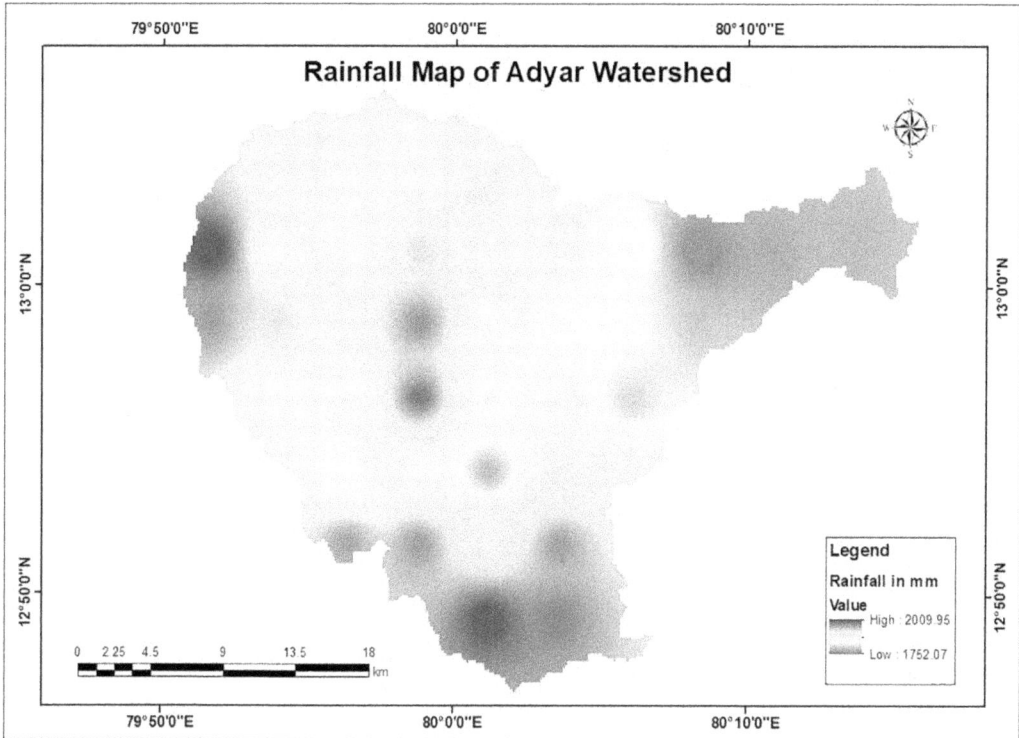

(F)

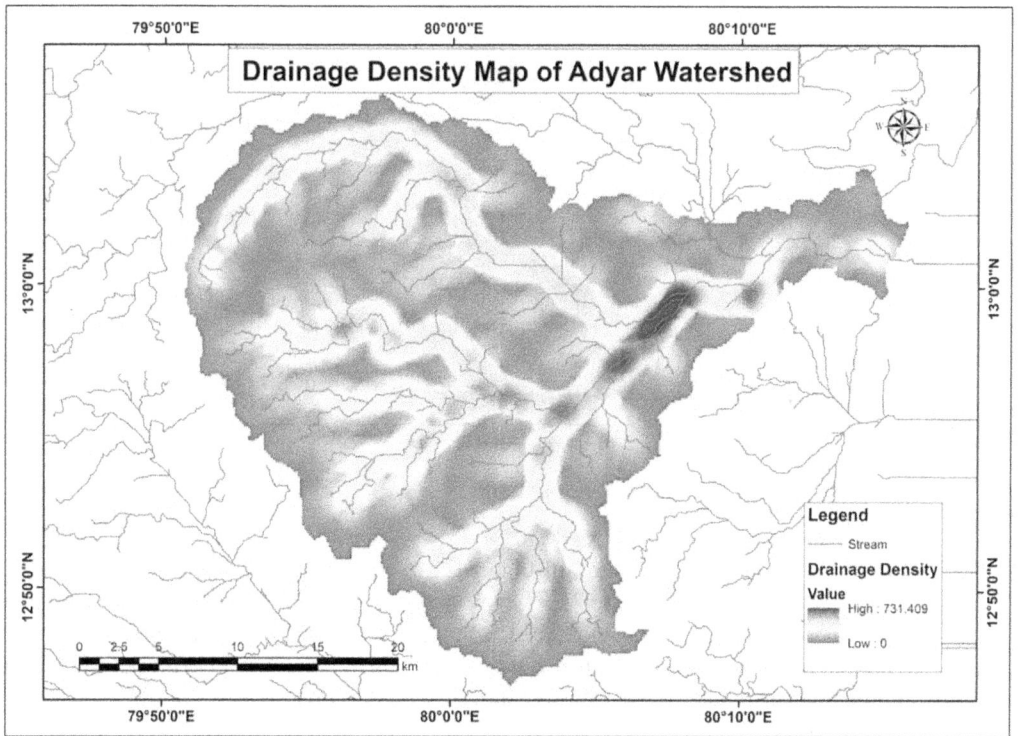

(G)

FIGURE 22.15 (Continued)

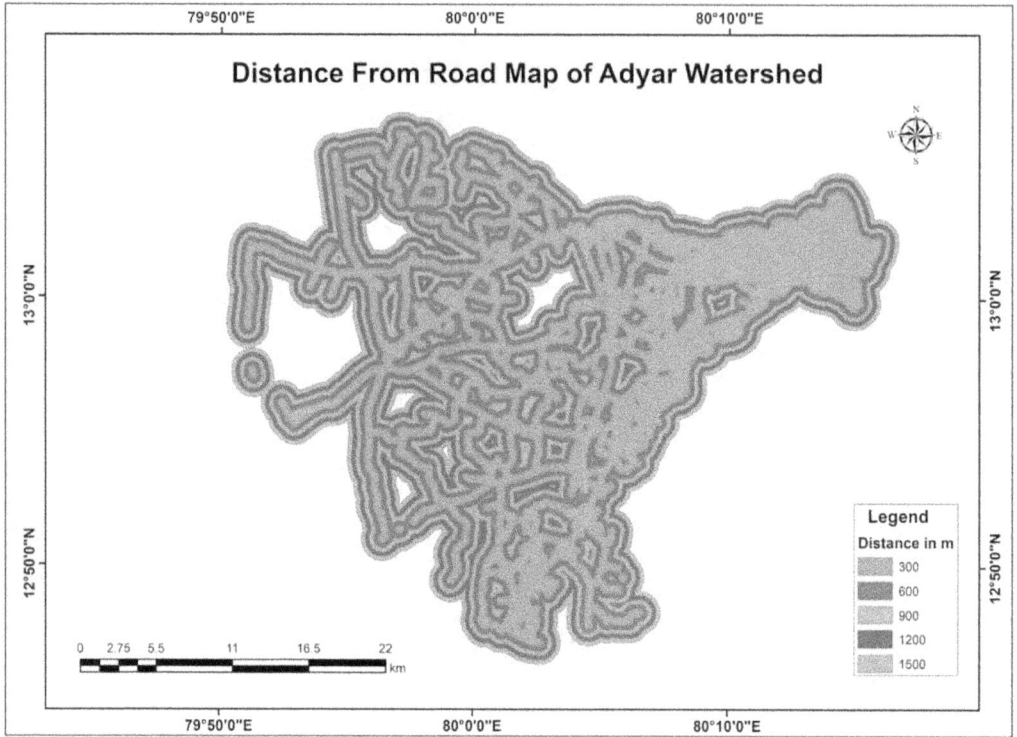

(H)

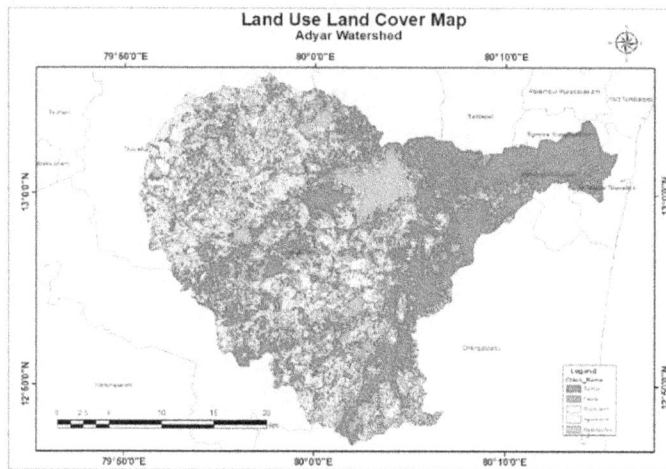

(I)

FIGURE 22.15 (Continued)

Flood susceptibility zones are depicted in Figure 22.16 and are divided into four categories: very high, high, medium, and low. Figure 22.16 shows that very high flood-vulnerable zones cover 50.85% of the watershed, high flood-vulnerable zones cover 47.38%, moderate flood-vulnerable zones cover 0.88%, and low flood-vulnerable zones covers 0.86%. The lower portion of the Adyar watershed, it might be established, is extremely vulnerable to flooding.

22.4.5 ONE-DIMENSIONAL SIMULATION MODELING

The expected flood levels were determined using the HEC-RAS model and the peak flood data. In addition to creating the stream's center line, bank line, flow routes, and xs cutline, it also offered

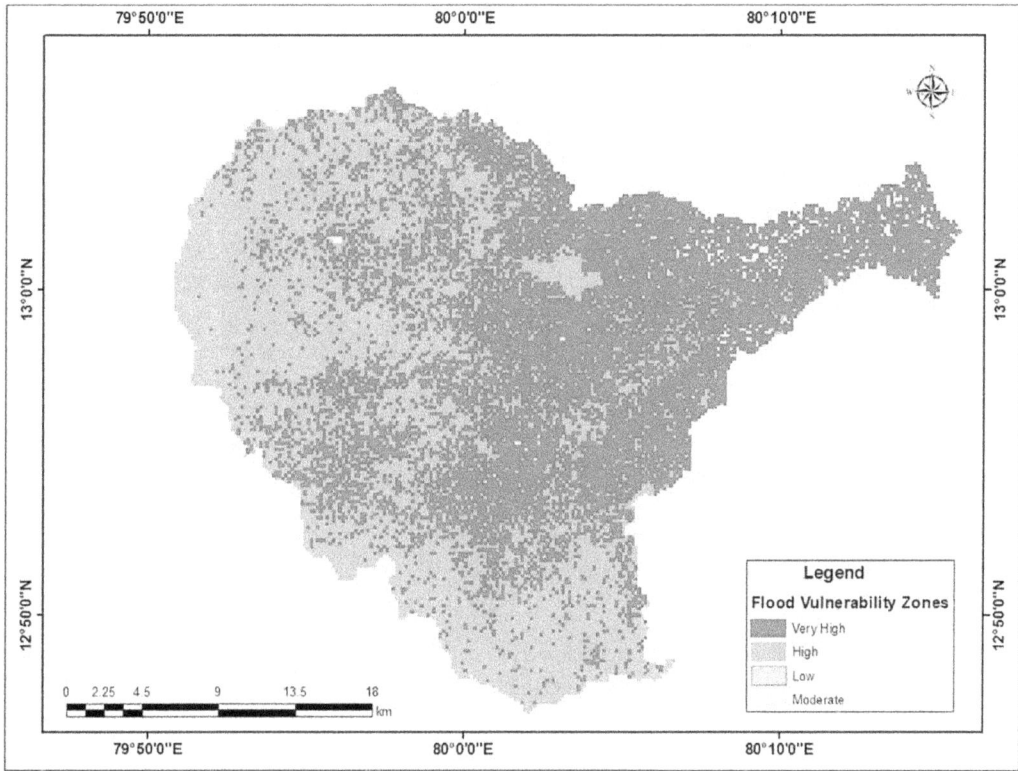

FIGURE 22.16   Flood vulnerability map of Adyar watershed.

FIGURE 22.17   Profile plot of Adyar River.

an ID for the center line and banks with channels in the flow path (Yerramilli, 2012). Then, a 1D steady flow analysis for the current river under investigation was carried out using the HEC-RAS version 5.0.7. With the parameters set to their appropriate cross-sectional regions and no class left untouched, the model was then ready to run and replicate the desired results. Several cross sections has been extracted from the ALOS PALSAR DEM for the flood modelling of Adyar watershed. Here, only one has shown (Figure 22.17).

Using the 1D hydraulic simulation method and the HEC-RAS 5.0.7 hydraulic model, flood-prone zones were identified, and the structural susceptibility to flood peaks with various exceedance probabilities was evaluated.

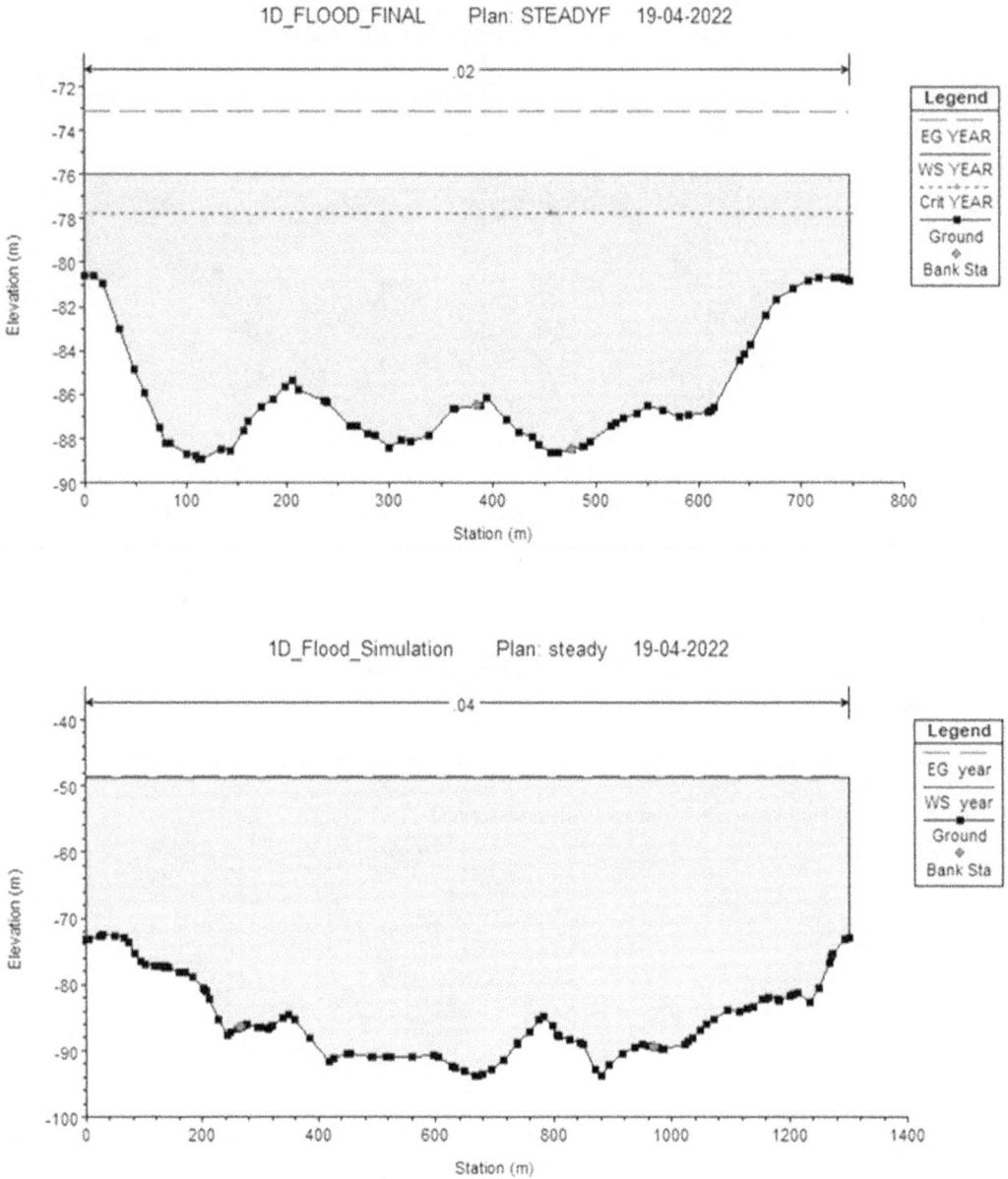

FIGURE 22.18 Representation of cross-sectional characteristics.

The output cross-sectional plots obtained demonstrate that the down part of the river, as well as the accessible bunds and surrounding roadways, is a flood-prone area that might inundate most of the settlement in that area. The integrative water surface profile illustrates the irregular flow behavior of the river from both two- and three-dimensional viewpoints (Figures 22.18a and 22.18b).

The river's average speed has been between 3.5 and 4.5 m/s. In contrast, a velocity of 7.5 to 8.6 m/s has been observed within the limits of Chennai, indicating that the flow through Chennai runs at a very high velocity in its higher reaches (Figure 22.19). This can be linked to the river's maximum spilling inside the lower reach, resulting in flooding of extensive areas next to the river through Chennai.

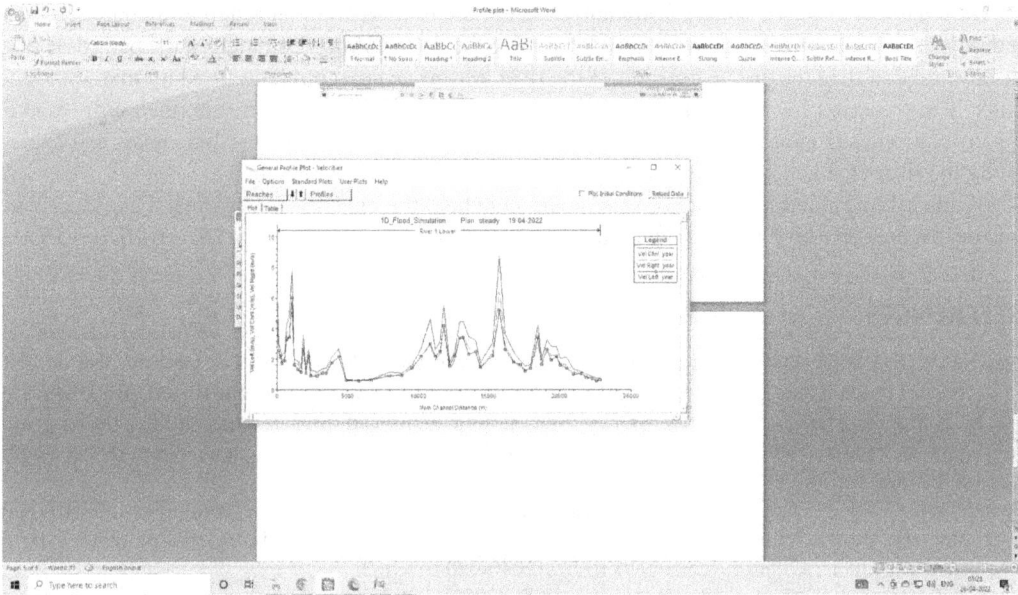

FIGURE 22.19   Velocity plot of river.

22.5   CONCLUSION

If the Adyar River floods again, it will be easier to predict how much of the area would be flooded. This would aid administrative personnel in taking required actions in the area and evacuating people from flood-prone areas to safe zones.

Susceptibility maps will give decision-makers a better visual depiction of vulnerability, allowing them to better understand risks and weaknesses and determine what resources are required to secure such regions. It helps them to make decisions on mitigation measures in advance of a disaster to prevent or lessen the loss of life, injuries, and environmental consequences. It can be used in all aspects of disaster management, including prevention, mitigation, readiness, operations, relief, and recovery, as well as lessons learned. Encroachment obstructs the normal flow of water, causing flooding. It was one of the primary reasons for the 2015 floods in Chennai. Streams, riverbeds, riverbanks, ponds, lakes, tanks, and other water bodies are commonly encroached upon and turned into settlements or commercial operations.

The Tamil Nadu state government has reintroduced the century-old Kudimaramathu plan to manage water bodies. It entails community participation in the desilting, restoration, and maintenance of existing water bodies, as well as the construction of new ones. Even if the research region contains a dense network of water bodies, it is possible to achieve this with the help of the local community. Such plans must be carried out systematically at regular periods.

In any future infrastructure design, the Chennai district should not be viewed as a single geographical region. The government should regard the districts of Thiruvallur, Kanchipuram, and Chennai as a single planning entity. Because rivers that run through these districts carry a lot of water after heavy rains and discharge it into Chennai, these districts are flooded.

One of the most beneficial techniques for reducing flood damage is the mapping of flood disaster-prone zones. The multi-parametric AHP-GIS methodology is a reliable alternative method for locating flood-prone areas in India's Adyar watershed. The final flood vulnerability map was divided into four different categories: low, moderate, high, and extremely high. Figure 22.16 demonstrates the identification of extremely highly vulnerable locations in the bottom stretch next to the river.

The results of using 1D steady flow analysis using the HEC-RAS model suggest that the research region was found to be more heavily flooded during the floods of the aforementioned period. It was discovered that the area on the left side of the river was more susceptible to flooding than the right bank. The scenario of water surface levels at various return periods together with their corresponding cross profiles demonstrates the requirement for engineering infrastructure like dykes and levees along the river channel. The river overflows the majority of its accessible buds.

By identifying flood-prone areas, this model may help the government, planners, and decision-makers execute effective management plans in the studied region and restrain the growth processes.

REFERENCES

Agarwal, C. S. (1998). Study of drainage pattern through aerial data in Naugarh area of Varanasi district, U.P. Journal of the Indian Society of Remote Sensing, 26(4), 169–175. https://doi.org/10.1007/BF02990795

Ali, R., Kuriqi, A., Abubaker, S., & Kisi, O. (2019). Long-term trends and seasonality detection of the observed flow in Yangtze River using Mann-Kendall and Sen's innovative trend method. Water, 11(9), 1855. https://doi.org/10.3390/w11091855

Anandharuban, P., & Elango, L. (2021). Spatio-temporal analysis of rainfall, meteorological drought and response from a water supply reservoir in the megacity of Chennai, India. Journal of Earth System Science, 130(1), 17. https://doi.org/10.1007/s12040-020-01538-2

Arnaud, P., Bouvier, C., Cisneros, L., & Dominguez, R. (2002). Influence of rainfall spatial variability on flood prediction. Journal of Hydrology, 260(1–4), 216–230. https://doi.org/10.1016/S0022-1694(01)00611-4

Asdak, C., Supian, S., & Subiyanto. (2018). Watershed management strategies for flood mitigation: A case study of Jakarta's flooding. Weather and Climate Extremes, 21, 117–122. https://doi.org/10.1016/j.wace.2018.08.002

Avand, M., Moradi, H. R., & Ramazanzadeh Lasboyee, M. (2021). Spatial prediction of future flood risk: An approach to the effects of climate change. Geosciences, 11(1), 25. https://doi.org/10.3390/geosciences11010025

Berhanu, B., & Bisrat, E. (2018). Identification of surface water storing sites using Topographic Wetness Index (TWI) and Normalized Difference Vegetation Index (NDVI). Journal of Natural Resources and Development, 8, 91–100. https://doi.org/10.5027/jnrd.v8i0.09

Beven, K. J., & Kirkby, M. J. (1979). A physically based, variable contributing area model of basin hydrology/ Un modèle à base physique de zone d'appel variable de l'hydrologie du bassin versant. Hydrological Sciences Bulletin, 24(1), 43–69. https://doi.org/10.1080/02626667909491834

Brody, S. D., Blessing, R., Sebastian, A., & Bedient, P. (2014). Examining the impact of land use/land cover characteristics on flood losses. Journal of Environmental Planning and Management, 57(8), 1252–1265. https://doi.org/10.1080/09640568.2013.802228

Brody, S. D., Highfield, W. E., & Blessing, R. (2015). An analysis of the effects of land use and land cover on flood losses along the Gulf of Mexico coast from 1999 to 2009. JAWRA Journal of the American Water Resources Association, 51(6), 1556–1567. https://doi.org/10.1111/1752-1688.12331

Bui, Q.-T., Nguyen, Q.-H., Nguyen, X. L., Pham, V. D., Nguyen, H. D., & Pham, V.-M. (2020). Verification of novel integrations of swarm intelligence algorithms into deep learning neural network for flood susceptibility mapping. Journal of Hydrology, 581, 124379. https://doi.org/10.1016/j.jhydrol.2019.124379

Choubin, B., Moradi, E., Golshan, M., Adamowski, J., Sajedi-Hosseini, F., & Mosavi, A. (2019). An ensemble prediction of flood susceptibility using multivariate discriminant analysis, classification and regression trees, and support vector machines. Science of the Total Environment, 651, 2087–2096. https://doi.org/10.1016/j.scitotenv.2018.10.064

Douben, K.-J. (2006). Characteristics of river floods and flooding: A global overview, 1985–2003. Irrigation and Drainage, 55(S1), S9–S21. https://doi.org/10.1002/ird.239

Duan, Y., Xiong, J., Cheng, W., Li, Y., Wang, N., Shen, G., & Yang, J. (2022). Increasing global flood risk in 2005–2020 from a multi-scale perspective. Remote Sensing, 14(21), 5551. https://doi.org/10.3390/rs14215551

Elkins, N. Z., Sabol, G. V., Ward, T. J., & Whitford, W. G. (1986). The influence of subterranean termites on the hydrological characteristics of a Chihuahuan desert ecosystem. Oecologia, 68(4), 521–528. https://doi.org/10.1007/BF00378766

El-Magd, I. A., Hermas, E., & Bastawesy, M. E. (2010). GIS-modelling of the spatial variability of flash flood hazard in Abu Dabbab catchment, Red Sea Region, Egypt. The Egyptian Journal of Remote Sensing and Space Science, 13(1), 81–88. https://doi.org/10.1016/j.ejrs.2010.07.010

Farhadi, H., & Najafzadeh, M. (2021). Flood risk mapping by remote sensing data and random forest technique. Water, 13(21), 3115. https://doi.org/10.3390/w13213115

Gao, B. (1996). NDWI: A normalized difference water index for remote sensing of vegetation liquid water from space. Remote Sensing of Environment, 58(3), 257–266. https://doi.org/10.1016/S0034-4257(96)00067-3

Ghosh, A., & Kar, S. K. (2018). Application of Analytical Hierarchy Process (AHP) for flood risk assessment: A case study in Malda district of West Bengal, India. Natural Hazards, 94(1), 349–368. https://doi.org/10.1007/s11069-018-3392-y

Hirabayashi, Y., Mahendran, R., Koirala, S., Konoshima, L., Yamazaki, D., Watanabe, S., Kim, H., & Kanae, S. (2013). Global flood risk under climate change. Nature Climate Change, 3(9), 816–821. https://doi.org/10.1038/nclimate1911

Hoque, M., Tasfia, S., Ahmed, N., & Pradhan, B. (2019). Assessing spatial flood vulnerability at Kalapara Upazila in Bangladesh using an analytic hierarchy process. Sensors, 19(6), 1302. https://doi.org/10.3390/s19061302

Horton, R. E. (1945). Erosional development of streams and their drainage basins: Hydrophysical approach to quantitative morphology. Geological Society of America Bulletin, 56(3), 275. https://doi.org/10.1130/0016-7606(1945)56[275:EDOSAT]2.0.CO;2

Jones, R. (2002). Algorithms for using a DEM for mapping catchment areas of stream sediment samples. Computers & Geosciences, 28(9), 1051–1060. https://doi.org/10.1016/S0098-3004(02)00022-5

Kumar Dubey, S. (2015). Morphometric analysis of the Banas River Basin using the geographical information system, Rajasthan, India. Hydrology, 3(5), 47. https://doi.org/10.11648/j.hyd.20150305.11

McFeeters, S. K. (1996). The use of the Normalized Difference Water Index (NDWI) in the delineation of open water features. International Journal of Remote Sensing, 17(7), 1425–1432. https://doi.org/10.1080/01431169608948714

McFeeters, S. K. (2013). Using the Normalized Difference Water Index (NDWI) within a geographic information system to detect swimming pools for mosquito abatement: A practical approach. Remote Sensing, 5(7), 3544–3561. https://doi.org/10.3390/rs5073544

Niemczynowicz, J. (1999). Urban hydrology and water management: Present and future challenges. Urban Water, 1(1), 1–14. https://doi.org/10.1016/S1462-0758(99)00009-6

Otukei, J. R., & Blaschke, T. (2010). Land cover change assessment using decision trees, support vector machines and maximum likelihood classification algorithms. International Journal of Applied Earth Observation and Geoinformation, 12, S27–S31. https://doi.org/10.1016/j.jag.2009.11.002

Rahmati, O., Pourghasemi, H. R., & Zeinivand, H. (2016). Flood susceptibility mapping using frequency ratio and weights-of-evidence models in the Golastan Province, Iran. Geocarto International, 31(1), 42–70. https://doi.org/10.1080/10106049.2015.1041559

Rai, P. K., Chandel, R. S., Mishra, V. N., & Singh, P. (2018). Hydrological inferences through morphometric analysis of lower Kosi river basin of India for water resource management based on remote sensing data. Applied Water Science, 8(1), 15. https://doi.org/10.1007/s13201-018-0660-7

Ralph, T. J., & Hesse, P. P. (2010). Downstream hydrogeomorphic changes along the Macquarie River, southeastern Australia, leading to channel breakdown and floodplain wetlands. Geomorphology, 118(1–2), 48–64. https://doi.org/10.1016/j.geomorph.2009.12.007

Rozalis, S., Morin, E., Yair, Y., & Price, C. (2010). Flash flood prediction using an uncalibrated hydrological model and radar rainfall data in a Mediterranean watershed under changing hydrological conditions. Journal of Hydrology, 394(1–2), 245–255. https://doi.org/10.1016/j.jhydrol.2010.03.021

Saint-Laurent, D., Arsenault-Boucher, L., & Berthelot, J.-S. (2019). Contrasting effects of flood disturbance on alluvial soils and riparian tree structure and species composition in mixed temperate forests. Air, Soil and Water Research, 12, 117862211987277. https://doi.org/10.1177/1178622119872773

Sanyal, J., & Lu, X. X. (2004). Application of remote sensing in flood management with special reference to monsoon Asia: A review. Natural Hazards, 33(2), 283–301. https://doi.org/10.1023/B:NHAZ.0000037035.65105.95

Sarkar, D., Mondal, P., Sutradhar, S., & Sarkar, P. (2020). Morphometric analysis using SRTM-DEM and GIS of Nagar river basin, Indo-Bangladesh Barind tract. Journal of the Indian Society of Remote Sensing, 48(4), 597–614. https://doi.org/10.1007/s12524-020-01106-7

Schumm, S. A. (1956). Evolution of drainage systems and slopes in badlands at Perth Amboy, New Jersey. Geological Society of America Bulletin, 67(5), 597. https://doi.org/10.1130/0016-7606(1956)67[597:EODSAS]2.0.CO;2

Settle, J. J., & Briggs, S. A. (1987). Fast maximum likelihood classification of remotely-sensed imagery. International Journal of Remote Sensing, 8(5), 723–734. https://doi.org/10.1080/01431168708948683

Sisodia, P. S., Tiwari, V., & Kumar, A. (2014). Analysis of supervised maximum likelihood classification for remote sensing image. International Conference on Recent Advances and Innovations in Engineering (ICRAIE-2014), 1–4. https://doi.org/10.1109/ICRAIE.2014.6909319

Smith, B. K., Smith, J. A., Baeck, M. L., Villarini, G., & Wright, D. B. (2013). Spectrum of storm event hydrologic response in urban watersheds: Spectrum of urban storm event hydrologic response. Water Resources Research, 49(5), 2649–2663. https://doi.org/10.1002/wrcr.20223

Soltani, K., Ebtehaj, I., Amiri, A., Azari, A., Gharabaghi, B., & Bonakdari, H. (2021). Mapping the spatial and temporal variability of flood susceptibility using remotely sensed normalized difference vegetation index and the forecasted changes in the future. Science of the Total Environment, 770, 145288. https://doi.org/10.1016/j.scitotenv.2021.145288

Strahler, A. H. (1980). The use of prior probabilities in maximum likelihood classification of remotely sensed data. Remote Sensing of Environment, 10(2), 135–163. https://doi.org/10.1016/0034-4257(80)90011-5

Strahler, A. N. (1957). Quantitative analysis of watershed geomorphology. Transactions, American Geophysical Union, 38(6), 913. https://doi.org/10.1029/TR038i006p00913

Sugianto, S., Deli, A., Miswar, E., Rusdi, M., & Irham, M. (2022). The effect of land use and land cover changes on flood occurrence in Teunom Watershed, Aceh Jaya. Land, 11(8), 1271. https://doi.org/10.3390/land11081271

Sundaram, S., Devaraj, S., & Yarrakula, K. (2023). Mapping and assessing spatial extent of floods from multitemporal synthetic aperture radar images: A case study over Adyar watershed, India. Environmental Science and Pollution Research, 30(22), 63006–63021. https://doi.org/10.1007/s11356-023-26467-7

Suriya, S., & Mudgal, B. V. (2012). Impact of urbanization on flooding: The Thirusoolam sub watershed: A case study. Journal of Hydrology, 412–413, 210–219. https://doi.org/10.1016/j.jhydrol.2011.05.008

Tehrany, M. S., Pradhan, B., & Jebur, M. N. (2014). Flood susceptibility mapping using a novel ensemble weights-of-evidence and support vector machine models in GIS. Journal of Hydrology, 512, 332–343. https://doi.org/10.1016/j.jhydrol.2014.03.008

Tian, X., Van Overloop, P.-J., Negenborn, R. R., & Van De Giesen, N. (2015). Operational flood control of a low-lying delta system using large time step model predictive control. Advances in Water Resources, 75, 1–13. https://doi.org/10.1016/j.advwatres.2014.10.010

Tripathi, G., Parida, B. R., & Pandey, A. C. (2019). Spatio-temporal rainfall variability and flood prognosis analysis using satellite data over North Bihar during the August 2017 flood event. Hydrology, 6(2), 38. https://doi.org/10.3390/hydrology6020038

Viglione, A., & Blöschl, G. (2009). On the role of storm duration in the mapping of rainfall to flood return periods. Hydrology and Earth System Sciences, 13(2), 205–216. https://doi.org/10.5194/hess-13-205-2009

Viglione, A., Merz, R., & Blöschl, G. (2009). On the role of the runoff coefficient in the mapping of rainfall to flood return periods. Hydrology and Earth System Sciences, 13(5), 577–593. https://doi.org/10.5194/hess-13-577-2009

Vojtek, M., & Vojteková, J. (2019). Flood susceptibility mapping on a national scale in Slovakia using the analytical hierarchy process. Water, 11(2), 364. https://doi.org/10.3390/w11020364

Yerramilli, S. (2012). A hybrid approach of integrating HEC-RAS and GIS towards the identification and assessment of flood risk vulnerability in the city of Jackson, MS. American Journal of Geographic Information System, 1(1), 7–16. https://doi.org/10.5923/j.ajgis.20120101.02

Ziegler, A. D., & Giambelluca, T. W. (1997). Importance of rural roads as source areas for runoff in mountainous areas of northern Thailand. Journal of Hydrology, 196(1–4), 204–229. https://doi.org/10.1016/S0022-1694(96)03288-X

Zope, P. E., Eldho, T. I., & Jothiprakash, V. (2016). Impacts of land use: Land cover change and urbanization on flooding: A case study of Oshiwara River Basin in Mumbai, India. CATENA, 145, 142–154. https://doi.org/10.1016/j.catena.2016.06.009

23 Assessment of Extreme Climatic Impact on Agricultural Livelihood in the Purulia District of West Bengal through Multi-Criteria Decision Model

Asutosh Goswami and Priyanka Majumder

23.1 INTRODUCTION

Severe droughts are occurring at an alarming rate and causing significant damage all over the world (Adams et al. 1998). In today's world, the increasing impact of climate disturbances like droughts and variations in rainfall is disproportionately putting the most vulnerable people at risk (Adhikari 2018) and putting their ability to earn a living in jeopardy (Agada and Obi 2015; Afodu et al. 2019). Climate change is making disasters worse all over the world and is expected to make them worse even more. Poor people are the ones who fall victim to this calamity, and 75% of them survive on agriculture. Extreme weather poses a significant threat to the livelihoods of vulnerable individuals who are dependent on agriculture, livestock, and fisheries and lack adaptability (Asha latha et al. 2012; Babar et al. 2015). It is estimated that there is insufficient food for a normal life for nearly a quarter of the world's population, and nearly a billion people go hungry each year. Since dry seasons diminish farming efficiency, it is one of the primary drivers of appetite and unhealthiness (Bello et al. 2012). Long-haul mechanically upgraded grain creation, which is the essential calculate food security, is diminishing (Bera et al. 2017), notwithstanding rural misfortunes brought about by yearly dry seasons. Poor country families are compelled to sell animals for food utilization because of the dry spell impacts, which additionally fuel neediness, movement, desperation, and social destabilization (Berhanu and Beyene 2015). One of the driest years in many years struck West Bengal provincial families in 2009, bringing about broad yield misfortunes and food deficiencies. This chapter looks at the moves made by agrarian families in the Purulia region of West Bengal against those climatic limits to address the elevated degrees of food frailty experienced by north of 0.2 million individuals consistently. We focus on how food procurement strategies, production adjustments, and livelihood outcomes are influenced by household resource access profiles and livelihood portfolios. The case in question is a good example of the difficulties that farming families face when it comes to their means of subsistence and production in places with a lot of climate risk and little natural resources. Human costs may be higher because many households operate near the margins of subsistence, despite the fact that absolute income loss may be lower than in regions with better endowment. The identification of factors that shape vulnerability as well as documentation of the type, range, and sequence of coping strategies can inform policies and programs for reducing risk exposure, assessing need and prioritizing interventions, relieving stress, and enabling affected individuals to recover quickly (Holman 2006; Iglesias et al. 2011; Huho

DOI: 10.1201/9781003377825-26

et al. 2012; Jain and Kumar 2012; Kar et al. 2012; Khoshnodifar et al. 2012; Chakraborty et al. 2013; Hashemi et al. 2013; Ji-kun et al. 2014; Iglesias and Garrote 2015; Chatterjee 2016; Kaur and Kaur 2017; Cherian and Khanna 2018; Hasan et al. 2019; Kasim 2019; Javadinejad et al. 2020). In West Bengal, groundwater is used to supply about 80% of drinking water sources. The groundwater in the state goes about as a cradle for long-haul precipitation and surface water deficiency apart from the fact of contamination in some patches of the Indo-Gangetic Plain. Hence, delayed dry spells (e.g. more prominent than a two-year term) or successive long stretches of diminished rainstorm precipitation seriously compromise occupations. In 2009, the district (Purulia) as well as the state experienced severe water scarcity as a result of insufficient monsoon rain. The present study looks at how frequent droughts are affecting dryland small and marginal landholders more and more, causing a lot of damage to agriculture and making it hard for them to have sustainable livelihoods and ways to adapt. The five asset categories of a livelihood approach that are frequently exacerbated by the dynamics of repeated droughts are economic, social, natural, physical, and human capital. Their findings are presented in this chapter. This study analyzes the serious impacts of dry seasons on animals and agronomic creation, which bring about huge monetary misfortunes and unforeseen vulnerability for smallholders' method for means. Various snags, including an absence of present-day procedures and information, an absence of agro-data, deficient credit, capital insufficiency, agronomic harms, financial misfortunes, and determined dry spell episodes, remain to undermine adaptive capacity, according to the findings. Drought is a climate-related hazard that occurs frequently and has significant effects on natural and production systems, particularly in rural households whose livelihoods are dependent on agriculture (Kima et al. 2015; Kolawole et al. 2016). Agriculture accounts for 26% of gross domestic product in the economies of many low-income developing nations, making it the primary source of income for 2.5 billion people worldwide. Through increased temperature and rainfall variability, as well as the frequency and severity of extreme weather events, climate change is, however, a global driver that has a negative impact on the sustainability of the agricultural production system (Korkmaz 1988; Kumar 1997; Kubicz et al. 2019). Numerous effects are anticipated, such as altering outbreaks of disease and crop pests, crop disappointment, reduction in yield, loss of the diversity of fishes, and a higher rate of livestock death. The age of composite files in view of sets of markers that mirror numerous components of the weakness idea, catching the openness, responsiveness, and versatile limit of agro-ecological frameworks, is being endeavored on the grounds that horticulture is one of the areas generally defenceless against environmental change (Kundzewicz and Kaczmarek 2000; Kumar and Nain 2013; Kumar and Gautam 2014; Lacombe and McCartney 2014; Goswami 2019). According to Goswami (2019), one of the greatest threats to our world's environment, society, and economy is climate change. As is currently evident from perceptions of increases in global average air and sea temperatures, widespread melting of snow and ice, and rising global mean ocean levels, the warming of the environment framework is acknowledged as unambiguous (Lei et al. 2016). Despite the significant challenges posed by climate change and drought, many questions remain unanswered. As a result of widespread poverty, heavy reliance on rain-fed agriculture, a lack of economic and technological resources, inadequate safety nets, and educational progress, the region's low human adaptive capacity to anticipated increases in extreme events had been the focus of much of the attention paid to Africa in relation to climate change (Lerner et al. 1990; Lindenberg 2002; Liu et al. 2011). The farming families utilize a large number of preventive measures against the climatic variabilities. Encouraging a good extent of both shortcoming and flexibility as such is the essential in supporting a variety of frameworks that decline the impact of climatic variability (drought, dry spell) among smallholder ranchers. It is difficult to develop such measures at the household level due to the lack of high-frequency panel data that can provide insights into the relationship between household-level variables and climate events (Longobardi and Villani 2009; Lobell and Gourdji 2012; Liu and Basso 2020). Understanding resilience can also help direct resources toward fundamental change, and determining the root causes of vulnerability can assist in addressing structural issues (Loon and Laaha 2015; Kumar et al. 2017). Smallholder vulnerability can impede sustainable development in one of India's most productive agricultural regions, the Indo-Gangetic Plains. As there are significant differences in financial turn of events, normal assets, and agrarian efficiency inside the area, we have applied the strengths, weaknesses, opportunities, and threats (SWOT) model to decide appropriate transformation systems under different circumstances. Both self-reported climate shocks and spatially interpolated drought data to reflect distinct aspects of climate exposure have been incorporated. In light of the fact that a system's ability to adapt depends on its political, cultural, technological, financial, and institutional capacities, local vulnerability assessments are crucial for comprehending the effects of climate change. Climate change will have a significant impact on food supply and security, water availability, infrastructure, and agriculture income for individuals and their communities. The effects of climate change are threatening the assets of sustainable livelihood—human, social, natural, physical, and economic—and forecasts for the future are extremely gloomy.

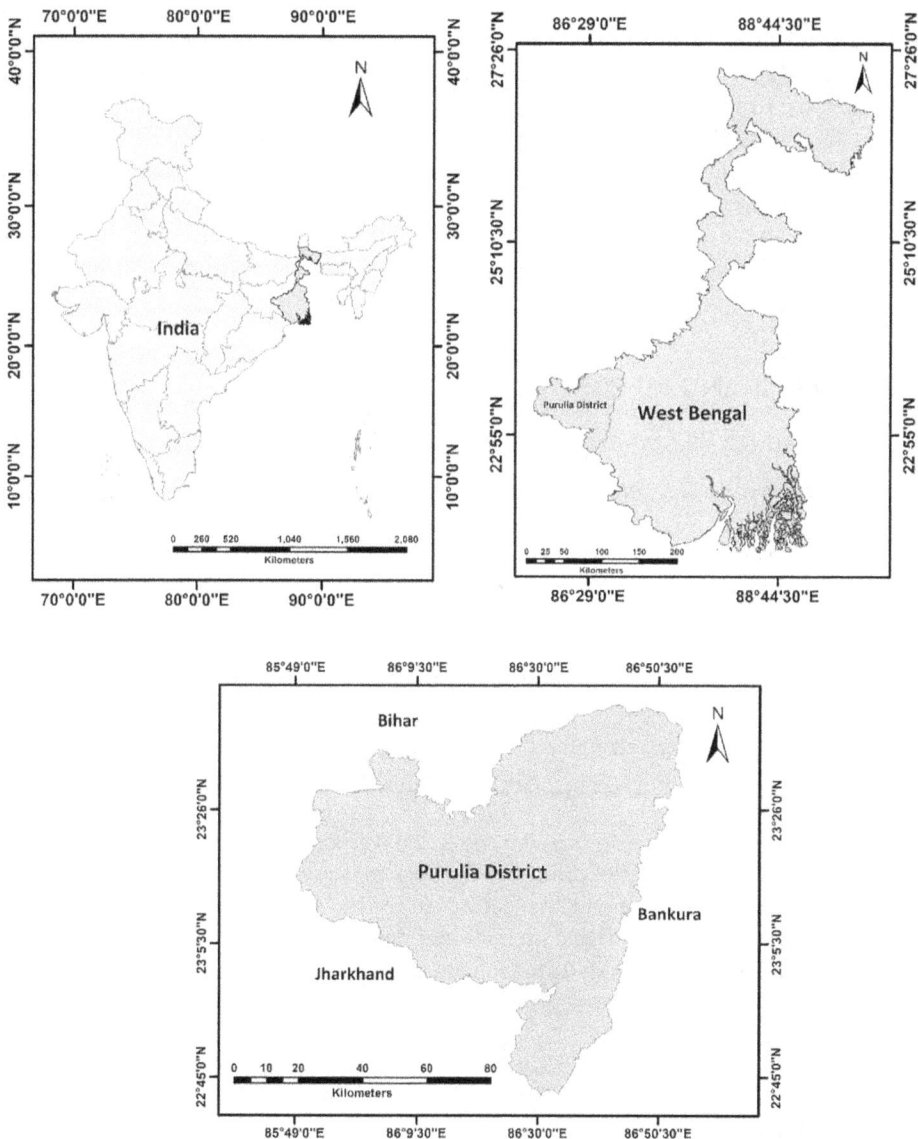

FIGURE 23.1 Location map of the study area.

In the Asia-Pacific region, climate change poses a significant threat to agriculture, food security, and rural livelihoods for billions of people, including the poor. Due to its high reliance on weather and climate, agriculture is the most vulnerable sector, and agricultural workers typically live in poorer conditions than urban residents. Our study area is not an exception, where more than 60% of the population relies on agriculture as a means of subsistence either directly or indirectly. Both the issue and the solution lie in agriculture. The agricultural sector in Asia is already confronted with numerous sustainability issues. Biophysical and socioeconomic factors increase community susceptibility to the effects of climatic stressors, making it more vulnerable to climate change. Numerous studies on climate change risk and adaptation have included vulnerability assessment. In most of these studies, vulnerability was viewed as an outcome rather than a pre-existing condition. The possibility of weakness has risen out of various fields of concentration throughout the long term. Various methodologies—human environment, political nature, actual science, and spatial examination—have prompted various translations of the expression "weakness". The impacts of environmental change have been felt in different areas, viz. energy, biodiversity, water resources, forestry, and agriculture The current review means to recognize the limit of climate and environment in the Purulia area of West Bengal (Figure 23.1) through the application of various indices and their impact on the livelihood of agrarian families in the district with the employment of both the primary and secondary observations. The study may be classified into two segments. The first part of the study highlights the climatic extremes of the district in terms of climatological, groundwater and consequent vegetative droughts. The other part of the study highlights the impact of these climatic extremes on the agricultural potential of the district as well as the adaptation techniques against those extremes through the computation of Garrett's ranking method and quantitative strategic planning matrix.

23.2 MATERIALS AND METHODS

23.2.1 STANDARDIZED PRECIPITATION INDEX

When a region experiences an abnormally high deficit in precipitation in comparison to the long-term average, this is known as a meteorological drought. The abnormalities, or deviations from the mean, of the noticed absolute precipitation are shown by the Standardized Precipitation Index (SPI) pointer for a specific area and collection time of interest. The magnitude of the deviation from the mean is a probabilistic indicator of the severity of a wet or dry event. The subsequent SPI pointers, which can be determined over an assortment of precipitation gathering periods (normally going from one to four years), make it conceivable to gauge various possible impacts of a meteorological dry spell.

When calculated for shorter accumulation periods (e.g. one to three months), SPI-1 to SPI-3 can be used as an indicator for immediate effects like decreased soil moisture, snowpack, and flow in smaller creeks.

SPI-3 to SPI-12 are determined for medium collection periods (like three months to a year). They tend to be utilized as an indication of diminished supply capacity and stream.

When calculated for longer accumulation periods (e.g. 12 to 48 months), SPI-12 to SPI-48 can be used as a sign of decreased reservoir and groundwater recharge.

SPI-3 has been calculated in this study to indicate a change in the moisture regime over a shorter accumulation period. The longer time period may generalize the study, which is why SPI-6 and SPI-12 were not selected. It is essential to keep in mind that both the natural environment (such as soils and geology) and human intervention (such as the existence of irrigation plans) influence the precise relationship between the accumulation period and the effects of the drought. To get a complete picture of the potential effects of a drought, the SPI needs to be calculated and compared for different accumulation periods. A comparison with other drought indicators is required in order to evaluate the actual effects on the vegetation cover and various economic sectors. The "gamma distribution", a two-parameter continuous probability distribution, can effectively represent a time series consisting of three months spread out over at least 30 years. One of two elective approximations of the "most extreme probability assessors" for the gamma circulation are utilized to fit the two boundaries (i.e. shape and size) of the gamma dissemination on the recurrence dispersion of the verifiable non-zero precipitation collections for the entire years in the accessible time series to figure the SPI for a noticed precipitation gathering for a time of interest. After adjusting for McKee and others, the cumulative probability of the observed rainfall is then transformed (converted) to the standard normal random variable Z with a mean of zero and a variance of 1, using an approximation. To watch out for dry season, the normalized precipitation record (SPI) was created in 1993.

The climatological precipitation time series has been found to be well-suited to the gamma distribution.

SPI is equal to the following:

$$(Xik\text{-}Xi)/ói \tag{23.1}$$

where $ói$ is the standardized deviation for the eighth station; Xik is the precipitation at the kth observation and the tenth station; Xi is the average precipitation at the eighth station. But the calculation of SPI is never free from some complexities which may arise due to non-homogenous data structure and gamma distribution of rainfall potentials. To minimize the role of non-homogenous data structure, the following equations may be taken into account:

$$G\ (r) = (1/\beta\ \alpha\ r\alpha) * a\text{-}1e\text{-}x/b \tag{23.2}$$

where $\alpha > 0$; α is considered as a shape factor and $\beta > 0$, β is denoted as a scale factor.

$$\Gamma(a) = \int\infty\ ya\text{-}1e\text{-}y$$

where $\Gamma(\alpha)$ denotes gamma distribution.

The SPI is calculated by fitting a station's precipitation total frequency distribution to a gamma probability density function. The maximum likelihood solutions are used to best estimate α and β:

$$\alpha^{\cdot} = 1/4A\ [1 + (\sqrt{1} + 4A/3)] \tag{23.3}$$
$$\beta^{\cdot} = \bar{x}/\alpha \tag{23.4}$$
$$A = In\ \bar{x} - \Sigma\ In\ (x)/n$$

where n is the number of precipitation observations.

The SPI is calculated by fitting a station's precipitation total frequency distribution to a gamma probability density function. The most extreme probability arrangements are utilized to ideally gauge α and β, where q is the likelihood of a zero. If m is the number of zeros in a precipitation time series, then m/n can be used to estimate q. The cumulative probability, $H(x)$, is then transformed into the SPI value, the standard normal random variable Z with a mean of zero and a variance of one.

23.2.2 Standardized Water-Level Index

For the measurement of hydrological drought, the Standardized Water-Level Index (SWI) has been implemented in the present study with the data of groundwater collected through primary survey of different wells during winter and monsoon, and secondary observations (central groundwater board [CGWB]). The water levels have been measured from the wells in the four community development (CD) blocks (Balarampur, Barabazar, Jhalda II, and Purulia II) during winter and monsoon. The differences of water levels during these two seasons are employed for the computation of SWI. The index is also proposed to analyze the status of groundwater in the district. Here the seasonal water level is normalized and divided by differential standard deviation between the seasonal water level and its long-term seasonal mean for the SWI computation. The groundwater table can be indirectly measured by SWI, which reveals a drop in water level. For ground surface as the basis for the groundwater-level measurement in the observation well, values of less than zero indicate no drought and larger than two indicate significant drought condition. A hydrological drought is characterized by a decrease in water levels in various water bodies. Groundwater levels from each of the four seasons—pre-monsoon, monsoon, post-monsoon, and winter—have been used to calculate SWI in order to comprehend the impacts of these seasons on water resources. In addition, primary observations were made in a few selected CD blocks in various parts of the district (Purulia) at 08:00 and 14:00 to determine the depth of the groundwater table during the two seasons—winter and monsoon. The pictorial representations have been made with straightforward bar graphs.

23.2.3 Soil Adjusted Vegetation Index

For the determination of agricultural drought in the four studied blocks of the different portions of the district, Soil Adjusted Vegetation Index (SAVI) has been employed. The main drawback of the Normalized Difference Vegetation Index (NDVI) is its non-linear relationship with biophysical characteristics and sensitivity to soil background. A transformation method is presented for minimizing the effects of external factors on spectral vegetation indices. The red and near-infrared (NIR) spectral bands are involved, and the change in origin of NIR and red reflectance in vegetated canopies is depicted graphically. The effects of soil on indices are eliminated by this transformation. The following formula is in use to calculate SAVI:

$$SAVI = \frac{NIR - RED}{NIR + RED + L} * (1 + L) \tag{23.5}$$

where L is the constant parameter, and NIR and RED are the spectral band reflectance, respectively. The SAVI calculation utilized three soil factor values. They are 0.2, 0.5, and 0.9.

23.2.4 SWOT Analysis in Agriculture

To identify the SWOT in the agricultural system of the district, the analytical method has been employed on the basis of the primary survey on 200 respondents in the four CD blocks of the district (Balarampur, Barabazar, Purulia II, and Jhalda II), selected through random sampling method.

23.2.4.1 Designing an Evaluation Matrix for External and Internal Factors

Both internal (strengths and weaknesses) and external (opportunities and threats) factors affecting farming systems are evaluated during this phase of the research. Everything is assessed as per the ranchers' idea. The concept of the farmers was used to rank each item and calculate its importance ratio coefficient.

23.2.4.1.1 External Factor Evaluation Matrix

In the first part of the SWOT analysis, we look at things outside of our farming practices that we cannot control but can improve or reduce their impact on our farming system. A strategic management tool that is frequently utilized to assess a company's current state is the external factor evaluation (EFE) matrix. It is a useful tool for imagining and focusing on potential hazards and openings in a system.

However, the EFE matrix procedure consists of the following five steps:

1. Give specifics: The first step is to create a list of external factors and categorize them into two groups: dangers and open doors.

2. Weights for attributes: Weight is given to each factor. If a 10 to 100 scale is used, the value of each weight should be between 10 and 100. Zero demonstrates that the variable is irrelevant, while one or a hundred shows that it is the most huge and persuasive. On the other hand, the sum of all weights ought to be 100.
3. Rate factors: The rating, which goes from one to four, is given to each factor. The rating indicates how effectively the factor is handled by the company's current strategies.
   A rating of an element shows whether it is a significant danger (rating = 1), a minor danger (rating = 2), a minor open door (rating = 3), or a significant open door (rating = 4). On the off chance that a scale from 1 to 4 is utilized, qualities should be given a rating of 4 or 3, and shortcomings should be given a rating of 1 or 2.
4. Divide ratings by weights: To determine each factor's weighted score, its rating is multiplied by its weight.
5. Weighted scores in total: The total weighted score is the sum of all the weighted scores for each factor.

23.2.4.1.2 Internal Factor Evaluation Matrix

A strategic management tool for determining a system's strengths and weaknesses within its functional areas is the internal factor evaluation (IFE) matrix. The IFE matrix and the EFE matrix, two strategies formulation tools, can be used to compare a company's performance to its own internal strengths and weaknesses. The following three stages can be utilized to construct the IFE grid:

1. Important internal aspects: Recognizing strengths and weaknesses is the first step.
2. Weights: The IFE matrix assigns a weight to each factor that ranges from 0.00 to 1.00. The weight given to a factor demonstrates its relative importance to other factors. Zero demonstrates that nothing is significant, while one shows that something is vital.
3. Rating: Practitioners typically use ratings of one to four on a scale of one to ten. The rating indicates whether the factor is a major weakness (rating = 1), minor weakness (rating = 3), or major strength (rating = 4).

The SWOT matrix contains four strategic groups:

1. Making the most of opportunities by utilizing one's strengths
2. How strengths can be overcome through opportunities
3. How strengths can be utilized to lessen the impact of threats
4. How to fix flaws that could cause these threats

23.2.4.2 Quantitative Strategic Planning Matrix

One of the fundamental tenets of the quantitative strategic planning matrix (QSPM) is that businesses must systematically evaluate their internal and external environments, conduct research, carefully consider the benefits and drawbacks of various alternatives, carry out analyses, and then select a specific course of action.

The objective of the QSPM method is to select the best business strategy objectively. Although the QSPM's left column contains both internal and external key factors, the matrix's list of factors is directly derived from the EFE and IFE. The viable alternate strategies derived from the SWOT matrix can be found in the top row. Nonetheless, factor loads are remembered for the main mathematical segment.

The QSPM engaging quality scores (AS) show how significant or engaging each variable is to every conceivable technique. Scores for allure fall into four classes: very less attractive, less attractive, moderate attractive, and very attractive.

The absolute allure scores, or TASs, address the general appeal of each vital component and the singular procedure that is connected with it. The QSPM's total attractiveness score, on the other hand, is calculated by adding the total attractiveness scores for each strategy column.

The sum of all attractiveness scores (STASs) reveals the most appealing strategy. At the point when all significant outer and inward basic factors that could influence the essential choice are thought about, higher scores show a procedure that is seriously engaging. The reach for engaging quality scores is 1 = not alluring, 2 = fairly appealing, 3 = sensibly alluring, and 4 = exceptionally appealing.

23.2.5 GARRETT RANKING METHOD

An effort is made to acknowledge the challenges that growers face when cultivating various crops. For the said purpose, 200 respondents have been surveyed. The respondents are selected through systematic sampling method from four CD blocks. From each CD block 50 respondents have been selected. Using Garrett's ranking method, the problems that ranchers have experienced are ranked. This method is also used to rank the respondents' preferences regarding various aspects of the cultivation process. It is used to determine the most significant factor in the farming practices. Based on Garrett's ranking, the study asked respondents to rank various problems and outcomes according to their impact, which is then converted into a score value and ranked using the following formula:

$$\text{Percentile position: } 100 \, (\text{Rank}_{in} - 0.5)/T_n \tag{23.6}$$

where $Rank_{in}$ is the assignment of rank to the ith variable by the nth respondents, and T_n is the total number of factors ranked by the nth respondents.

The individual percentile location is then transformed into Garrett's score value with the employment of Garrett's ranking table. The scores of all the rankings are then summed up to get the importance of the factors.

The problems identified through personal interviews and questionnaires are categorized at random in the table. The individual scores for each factor are then added together, and the total scores and mean scores are then calculated. The most significant factor is thought to be the one with the highest mean value.

23.2.6 ANALYTICAL HIERARCHY PROCESS MODELLING

The present study employs analytical hierarchy process (AHP) modelling to detect the best supportive policy to boost up the agro-industry in the study area. An overall estimation hypothesis is the AHP. Correlations based on mathematical decisions are used to determine them from a massive numbers. Here both objective elements should be considered to be in a pool that is similar. AHP provides a method for deliberately determining and incorporating family scales. The component and sub-factor addressed in these designs estimate the various variables, which are arranged in a pecking order. This chapter examines the current state of the AHP's approach in light of its expanding application. It is used to obtain proportion scales from both continuous and discrete matched correlations. The method is for accurately reflecting each member's perspective and determining which instrument is best for each choice situation. The method of thinking about each component at the relational level and aligning them with the mathematical size of numbers that demonstrates how frequently one component is more significant or prevailing than another component with recognition of the factor that they are examined. Local evaluations of each variable are obtained by adding the loads of the subfactors to each option rating. The loads are then added to nearby evaluations, which are then compiled for combined efficiency. The AHP generates weight values for each option in order to determine the significance of one option over another in relation to a standard. The AHP and other multi-criteria decision model devices are used in this investigation. AHP is used to connect a staggered order design of objectives, models, sub-criteria, and choices to explain various decision contentions. Identifying the models for advisor selection is one way to get the best professional measures right away. The fundamental component of the AHP cycle is the evaluation of the standards and subrules by relevant experts in their respective fields. One of the main promises is that it can choose the most prevalent measures and sub-models for each objective. The main step in AHP is to create a different levelled structure with goals, measures, and sub-models. Then, the AHP model is made through a couple wise review that will be assessed by the trained professionals. As a result, the survey results will also be recorded, and the mathematical mean will be calculated before the framework structure and the weight of each standard are acquired. Through the consistency record, the AHP strategy, in particular, can determine to what extent the reactions are reliable or contradictory.

For the correlation, a mathematical scale with nine points is used. The power besides, the implications of the pair-wise assessment are according to the following:

- 1 indicates the same significance;
- 2 means powerless or minuscule;
- 3 means there is some significance;
- 4 mean moderate notwithstanding;
- 5 indicates solid significance;
- 6 means solid as well as significant;
- 7 indicates incredibly amazing;
- 8 is extremely sturdy;
- 9 is of very high significance, or
- 1.1–1.9 if the exercises are very close.

23.3   RESULTS AND DISCUSSION

23.3.1   SPI

The computation of SPI-3 for some selected stations in the district indicates that the meteorological drought is not a very much recurrent phenomena over the district. The number of years with negative SPI values is higher than the years with positive SPI values. Compared to monsoon, the station Hatwara (for Purulia II CD block) has experienced fewer drought events during the winter, pre-monsoon, and post-monsoon (Figure 23.2), where the total number of drought events are found to be 9, 13, and 24, respectively. The station has reported extremely wet conditions with SPI values of 2.0 or higher in the years 1979, 1987, 1991, 1998, 2007, 2013, 2015, 2016, and 2017. Additionally, the station Jhalda (for Jhalda II CD block) experiences seasonal shifts in the spells of dry and wet weather (Figure 23.2). Compared to the monsoon season, this station has experienced fewer drought events during the winter, pre-monsoon, and post-monsoon seasons. The absolute quantities of dry spell occasions are distinguished as 25, 21, and 6 during rainstorm, pre-storm, and post-rainstorm seasons, respectively. During the monsoon season, the station Balarampur has only experienced three extreme drought events (Figure 23.2). Compared to the pre-monsoon, winter, and post-monsoon seasons, this station has reported a higher number of moderate drought events during the monsoon. There have been fewer normal to near-normal precipitation events reported in the monsoon months of July and August than in the other months. In comparison to the other three seasons, the station Barabazar has reported a higher number of moderate, severe, and extreme drought events during the monsoon. According to month-by-month analysis of SPI, the post-monsoon month of October has seen a greater number of moderate drought events than the other months. The station Hatwara has only experienced two severe droughts, both of which occurred exclusively during the monsoon. In contrast, there is no indication of an extreme drought during other seasons. Additionally, the station Jhalda has only experienced two extreme droughts, both of which occurred exclusively during the monsoon. It is also evident that there have not been any severe or extreme drought events in the post-monsoon and winter seasons for the time span of 42 years, i.e. from 1976 to 2017.

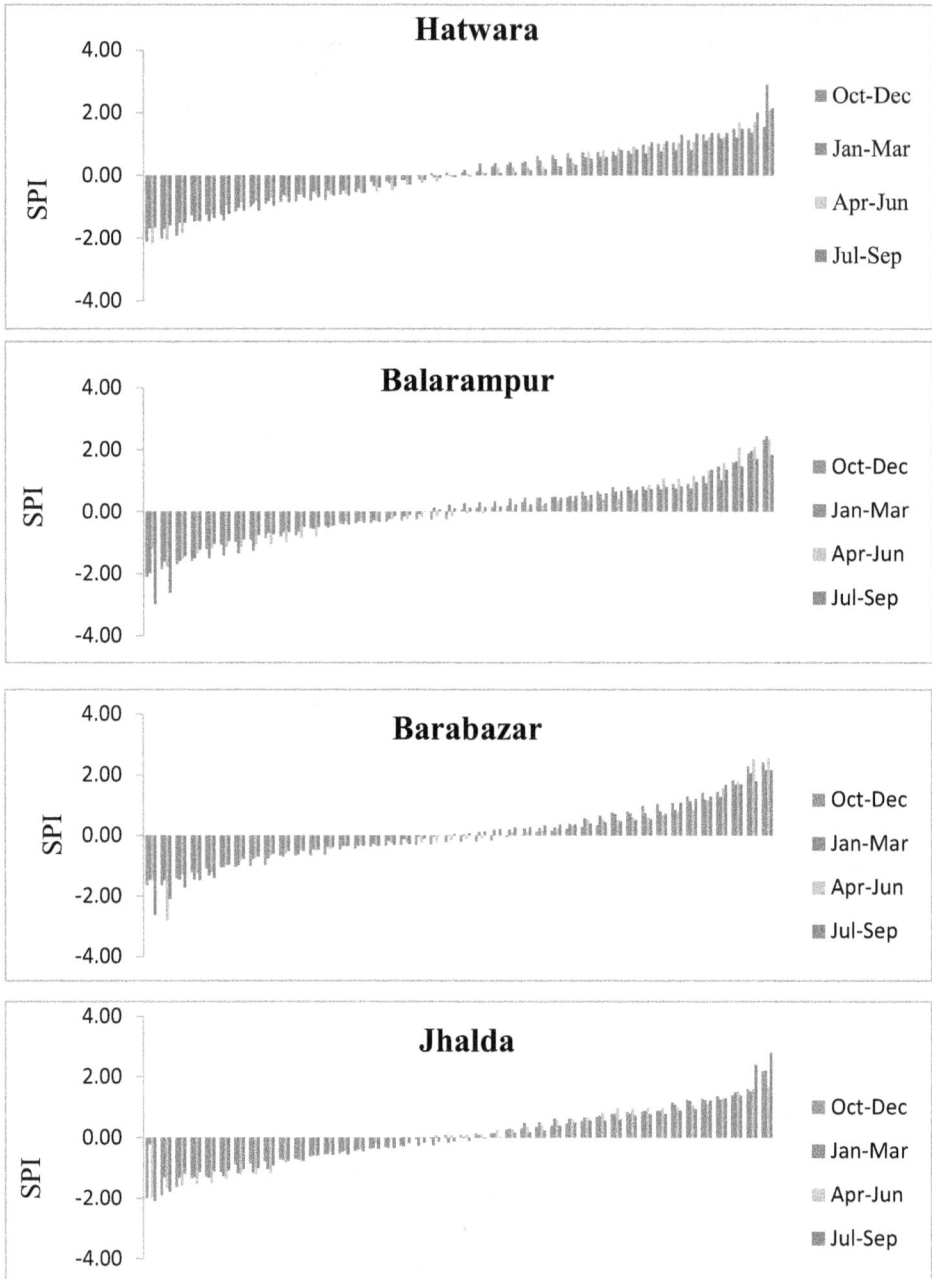

FIGURE 23.2   SPI- 3 for some selected stations of Purulia district.

23.3.2   SOIL ADJUSTED VEGETATION INDEX

The study indicates that the agricultural drought is a regular phenomenon in the study area associated with the higher variability of rainfall mainly during pre-monsoon and winter and less surface water retention capacity by the soil. The SAV index has been computed for the four CD blocks lying in the different geo-environmental spectrums of the district for the two separate years, 2011 and 2020 to visualize the change of moisture regime within vegetation (Figure 23.3–23.7). The western

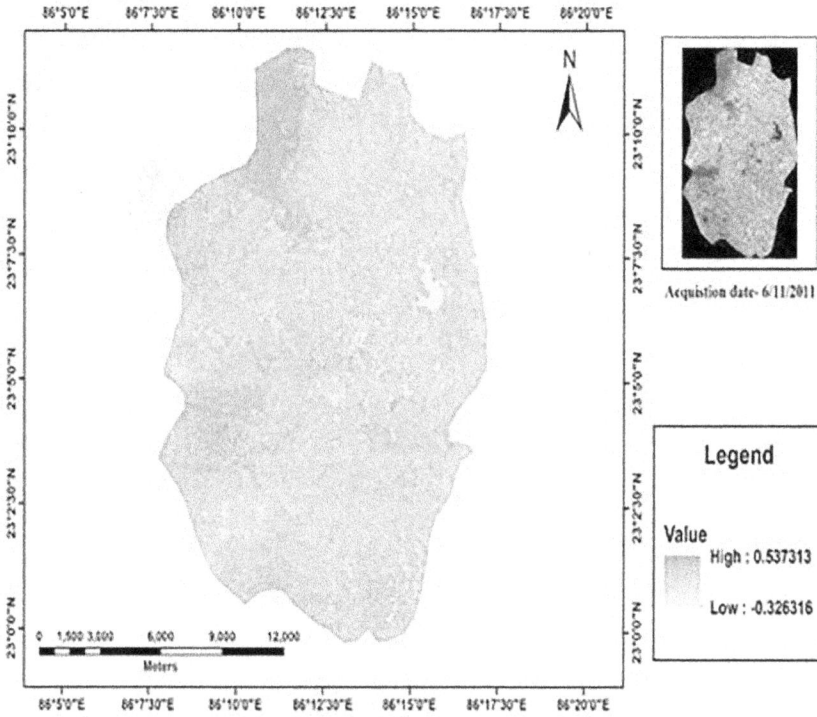

FIGURE 23.3   SAVI map of Balarampur CD block of Purulia in 2011.

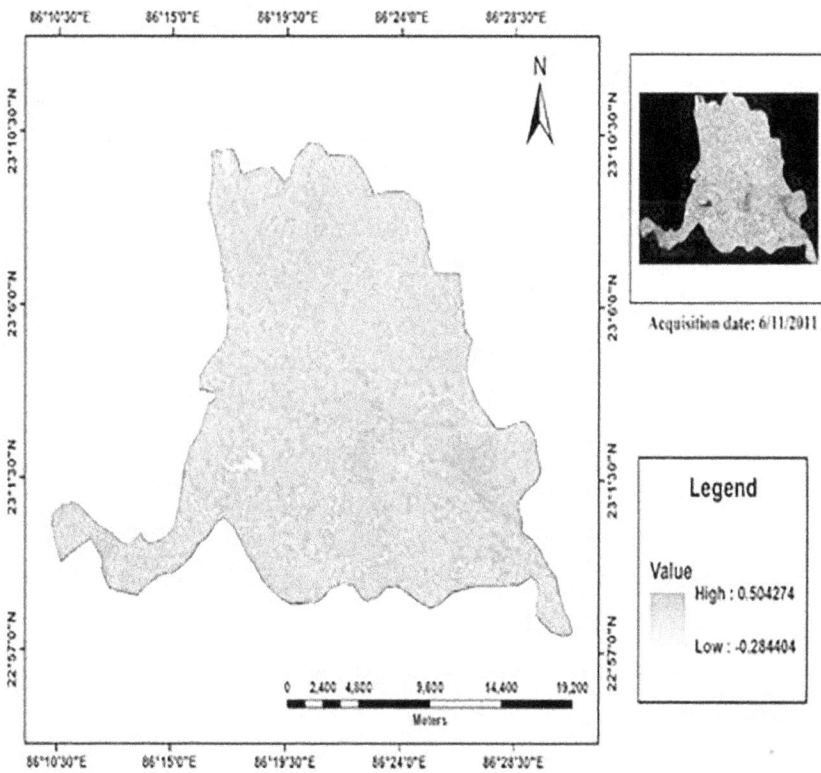

FIGURE 23.4   SAVI map of Barabazar CD block of Purulia in 2011.

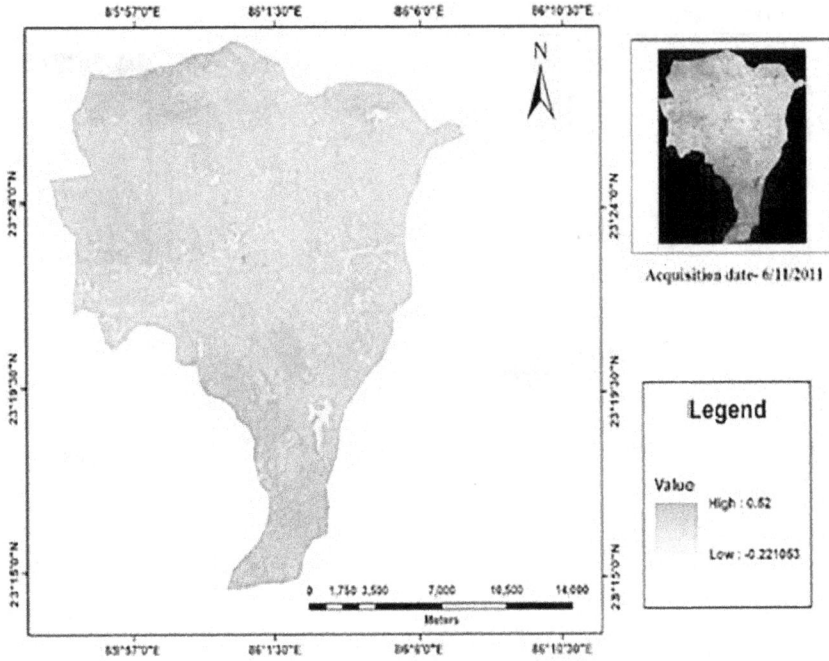

FIGURE 23.5   SAVI Map of Jhalda II CD block of Purulia in 2011.

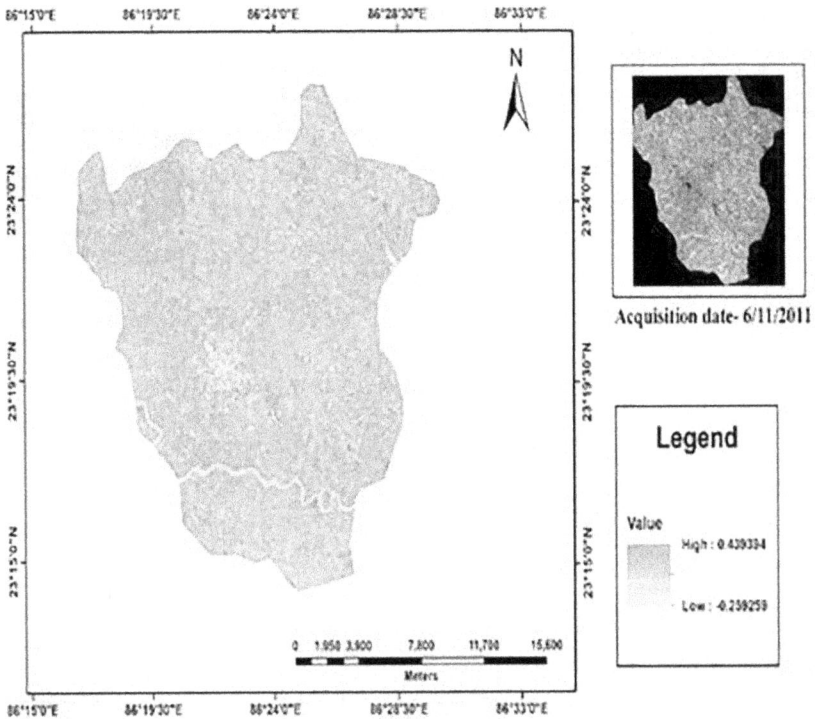

FIGURE 23.6   SAVI map of Purulia II CD block of Purulia in 2011.

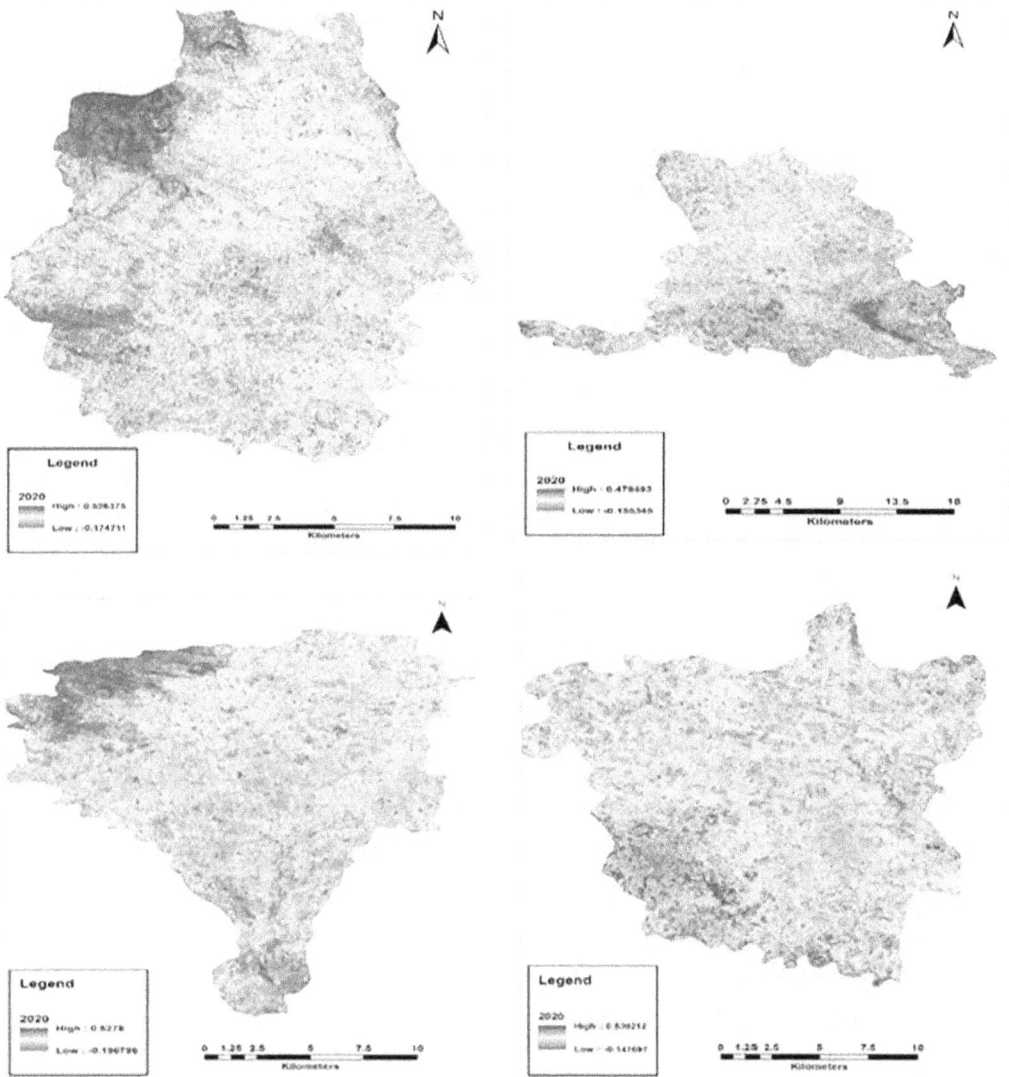

FIGURE 23.7   SAVI maps of the four CD blocks of Purulia.

part of the district indicates a worse scenario of vegetation moisture status due to exposed rocky terrain coupled with thin soil cover marked by low water holding capacity. The values of SAVI are much closer to 1 in CD block Barabazar, indicating a very sparse vegetation cover. On its contrary, the situation of Purulia II is satisfactory due to thick soil cover and less gradient of topography. The situation is getting worse for Barabazar which may also be validated with the values of SPI-3. On its contrary, the CD block Purulia II has maintained a stable condition in this regard.

23.3.3   STANDARDIZED WATER-LEVEL INDEX

The longest periods of normal to near-normal water levels in the CD block Balarampur are found from 1996 to 1998 and from 2013 to 2015. From 1996 to 2017, the monsoon season experienced eight mild hydrological droughts. A year of extreme hydrological drought has followed a year of moderate hydrological drought in 2010. This situation appears to be getting better before the monsoon because this season has seen water levels that are normal or close to normal for the first time in 15 years. Despite the fact that this season has experienced four moderate hydrological droughts in 1999, 2005, 2009, and 2015, there is no evidence of a severe hydrological drought. This season has been marked by a severe hydrological drought in 2010 for all the CD blocks. The coherence of the typical to approach ordinary water levels from 1996 to 2004 is broken by a time of gentle hydrological dry spell in 2001. Despite the absence of post-monsoon evidence of extreme hydrological drought, three severe hydrological droughts are observed in 2010, 2014, and 2015. Nine mild hydrological droughts occurred in the winter; which covers the most time from 2006 to 2008. Although there is no evidence of a moderate hydrological drought, 2011 and 2016 are found to have experienced severe and extreme hydrological droughts. In Jhalda-II CD block, the number of years with normal to near-normal water levels is highest (13 years) in the winter, followed by 12 years in monsoon and post-monsoon seasons (Figure 23.8). Although severe hydrological drought events are observed during the monsoon (2005), pre-monsoon (2010), and winter (2001 and 2011), there is no evidence of occurrence during post-monsoon. Then again, outrageous hydrological dry spell occasions are found to have happened in 2010, 2005, and 2002 during rainstorm, post-storm, and winter seasons respectively. The hydrological drought matrix demonstrates that from 2011 to 2017, the monsoon season experienced the longest stretch of normal to near-normal water levels for the CD block Barabazar (Figure 23.8). Extreme hydrological droughts have not occurred for Purulia II during the monsoon or post-monsoon seasons, while severe hydrological droughts are observed in 2010, 2014, and 2010, respectively, during post-monsoon and monsoon conditions.

23.3.3.1 Field-Based Perceptions of Hydrological Droughts

Field-based perceptions of hydrologically dry seasons from the year 2015 to 2019 at 8:00 AM and 2:00 PM additionally demonstrate comparative outcomes. Pre-monsoon and winter, when groundwater extraction is significantly greater than precipitation-based groundwater recharge, complicates the situation (Figure 23.9). Even during the monsoon, all of the selected CD blocks have experienced hydrological droughts at 2:00 PM (Figure 23.9) as a result of the excessive use of groundwater for irrigation.

23.3.4 SWOT Analysis in Agriculture

The strengths and opportunities indicate the potentialities of the district to minimize the negative issues that may arise from environmental extremeness. Random sampling technique has been employed to select 200 farmers and respondents for the sample. The present study looks into the various problems in agriculture using SWOT analysis. The technique for SWOT examination has been used in the current research to investigate the different open doors and dangers in agribusiness. With a weighted score of 0.32, the factor of market facilities is identified as the primary strength for the district (Table 23.1). The majority of the respondents have highlighted increasing support from the government in agri-business as one of the most crucial strengths with the weighted score of 0.27. Additionally, this region receives a significant amount of bright sunlight and radiation. With a weighted score of 0.24, this factor is also regarded as a major strength in the study area for the proper agricultural prospect. Even though the study area receives plenty of rain, droughts are a common occurrence there. As a result, majority of the CD blocks primarily those are located at the eastern fringe of Chhota Nagpur plateau witness drought as a major weakness (Table 23.1). The undulating terrain coupled with coarse-grained soil takes the most pivotal role for the water scarcity even after a considerable amount of rainfall during monsoon. On the other hand, for majority of the CD blocks a number of weakness factors such as excessive reliance on rainfall, weak agro-based industry, and a lack of water conservation practices have been recorded. Additionally, transportation, credit support, and market infrastructure are also found to be immature in the study area. Opportunities for horticulture and other cash crops, a rise in governmental support for agricultural development, the introduction of new agricultural technologies are some of the most significant opportunities for the

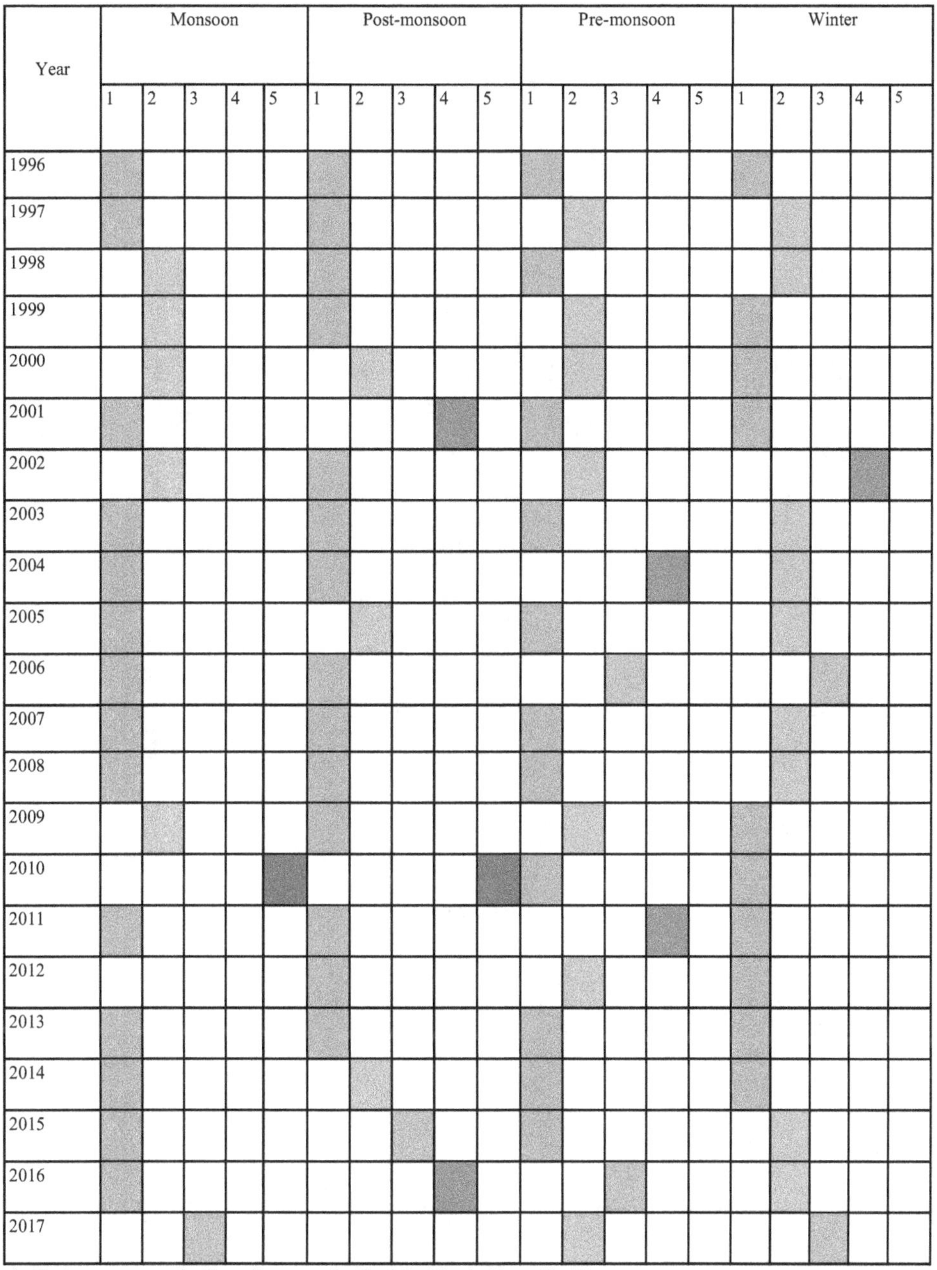

Year	Monsoon					Post-monsoon					Pre-monsoon					Winter				
	1	2	3	4	5	1	2	3	4	5	1	2	3	4	5	1	2	3	4	5
1996																				
1997																				
1998																				
1999																				
2000																				
2001																				
2002																				
2003																				
2004																				
2005																				
2006																				
2007																				
2008																				
2009																				
2010																				
2011																				
2012																				
2013																				
2014																				
2015																				
2016																				
2017																				

1: Balarampur; 2: Barabazar; 3: Jhalda II; 4: Purulia II

FIGURE 23.8 Hydrological drought scenario of Purulia.

CD blocks in their agricultural potentiality. The future of agriculture in the study area will unquestionably depend on reducing threats. The primary threats to the area under investigation have been identified as the growing climatic variability and market price swings for crops.

23.3.4.1 Quantitative Strategic Planning Matrix Approach

The QSPM approach is intended here for evaluating the internal and external environments of a farming business system to indicate the best ideal strategy to combat the loopholes that may arise from climatic disasters. In order to identify and evaluate the various options, the matrix conducts

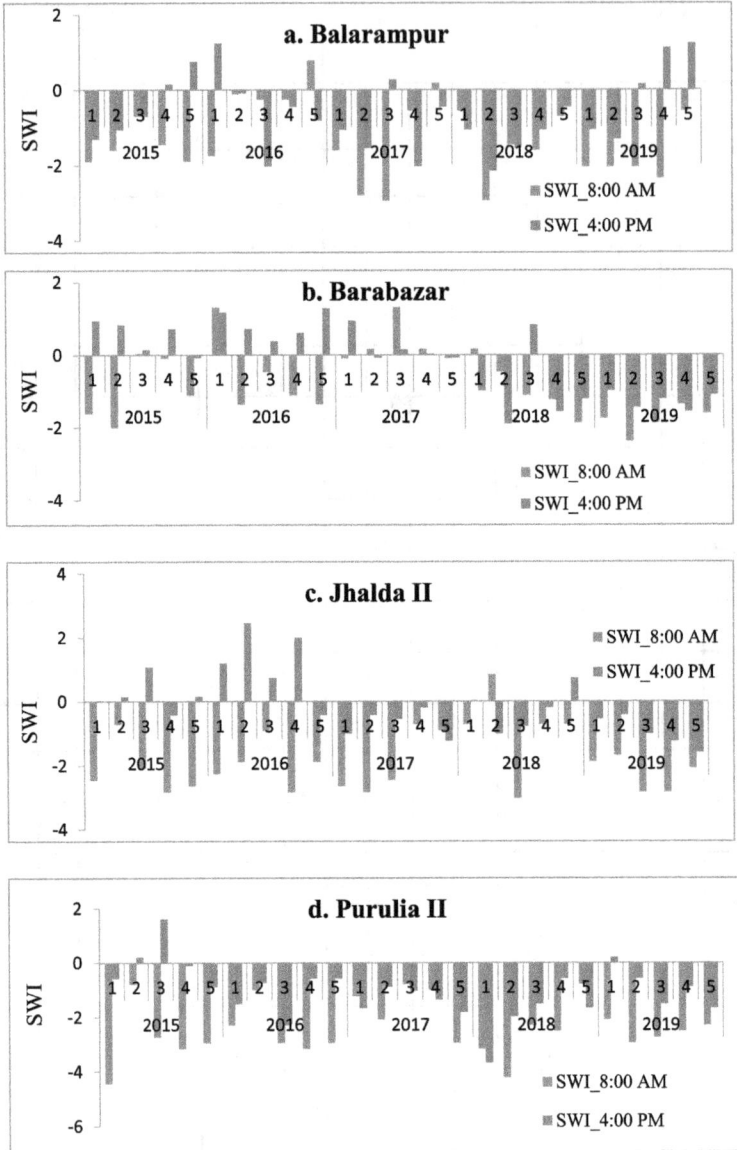

FIGURE 23.9 Elevation-wise observations of hydrological drought in pre-monsoon.

in-depth research that determines a particular course of action. Seven weakness-threat (WT) strate-gies have been listed for the study area (Table 23.2). With the total attractiveness scores (STASs) of 5.4, the strategy for increasing diversification of crops has acquired the rank 1 strategy to combat the loopholes in agriculture for the district (Table 23.3). The state as well as the district gets plenty of rainfall in the monsoon. But due to undulating nature of terrain, water moves as surface runoff following the general gradient of the topography. During pre-monsoon and winter, there is basically no reliable source of rainfall except some sporadic thunderstorms. The QSPM model suggests to make the cropping pattern more diverse instead of over-reliance on rice farming throughout the year more specifically during pre-monsoon and winter months.

TABLE 23.1

SWOT Analysis in Agriculture

Strengths	Weight	Score	Weighted Score
Sufficient rainfall	0.12	4	0.48
Adequate hours of sunshine and radiation	0.06	4	0.24
High manpower	0.04	3	0.12
Introduction of new technologies in agricultural practices	0.03	3	0.09
Wide variety of natural resources	0.02	3	0.06
Agroecological practices	0.06	4	0.24
Strong governmental support	0.07	3	0.21
Liaison with many development organizations	0.02	3	0.06
Total			**1.5**
Weaknesses	**Weight**	**Score**	**Weighted score**
Decreased crop diversification	0.09	1	0.09
Excessive reliance on precipitation	0.04	2	0.08
Weaknesses of agriculture-related industries	0.09	1	0.09
Small and dispersed landholdings	0.04	1	0.04
Vulnerabilities in market infrastructure, storage, transportation, credit support, etc.	0.03	1	0.03
Lack of research facilities in agriculture	0.09	1	0.09
Lack of water conservation practices	0.04	2	0.08
Greater surface runoff	0.05	2	0.1
High intensity of drought	0.03	1	0.03
Reducing private sector investment in agriculture	0.02	2	0.04
Presence of a middleman	0.06	1	0.06
Total			**0.73**

Internal Factors Evaluation Matrix (IFEM)

Total Weighted Score 2.28

(*Continued*)

TABLE 23.1   (Continued)
WOT Analysis in Agriculture

Opportunities	Weight	Score	Weighted Score
Horticultural and other cash crop opportunities	0.1	4	0.4
Increase in agriculture-related industries	0.04	3	0.12
Development of new technology	0.06	4	0.24
Increased government support for agriculture	0.07	4	0.28
Diversification of agriculture	0.07	3	0.21
Organic farming opportunities	0.04	3	0.12
Food processing sector outlook	0.04	3	0.12
Opportunities for agro-ecology practices	0.03	4	0.12
Fostering market facilities for alternative crops	0.05	3	0.15
Total			**1.76**

Threats	Weight	Score	Weighted Score
Crop production hazards	0.05	2	0.1
Crop price fluctuations in the market	0.09	1	0.09
Increased climate change	0.1	1	0.1
Low prices for some crops	0.06	1	0.06
Soil degradation	0.07	1	0.07
Participation scale	0.05	2	0.1
Declining interest in agriculture	0.08	1	0.08
Total			**0.6**

External Factors Evaluation Matrix (EFEM)

Total Weighted Score   2.37

TABLE 23.2

Weakness-Threat Strategies

WT 1	Promote crop diversification
WT 2	Applying sustainable water resource management practices
WT 3	Promotion of agricultural industry
WT 4	Encouraging private investment in agriculture
WT 5	Promotion of agricultural research institutes
WT 6	Development of market infrastructure, storage, transportation, credit support, etc.

23.3.5 RANKING OF ADAPTATION OPTIONS AND HINDRANCES

The majority of respondents gave livelihood diversification a total score of 5,816, placing it ahead of storage of fodder (rank 2) and shifting planting times (rank 3) with total scores of 5,408 and 5,333, respectively, in the study area. With the scores of 5,108, 5,045, 4,994, and 4,867, respectively, water conservation measures (rank 4), rainwater harvesting (rank 5), diversifying livestock sources (rank 6), and the introduction of drought-tolerant rice varieties (rank 7) have also appeared as significant adaptation options in the study area (Table 23.4).

There are some limitations and restrictions on agricultural activities in the area under investigation. Table 23.5 contains the calculated scores for a total of 15 constraints. Some of the most significant agricultural constraints in the study area include the use of traditional farming methods, a lack of water storage, an excessive reliance on rice cultivation, and frequent droughts. Garrett's ranking method has been utilized to rank the constraints. With a total score of 5,748, the majority of respondents identified lack of irrigation as the primary agricultural constraint (rank 1), followed by overreliance on rice production, frequent droughts, traditional agricultural practices, and overreliance on rain-fed agriculture (Table 23.5), with scores of 5,697, 5,689, 5,687, and 5,542, respectively. With the total scores of 5,522, 5,436, and 5,236, respectively, respondents cited lack of water conservation measures, a lack of alternative crops (rank 6), a limited supply of fertilizer and plant protection chemicals (rank 7), and a lack of alternative crops (rank 8) as crucial agricultural constraints in the study area.

23.3.6 ANALYTICAL HIERARCHY PROCESS MODEL FOR SELECTING APPROPRIATE DECISION SUPPORT SYSTEM

AHP is used in the study to discuss the various alternatives that influence the farming business system of the district. Based on study, five criteria have been chosen. The impact of each parameter on the agricultural prospect has been highlighted using scale values from 1 to 9. Water conservation has always played a significant role in determining the agricultural status in the district. When compared to the rainfall factor, indicators that are based on the alternative income sources typically receive significant values of three. Strong governmental support is given the appropriate significance to combat the impact of climatic extremes on the agro-systems. The factor of cultivating short duration rice varieties has been given the utmost significance to minimize the role of climatic creeping disaster on the farming system.

Figure 23.10 emphasizes the proportional weighting of various controlling parametric values. With 27.6% given to each, the strong governmental support and agri-business factors gain the most weight. The lack of rainwater storage facility is without a doubt a controlling variable for limiting the agricultural prospect in the study area. If the impact of droughts is to be minimized, the role of water conservation can be interpreted as a systematic and dependable factor once more. The graph

TABLE 23.3

Quantitative Strategic Planning Matrix Model for Appropriate Strategy Selection

SWOT	Key Factor	Weight	WT 1		WT 2		WT 3		WT 4		WT 5		WT6		Key Factor
			AS	TAS	AS	TAS	AS	TAS	AS	TAS	AS	TAS	AS	TAS	
Sufficient rainfall	S1	0.12	4	0.48	4	0.48	3	0.36	3	0.36	3	0.36	3	0.36	S1
Adequate hours of sunshine and radiation	S2	0.06	2	0.12	3	0.18	2	0.12	1	0.06	2	0.12	1	0.06	S2
High manpower	S3	0.04	2	0.08	2	0.08	2	0.08	2	0.08	2	0.08	2	0.08	S3
Introduction of new technologies in agricultural practices	S4	0.02	3	0.06	3	0.06	4	0.08	3	0.06	3	0.06	3	0.06	S4
Wide variety of natural resources	S5	0.03	3	0.09	2	0.06	2	0.06	2	0.06	3	0.09	2	0.06	S5
Agro-ecological practices	S6	0.02	2	0.04	2	0.04	3	0.06	2	0.04	3	0.06	3	0.06	S6
Strong governmental support	S7	0.06	3	0.18	3	0.18	4	0.24	3	0.18	2	0.12	3	0.18	S7
Liaison with many development organizations	S8	0.07	4	0.28	3	0.21	3	0.21	3	0.21	3	0.21	3	0.21	S8
Decreased crop diversification	W1	0.06	3	0.18	2	0.12	2	0.12	1	0.06	2	0.12	3	0.18	W1
Excessive reliance on precipitation	W2	0.09	2	0.18	2	0.18	2	0.18	2	0.18	2	0.18	2	0.18	W2
Weaknesses of agriculture-related industries	W3	0.04	3	0.12	3	0.12	2	0.08	3	0.12	3	0.12	3	0.12	W3
Small and dispersed landholdings	W4	0.09	3	0.27	3	0.27	3	0.27	3	0.27	2	0.18	2	0.18	W4
Vulnerabilities in market infrastructure, storage, transportation, credit support, etc.	W5	0.04	3	0.12	2	0.08	2	0.12	2	0.08	3	0.12	3	0.12	W5

Code	Description														Code
W6	Lack of research facilities in agriculture	0.03	2	0.06	3	0.09	3	0.09	2	0.06	3	0.09	2	0.06	W6
W7	Lack of water conservation practices	0.09	2	0.18	2	0.18	2	0.18	2	0.18	2	0.18	2	0.18	W7
W8	Greater surface runoff	0.04	2	0.08	3	0.12	2	0.08	2	0.08	3	0.12	1	0.04	W8
W9	High intensity of drought	0.05	3	0.15	3	0.15	2	0.1	3	0.15	2	0.1	2	0.1	W9
W10	Reducing private sector investment in agriculture	0.03	2	0.06	2	0.06	2	0.06	2	0.06	2	0.06	3	0.09	W10
W11	Presence of a middleman	0.02	3	0.06	1	0.02	3	0.06	3	0.06	2	0.04	3	0.06	W11
O1	Horticultural and other cash crop opportunities	0.1	3	0.3	2	0.2	2	0.2	3	0.3	2	0.2	2	0.2	O1
O2	Increase in agriculture-related industries	0.04	3	0.12	2	0.08	3	0.12	3	0.12	3	0.12	3	0.12	O2
O3	Development of new technology	0.06	4	0.24	3	0.18	3	0.18	3	0.18	2	0.12	3	0.18	O3
O4	Increased government support for agriculture	0.07	3	0.21	3	0.21	3	0.21	3	0.21	3	0.21	3	0.21	O4
O5	Diversification of agriculture	0.07	2	0.14	3	0.21	3	0.21	3	0.21	3	0.21	2	0.14	O5
O6	Organic farming opportunities	0.04	2	0.08	2	0.08	2	0.08	2	0.08	3	0.12	1	0.04	O6
O7	Food processing sector outlook	0.04	1	0.04	1	0.04	2	0.08	3	0.12	2	0.08	2	0.08	O7
O8	Opportunities for agro-ecology practices	0.03	2	0.06	2	0.06	2	0.06	2	0.06	3	0.09	2	0.06	O8
O9	Fostering market facilities for alternative crops	0.05	3	0.15	2	0.1	3	0.15	3	0.15	3	0.15	3	0.15	O9

(Continued)

TABLE 23.3 (Continued)

Quantitative Strategic Planning Matrix Model for Appropriate Strategy Selection

SWOT	Key Factor	Weight	WT 1 AS	WT 1 TAS	WT 2 AS	WT 2 TAS	WT 3 AS	WT 3 TAS	WT 4 AS	WT 4 TAS	WT 5 AS	WT 5 TAS	WT6 AS	WT6 TAS	Key Factor
Crop production hazards	T1	0.05	2	0.1	2	0.1	2	0.1	2	0.1	2	0.1	2	0.1	T1
Crop price fluctuations in the market	T2	0.09	3	0.27	1	0.09	3	0.27	2	0.18	3	0.27	3	0.27	T2
Increased climate change	T3	0.1	3	0.3	2	0.2	2	0.2	3	0.3	2	0.2	2	0.2	T3
Low prices for some crops	T4	0.06	2	0.12	2	0.12	2	0.12	2	0.12	3	0.18	3	0.18	T4
Soil degradation	T5	0.07	2	0.14	2	0.14	2	0.14	2	0.14	3	0.21	2	0.14	T5
Participation scale	T6	0.05	2	0.1	3	0.15	2	0.1	2	0.1	2	0.1	3	0.15	T6
Declining interest in agriculture	T7	0.08	3	0.24	3	0.24	3	0.24	3	0.24	3	0.24	3	0.24	T7
STAS				5.4		4.88		5.01		4.96		5.01		4.84	
Priority				1		4		2		3		2		5	

TABLE 23.4
Ranking of Adaptation Options

Adaptation Options	Total	Total Score/100	Rank
Short-lived rice variety	4848	48.48	8
Surface water–based irrigation	4678	46.78	10
Postponement of planting	5333	53.33	3
Drought-tolerant rice cultivars	4867	48.67	7
Collection of rainwater	5045	50.45	5
Feed storage	5408	54.08	2
Development of water protection infrastructure	5108	51.08	4
Diversification of livelihood means	5816	58.16	1
Disease-resistant plant species	4806	48.06	9
Livestock diversification	4994	49.94	6

TABLE 23.5
Ranking of Agricultural Constraints

Agricultural Cnstraints	Total	Total Score/100	Rank
Traditional technique	5687	56.87	4
Limited land area	5176	51.76	10
Limited market access	4822	48.22	15
Water storage	5236	52.36	8
Low price of crops	5218	52.18	9
Missing market information	5042	50.42	11
Lack of watering	5748	57.48	1
Pests and diseases	4918	49.18	14
Low-yielding varieties	4923	49.23	13
Relying too much on rice	5697	56.97	2
Lack of alternative earning sources	5522	55.22	6
Lack of fertilizers and pesticides	5436	54.36	7
Repeated droughts	5689	56.89	3
Excessive reliance on rain-fed agriculture	5542	55.42	5
Improper land allocation	4984	49.84	12

shows that, out of the five factors, cultivation of short breeding rice varieties plays the most important role in neutralizing the drought factor.

23.4   CONCLUSION

Certain essential factors, such as adequate rainfall, favourable temperature conditions, bright sunlight, fertile soil, and so forth, are necessary for the successful expansion of agriculture in a region. This district cannot be classified as drought-prone if we adhere to the IMD criteria. Even though the study area receives a lot of rain, the coarse-grained soil has little ability to hold moisture, and the upper layer of the soil becomes dry quickly after a period of more intense rain, which moves as

Matrix		Short duration rice varieties	Promotion of agri-business	Water conservation	Labour migration	Strong governmental support	○	○	○	○	○	normalized principal Eigenvector
		1	2	3	4	5	6	7	8	9	10	
Short duration rice varieties	1	1	1	3	2	1	0	0	0	0	0	25.59%
Promotion of agri-business	2	1	1	3	3	1	0	0	0	0	0	27.59%
Water conservation	3	1/3	1/3	1	1	1/3	0	0	0	0	0	9.20%
Labour migration	4	1/2	1/3	1	1	1/3	0	0	0	0	0	10.05%
Strong governmental support	5	1	1	3	3	1	0	0	0	0	0	27.59%
0	6	0	0	0	0	0	1	0	0	0	0	0.00%
0	7	0	0	0	0	0	0	1	0	0	0	0.00%
0	8	0	0	0	0	0	0	0	1	0	0	0.00%
0	9	0	0	0	0	0	0	0	0	1	0	0.00%
0	10	0	0	0	0	0	0	0	0	0	1	0.00%

FIGURE 23.10 Analytical hierarchy process model for selecting the best criteria.

surface runoff along the slope of the area. Some of the factors that make it unsuitable to cultivate long-term crop varieties include the regional harsh climate, highly diversified soil, and undulating terrain. Therefore, a significant amount of surface runoff must be stopped for the area's proper agricultural development, and new short-term crop varieties must be introduced. Giving climatic data to the cultivating communities is extremely fundamental; otherwise, it is impossible to attain the desired level of productivity. This chapter is exceptionally urgent in that sense as it distinguishes farming advancement of the area according to eco-natural perspective. The occurrence of heat waves and droughts is a natural occurrence that cannot be avoided. However, droughts and heat waves can be somewhat mitigated by conserving excess water during the monsoon and using it in the fields during non-monsoon seasons. The study area is not found to be potential for the cultivation of long duration rice varieties; but some of the crucial steps like water conservation, cultivation of short duration crop varieties may provide the backbone to the overall economic boost up for the district.

REFERENCES

Adams, R. M., Hurd, B. H., Lenhart, S., & Leary, N. (1998). Effects of global climate change on agriculture: An interpretative review. *Climate Research*, 11, 19–30.

Adhikari, S. (2018). Drought impact and adaptation strategies in the mid-hill farming system of Western Nepal. *Environments*, 5, 101.

Afodu, O. J., Afolami, C. A., Akinboye, O. E., Ndubuisi-Ogbonna, L. C., Aye-Bello, T. A., Shobo, B. A., & Ogunnowo, D. M. (2019). Livelihood diversification and it's determinants on rice farming households in Ogun State, Nigeria. *African Journal of Agricultural Research*, 14 (35), 2104–2111.

Agada, B. I., & Obi, M. E. (2015). Rainfall characteristics at Makurdi, North-Central Nigeria. *Researchjournali's Journal of Agriculture*, 2 (8), 1–6.

Asha latha K. V., Gopinath, M., & Bhat, A. R. S. (2012). Impact of climate change on rainfed agriculture in India: A case study of Dharwad. *International Journal of Environmental Science and Development*, 3 (4), 368–371.

Babar, S., Gul, S., Amin, A., & Mohammad, I. (2015). Climate change: Region and season specific agriculture impact assessment (thirty year analysis of Khyber Pakhtunkhwa i.e. 1980–2010). *FWU Journal of Social Sciences*, 9 (1), 89–98.

Bello, O. B., Ganiyu, O. T., Wahab, M. K. A., Afolabi, M. S., Oluleye, F., Ig, S. A., Mahmud, J., Azeez, M. A., & Abdulmaliq, S. Y. (2012). Evidence of climate change impacts on agriculture and food security in Nigeria. *International Journal of Agriculture and Forestry*, 2 (2), 49–55.

Bera, S., Ammad, M., & Suman, S. (2017). Land suitability analysis for agricultural crop using remote sensing and GIS: A case study of Purulia district. *IJSRD International Journal for Scientific Research & Development*, 5 (6), 999–1004.

Berhanu, W., & Beyene, F. (2015). Climate variability and household adaptation strategies in Southern Ethiopia. *Sustainability*, 7, 6353–6375.

Chakraborty, S., Pandey, R. P., Chaube, U. C., & Mishra, S. K. (2013). Trend and variability analysis of rainfall series at Seonath River Basin, Chhattisgarh (India). *International Journal of Applied Sciences and Engineering Research*, 2 (4), 425–434.

Chatterjee, D. (2016). Strenghts-Weaknesses-Opportunities-Threats (SWOT) analysis of conservation agriculture. *Indian Journal of Hill Farming*, 29 (1), 18–23.

Cherian, B., & Khanna, V. K. (2018). Impact of climate change in Indian agriculture: Special emphasis to soybean (Glycine max (L.) merr.). *Open Access Journal of Oncology and Medicine*, 2 (4), 179–185.

Goswami, A. (2019). Identifying the trend of meteorological drought in Purulia District of West Bengal, India. *Environment and Ecology*, 37 (1B), 387–392.

Hasan, H. H., Razali, S. F. M., Muhammad, N. S., & Ahmad, A. (2019). Research trends of hydrological drought: A systematic review. *Water*, 11, 2252.

Hashemi, H., Berndtsson, R., Kompani-Zare, M., & Persson, M. (2013). Natural vs. artificial groundwater recharge, quantification through inverse modeling. *Hydrology and Earth System Sciences*, 17, 637–650.

Holman, I. P. (2006). Climate change impacts on groundwater recharge-uncertainty, shortcomings and the way forward? *Hydrogeology Journal*, 14 (5), 637–647.

Huho, J. M., Ngaira, J. K. W., Ogindo, H. O., & Masayi, N. (2012). The changing rainfall pattern and the associated impacts on subsistence agriculture in Laikipia East District, Kenya. *Journal of Geography and Regional Planning*, 5 (7), 198–206.

Iglesias, A., & Garrote, L. (2015). Adaptation strategies for agricultural water management under climate change in Europe. *Agricultural Water Management*, 155 (2015), 113–124.

Iglesias, A., Mougou, R., Moneo, M., & Quiroga, S. (2011). Towards adaptation of agriculture to climate change in the Mediterranean. *Regional Environmental Change*, 11 (Suppl 1), 159–166.

Jain, S. K., & Kumar, V. (2012). Trend analysis of rainfall and temperature data for India. *Current Science*, 102 (1), 37–49.

Javadinejad, S., Dara, R., & Jafary, F. (2020). Evaluation of hydro-meteorological drought indices for characterizing historical and future droughts and their impact on groundwater. *Resources Environment and Information Engineering*, 2 (1), 71–83.

Ji-kun, H., Jing, J., Jin-xia, W., & Ling-ling, H. (2014). Crop diversification in coping with extreme weather events in China. *Journal of Integrative Agriculture*, 13 (4), 677–686.

Kar, B., Saha, J., & Saha, J. D. (2012). Analysis of meteorological drought: The scenario of West Bengal. *Indian Journal of Spatial Science*, 3 (2), 1–11.

Kasim, Y. (2019). Impacts of livelihood assets on wellbeing of rural households in Northern Nigeria. *International Transaction Journal of Engineering, Management & Applied Sciences & Technologies*, 10 (13), 1–12.

Kaur, H., & Kaur, S. (2017). Climate change impact on agriculture and food security in India. *Journal of Business Thought*, 7, 35–62.

Khoshnodifar, Z., Sookhtanlo, M., & Gholami, H. (2012). Identification and measurement of indicators of drought vulnerability among wheat farmers in Mashhad Country, Iran. *Annals of Biological Research*, 3 (9), 4593–4600.

Kima, S. A., Okhimamhe, A. A., Kiema, A., Zampaligre, N., & Sule, I. (2015). Adapting to the impacts of climate change in the sub-humid zone of Burkina Faso, West Africa: Perceptions of agro-pastoralists. *Pastoralism: Research, Policy and Practice*, 5, 16.

Kolawole, O. D., Motsholapheko, M. R., Ngwenya, B. N., Thakadu, O., Mmopelwa, G., & Kgathi, D. L. (2016). Climate variability and rural livelihoods: How households perceive and adapt to climate shocks in the Okavango Delta, Botswana. *Weather, Climate and Society*, 8, 131–145.

Korkmaz, N. (1988). The estimation of groundwater recharge from water level and precipitation data. *Journal of Islamic Academy of Sciences*, 1 (2), 87–93.

Kubicz, J., Kajewski, I., Kajewska-Szkudlarek, J., & Dąbek, P. B. (2019). Groundwater recharge assessment in dry years. *Environmental Earth Sciences*, 78, 555.

Kumar, C. P. (1997). Estimation of natural ground water recharge. *ISH Journal of Hydraulic Engineering*, 3 (1), 61–74.

Kumar, P., & Nain, M. S. (2013). Agriculture in India: A SWOT analysis. *Indian Journal of Applied Research*, 3 (7), 4–6.

Kumar, R., & Gautam, H. R. (2014). Climate change and its impact on agricultural productivity in India. *Journal of Climatology & Weather Forecasting*, 2 (1), 1–3.

Kumar, S., Subash, S., Baindha, A., & Jangir, R. (2017). Perceived constraints of farmers in indigenous cattle dairy farming in Rajasthan, India. *Journal of Animal Health and Production*, 5 (4), 172–175.

Kundzewicz, Z. W., & Kaczmarek, Z. (2000). Coping with hydrological extremes. *IWRA Water International*, 25 (1), 66–75.

Lacombe, G., & McCartney, M. (2014). Uncovering consistencies in Indian rainfall trends observed over the last half century. *Climate Change*, 123, 287–299.

Lei, Y., Liu, C., Zhang, L., & Luo, S. (2016). How smallholder farmers adapt to agricultural drought in a changing climate: A case study in southern China. *Land Use Policy*, 55 (2016), 300–308.

Lerner, D. N., Issar, A. S., & Simmers, I. (1990). Groundwater recharge: A guide to understanding and estimating natural recharge. *International Association of Hydrogeologists, Kenilworth*, 8, 345.

Lindenberg, M. (2002). Measuring household livelihood security at the family and community level in the developing world. *World Development*, 30 (2), 301–318.

Liu, L., & Basso, B. (2020). Impacts of climate variability and adaptation strategies on crop yields and soil organic carbon in the US Midwest. *Plos One*, 15 (1), e0225433.

Liu, T. T., McConkey, B. G., Ma, Z. Y., Liu, Z. G., Li, X., & Cheng, L. L. (2011). Strengths, weaknesses, opportunities and threats analysis of bioenergy production in marginal land. *Energy Procedia*, 5 (2011), 2378–2386.

Lobell, D. B., & Gourdji, S. M. (2012). The influence of climate change on global crop productivity. *Plant Physiology*, 160, 1686–1697.

Longobardi, A., & Villani, P. (2009). Trend analysis of annual and seasonal rainfall time series in the Mediterranean area. *International Journal of Climatology*, 30 (10), 1538–1546.

Loon, A. F. V., & Laaha, G. (2015). Hydrological drought severity explained by climate and catchment characteristics. *Journal of Hydrology*, 526 (2015), 3–14.

24 Assessment and Possible Mitigation Strategy of Climate Change and Urban Heat Island Using GIS and Remote Sensing
A Case Study of Agartala City

Sabirul Sk, Surajit Debarma, Asraful Alam,
Lakshminarayan Satpati, and R. Jagannathan

24.1 INTRODUCTION

Urban regions are facing a serious problem as a result of the worldwide phenomena of climate change, which is affecting their social, economic, and environmental systems (Corburn, 2009). One of the major effects of climate change is the urban heat island (UHI) effect, which is characterized by hotter temperatures in cities than in the nearby rural areas (Golden, 2004). The health, happiness, and sustainability of metropolitan populations are seriously impacted. India's northern city of Agartala is not exempt from these difficulties. The city is becoming more susceptible to the effects of climate change and the UHI effect as it rapidly urbanizes. A thorough examination is essential to properly address and minimize these problems. The purpose of this study is to analyze the patterns of climate change and the impact of UHI in Agartala City using geographic information systems (GIS) and remote sensing technologies, as well as to suggest suitable mitigating measures.

Large or semi-large urban areas like cities attract more population because of the opportunity the city provides (Emmanuel & Krüger, 2012). It means an increase in human population, urban expansion, increase in pollution, energy use, etc. all of which highly contributes to the temperature of the area causing urban heat effect and in turn huge environmental and human health issues (Zhou & Chen, 2018). A UHI is a much warmer location in a major urban area or metropolitan city caused by human activities, with surrounding rural areas remaining at normal temperatures (Zhou & Chen, 2018). At night, the temperature differential is greater than during the day (Goward, 1981). This is especially noticeable when the breezes in the vicinity are weak. This temperature differential is particularly visible in the summer and winter. The conversion of land surfaces from dirt to concrete metal is the primary cause of the UHI effect (Chatterjee et al., 2019).

The UHI effect can be caused by a variety of factors. Dark surfaces, for example, absorb more solar heat and radiation, which is why urban roads, pavements, and building roofs experience higher temperatures than their counterparts in suburban and rural locations (Peng et al., 2012). The reason for this is that common urban pavement and roof materials, such as concrete and asphalt, have vastly differing thermal bulk qualities such as heat capacity and thermal conductivity (Mohajerani et al., 2017). As a result, urban regions frequently experience higher temperatures than rural areas (Pandey et al., 2012). Land surface temperature (LST) is the radiative surface temperature of the land that is acquired from solar radiation (Li et al., 2013). LST is a measurement of the thermal radiation emitted by the land surface, where incoming solar energy interacts with objects on the

DOI: 10.1201/9781003377825-27

ground and heats the surface, or the surface of the canopy in vegetated regions (Yue et al., 2019). LST is composed of temperatures of bare soil and vegetation (Li et al., 2013).

To study UHIs, the land surface temperature is required to associate the temperature variation in vegetation-covered areas and non-vegetation-covered areas, and also areas that are highly urbanized and less urbanized (Arnfield, 2003). Based on this temperature variation UHI is understood (Kalma et al., 2008). The twenty-first century marked one of the major global challenges, especially with the recent pandemic affecting the livelihood of the entire population of the globe (Alqasemi et al., 2021). A huge change was seen environmentally with less pollution, clear air because of few vehicles on the road, and less stress on fossil fuels, etc. that came with more than a year of lockdown protocol because of COVID-19 (Chakraborty et al., 2021). This pandemic brought a nationwide crisis, with chaos and loss of life, etc. Most people would argue that it rejuvenated the weather and climate of the area, which is good from an environmental perspective (Ali et al., 2021). Since 2022, things started to get back to normal and the rate of temperature has hiked up again causing irritation, heat rashes, and even a few deaths. If this keeps on going unapproached, this urban heat will become one of the major crises after COVID (Liu et al., 2022).

A UHI occurs when cities replace natural land surfaces with dense concentrations of pavements, buildings, and other objects that absorb and retain heat for a longer period (Gago et al., 2013; Mohajerani et al., 2017). This causes a huge abnormal temperature variation between the highly built-up area and outskirt vegetated area. Rising temperature is an inevitable result of urbanization (Argüeso et al., 2014). It is highly observed when an area is remodeled to fit in with the growing urban development (Mohajerani et al., 2017). Urban heat is not going anywhere and based on the current trend and it is only going to get worst (Yang et al., 2020). The temperature in urban areas is just going to rise over time (Yang et al., 2020). This begs for the need to study the area for its temperature rise, population changes, loss of vegetation cover, etc. Only then can we take effective steps to mitigate it and develop new strategies that will be helpful to regulate urban heat (Hu et al., 2020).

This study aims to analyse UHI to understand the heat island distribution effect of the study area using an open-source platform and provide an effective mitigation strategy. The objectives of this study are to analyze UHI in the study area, to identify the temporal temperature variation of the study area, and to provide a mitigation strategy based on the current scenario of the urban city. Software used for the study QGIS and Google Earth Pro. GIS allows for the integration, visualization, and analysis of spatial data, enabling researchers to assess the complex interplay between climate variables, land-use patterns, and urban development (Tran et al., 2017). Remote sensing, on the other hand, provides valuable data on surface temperature, vegetation cover, and other relevant parameters by utilizing satellite imagery and aerial photographs (Yang & Lo, 2002).

24.2 STUDY AREA

Agartala is the only major city in Tripura. It extends from 23°46′41.00″ N to 23°54′36.00″ N latitude, and 91°13′45.00″ E to 91°20′33.00″ E longitude (Figure 24.2). The elevation of the city is 12.8 m above mean sea level (Mitra et al., 2022). One of the largest cities in north-east India is Agartala, which serves as the capital of the Indian state of Tripura (Mitra et al., 2022). Agartala is made up of the terms *agar*, a priceless tree from the species *Aquilaria* that is used in perfume and incense, and the suffix *tala*, which describes the abundance of agarwood trees in the region. The legend of King Raghu, who tied the feet of his elephant to an agar tree on the banks of the Lauhitya River, refers to the agar tree. Tripura's government is headquartered in Agartala. The Agartala Municipal Corporation governs the city, which spans 76.504 km2 and has a population density of 5,200 people per square kilometer (Mallik et al., 2021). It is close to the Bangladesh border, on the banks of the Howrah River. The peoples of Bangladesh and Tripura have a considerable mutual economic and social trade. Every year, an extensive number of populations migrate to Tripura from the Sylhet Valley, Bangladesh, generally settling in

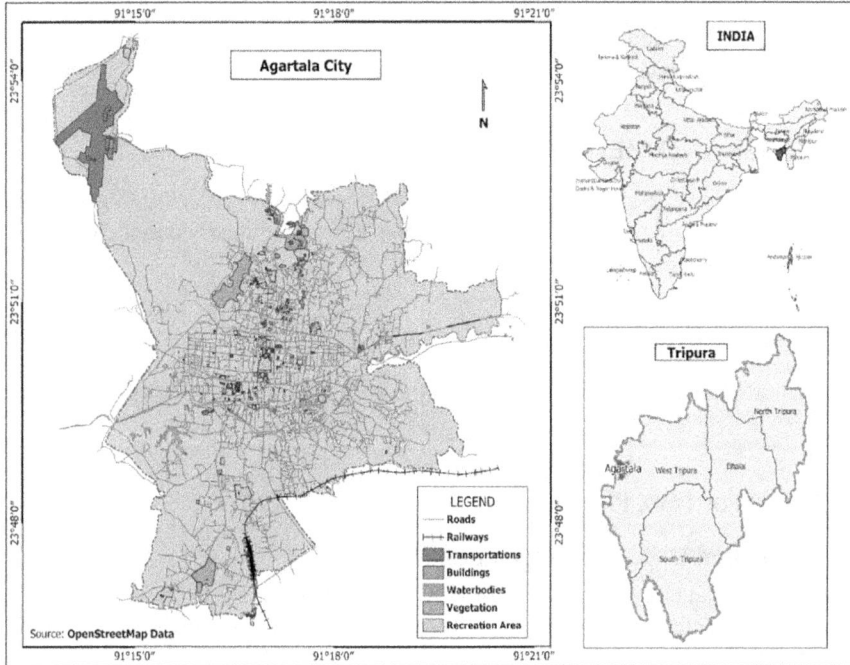

FIGURE 24.1 Study area map.

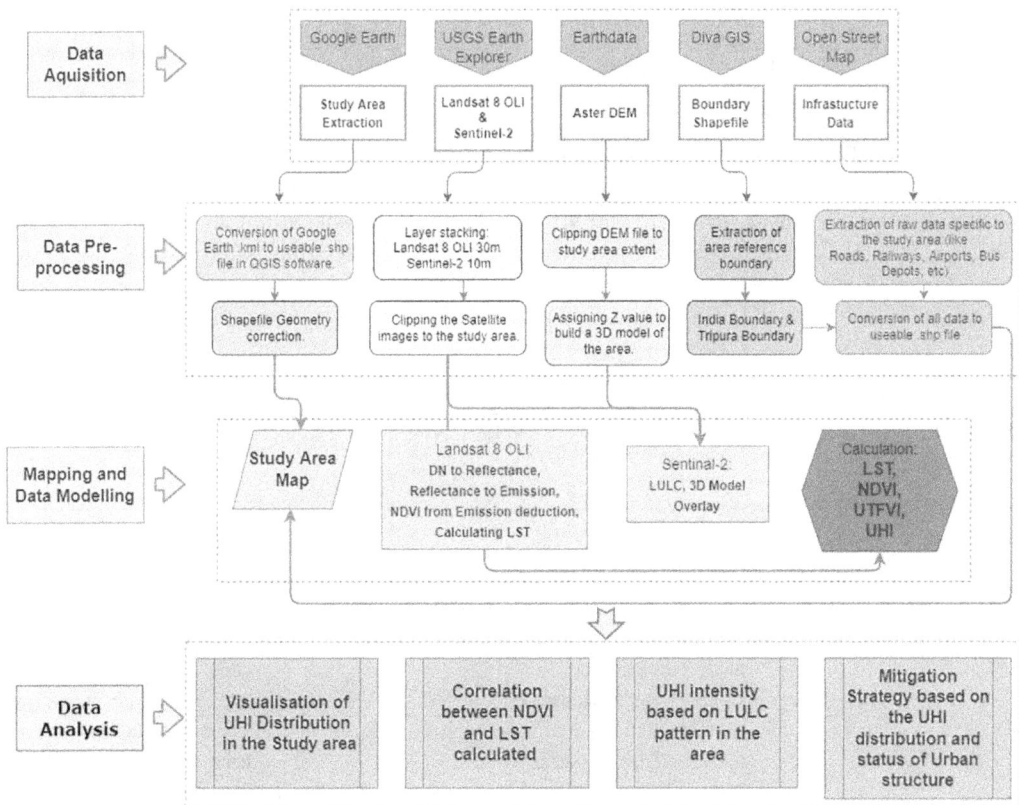

FIGURE 24.2 Flowchart of methodology.

the metropolitan area and bringing with them a large chunk of their culture and heritage. This region has a subtropical, hot, and humid environment with average maximum summertime temperatures of 29°C and average lowest wintertime temperatures of 10°C (Mallik et al., 2022).

24.3 DATABASE AND METHODOLOGY

The methodological flowchart for the assessment and possible mitigation strategy of UHIs using GIS and remote sensing in Agartala City would involve a series of clearly defined steps. Here is the flowchart that defined the overall steps of the study.

24.3.1 DATA COLLECTION

Different types of data have been collected for the study. Table 24.1 shows the sources of the database.

24.3.2 METHODS OF DATA PROCESSING

24.3.2.1 Normalized Difference Vegetation Index

Using remote sensing data, the Normalized Difference Vegetation Index (NDVI) was used to determine the amount of vegetation on the earth's surface and to determine whether or not there was any urban vegetation (Weng & Lo, 2001). NDVI has been calculated using Equation 24.2:

$$NDVI = \frac{(\text{NIR} - \text{Red})}{(\text{NIR} + \text{Red})} \tag{24.2}$$

where *NIR* is the reflection in the near-infrared spectrum, and *Red* is the reflection in the red range of the spectrum.

24.3.2.2 Normalized Difference Built-up Index

The Normalized Difference Built-up Index (NDBI) is a spectral index frequently used in remote sensing and image analysis to locate and measure built-up regions within an image or a region (Zha et al., 2003). It is estimated using remotely sensed data, including satellite imaging, in the near-infrared (NIR) and shortwave infrared (SWIR) bands.

The NDBI is determined using Equation 24.3:

$$NDVI = \frac{(\text{SWIR} - \text{NIR})}{(\text{SWIR} + \text{NIR})} \tag{24.3}$$

TABLE 24.1
Database and source

Sl. No.	Data Type	Sources
1	Study Area Shapefile	https://earth.google.com
2	Landsat 8 and 9, Sentinel-2	https://earthexplorer.usgs.gov
3	ASTER DEM	www.earthdata.nasa.gov
4	India and Tripura boundary	https://diva-gis.org
5	Roads, rivers, and transportation	www.openstreetmap.org
6	Population data	www.macrotrends.net

In this equation, *SWIR* refers to the reflectance or radiance values in the shortwave infrared band, while *NIR* represents the values in the near-infrared band.

24.3.2.3　Land Surface Temperate Calculation

1　Conversion from DN to TOA radiance

It is calculated (Twumasi et al., 2021) using Equation 24.4:

$$L_\lambda = M_L Q_{cal} + A_L - O_i \tag{24.4}$$

where
　　M_L = the band-specific multiplicative rescaling factor
　　Q_{cal} = the Band 10 image
　　A_L = the band-specific additive rescaling factor
　　O_i = the correction for Band 10
For
　　Landsat 8 Data, Formula: Float(0.00033420 * "B10.tif" + 0.1)
　　Landsat 9 Data, Formula: Float(0.00038* "B10.tif" + 0.1)

2　Conversion of TOA radiance to at-sensor temperature

It is determined (Twumasi et al., 2021) by Equation 24.5 that

$$BT = \frac{K_2}{In\left[\left(\dfrac{k_1}{L\lambda}\right)+1\right]} - 273.15 \tag{24.5}$$

where
　　k_1 and K_2 = specific thermal conversion constants from the metadata.
　　(For obtaining the results in Celsius, the radiant temperature is reserved by adding the absolute zero [approximately −273.15°C].)
For, Landsat 8 Data, Formula: Float([1321.0789/Ln{774.8853/"rad"} + 1] − 273.15)
　　　Landsat 9 Data, Formula: Float([1329.2405/Ln{799.0284/"rad"} + 1] − 273.15)

3　Land surface emissivity

The term "surface emissivity" refers to the capacity of a surface to convert thermal energy into radiant radiation. Land surface emissivity (LSE) (ε) is one of the crucial variables for accurately obtaining LST from remotely sensed images. It is calculated (Sekertekin & Bonafoni, 2020) using Equation 24.6:

$$\varepsilon = 1.009 + 0.0047 \times (NDVI) \tag{24.6}$$
$$\text{Formula: Float}(1.009 + [0.0047 * Ln\{\text{"NDVI"}\}])$$

4　Retrieving the emissivity-corrected LST (land surface temperature) is computed (Twumasi et al., 2021) as follows:

$$T_s = \frac{BT}{\left\{1+\left[\left(\dfrac{\lambda BT}{p}\right)In\varepsilon_\lambda\right]\right\}} \tag{24.7}$$

where
   T_s = LST in Celsius (°C)
   BT = at-sensor BT (°C)
   λ = the wavelength of emitted radiance

(For which the peak response and the average of the limiting wavelength [λ= 10.895 will be used] are the emissivity calculated.)

Formula: ("bt.tif"/[1+{0.00115 * "bt.tif"/1.4388} * Ln{"E"}])

5 Convert LST data from Celsius (Semenza et al., 1996) to Kelvin

Formula: Float("LST" + 273.15)   (24.8)

6 Extraction of UTFVI and UHI temperature range

   a. The Urban Field Thermal Variance Index (Kafy et al., 2021)

$$\text{UTFVI} = \frac{T_s + T_{mean}}{T_{mean}}$$ (24.9)

where
   $UTFVI$ = Urban Thermal Field Variance Index
   T_s = LST (°k)
   T_0 = Mean LST (°k)

Formula: Float(["K"—T_{mean}]/T_{mean})

   b. The Urban Heat Island (Halder et al., 2021)

$$\text{UHI} = \frac{T_s + T_{mean}}{SD}$$ (24.10)

where
   UHI = urban heat island
   T_s = LST (°K)
   T_0 = Mean LST (°K)

The UHI analysis of Agartala is carried out using a modified version of the USGS algorithm for automated mapping of LST using LANDSAT satellite data.
   The process is synthesized in six steps and two more to extract the UHI value as follows:

1 Calculation of TOA (Top of Atmospheric) spectral radiance.

TOA = ML * Qcal + AL

where

M_L = the band-specific multiplicative rescaling factor

Qcal = corresponds to band 10

A_L = the band-specific additive rescaling factor

TOA = 0.0003342 * "Band 10" + 0.1

2 TOA to brightness temperature conversion

$$BT = (K2/[\ln \{K1/TOA\} + 1]) - 273.15$$
$$BT = (1321.0789/\ln [\{774.8853/\text{"TOA"}\} + 1]) - 273.15$$

3 Calculate emissivity (ε)

$$\varepsilon = 1.009 + (0.0047 * \ln[\text{"NDVI"}])$$

4 Calculate the LST

$$LST = (BT/[1 + \{0.00115 * BT/1.4388\} * \ln\{\varepsilon\}])$$

5 Convert the LST data from °C to K

$$K = LST + 273.15$$

6 Get K Mean and SD

Open Processing Toolbox > Raster Analysis > Raster Layer Statistics

7. Calculate the UHI

$$UHI = (K - Tmean)/SD$$

24.4 RESULTS AND DISCUSSION

Global temperature rise is not only an academic debate anymore but an inevitable natural trend. In urban cities especially, new developing ones tend to have less vegetation cover due to clearance for urban expansion making them vulnerable to urban heat (Corburn, 2009). Although vegetation cover is the major regulator of urban heat, many other factors are responsible for increasing the heat (Loughner et al., 2012). This global warming and urban heat already pose serious health and environmental challenges (Alcoforado & Andrade, 2008). Currently, with growing urbanization and the population migrating from rural to city life, these temperature issues have gotten from bad to worst (Veena et al., 2020).

24.4.1 LAND USE AND LAND COVER

The study area has been classified using the supervised classification method of 2022 Sentinel-2 data, categorized into five LULC categories (agriculture land, barren land, built-up, vegetation, and water body). The study area is approximately 16,621 ha, of which the built-up area covers the highest, about 11,618 ha, i.e. nearly, 70% of the total study area. Agricultural land covers approximately, 3,377 ha which are about 20.32%, and water bodies cover about 129.31 ha, i.e. nearly 0.78% (Table 24.2).

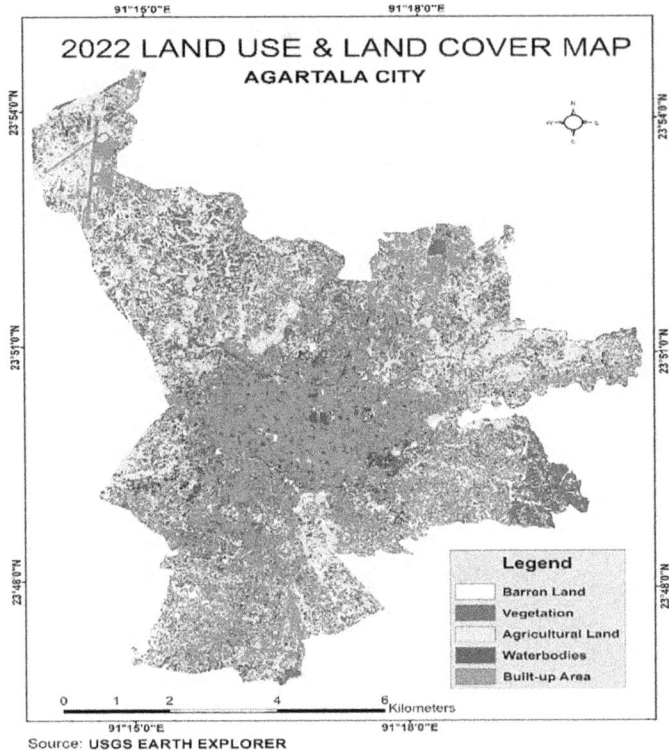

FIGURE 24.3   Land-use/land-cover map of the year 2022.

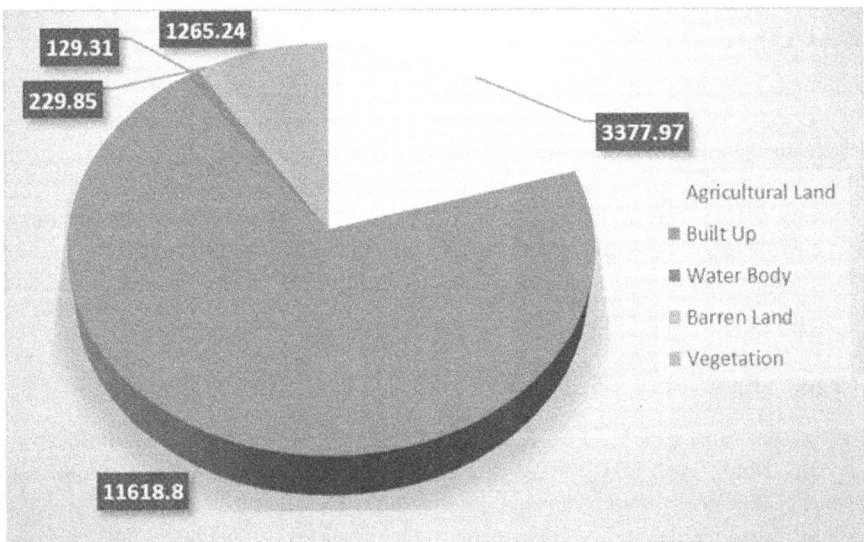

FIGURE 24.4   Piechart showing land-use/land-cover category.

TABLE 24.2
2022 Land Use/Land Cover in Percentage.

Sl. No.	Class Name	Area	
		In Hectares	In Percentage (%)
1	Agriculture land	3,377.97	20.32
2	Barren land	229.85	1.38
3	Built-up	11,618.8	69.90
4	Vegetation	1,265.24	7.61
5	Water body	129.31	0.78
Total area =		16,621.17	

24.4.2 Temperature and Humidity

The recorded temperature of the area averages about 30°C, highest and about 16°C, lowest, where, the highest temperature recorded is 38°C in September 2016, and the lowest is 4°C in January 2013. The highest temperature is recorded in April, May, June, and September averaging about 35°C. During this period the area is met with sunny skies and sudden rainfall showers for a short period (Table 24.3). The humidity also averages 70% to 85% based on the time of the day as well as other weather phenomena. The lowest temperature, on the other hand, is recorded starting from the end of October to the first week of March with December, January, and February recording the lowest temperatures of all.

The temperature variation and humidity of the years under study are shown using a line graph and bar graph, respectively, in Figures 24.5 and 24.6. Although the temperature and humidity graphs are shown monthly for four years, the graphs for 2022 are not complete due to a lack of data. It only shows data up to May.

A UHI is an urban climate effect wherein urban areas have higher surface or air temperatures than nearby rural areas. This phenomenon is receiving attention from all over the world due to its detrimental effects on both human health and the environment, such as cardiovascular mortality and morbidity, vegetation phenology, and energy consumption (Li et al., 2020). The heat island effect, which is also anticipated to get worse in the future, makes air temperatures brought on by climate change worse in metropolitan areas (Watkins et al., 2007).

The LST of the four years analyzed shows the lowest and the highest temperature readings in April (Dey et al., 2013). Based on visual inspection it is visible that the LST distribution of the year 2022 is more spread over the main built-up area. A huge temperature rise is also observed at the top of the map from the airport of the state. The map clearly shows that the temperature gradually rises and expands slowly from 2013 to 2022 indicating the urban expansion and because of which the temperature rise occurred. Because of increasing temperature global warming is taking place (Shaver et al., 2000). Changing climate is a very challenging issue of the increasing urban heat (Mirzaei & Haghighat, 2010). Increasing temperatures and heat islands help to create severe health issues in this region (Li & Bou-Zeid, 2013).

Figure 24.8 shows the NDVI distribution in the area from 2013 to 2022. Although there is a minor change in the reduction of vegetation in the area, if observed, the NDVI of 2022 has reduced in the center portion of the map indicating the removal of vegetation in the area for built-up. As seen, the map 2013 shows a distinct NDVI value indicating the availability of vegetation in the area compared to other years.

TABLE 24.3
Monthly Temperature and Humidity Data (three-year interval, 2013–2022)

2013

Sl. No.	Month	Temperature (in °C)			Humidity (in %)		
		Highest	Lowest	Average Temperature	Highest	Lowest	Average Humidity
1	January	28	4	16	100	33	76
2	February	32	10	22	100	23	69
3	March	35	13	27	95	17	66
4	April	36	19	28	98	17	70
5	May	35	20	27	100	55	82
6	June	36	25	30	97	51	81
7	July	35	25	29	97	58	81
8	August	34	25	28	98	62	85
9	September	36	24	28	99	51	85
10	October	35	20	27	100	51	85
11	November	32	13	23	98	27	76
12	December	30	9	19	98	31	79

2016

Sl. No.	Month	Temperature (in °C)			Humidity (in %)		
		Highest	Lowest	Average Temperature	Highest	Lowest	Average Humidity
1	January	29	8	18	100	27	77
2	February	34	10	24	100	28	73
3	March	35	16	27	100	24	71
4	April	36	20	30	98	52	78
5	May	37	21	28	100	1	81
6	June	36	24	30	98	48	80
7	July	35	24	28	100	63	86
8	August	36	24	30	100	54	81
9	September	38	25	29	98	50	85
10	October	35	20	28	100	47	83
11	November	34	15	23	100	5	82
12	December	30	12	21	100	47	83

2019

St. No.	Month	Temperature (in °C)			Humidity (in %)		
		Highest	Lowest	Average Temperature	Highest	Lowest	Average Humidity
1	January	30	9	18	100	33	75
2	February	32	10	21	100	29	74

2022

Sl No.	Month	Temperature (in °C)			Humidity (in %)		
		Highest	Lowest	Average Temperature	Highest	Lowest	Average Humidity
1	January	28	10	18	100	43	84
2	February	31	10	20	100	28	75

3	March	36	14	26	98	32	71
4	April	37	15	30	98	49	74
5	May	37	22	28	100	30	82

3	March	35	11	25	100	30	73
4	April	37	18	28	100	35	75
5	May	37	21	29	100	54	80
6	June	37	23	30	100	59	83
7	July	36	18	29	98	59	85
8	August	37	26	30	100	55	83
9	September	35	23	29	98	56	84
10	October	34	22	27	98	49	86
11	November	32	16	24	100	48	83
12	December	30	10	19	100	45	84

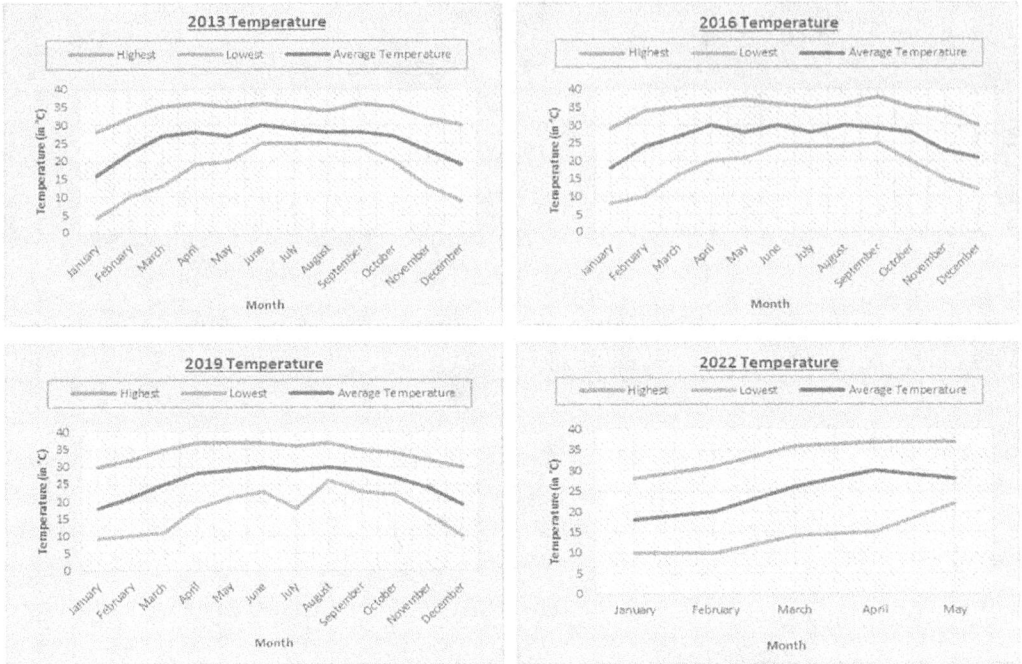

FIGURE 24.5   Line graph showing temperature variation.

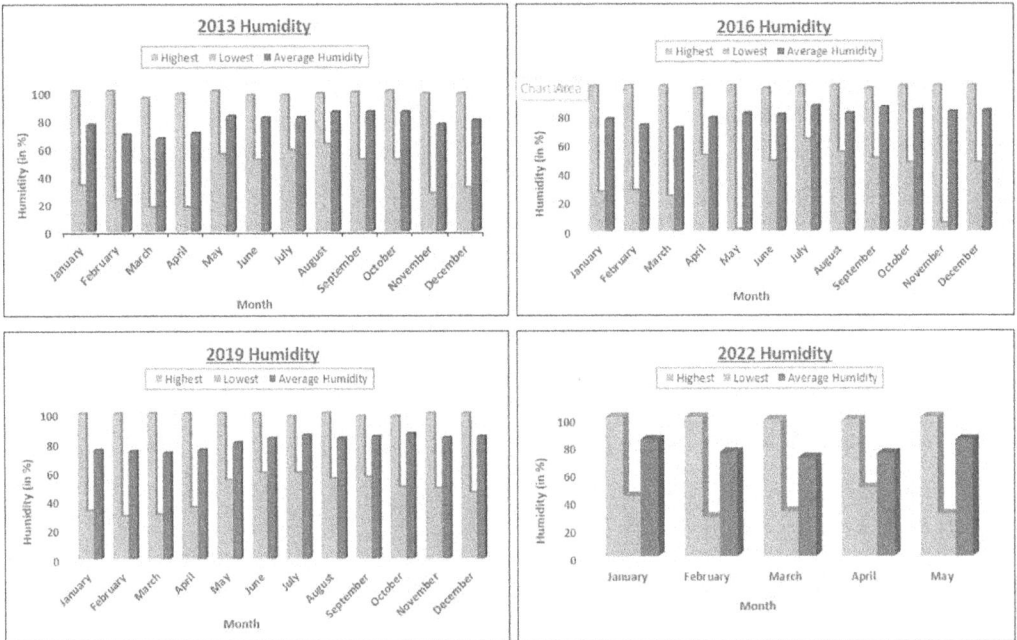

FIGURE 24.6   Bar graph showing humidity variation.

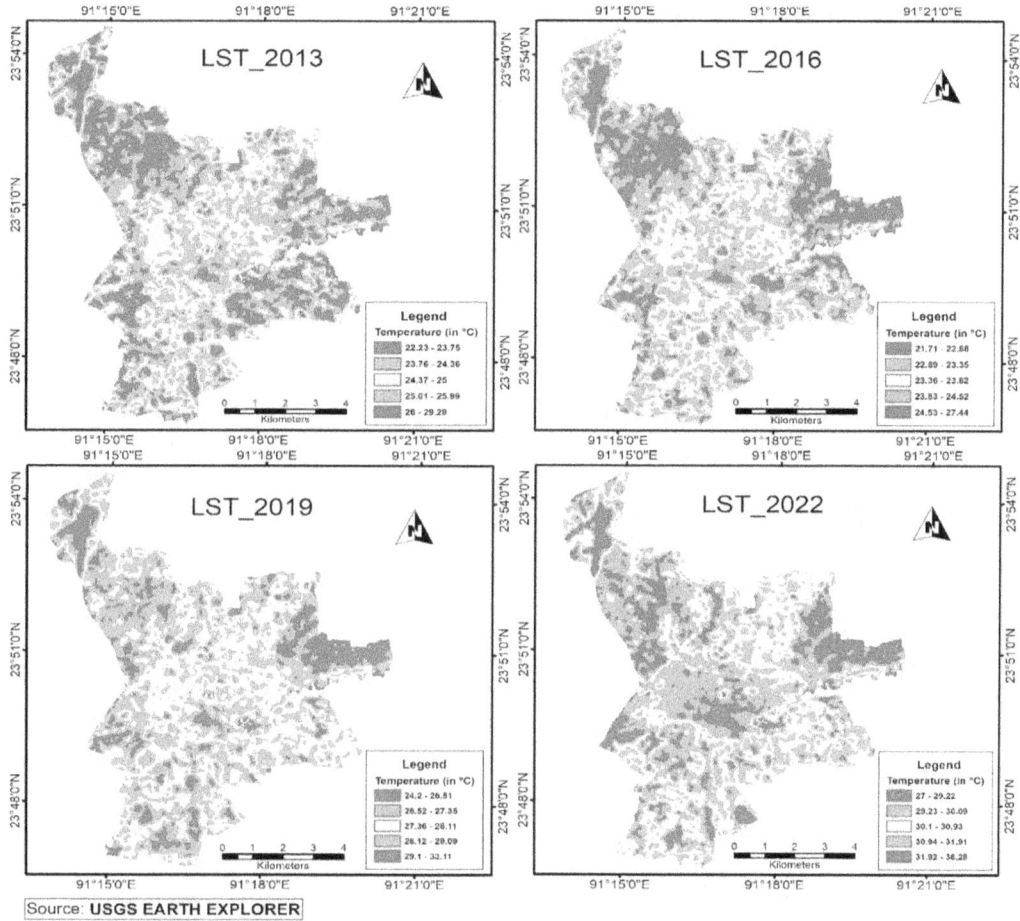

FIGURE 24.7 Temporal land surface temperature comparison map (three-year interval).

After preparation of the correlation trend between LST and NDVI, it shows that the LST has an inverse correlation trend with LST in all four years. This means that the temperature rises with the lessening of NDVI, and vice versa. The four correlation trend is shown using different color. The trend of the correlation is plotted over the scatter plot of every year.

The NDBI time series data are calculated using Landsat data bands 5 and 6. Figure 24.10 shows the NDBI distribution. The year 2022 has the most densely proportionate built-up area. These densely built buildings are trapping the air and block the airflow creating huge temperatures with high humidity. Based on this map it is clear that as long as the built-up areas keep expanding, the temperature will also increase with it. To statistically analyze LST across the seasons for UHI investigations, NDBI can be employed as a companion to the commonly utilized NDVI measure. NDBI is a reliable indication of UHI impacts (Li & Liu, 2008). The LST and NDBI relationship's trend varies according to the season, which amply proves that the two variables' relationships rely on one another (Mathew et al., 2017).

Here, the scatter plot clearly shows that the correlation between the LST and NDBI is of positive correlation. This means with the increase in built-up area, the temperature of the area will also increase. The expansion of built-up areas is one of the main reasons for UHI.

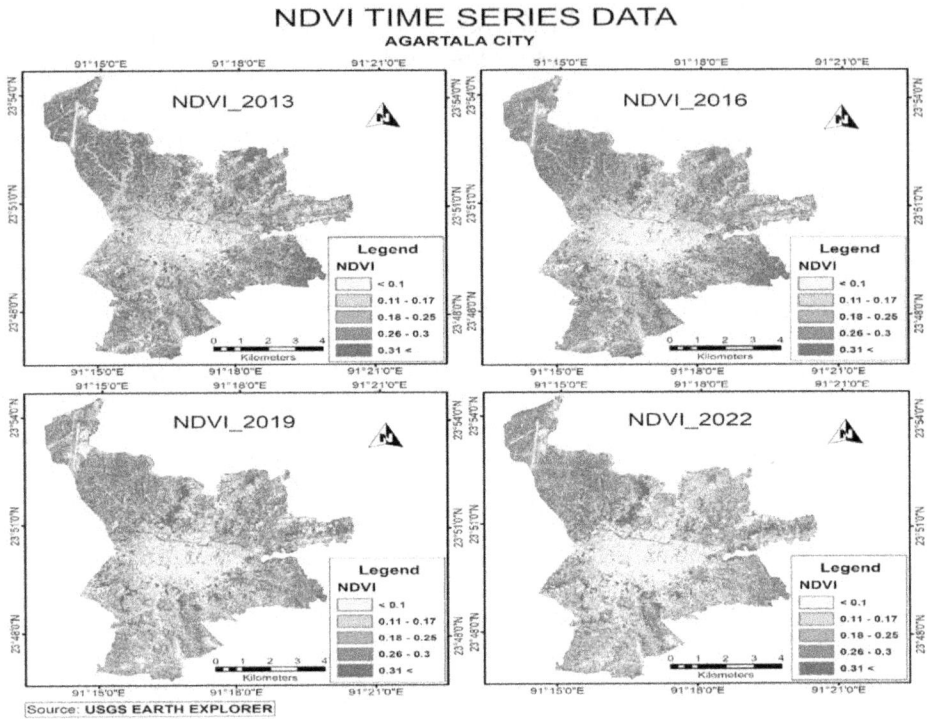

FIGURE 24.8   Temporal Normalized Difference Vegetation Index comparison map (three-year interval).

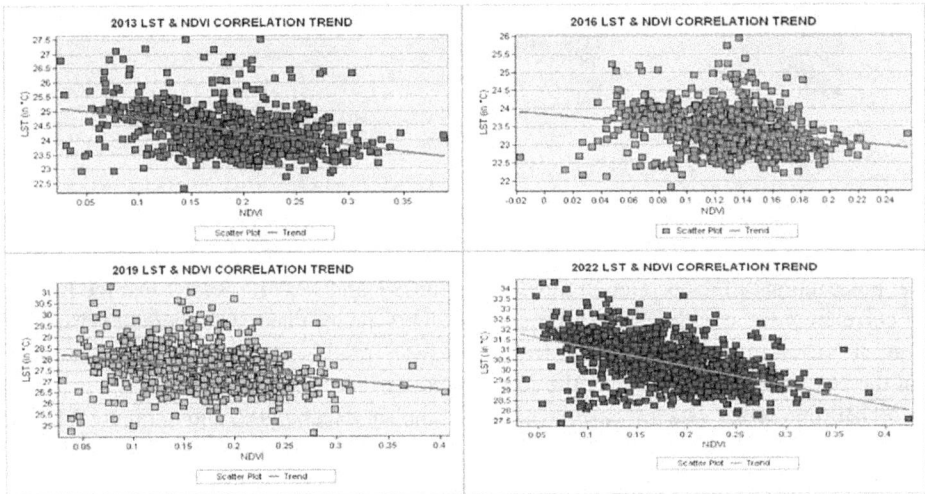

FIGURE 24.9   Temporal correlation trend between land surface temperature and Normalized Difference Vegetation Index (three-year interval).

In the time series map in Figure 24.12, four satellite images have been taken of different time frames (three years apart), and UHI is calculated from it using a modified US Geological Services thermal equation. Upon inspection, it is clear that the heat island was a small issue back in 2013. As it progressed toward further years the UHI seems to increase its radius, and the rate of heat distribution is massive (Dos Santos et al., 2017). Looking at the 2022 UHI map, it is clear that we are not so far from a massive crisis if this issue is not tackled properly.

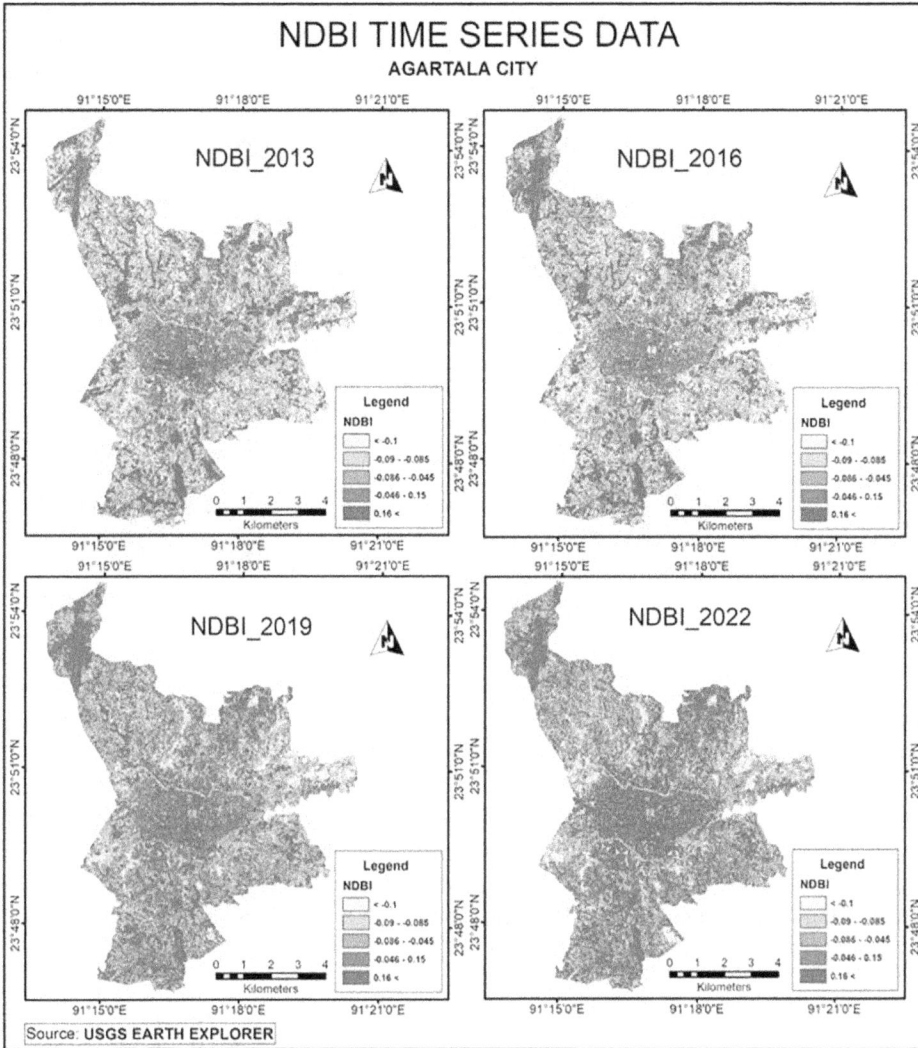

FIGURE 24.10 Temporal Normalized Difference Built-up Index comparison map (three-year interval).

In Figure 24.13, 2022 UHI and LULC, a clear distinction can be made as the map shows the very high-temperature recording falls under an area where there are more human activities. The airport area at the top of the map is also shown radiating high temperature because the large runway made of concrete materials is radiating a high amount of heat, while the periphery areas around the main city are showing low to very low-temperature change. That is because these areas are covered with more vegetation, and many uncultivated wetlands help the outskirt area to cool down (Du et al., 2016).

Based on this analysis, it is clear that vegetation plays its part in keeping the temperature of the area cool while, city areas, areas where the most topsoil is removed and replaced with hardened concrete, are causing the urban temperature rise (Mohajerani et al., 2017).

24.4.3 POSSIBLE MITIGATION STRATEGY FOR THE STUDY AREA

24 4.3.1 Green Roofing and Wall

A green roof and wall are a layer of vegetation planted over a waterproofing system that is installed on top of a flat or slightly sloped roof. Green roofs are also known as vegetative or eco-roofs

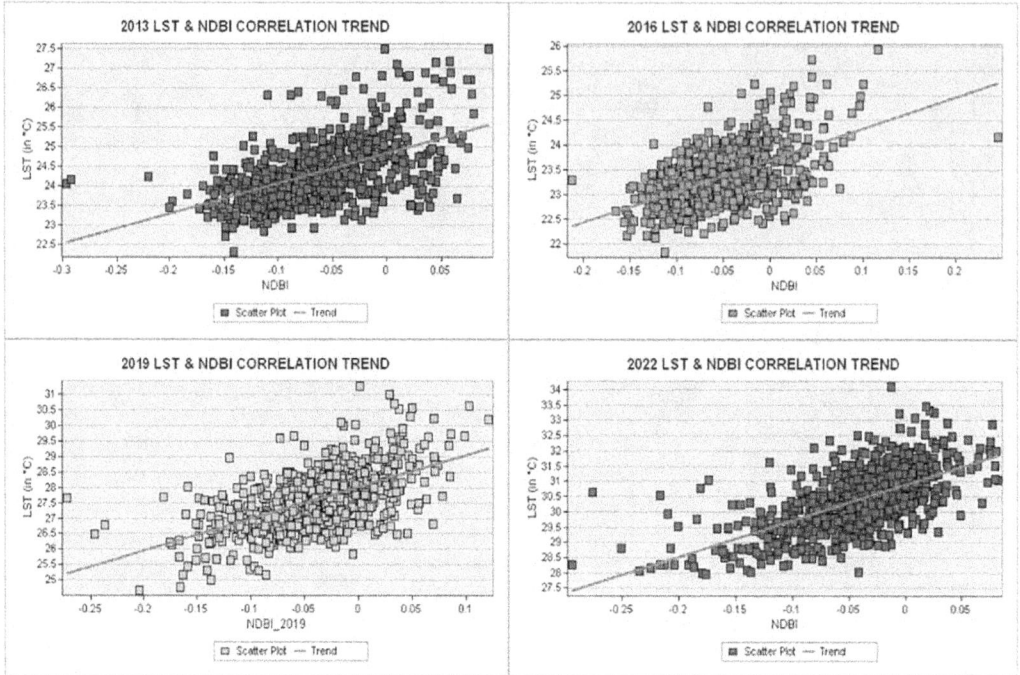

FIGURE 24.11 Temporal correlation trend between land surface temperature and Normalized Difference Built-up Index (three-year interval).

(Alexandri & Jones, 2008). This method is cost-optimized but requires constant maintenance and is most opted for because it gives the building a distinct aesthetic look (Santamouris, 2013). The greens need to be removed once in a while after several years and the concrete walls need to be inspected for damage from the roots. The walls are then maintained and left uncovered for several months and then new greens are attached again.

24.4.3.2   Architecture Based on Termite Mound

This is an architecture model taken from termite and ant mounds. This building model adopts viaducts that help regulate temperature. Latest skyscrapers or very tall buildings mostly adopt this method for keeping the tower cool. This process requires lots of planning, and the designs change based on the architect's perspective (Wang et al., 2016).

24.4.3.3   Urban River Restoration and Revitalization

The phrase "river restoration" covers a wide range of biological, physical, spatial, and managerial practices. These are intended to bring back the river system's natural functioning and status to promote landscape development, recreation, flood control, and biodiversity (Hathway & Sharples, 2012).

The research area is split roughly in half by two rivers that cross through it. Locals now frequently throw waste goods into these rivers as a result of local markets and enterprises, which harms the river's environment and pollutes the river. If the river is restored, it will flow, transport cold air, and lower the temperature in the area (Blau et al., 2018).

24.4.3.4   Whitening of Flat Concrete Surface

Countries like the United States and other European states adopt the method of coating flat concrete surfaces like roofs, pavements, and roads with lighter colors having special mold for heat reflection

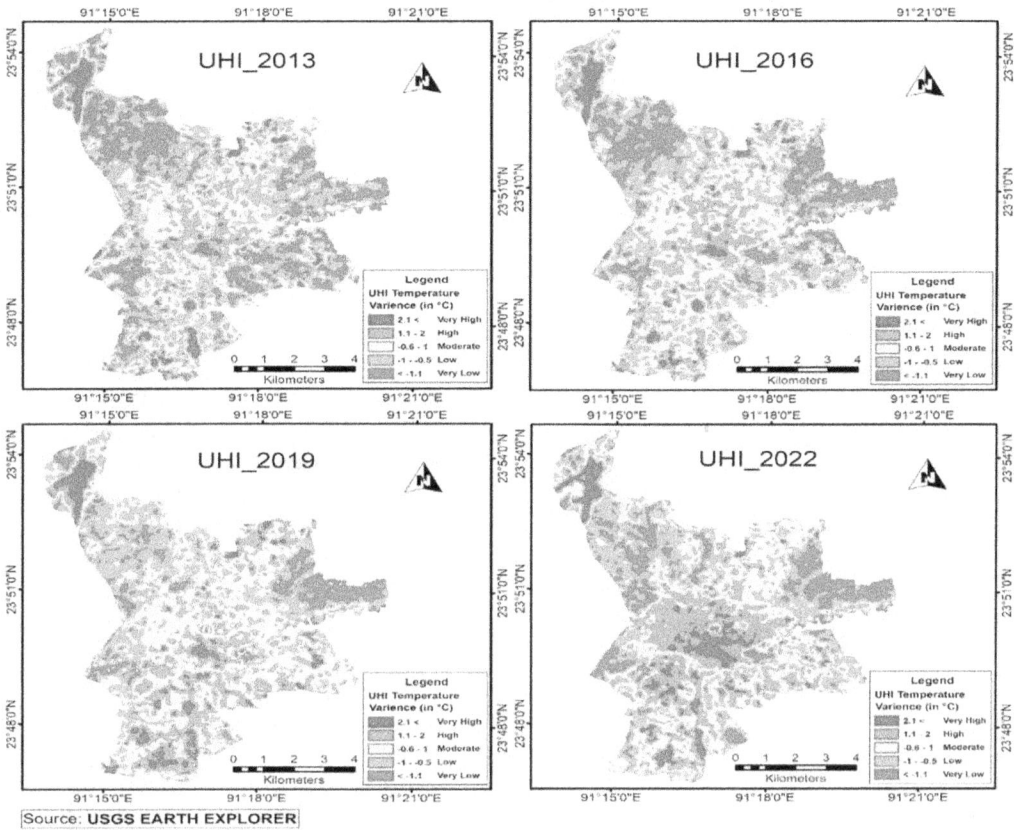

FIGURE 24.12 Temporal urban heat island comparison map (three-year interval).

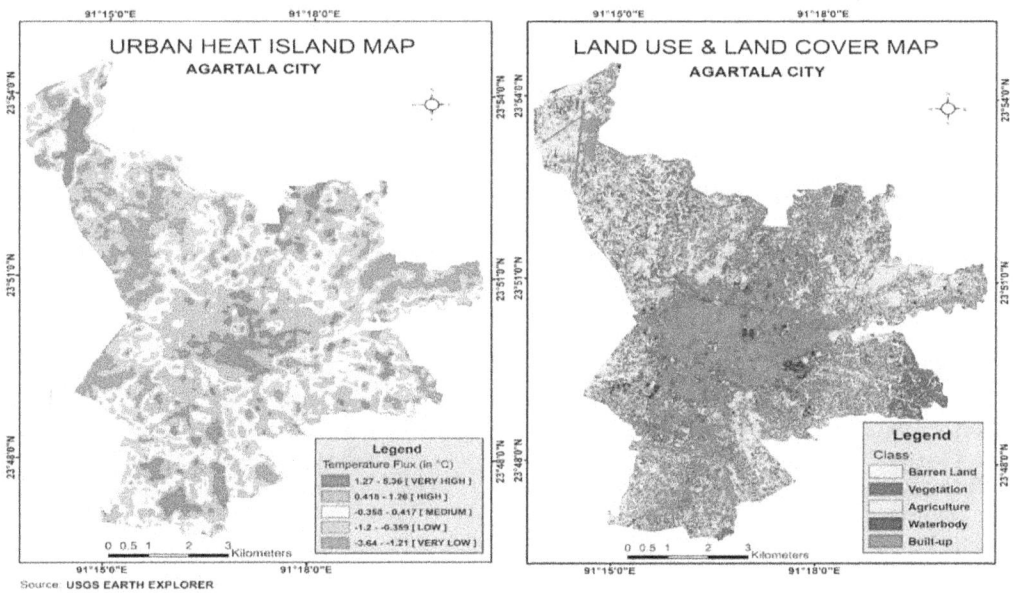

FIGURE 24.13 Urban heat island association based on land-use/land-cover status.

(Rosenfeld et al., 1995). This is another mitigation strategy against the UHI effect that is cost-effective and showing definitive results. This method can also be adopted in the study area as it does not restrict its use based on the size of the area (Akbari et al., 2011).

24.4.3.5 Underground Cooling System

This one is the most effective of all where a huge mechanical cooling system is installed cooling the surface water that flows through the area keeping the area cool. To utilize this strategy, there has to be a huge space, a large number of skilled personnel, and an abundance of funding (Zhu et al., 2010).

24.5 CONCLUSION

A UHI is no minor concept anymore. Although most developing countries still overlook this issue, many developed countries like the United States, United Kingdom, Switzerland, Malaysia, etc. are now focusing on the topic of reducing or managing urban heat distribution. Based on their results, it is clear that if there is enough funding and the right technology, this urban heat rate can be reduced by a substantial amount. For example, the world's largest underground district cooling system in Singapore requires a large underground area and highly trained technical personnel to operate this cooling system. Since all geographical areas have unique structures, not every strategic method for this issue is practical for sustainable use. To best come up with the information on the UHI of the study area, LST data are extracted from satellite images of the study area and associated with the NDVI and NDBI data of the area. The study area was small in size, densely populated, highly polluted, and had very small gaps between buildings for airflow providing challenges to finding suitable mitigation methods that have already been developed.

After careful evaluation and field validation, the study concluded that the study area is not big enough for a sophisticated cooling system like that in Singapore as it is just a small city of the third smallest state in India, nor there is a possibility to build air corridors. However, green walls and roofs can be added, rivers can be revitalized with strict rules against trash dumping, and every new building that is to be built can be regulated to have air ducts taking as an example the termite mounds to give the infrastructure better airflow, along with other minor mitigation strategies that are possible as they do not require significant costs and ongoing maintenance once they are implemented. This updated approach will effectively control the temperature and preserve cost efficiency. This study can also be used as a guide for new urban planners, working with small areas to come up with designs that will help reduce the urban heat while being cost-effective without the need for a very large area, or high financial expenses.

As a result, urban planners and decision-makers may benefit from the combination of GIS and remote sensing techniques in monitoring and mitigating climate change and the UHI in Agartala City. It also gives useful insights and workable solutions. To ensure the well-being of its citizens and the preservation of its environment, Agartala City may strive towards a more sustainable and resilient future by implementing suitable mitigation methods based on these results.

REFERENCES

Akbari, H., Xu, T., Taha, H., Wray, C., Sathaye, J., Garg, V., Tetali, S., Babu, M. H., & Reddy, K. N. (2011). Using cool roofs to reduce energy use, greenhouse gas emissions, and urban heat-island effects: Findings from an India experiment (LBNL-4746E, 1026804; p. LBNL-4746E, 1026804). https://doi.org/10.2172/1026804

Alcoforado, M. J., & Andrade, H. (2008). Global warming and the urban heat island. In J. M. Marzluff, E. Shulenberger, W. Endlicher, M. Alberti, G. Bradley, C. Ryan, U. Simon, & C. ZumBrunnen (Eds.), Urban ecology (pp. 249–262). Springer US. https://doi.org/10.1007/978-0-387-73412-5_14

Alexandri, E., & Jones, P. (2008). Temperature decreases in an urban canyon due to green walls and green roofs in diverse climates. Building and Environment, 43(4), 480–493. https://doi.org/10.1016/j.buildenv.2006.10.055

Ali, G., Abbas, S., Qamer, F. M., Wong, M. S., Rasul, G., Irteza, S. M., & Shahzad, N. (2021). Environmental impacts of shifts in energy, emissions, and urban heat island during the COVID-19 lockdown across Pakistan. Journal of Cleaner Production, 291, 125806. https://doi.org/10.1016/j.jclepro.2021.125806

Alqasemi, A. S., Hereher, M. E., Kaplan, G., Al-Quraishi, A. M. F., & Saibi, H. (2021). Impact of COVID-19 lockdown upon the air quality and surface urban heat island intensity over the United Arab Emirates. Science of the Total Environment, 767, 144330. https://doi.org/10.1016/j.scitotenv.2020.144330

Argüeso, D., Evans, J. P., Fita, L., & Bormann, K. J. (2014). Temperature response to future urbanization and climate change. Climate Dynamics, 42(7–8), 2183–2199. https://doi.org/10.1007/s00382-013-1789-6

Arnfield, A. J. (2003). Two decades of urban climate research: A review of turbulence, exchanges of energy and water, and the urban heat island. International Journal of Climatology, 23(1), 1–26. https://doi.org/10.1002/joc.859

Chakraborty, T., Sarangi, C., & Lee, X. (2021). Reduction in human activity can enhance the urban heat island: Insights from the COVID-19 lockdown. Environmental Research Letters, 16(5), 054060. https://doi.org/10.1088/1748-9326/abef8e

Chatterjee, S., Khan, A., Dinda, A., Mithun, S., Khatun, R., Akbari, H., Kusaka, H., Mitra, C., Bhatti, S. S., Doan, Q. V., & Wang, Y. (2019). Simulating micro-scale thermal interactions in different building environments for mitigating urban heat islands. Science of the Total Environment, 663, 610–631. https://doi.org/10.1016/j.scitotenv.2019.01.299

Corburn, J. (2009). Cities, climate change and urban heat island mitigation: Localising global environmental science. Urban Studies, 46(2), 413–427. https://doi.org/10.1177/0042098008099361

Dey, S., Mukherjee, G., & Paul, S. (2013). Imaging and visualizing the spectral signatures from Landsat TM and 'τ' value-based surface soil microzonation mapping at and around Agartala (India). Geocarto International, 28(2), 144–158. https://doi.org/10.1080/10106049.2012.662528

Dos Santos, A. R., De Oliveira, F. S., Da Silva, A. G., Gleriani, J. M., Gonçalves, W., Moreira, G. L., Silva, F. G., Branco, E. R. F., Moura, M. M., Da Silva, R. G., Juvanhol, R. S., De Souza, K. B., Ribeiro, C. A. A. S., De Queiroz, V. T., Costa, A. V., Lorenzon, A. S., Domingues, G. F., Marcatti, G. E., De Castro, N. L. M., . . . Mota, P. H. S. (2017). Spatial and temporal distribution of urban heat islands. Science of the Total Environment, 605–606, 946–956. https://doi.org/10.1016/j.scitotenv.2017.05.275

Du, H., Wang, D., Wang, Y., Zhao, X., Qin, F., Jiang, H., & Cai, Y. (2016). Influences of land cover types, meteorological conditions, anthropogenic heat and urban area on surface urban heat island in the Yangtze River Delta Urban Agglomeration. Science of the Total Environment, 571, 461–470. https://doi.org/10.1016/j.scitotenv.2016.07.012

Emmanuel, R., & Krüger, E. (2012). Urban heat island and its impact on climate change resilience in a shrinking city: The case of Glasgow, UK. Building and Environment, 53, 137–149. https://doi.org/10.1016/j.buildenv.2012.01.020

Gago, E. J., Roldan, J., Pacheco-Torres, R., & Ordóñez, J. (2013). The city and urban heat islands: A review of strategies to mitigate adverse effects. Renewable and Sustainable Energy Reviews, 25, 749–758. https://doi.org/10.1016/j.rser.2013.05.057

Golden, J. S. (2004). The built environment induced urban heat island effect in rapidly urbanizing arid regions: A sustainable urban engineering complexity. Environmental Sciences, 1(4), 321–349. https://doi.org/10.1080/15693430412331291698

Goward, S. N. (1981). Thermal behavior of urban landscapes and the urban heat island. Physical Geography, 2(1), 19–33. https://doi.org/10.1080/02723646.1981.10642202

Halder, B., Bandyopadhyay, J., & Banik, P. (2021). Monitoring the effect of urban development on urban heat island based on remote sensing and geo-spatial approach in Kolkata and adjacent areas, India. Sustainable Cities and Society, 74, 103186. https://doi.org/10.1016/j.scs.2021.103186

Hu, Y., Dai, Z., & Guldmann, J.-M. (2020). Modeling the impact of 2D/3D urban indicators on the urban heat island over different seasons: A boosted regression tree approach. Journal of Environmental Management, 266, 110424. https://doi.org/10.1016/j.jenvman.2020.110424

Kafy, A.-A., Abdullah-Al-Faisal, Rahman, M. S., Islam, M., Al Rakib, A., Islam, M. A., Khan, M. H. H., Sikdar, M. S., Sarker, M. H. S., Mawa, J., & Sattar, G. S. (2021). Prediction of seasonal urban thermal field variance index using machine learning algorithms in Cumilla, Bangladesh. Sustainable Cities and Society, 64, 102542. https://doi.org/10.1016/j.scs.2020.102542

Kalma, J. D., McVicar, T. R., & McCabe, M. F. (2008). Estimating land surface evaporation: A review of methods using remotely sensed surface temperature data. Surveys in Geophysics, 29(4–5), 421–469. https://doi.org/10.1007/s10712-008-9037-z

Li, D., & Bou-Zeid, E. (2013). Synergistic interactions between urban heat islands and heat waves: The impact in cities is larger than the sum of its parts. Journal of Applied Meteorology and Climatology, 52(9), 2051–2064. https://doi.org/10.1175/JAMC-D-13-02.1

Li, H., & Liu, Q. (2008). Comparison of NDBI and NDVI as indicators of surface urban heat island effect in MODIS imagery (D. Li, J. Gong, & H. Wu, Eds., p. 728503). International Society for Optics and Photonics. https://doi.org/10.1117/12.815679

Li, L., Zha, Y., & Wang, R. (2020). Relationship of surface urban heat island with air temperature and precipitation in global large cities. Ecological Indicators, 117, 106683. https://doi.org/10.1016/j.ecolind.2020.106683

Li, Z.-L., Tang, B.-H., Wu, H., Ren, H., Yan, G., Wan, Z., Trigo, I. F., & Sobrino, J. A. (2013). Satellite-derived land surface temperature: Current status and perspectives. Remote Sensing of Environment, 131, 14–37. https://doi.org/10.1016/j.rse.2012.12.008

Liu, Z., Lai, J., Zhan, W., Bechtel, B., Voogt, J., Quan, J., Hu, L., Fu, P., Huang, F., Li, L., Guo, Z., & Li, J. (2022). Urban heat islands significantly reduced by COVID-19 lockdown. Geophysical Research Letters, 49(2). https://doi.org/10.1029/2021GL096842

Loughner, C. P., Allen, D. J., Zhang, D.-L., Pickering, K. E., Dickerson, R. R., & Landry, L. (2012). Roles of urban tree canopy and buildings in urban heat island effects: Parameterization and preliminary results. Journal of Applied Meteorology and Climatology, 51(10), 1775–1793. https://doi.org/10.1175/JAMC-D-11-0228.1

Mallik, S., Chakraborty, A., Mishra, U., & Paul, N. (2022). Prediction of irrigation water suitability using geospatial computing approach: A case study of Agartala city, India. Environmental Science and Pollution Research. https://doi.org/10.1007/s11356-022-21232-8

Mallik, S., Mishra, U., & Paul, N. (2021). Groundwater suitability analysis for drinking using GIS based fuzzy logic. Ecological Indicators, 121, 107179. https://doi.org/10.1016/j.ecolind.2020.107179

Mathew, A., Khandelwal, S., & Kaul, N. (2017). Investigating spatial and seasonal variations of urban heat island effect over Jaipur city and its relationship with vegetation, urbanization and elevation parameters. Sustainable Cities and Society, 35, 157–177. https://doi.org/10.1016/j.scs.2017.07.013

Mirzaei, P. A., & Haghighat, F. (2010). Approaches to study urban heat island: Abilities and limitations. Building and Environment, 45(10), 2192–2201. https://doi.org/10.1016/j.buildenv.2010.04.001

Mitra, S., Roy, S., & Hore, S. (2022). Assessment and forecasting of the urban dynamics through lulc based mixed model: Evidence from Agartala, India. GeoJournal, 88(2), 2399–2422. https://doi.org/10.1007/s10708-022-10730-4

Mohajerani, A., Bakaric, J., & Jeffrey-Bailey, T. (2017). The urban heat island effect, its causes, and mitigation, with reference to the thermal properties of asphalt concrete. Journal of Environmental Management, 197, 522–538. https://doi.org/10.1016/j.jenvman.2017.03.095

Pandey, P., Kumar, D., Prakash, A., Masih, J., Singh, M., Kumar, S., Jain, V. K., & Kumar, K. (2012). A study of urban heat island and its association with particulate matter during winter months over Delhi. Science of the Total Environment, 414, 494–507. https://doi.org/10.1016/j.scitotenv.2011.10.043

Peng, S., Piao, S., Ciais, P., Friedlingstein, P., Ottle, C., Bréon, F.-M., Nan, H., Zhou, L., & Myneni, R. B. (2012). Surface urban heat island across 419 global big cities. Environmental Science & Technology, 46(2), 696–703. https://doi.org/10.1021/es2030438

Rosenfeld, A. H., Akbari, H., Bretz, S., Fishman, B. L., Kurn, D. M., Sailor, D., & Taha, H. (1995). Mitigation of urban heat islands: Materials, utility programs, updates. Energy and Buildings, 22(3), 255–265. https://doi.org/10.1016/0378-7788(95)00927-P

Santamouris, M. (2013). Using cool pavements as a mitigation strategy to fight urban heat island: A review of the actual developments. Renewable and Sustainable Energy Reviews, 26, 224–240. https://doi.org/10.1016/j.rser.2013.05.047

Sekertekin, A., & Bonafoni, S. (2020). Land surface temperature retrieval from Landsat 5, 7, and 8 over rural areas: Assessment of different retrieval algorithms and emissivity models and toolbox implementation. Remote Sensing, 12(2), 294. https://doi.org/10.3390/rs12020294

Semenza, J. C., Rubin, C. H., Falter, K. H., Selanikio, J. D., Flanders, W. D., Howe, H. L., & Wilhelm, J. L. (1996). Heat-related deaths during the July 1995 heat wave in Chicago. New England Journal of Medicine, 335(2), 84–90. https://doi.org/10.1056/NEJM199607113350203

Shaver, G. R., Canadell, J., Chapin, F. S., Gurevitch, J., Harte, J., Henry, G., Ineson, P., Jonasson, S., Melillo, J., Pitelka, L., & Rustad, L. (2000). Global warming and terrestrial ecosystems: A conceptual framework for analysis. BioScience, 50(10), 871. https://doi.org/10.1641/0006-3568(2000)050[0871:GWATEA]2.0.CO;2

Tran, D. X., Pla, F., Latorre-Carmona, P., Myint, S. W., Caetano, M., & Kieu, H. V. (2017). Characterizing the relationship between land use land cover change and land surface temperature. ISPRS Journal of Photogrammetry and Remote Sensing, 124, 119–132. https://doi.org/10.1016/j.isprsjprs.2017.01.001

Twumasi, Y. A., Merem, E. C., Namwamba, J. B., Mwakimi, O. S., Ayala-Silva, T., Frimpong, D. B., Ning, Z. H., Asare-Ansah, A. B., Annan, J. B., Oppong, J., Loh, P. M., Owusu, F., Jeruto, V., Petja, B. M., Okwemba, R., McClendon-Peralta, J., Akinrinwoye, C. O., & Mosby, H. J. (2021). Estimation of land surface temperature from Landsat-8 OLI thermal infrared satellite data: A comparative analysis of two cities in Ghana. Advances in Remote Sensing, 10(4), 131–149. https://doi.org/10.4236/ars.2021.104009

Veena, K., Parammasivam, K. M., & Venkatesh, T. N. (2020). Urban heat island studies: Current status in India and a comparison with the international studies. Journal of Earth System Science, 129(1), 85. https://doi.org/10.1007/s12040-020-1351-y

Wang, Y., Berardi, U., & Akbari, H. (2016). Comparing the effects of urban heat island mitigation strategies for Toronto, Canada. Energy and Buildings, 114, 2–19. https://doi.org/10.1016/j.enbuild.2015.06.046

Watkins, R., Palmer, J., & Kolokotroni, M. (2007). Increased temperature and intensification of the urban heat island: Implications for human comfort and urban design. Built Environment, 33(1), 85–96. https://doi.org/10.2148/benv.33.1.85

Weng, Q., & Lo, C. P. (2001). Spatial analysis of urban growth impacts on vegetative greenness with Landsat TM data. Geocarto International, 16(4), 19–28. https://doi.org/10.1080/10106040108542211

Yang, C., Yan, F., & Zhang, S. (2020). Comparison of land surface and air temperatures for quantifying summer and winter urban heat island in a snow climate city. Journal of Environmental Management, 265, 110563. https://doi.org/10.1016/j.jenvman.2020.110563

Yang, X., & Lo, C. P. (2002). Using a time series of satellite imagery to detect land use and land cover changes in the Atlanta, Georgia metropolitan area. International Journal of Remote Sensing, 23(9), 1775–1798. https://doi.org/10.1080/01431160110075802

Yue, W., Liu, X., Zhou, Y., & Liu, Y. (2019). Impacts of urban configuration on urban heat island: An empirical study in China mega-cities. Science of the Total Environment, 671, 1036–1046. https://doi.org/10.1016/j.scitotenv.2019.03.421

Zha, Y., Gao, J., & Ni, S. (2003). Use of normalized difference built-up index in automatically mapping urban areas from TM imagery. International Journal of Remote Sensing, 24(3), 583–594. https://doi.org/10.1080/01431160304987

Zhou, X., & Chen, H. (2018). Impact of urbanization-related land use land cover changes and urban morphology changes on the urban heat island phenomenon. Science of the Total Environment, 635, 1467–1476. https://doi.org/10.1016/j.scitotenv.2018.04.091

Zhu, K., Blum, P., Ferguson, G., Balke, K.-D., & Bayer, P. (2010). The geothermal potential of urban heat islands. Environmental Research Letters, 5(4), 044002. https://doi.org/10.1088/1748-9326/5/4/044002

Index

For Product Safety Concerns and Information please contact our EU
representative GPSR@taylorandfrancis.com
Taylor & Francis Verlag GmbH, Kaufingerstraße 24, 80331 München, Germany